COLLOQUIA MATHEMATICA
SOCIETATIS JÁNOS BOLYAI, 48

INTUITIVE GEOMETRY

Edited by

K. BÖRÖCZKY and G. FEJES TÓTH

NORTH-HOLLAND PUBLISHING COMPANY
AMSTERDAM–OXFORD–NEW YORK

Budapest, Hungary, 1987

ISBN North-Holland: 0444 879 33 1

ISBN Bolyai: 963 8022 22 1

ISSN Bolyai: 0139 3383

Joint edition published by

JÁNOS BOLYAI MATHEMATICAL SOCIETY

and

ELSEVIER SCIENCE PUBLISHERS B. V.

P. O. Box 1991

1000 BZ Amsterdam, The Netherlands

In the U.S.A. and Canada:

ELSEVIER SCIENCE PUBLISHING COMPANY INC.

52 Vanderbilt Avenue

New York, N.Y. 10017

U.S.A.

Editorial assistant: É. VÁSÁRHELYI

Printed in Hungary
Szegedi Nyomda
Szeged

CONTENTS

PREFACE

The János Bolyai Mathematical Society held an International Conference on Intuitive Geometry in Siófok from 12 to 18 May 1985. The meeting was devoted to discussing problems in geometry which were described by Hilbert as being explainable to the man in the street. Naturally, the same was not expected from the solutions of the problems.

At the conference 83 mathematicians representing 15 countries joined the 38 Hungarian participants. It consisted of five 50 minutes plenary talks by outstanding specialists, and 103 talks given in three parallel sections. There was also a plenary session where research problems were presented and discussed. The conference emphasised discrete geometry, convexity and combinatorial geometry in accordance with the Hungarian traditions of research. Several participants presented results in crystallography, classical differential geometry and rigidity theory. Some talks discussed the connections of geometry with biology, architecture, arts, and even psychology.

This book presents a collection of the papers and problems outlined at the conference.

We wish to thank all the authors for their contributions. Thanks are due also to the referees for their work.

The Editors

SCIENTIFIC PROGRAM

May 13, Monday

PLENARY SESSION

10.00-10.50 A. FLORIAN: Packing and covering with convex discs

SECTION A

Chairman: L. FEJES TÓTH

11.20-11.50 W. KUPERBERG: On packing the plane with congruent copies of a convex body

12.00-12.20 I. VINCZE: On system of circles surounding a circle

Chairman: E. JUCOVIČ

14.00-14.30 J.E. GOODMAN-R. POLLACK: Upper bounds for configurations and polytopes in R^d

14.40-15.20 W. WHITELEY: When is a cylindrical tower rigid?

15.30-15.50 H. HARBOTH: Regular point sets with unit distances

16.00.16.20 J. PALÁSTI: A construction for arrangements of lines with vertices of large multiplicity

16.30-17.00 A. BARAGAR: A combinatorial problem with convex polygons

17.00-18.30 P. ERDÖS: Open problems

SECTION B

Chairman: R. SCHNEIDER

11.20-11.50 C. BUCHTA: On the number of vertices of a polyhedron with a given number of facets

12.00-12.20 J. MÖLLER: Random intersections of the unit ball, random McKinney-simplices

Chairman: J. BÖHM

14.00-14.30 J. LINHART: An inequality for zonotopes

14.40-15.20 I. SZEPESVÁRI: Semi-regular star-polytopes in four dimension

15.30-15.50 E. SCHULTE: Can all edges be tangent to a sphere?

16.00-16.20 D.G. LARMAN - A. HILL: Interchanging facets

16.30-17.00 B. MONSON: A family of uniform polytopes with 5-fold symmetric shadows

SECTION C

Chairman: L.W. DANZER

11.20-11.50 C.W. LEE: Antagonistic knight placement problems

12.00-12.20 A. KEMNITZ: Diameters of integral point sets

Chairman: H. SACHS

14.00-14.30 P. McMULLEN: Ange-sum relations for polyhedral sets

14.40-15.20 A. VOLČIČ: Tomography of convex bodies

15.30-15.50 A. KUBA-A. VOLČIČ-D. KÖLCOW: An algorithm to reconstruct convex bodies from their projections

16.00-16.20 P. MANI: An intuitive notion of surface area?

16.30-17.00 B. UHRIN: Sharpenings and extensions of Brunn-Minkowski inequality

May 14, Tuesday

PLENARY SESSION

8.45-9.35 J. BÖHM: Shlaefli's differential form in the non-Euclidean geometry of polytopes and its applications

SECTION A

Chairman: J.J. SEIDEL

10.00-10.30 H.C. IM HOF: Generalized orthoschemes, frieze patterns, and the Coxeter-Bennet configuration

10.40-11.10 H. KAISER: Typisierung der elliptischen Dreiecke nach der Qualität ihrer Winkel und Seiten

11.20-11.50 P. NAGY: Free mitoin of rigid bodies in hyperbolic space

12.00-12.20 I. VERMES: Unterdeckungen und Überdeckungen in der byperbolischen Ebene

Chairman: W. NOWACKI

14.45-15.15 A.W.M. DRESS: How to classify local distortions in globally regular patterns

15-25-15-55 A.L. MACKAY: Periodical minimal surfaces

16.20-16.40 Sz. BÉRCZI: Symmetries in the plant surface lattice systems

16.50-17.10 P. ENGEL: Geometric crystallography in higher dimension

SECTION B

Chairman: K. BÖRÖCZKY

10.00-10.30 R. CONNELLY: Rigid packings

10.40-11.10 W. MÖGLING: Über Punktverteilungen in einem Quadrat und in einem Kreis mit Speziellem Abstandsbegriff

11.20-11.50 K. BEZDEK: The thinnest holding-lattice of a set

12.00-12.20 G. FEJES TÓTH: Totally separable packing and covering with circles

Chairman: A. FLORIAN

14.45-15.15 K. BÖRÖCZKY: Befreundete Punktsysteme und Kreisanordnungen

15.25-15.55 J. MOLNÁR: Sur la densitè d'empilement des sphères incongruents

16.20-16.40 U. BOLLE: K-fold lattice packing of translates of a convex disc

16.50-17.10 E. JUCOVIČ: On certain circle packings in the plane

17.20-17.40 K. BOGNÁR-MÁTHÉ: Über Kugelsysteme unter Geräumigkeitsbedingungen

SECTION C

Chairman: H. HARBORTH

10.00-10.30 T. BISZTRICZKY-J. SCHAER: Linearly related convex sets

10.40-11.10 J. SCHAER- T. BISZTRICZKY: Affinely embeddable families of convex sets

11.20-11.50 G. WEGNER: Intersecting rectangles

12.00-12.20 J. KINCSES: The classification of 3 and 4 Helly dimensional convex bodies

ChairmanÉ D.G. LARMAN

14.45-15.15 J.M. WILLS: Platonic manifolds

15.25-15.55 S. BILINSKI: Die quasiregularen Polyeder der Kubooktaeder-Familie

16.20-16.40 G. WEIß: Isoperimetrische tetraeder

16.50-17.10 Z. DZIECHCINSKA-HALAMODA-W.SZWIEC: On Critical sets of convex polyhedra

17.20-17.40 B. WEIßBACH: Polyedrische Deckel

May 15, Wednesday

PLENARY SESSION

8.45-9.35 P. ERDŐS: Problems and results in combinatotial and metric geometry

SECTION A

Chairman: P. GRUBER

10.00-10.30 J. VAN DE CRAATS: Varying two tangent hyperbolas

10.40-11.10 J. SIMONIS -J. VAN DE CRAATS: Coordinate free definitions of cross ratio

11.20-11.50 A. BLOKHUIS-K.A. POST-C.L.M. VAN PAUL: Note on a theorem of Wik

12.00-12.20 C. LOZANOV: The theorem of Miquel in finite non-Miquelian inversive planes

SECTION B

Chairman: S. BILINSKY

10.00-10.30 M.N. BLEICHER: Isoperimetric partitions of the plane

10.40-11.10 R.K. GUY: Tiling the square with rational triangles

11.20-11.50 É. VÁSÁRHELYI: Ein isoperimetrisches Problem für Mosaike

12.00-12.20 G. BARON: Maximale k-Pflasterungen von Rechtecken

SECTION C

Chairman: Gy. STROMMER

10.00-10.30 W. WUNDERLICH: Spiegelprobleme

10.40-11.10 H. SACHS: Merkwürdige Kennzeichnungen ebener Kurven

11.20-11.50 H. STACHEL: Pairs of curves with a certain distance-property

12.00-12.30 M. HUSTY: Eine anschauliche Erzeugung von Flachen vierter Ordnung mit zerfallendem Doppelkegelschnitt

May 16, Thursday

PLENARY SESSION

8.45-9.35 R. SCHNEIDER: Equidecomposable polyhedra

SECTION A

Chairman: P. McMULLEN

10.00-10.30 P.M. GRUBER: An asimptotic formula for the approximation of a convex body by polytopes

10.40-11.10 T. ZAMFIRESCU: Nearly all convex bodies are smooth and strictly convex

11.20-11.50 I. BÁRÁNY-K.BÖRÖCZKY-E.MAKAI-J.PACH: Body lattices in R^n

12.00-12.20 A.C. THOMPSON: On the isoperimetric mapping for Minkowski spaces

Chairman: M.N. BLEICHER

14.45-15.15 L. FEJES TÓTH: Densest packing of transletes of the union of two circles

15-25-15.55 J. PACH: How to build a barricade

16.20-16.40 M. LASSAK: Covering plane convex bodies with smaller homothetical copies

16.50-17.10 T. TARNAI-Zs. GÁSPÁR: Covering the sphere with equal circles

17.20-17.40 S. SZABÓ: A star polyhedron that tiles but not as a fundamental domain

SECTION B

Chairman: H. STACHEL

10.00-10.30 B. KLOTZEK: Diskrete Gruppen in normierten Räumen

10.40-11.10 D.G. EMMERICH: Selftensioning structures

11.20-11.50 C. CETKOVIC: The classification of equi-affinities of the affine plane

12.00-12.20 B. WERNICKE: Topologische Möbiusebenen in spiegelungsgeometrischer Darstellung

Chairman: B. KLOTZEK

15.25-15.55 G. CSÓKA: There exists a basis of minimal vectors in any 8-dimensional perfect lattice

16.20-16.40 Z. MAJOR: Sur l'estremum de la somme des puissances des distances

16.50-17.10 D. NAGY: (Dis) symmetry and (neg)entropy of geometric figure-systems

SECTION C

Chairman: A.W.M. DRESS

10.00-10.30 W. NOVACKI: 50 years of the conceit of geometric and arithmetic crystal classes.- The analogy between elements of symmetry in science and archetypes in psychology

10.40-11.10 C.E. LINDERHOLM: An inequality for simplices

11.20-11.50 J. FLACHSMAYER: Cubes and cross-polytopes in 3 and 4 dimensions: Symmetries and their groups

12.00-12.20 E. MOLNÁR: Minimal presentation of Euclidean space forms by fundamental domains

Chairman: R.K. GUY

14.45-15.15 U. BREHM: Symmetric weakly neighborly polyhedral maps

15.25-15.55 C. SCHULZ: Embedding of neighborly 2-manifolds into sewn 4-polytopes

16.20-16.40 P. GOOSSENS: Hormology in polyhedral manifolds

16.50-17.10 I. HERBURT: On intrinsic isometries of manifolds

17.20-17.40 J. BOKOWSKI: Realisierung kombinatorischer Mannigfaltigkeiten im R^3

May 17, Friday

SECTION A

Chairman: A. FLORIAN

8.45- 9.15 L. DANZER: Über ebene Pflasterungen bei denen jeder Stein D_5 symmetrisch ist

9.25- 9.55 L. STRAMMLER: Classification of triangles by relations between distances

10.05-10.25 A. IVIC WEISS: Isoclinal sequences of spheres

10.50-11.10 J.G. TETERIN: Distribution of lattice points on a sphere

11.20-12.00 G.J. RIEGER: Geometry you can touch

Chairman: J.M. WILLS

14.45-15.15 J.B. WILKER: Periodic isoclinal sequences

15.25-15.55 W. JANK: Über eine Verallgemeinung der Drehkegel

16.05-16.25 H.S.M. Coxeter: (presented by J.B. Wilker) A packing of 840 balls of radius $9^{0}0'19''$ on the 3-sphere

SECTION B

Chairman: J. BARACS

8.45 - 9.15 V.F. IGNATENKO: On the geometrical theory of group invariants generated by reflections

9.25 - 9.55 D. LJUBIC: Classification of isometries of n-dimensional spaces

10.05-10.25 L. TAMÁSSY: Problem of the existence of a body with given section areas

10.50-11.10 S. FUDALI: Maximum-point primitive fixing system in a plane

11.20-11.40 Z. LUCIC: Uniform tessalations of the hyperbolic plane

Chairman: Gy. STROMMER

14.45-15.15 J. BARACS: Recognition of polyhedral projections

15.25-15.55 H. MARTINI: New results around zonotopes

16.05-16.25 W. KUPERBERG: An inequality linking packing- with covering densities in the plane

SECTION C

Chairman: L. TAMÁSSY

14.45-15.15 N.V. ZHIVKOV: A necessary condition for best Hausdorff approximation of plane convex compacta by polygons

15.25-15.55 E. MAKAI: Maximal value enclosed by plates

16.05-16.25 J.M. SMYT'KO: Geometrical aspects of x-ray divergent beam technique

PLENARY SESSION

17.00-17.50 J.J. SEIDEL: On the volume of a hyperbolic simplex

COLLOQUIA MATHEMATICA SOCIETATIS JÁNOS BOLYAI
48. INTUITIVE GEOMETRY, SIÓFOK, 1985.

COVERING ALL SECANTS OF A SQUARE

I. BÁRÁNY*- Z. FÜREDI*

Suppose that n points are given in the unit square. Then there exists an intersecting line whose L_∞-distance is at least 2/3(n+1) from each point. This is a slight improvement on the trivial lower bound 1/2n but it is still far from the best possible value 1/(n+1) conjectured by L. Fejes Tóth.

1. INTRODUCTION

Let S be a square on the plane with side length n (≥ 1), and let $\mathcal{S} = \{S_1, S_2, \ldots, S_t\}$ be a collection of unit squares whose sides are parallel to those of S. We say that $\mathcal{S}$ *covers* the lines intersecting S if for every line L (on the plane) which intersects S intersects some of the S_i's (i.e., $L \cap S \neq \emptyset$ implies $L \cap S_i \neq \emptyset$ for some i). Let $\tau(n) = \tau(n, S)$ denote the minimum cardinality of a cover, and let $\tau_{in}(n)$ denote the minimum cardinality of a covering system whose members are located inside S.

* Research supported by Hungarian National Foundation for Scientific Research no. 1238 and 1812.

L. Fejes Tóth [3,7] conjectured that for an odd integer n

$$\tau_{in}(n) = 2n-1$$

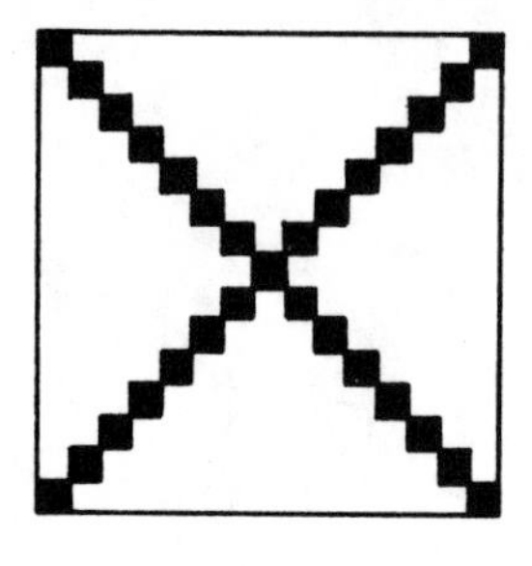

Figure 1

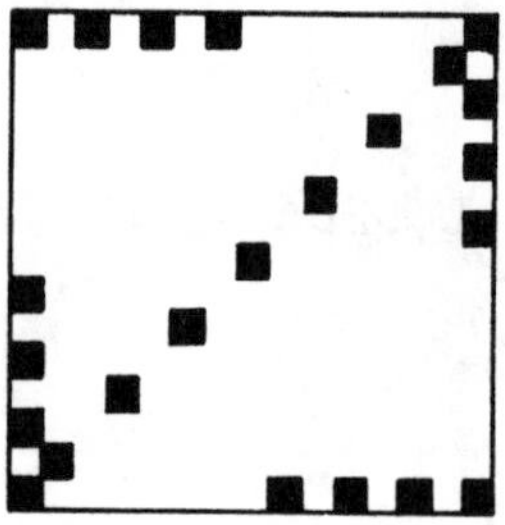

Figure 2

(see Figure 1.). Clearly, $\tau(n) \le \tau_{in}(n) \le 2\lceil n \rceil$ where $\lceil x \rceil$ denotes the upper integer part of the real x. The aim of this note is to improve on the trivial lower bound $\tau(n) \ge \lceil n \rceil$. Namely, we will prove $\tau(n) > (13n-1)/12$ (Theorem 2.1) and $\tau_{in}(n) > (4n-1)/3$ (Theorem 2.3).

The exact results are stated in Section 2. That section also contains examples showing the limit of our methods. Section 3 is devoted to the proof of the lower bounds. These proofs use weight functions, actually we calculate the fractional covering number of a hypergraph. In Section 4 we mention related problems and results.

2. INTERSECTING LINES PARALLEL TO THE SIDES OR THE DIAGONALS

THEOREM 2.1. *Let* S *be a square with side length* n ($n \ge 1$, *real*) *and let* $\mathcal{S} = \{S_1, \ldots, S_t\}$ *be a collection of unit squares in* S *whose sides are parallel to those of* S. *If* $t \le (4n-1)/3$ *then there exists a line parallel to either a side or a diagonal of* S, *which intersects* S *and avoids every* S_i.

The Example 2.2 shows that for $t \ge (3n+1)/2$, Theorem 2.1 does not remain true.

EXAMPLE 2.2. Let k be a positive integer, $n = 4k-1$. Suppose that the four vertices of S are given by their coordinates: (0,0), (0,n), (n,0) and (n,n). We will denote by S(i,j) the unit square $\{(x,): i \leq x \leq i+1, j \leq y \leq j+1\}$. Then the following set of squares, $\mathcal{S}$ covers every intersecting line of S with slope 0, 45^0, 90^0 or 135^0.

$\mathcal{S} = \{S(i,j)$: where $i,j \geq 0$ integers such that $i = 0$, $j = 2t$, $0 \leq t \leq k-1$ or $j = 0$, $i = 2(k+t)$, $0 \leq t \leq k-1$ or $i = 2k-2$, $j = 2(k+t)$, $0 \leq t \leq k-1$ or $j = 2k-2$, $i = 2t$, $0 \leq t \leq k-1$ or finally $i = j = 2t+1$, $0 \leq t \leq 2k-2\}$. See Figure 2.

If n is not an integer of the form 4k-1, then a minor modification of the above example (e.g., let $k = \lfloor (n-1)/4 \rfloor$) demands less than $(3n+9)/2$ unit squares. Denote by $t_{in}(n)$ the minimum value of th for which 2.1 does not hold. Similarly, let t(n) denote the minimum t such that there exists a cover consisting of t unit squares (located arbitrarily, not only inside S) which meets every intersecting line with slope 0, 45^0, 90^0 or 135^0.

THEOREM 2.3 $\quad \frac{13}{12}n - \frac{1}{12} < t(n) < \frac{4}{3}n + O(1)$.

The upper bound follows from the following example.

EXAMPLE 2.4. Suppose $n = 6k+3$, where k is an integer. Let $\mathcal{S} = \{S(i,j)$: where i,j are integers and either $i = 3j$, $0 \leq j \leq 3k+1$ or $j = 3i-2$, $1 \leq i \leq 3k+1$ or $(i,j) = (3k+2, 6k+2)$ or $j = i-2$, $i = 3k+3+t$, $0 \leq t \leq 3k-1$, $t \not\equiv 2 \pmod 3\}$. Then $|\mathcal{S}| = 8k+4$. See Figure 3.

These examples show that our method, i.e., to consider only 4 directions, can not lead to the proof of Fejes Tóth's conjecture.

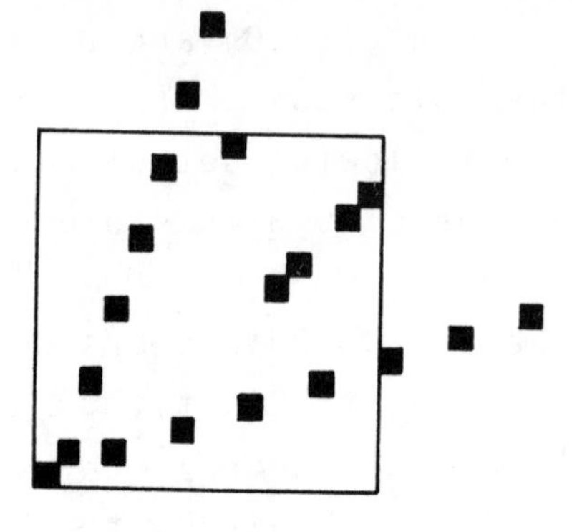

Figure 3

3. PROOFS

Suppose that $S_1, S_2, \ldots, S_t$ meet every line intersecting S with angle 0, 45°, 90° or 135°. We will show that $t > (4n-1)/3$. Consider a coordinate-system whose axes are parallel to the sides of S. Choose the unit and the origin of this system in such a way that the vertices of S have the coordinates $(\pm 1, \pm 1)$. Then the side length of a square S_i is $2/n$ denoted by 2ε. We define a *weight* function $w(L)$ on the set of intersecting lines L with slopes 0, 45°, 90° or 135° as follows. Actually, this weight-function is a measure on the set of these lines. If the equation of the line L is $y = c$ or $x = c$ then

$$w(L) = \frac{1}{2} - \frac{1}{2}c^2$$

and, if the form of the line L is $y = x+h$ or $y = -x+h$ then

$$w(L) = \frac{1}{8}h^2.$$

As for an intersecting line $|c| \leq 1$, $|h| \leq 2$ hold we have $\frac{1}{2} \geq w(L) \geq 0$. The total weight of the lines in these four directions is:

$$(1) \qquad 2\int_{-1}^{+1} \left(\frac{1}{2} - \frac{1}{2}c^2\right) dc + 2\int_{-2}^{2} \frac{1}{8}h^2 \, dh = \frac{8}{3}.$$

Now consider a square $Q = Q(a,b)$ with center (a,b) ($|a|$, $|b| \leq 1-\varepsilon$) and side length 2ε.

We will show that the weight of the lines intersecting Q is

$$(2) \qquad 2\varepsilon + \frac{2}{3}\varepsilon^3.$$

Hence (1) and (2) yield that for $n > 1$

$$t \geq \frac{8}{3} / \left(2\varepsilon + \frac{2}{3}\varepsilon^3\right) = \frac{4}{3}n - \frac{4}{9n + (3/n^2)} > \frac{4n-1}{3}$$

proving Theorem 2.1. The proof of (2) is simple because the weight of the lines intersecting Q and parallel to the axis $x = 0$ is

$$(3) \qquad \int_{a-\varepsilon}^{a+\varepsilon} \left(\frac{1}{2} - \frac{1}{2}c^2\right) dc = \varepsilon - a^2\varepsilon - \frac{1}{3}\varepsilon^3.$$

See Figure 4. Similarly the weights of the lines intersecting Q and parallel to the lines $y = 0$, $y = x$, $y = -x$ are

$$(4) \qquad \int_{b-\varepsilon}^{b+\varepsilon} \left(\frac{1}{2} - \frac{1}{2}c^2\right) dc = \varepsilon - b^2\varepsilon - \frac{1}{3}\varepsilon^3,$$

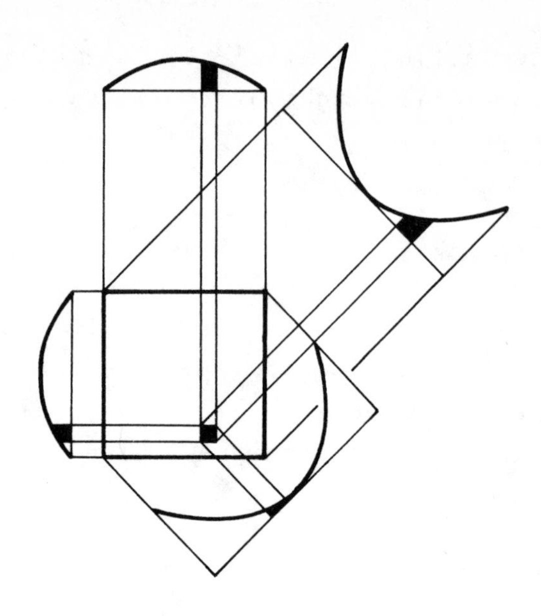

Figure 4

$$(5) \qquad \int_{b-a-2\varepsilon}^{b-a+2\varepsilon} \tfrac{1}{8}h^2 dh = \tfrac{1}{2}\varepsilon(b-a)^2 + \tfrac{2}{3}\varepsilon^3,$$

$$(6) \qquad \int_{a+b-2\varepsilon}^{a+b+2\varepsilon} \tfrac{1}{8}h^2 dh = \tfrac{1}{2}\varepsilon(a+b)^2 + \tfrac{2}{3}\varepsilon^3,$$

Summing up (3) - (6) we get (2).

The proof of 2.3 is analogous to the above. We modify the weight functions of the lines, because in the previous case a small square outside S, e.g., Q(0,2) could get too much weight.

If $y = c$ or $x = c$ then $w(L) = \begin{cases} \frac{1}{2} - \frac{1}{8}c^2 & \text{for } |c| \leq 1 \\ 0 & \text{otherwise} \end{cases}$

and if $y = \pm x + h$ then $w(L) = \begin{cases} \frac{1}{32}h^2 & \text{for } |h| \leq 2, \\ 0 & \text{otherwise.} \end{cases}$

Then the total weight of the lines is 13/6 and every small square covers lines with weight at most $2\varepsilon + \frac{1}{6}\varepsilon^3$. Hence $t < 13/(12\varepsilon + \varepsilon^3) = \frac{13}{12}n - 1/12(12n^2 + 1)$.

4. RELATED PROBLEMS AND RESULTS

We have the following conjectures:

$$t(n) = \frac{4}{3}n + O(1),$$

$$t_{in}(n) = \frac{3}{2}n + O(1).$$

We could not even prove that $\lim_{n\to\infty} t(n)/n$ exists (or $\lim t_{in}(n)/n$, or $\lim \tau(n)/n$ or $\lim \tau_{in}(n)/n$.) The only result we have is if we consider 8 directions of the lines, and define a more sophisticated weight-function, then we obtain

THEOREM 4.1. $\tau_{in}(n) > 1.43n - O(1)$.

Paul Endös asked what is the minimum number of covering unit squares *outside* S? It is very likely $3n + O(1)$.

Our problem is a particular case of a problem of Fejes Tòth [2]. Assume K is a convex body on the plane and $\lambda > 0$. Consider a set $\mathcal{S}$ of λ-homothetic copies of K having the property that each line intersecting K intersects at least one member of $\mathcal{S}$. What is the minimum cardinality of such a set? Fejes Tòth [3] points out further that this question is closely related to the dual of Tarski's plank problem (see Bang [1] or Fenchel [4]).

Another related problem is the following, considered by Makai and Pach [6]. Let $\mathcal{F}$ be a class of functions $f : R \to \mathbb{R}^d$. A set of points $\{(x_i, y_i) \in R \times \mathbb{R}^d,\ i = 1,2,\dots\}$

is said to be $\mathcal{F}$-controlling system if for each $f \in \mathcal{F}$ there is an i with $\|f(x_i) - y_i\| \leq 1$. So an $\mathcal{F}$-controlling system is a set of points P in $\mathbb{R}^1 \times \mathbb{R}^d$ with the property that for each $f \in \mathcal{F}$ one can find a point in P sufficiently close to the graph of f. The problem is to find an $\mathcal{F}$-controlling system with "few" points (or with small density if P must be inifinite). Makai and Pach [6], and Groemer [5] prove several results concerning this problem. In their case the norm is always the Euclidean norm.

When we take in the above formulation $d = 1$, $\mathcal{F}$ to be the class of all linear functions whose graphs intersect the square S, and $\| \ \|$ to be the L_∞ norm, then what we arrive to is exactly our problem about $\tau(n,S)$.

We end this paper by mentioning a question of Fejes Tóth [2] which we find very appealing and which belongs to the sort of questions considered here. A *zone* of widht w is defined as the parallel domain of a great circle (of the sphere) with angular distance w/2. Prove (or disprove) that the total width of any set of zones covering the sphere is at least π.

REFERENCES

[1] TH. BANG, *On covering by parallel-strips*, Mat. Tidsskrift B, (1950) 49-53./ A solution of the "plank problem," Proc. Amer. Math. Soc. 2 (1951), 990-993.

[2] L. FEJES TÓTH, *Exploring a planet*, Amer. Math. Monthly 80 (1973), 1043-1044.

[3] L. FEJES TÓTH, *Remarks on a dual of Tarski's plank problem*, Mat. Lapok 25 (1974), 13-20.

[4] W. FENCHEL, *On Th. Bang's solution of the plank problem*, Mat. Tidsskrift B, (1951), 49-51.

[5] H. GROEMER, *Covering and packing by sequences of convex sets*, Mathematika 29 (1982), 18-31.

[6] E. MAKAI, Jr. and J. PACH, *Controlling function classes and covering Euclidean space*, Studia Sci. Math. Hungar. 18 (1983), 435-459.

[7] W.O. MOSER and J. PACH, *Research Problems in Discrete Geometry*, Problem 84, Montreal 1985, mimeographed.

I. BÁRÁNY
Math. Inst. Hungar. Acad. Sci.,
Budapest 1364 P.O.B. 127,
Hungary

Z. FÜREDI
RUTCOR, RUTGERS
University,
New Brunswick,
NJ 08903, USA

COLLOQUIA MATHEMATICA SOCIETATIS JÁNOS BOLYAI
48. INTUITIVE GEOMETRY, SIÓFOK, 1985.

COUNTER-EXAMPLES TO A PACKING PROBLEM OF L. FEJES TÓTH

A. BEZDEK* and G. KERTÉSZ

Let S be an open set of points in the Euclidean plane. Let d(S) be the density of the densest packing of translates of S and d*(S) be the density of the densest lattice-packing of S. One natural question is the following. Under what conditions on S can we claim that the equality

$$(1) \qquad d(S) = d^*(S)$$

holds? Rogers [4] proved that if S is convex then (1) holds (see also [1] and [2]). Another class of domains {S} for which (1) holds was given by L. Fejes Tóth [3]. Following L. Fejes Tóth we say that an open connected domain S is *semi-convex* if there exists a pair of parallel supporting lines of S with touching points A and B such that one of the two arcs into which A and B divide the boundary of S is convex. Similarly we say that S is *limited* semi-convex, if choosing an arbitrary

*Research supported by Hungarian National Foundation for Scientific Research no. 1238.

point C on the convex arc AB and translating AB through the vectors $\vec{CA}$ and $\vec{CB}$ the original arc and its translates enclose a region which contains S. L. Fejes Tóth proved [3] that, if S is limited semi-convex then (1) holds, and asked, whether the semi-convexity of S also implies (1). We shall see that the answer is "no".

In 1. we shall construct a domain with bilateral symmetry for which (1) does not hold. By modifying this domain we shall obtain in 2. a semi-convex domain which will yield the desired counter-example.

We still note that altough the class {S} of limited semi-convex domains does not comprise the class of convex domains, the fact that for any $S \in \{S\}$ (1) holds implies the same for any convex domain (see [2]).

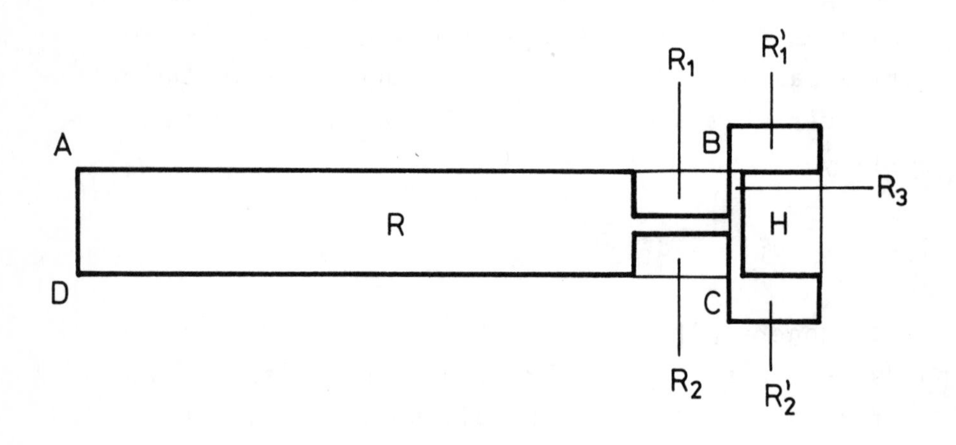

Figure 1 The disc S_1

1. Consider the rectangle R with vertices A,B,C,D and sides AD = 1, AB = n (n > 5). Let R_1, R_2 be two disjoint rectangles at the vertices B and C inside R each having side lengths 1 and $\frac{1-\varepsilon}{2}$ $(0 < \varepsilon \ll 1)$. Let R_1', R_2' be the images of R_1 and R_2 reflected in the vertices B and C respectively. Let us shift the side BC in the direction

of $\vec{AB}$ by the distance ε and denote the rectangle which was covered during this motion by R_3. Let

$$S_1 = R \cup R_2' \cup R_2' \cup R_3 \setminus (R_1 \cup R_2).$$

Denote the rectangle which fills out the gap next to S_1 between R_1' and R_2' of S_1 by H. Consider the packings P_1 and P_2 shown in Figure 2 and Figure 3.

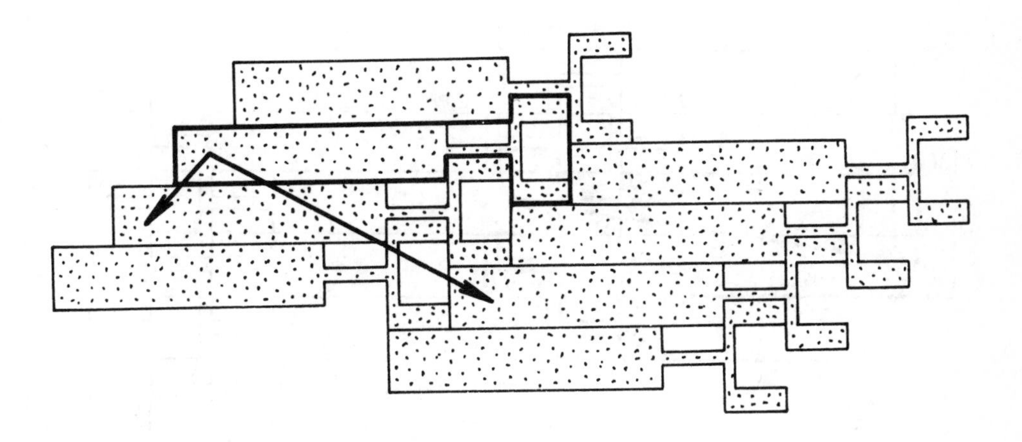

Figure 2 The packing P_1

Both consist of translates of S_1. Obviously P_1 is a lattice-packing, while P_2 is not. The required inequality $d(S_1) > d^*(S_1)$ follows from the next two lemmas:

LEMMA 1. *The density of the packing* P_1 *is less than the density of the packing* P_2.

LEMMA 2. *The packing* P_1 *is the densest lattice packing of translates of* S_1.

PROOF OF LEMMA 1.

In the Figure 1 and Figure 2 there are shown patterns and basic vectors which determine lattice-like tilings of the plane such that each tile in the case of P_1 is the union of one copy of S_1 and two gaps of total area $\frac{3(1-\varepsilon)}{2}$. On the other hand each tile in the case of the packing P_2 is the union of two copies of S_1 and three gaps of total area 2. Since $\frac{3(1-\varepsilon)}{2} > 1$ the packing P_2 has a larger density than that of P_1.

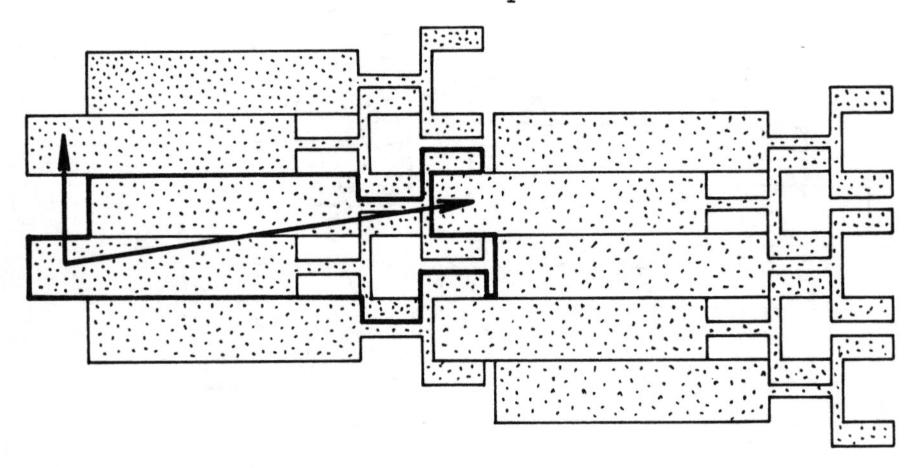

Figure 3 The packing P_2

PROOF OF LEMMA 2.

Let P be an arbitrary lattice-packing of S_1. It is enough to assign to each disc S in P a part f(S) of the gaps such that i/ the area of f(S) is $\geq \frac{3(1-\varepsilon)}{2}$, ii/ if S, S' are two different discs of P, then f(S) and f(S') are disjoint, iii/ for each S the union of S and f(S) can be covered by a circle of fixed radius. We will use the notations of Figure 1 in order to identify certain parts of the plane. For example the term "R_1 of S" means the image of the rectangle R_1 at the isometry which takes the disc S_1 into the disc S. We need

LEMMA 3. *At least one of the rectangles* R_1 *and* R_2 *of* S *is not overlapped by any other disc of* P.

PROOF OF LEMMA 3.

Suppose that R_1 of S is overlapped by another disc say S'. Only R_2' of S' can overlap R_1 of S. Since P is lattice-packing, S" the image of S at the translation which takes S' into S also belongs to P. But S" clearly prevents any disc of P from overlapping R_2 of S.

We may suppose that H of S is overlapped by another disc say S''', because otherwise the assignment where f(S) is the union of R_1 (or R_2) of S and H of S satisfies the conditions i/ - iii/ and we are done. The images of S at the translation $\pm kT$ (where T denotes the translation which takes S into S''' and k is an integer) form a row of P. One can easily check that between two rows there is a strip of width at least $\frac{1-\varepsilon}{2}$, which is overlapped by at most one R_1'-type (or R_2'-type) parts in each period of length n. This allows us to assign to each disc gaps right above it of area at least $(n-2)\frac{1-\varepsilon}{2}$. If n is large enough $(n > 5)$, then $(n-2)\frac{1-\varepsilon}{2} > \frac{3(1-\varepsilon)}{2}$ and we are done.

2. First we need to extend the notations of Figure 1. Suppose the disc S_1 is embedded in the coordinate system XY, such that the X axes is parallel to the edge AB and the vectors $\vec{AB}$ and $\vec{AD}$ point to the positive directions (Figure 4).

If P is a point of coordinates x,y and δ,ν are real numbers, then
$P(\delta,\nu)$ denotes the point of coordinates $x+\delta$ and $y+\nu$. Let EF and KL be the right vertical edges of the rectangles R_1' and R_2'. After modifying the boundary of S_1 by replacing

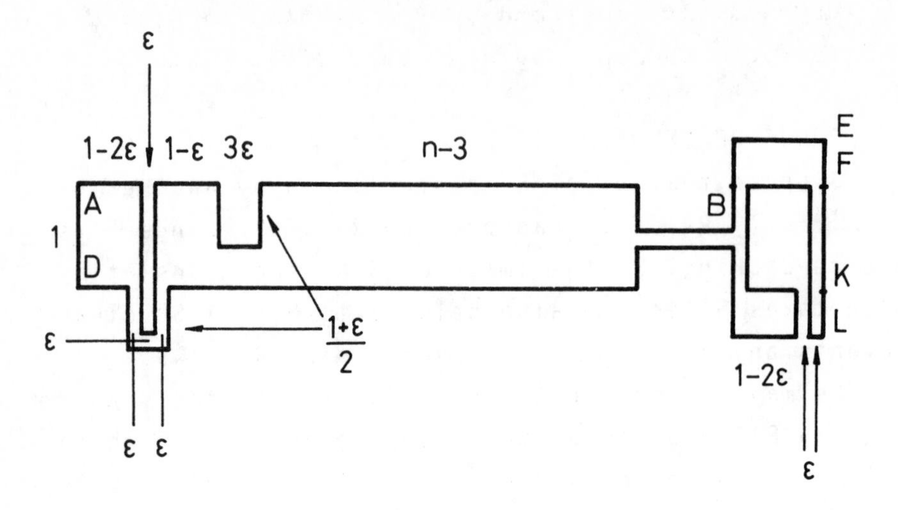

Figure 4 The disc S_2

the segment $A(1-2\varepsilon,0)A(1-\varepsilon,0)$ by the arc

$$A(1-2\varepsilon,0)A(1-2\varepsilon,-\frac{3-\varepsilon}{2})A(1-\varepsilon,-\frac{3-\varepsilon}{2})A(1-\varepsilon,0),$$

the segment $D(1-3\varepsilon,0)D(1,0)$ by the arc

$$D(1-3\varepsilon,0)D(1-3\varepsilon,-\frac{1+\varepsilon}{2})D(1,-\frac{1+\varepsilon}{2})D(1,0)$$

the segment $A(2-3\varepsilon,0)A(2,0)$ by the arc

$$A(2-3\varepsilon,0)A(2-3\varepsilon,-\frac{1+\varepsilon}{2})A(2,-\frac{1+\varepsilon}{2})A(2,0),$$

the segment $F(-\varepsilon,0)F$ by the arc $F(-\varepsilon,0)L(-\varepsilon,0)LF$, the arc $K(-2\varepsilon,0)KLL(-2\varepsilon,0)$ by the segment

$$K(-2\varepsilon,0)L(-2\varepsilon,0)$$

we obtain the region S_2. The pair of parallel supporting lines of slope μ will touch S_2 at the vertices $E(-1,0)$ and L, if $\frac{\varepsilon}{n} < \mu < \frac{1-\varepsilon}{2n}$. Since the arc $E(-1,0)EL$ of the boundary of S_2 is convex, S_2 is semi-convex.

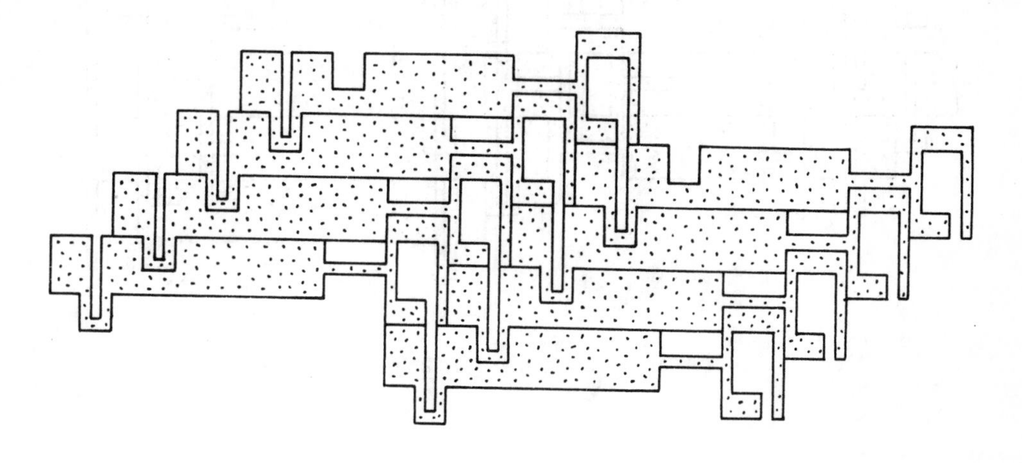

Figure 5 The packing P_1'

Derive the packings P_1' and P_2' by replacing each copy of S_1 in P_1 and P_2 by a copy of S_2 (Fig. 5 and 6.). If ε is very small, then the areas of S_1 and S_2 are very close to each other. Therefore the lemmas, analogous to the Lemmas 1, 2 and 3 can be proved in the same way, which means that $d^*(S_2) < d(S_2)$.

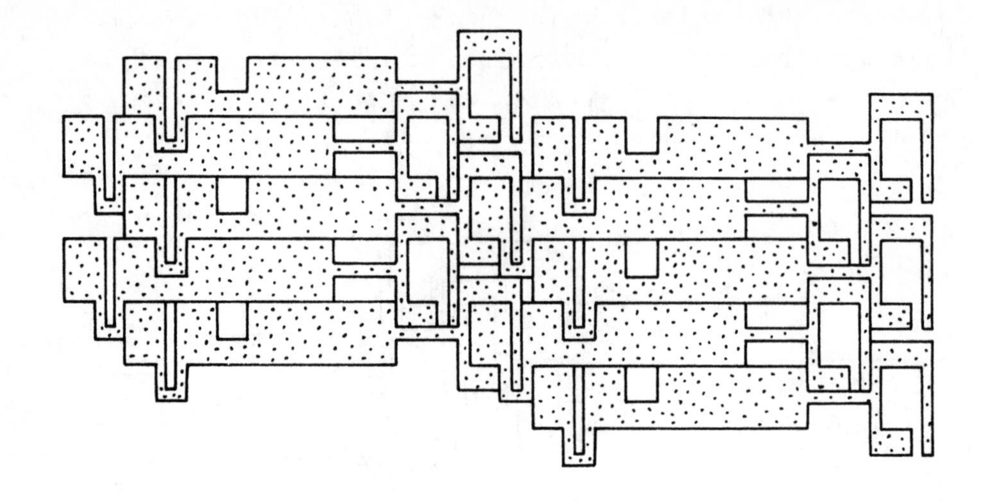

Figure 6 The packing P_2'

REFERENCES

[1] L. FEJES TÓTH, *Some packing and covering theorems* Acta Sci. Math. Szeged, 12/A (1950), 62-67.

[2] L. FEJES TÓTH, *On the densest packing of convex discs,* Mathematika 30(1983), 1-3.

[3] L. FEJES TÓTH, *Densest packing of translates of a domain,* Acta Math. Hung., 45(3-4) (1985), 437-440.

[4] C.A. ROGERS, *The closest packing of convex two-dimensional domains,* Acta Math., 86 (1951), 309-321.

A. BEZDEK
Mathematical Institute of the
Hungarian Academy of Sciences
Budapest, Reáltanoda u. 13-15.
H-1053.

G. KERTÉSZ
Technical Univ. Budapest
Faculty of Civil Eng.
Dep. of Mathematics
Budapest, Stoczek u. 2.
H-1152.

COLLOQUIA MATHEMATICA SOCIETATIS JÁNOS BOLYAI
48. INTUITIVE GEOMETRY, SIÓFOK, 1985.

ON THE AVERAGE NUMBER OF NEIGHBORS IN A SPHERICAL PACKING OF CONGRUENT CIRCLES

K. BEZDEK*, R. CONNELLY*, G. KERTÉSZ

1. INTRODUCTION

Recently L. Fejes Tóth [1] raised the following question: Consider a packing[1)] of circles of radius $r(>0)$ on the unit sphere. The neighbors of a circle are the circles that touch it, so we can compute the average number[2)] of neighbors for a given packing and also we can compute the maximum value of these average numbers for the given r. What is the lim sup of these maximal averages as r tends to zero? Is the lim sup smaller than five?[3)]

In the first part of this note we given an affirmative answer to the second question. The main theorem of this section computes an $\varepsilon > 0$ such that in any packing of circles of radius r, on the unit sphere where r is small enough (i.e. $r \leq r_0$ for a given real number r_0, which can

---

*) Supported by Hung. Nat. Found. for Sci. Research no. 1238.
*) Partially supported by N.S.F. grant, number MCS-790251.
1) Packing means collection of non-overlapping regions.
2) The average number is the arithmetic mean.
3) The lim sup is obviously smaller than or equal to five.

be computed) the average number of neighbors is $\leq 5 - \varepsilon$. See the end of section 2. The technique of our proof is strongly motivated by an article of R.M. Robinson [2], in which he investigated those packings of congruent circles on the unit sphere where each circle has exactly five neighbors. In the second part of our note we give constructions for packings of artibrarily small congruent circles on the unit sphere, in which the average number of neighbors is $4\frac{2}{5}$. Some of our remarks there support the conjecture that $4\frac{2}{5}$ might be the largest possible average number of neighbors in a packing of circles of radius approaching 0.

We also ask the following. Consider a packing of circles of radius r in the hyperbolic plane. As before we compute the average number of neighbors of the circles in the given packing, and we look for their maximum for a given r. Finally we try to determine the lim sup of these maximums as r tends to 0. Is this lim sup smaller than five? We conjecture the answer is "yes" to this question. An easy argument shows that the lim sup above is not greater than $5\frac{1}{2}$, so the real difficulty of our problem is to show that the lim sup is near five. Also, we refer the reader a good hyperbolic construction in the last section.

2. THE MAIN THEOREM

Suppose that we have circles of radius $r(>0)$ on the unit sphere[4)] forming a packing $\mathscr{P}$. We consider the graph $G(\mathscr{P})$ of this packing (the vertices are the centers of our circles, and the edges are the segments of length

---

[4)] As usual, in the spherical geometry the lines are greatcircles, the segments are their arcs, and we measure the length, the angles, the areas in degrees.

2r connecting the centers of two touching circles) which divides the sphere into connected regions, forming a tessallation $\mathcal{T}(\mathcal{P})$. Let us denote the average degree of $G(\mathcal{P})$ (or of $\mathcal{T}(\mathcal{P})$) by deg ($\mathcal{P}$). We are going to show that

$$(*) \qquad \limsup_{r\to 0}[\sup_{\mathcal{P}} \deg(\mathcal{P})] < 5.$$

In other words (*) means that there exist an $\varepsilon > 0$ and an $r_0 > 0$ such that in any packing of congruent circles of radius $r \leq r_0$ on the unit sphere the average number of neighbors of the circles in the packing is at most $5 - \varepsilon$.

A simple computational argument shows that the degree of any vertex of the graph $G(\mathcal{P})$ of $\mathcal{P}$ is at least 3 otherwise we can remove a vertex of $G(\mathcal{P})$ while the average degree of $G(\mathcal{P})$ increases except in the case when the average degree of $G(\mathcal{P})$ is less than four. (But in this case we do not have to look any further.) Thus the vertices of $G(\mathcal{P})$ are identical with the vertices of $\mathcal{T}(\mathcal{P})$. Now choose a face F of $\mathcal{T}(\mathcal{P})$. This is not necessarily simply connected. (Roughly speaking this could be a "ring" also.) The nth neighborhood of F (n is a positive integer) with respect to the faces of $\mathcal{T}(\mathcal{P})$ (with respect to the edges of $\mathcal{T}(\mathcal{P})$, respectively) is the union $\mathcal{N}_f(n,F)$ ($\mathcal{N}_e(n,F)$, respectively) of those faces (edges, respectively) of $\mathcal{T}(\mathcal{P})$, which can be reached from F using at most n consecutive faces (edges, respectively) of $\mathcal{T}(\mathcal{P})$. (Two elements are consecutive if they share a common point. Here we suppose also that the edges and the faces of $\mathcal{T}(\mathcal{P})$ are closed sets.) We denote the average degree of the vertices of the graph $G(\mathcal{P})$ belonging to $\mathcal{N}_f(n,F)$ ($\mathcal{N}_e(n,F)$, respectively) by $\deg_f(n,F)$ ($\deg_e(n,F)$, respectively).

($\deg_f(n,F) \le 5$, $\deg_e(n,F) \le 5$ and equality holds if and only if all of the vertices have degree 5.)

LEMMA 1. *If* $\deg_e(2(k+1),F) = 5$ $(k \ge 1)$ *and* r *is small enough, then*

(1) $\begin{cases} \deg_f(k+1,F) = 5 \text{ *and any face of* } \mathcal{N}_f(k+1,F) \\ \text{*is a regular triangle or a rhombus or a convex*} \\ \text{*equilateral pentagon, with angles*} < 120^0\text{, *of*} \\ \text{*side-length* } 2r\text{, *and at most two non-triangular*} \\ \text{*faces of* } \mathcal{T}(\mathcal{P}) \text{ *can meet at a vertex of*} \\ \mathcal{N}_f(k,F) \text{ *and if two do meet, then the three*} \\ \text{*regular triangles at the vertex are split*} \\ \text{*into groups of one and two.*} \end{cases}$

PROOF. First we show that if $k \ge 0$ and $\deg_e(2(k+1),F)=5$ and r is small enough, then $\deg_f(k+1,F) = 5$, and the faces of $\mathcal{N}_f(k+1,F)$ are regular triangles, rhombuses or convex equilateral pentagons (with angles $< 120^0$) of side-length 2r. We use induction on k. Suppose $k = 0$. Let $\bar{F}$ be an arbitrary face of $\mathcal{N}_f(1,F)$. Then there is a common vertex V of $\bar{F}$ and F (Fig.1.) and the consecutive edges VU', U'W' and VU, UW of $\bar{F}$ lie

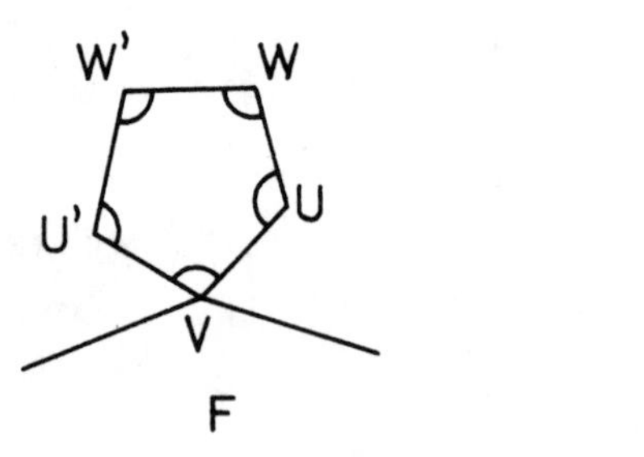

Figure 1

in $\mathcal{N}_e(2,F)$. Since $\deg_e(2,F) = 5$ the degree of the vertices V, U, W, U', W' is five and consequently the angles of $\bar{F}$ at these vertices are smaller than or equal to $360^0 - 4\alpha$, where α is the angle of the regular triangle of side $2r$ ($360^0 - 4\alpha < 120^0$). Then $\bar{F}$ is at most a five-gon and of course is convex and equilateral of side-length $2r$. Obviously $\deg_f(1,F) = 5$, so we are done with the case $k = 0$.

Now suppose that for k we know our assertion and let $\deg_e(2(k+1),F) = 5$. Then of course $\deg_e(2k,F) = 5$ so by induction $\deg_f(k,F) = 5$ and any face of $\mathcal{N}_f(k,F)$ is a regular triangle or a rhombus or a convex equilateral pentagon (with angles $< 120^0$) of side-length $2r$. Let $\bar{\bar{F}}$ be an arbitrary face of $\mathcal{N}_f(k+1,F)$ which shares a common vertex V with a face $\bar{F}$ of $\mathcal{N}_f(k,F)$. From F we can get to V using at most $2k$ consecutive edges of $\mathcal{T}(\mathcal{P})$ (using the inductive step) which means that we can repeat the logic of the case $k = 0$ for the faces $\bar{\bar{F}}$, $\bar{F}$, proving the first part of our Lemma 1. In order to prove the second part of our Lemma we observe that the sum of two adjacent angles in a rhombus or in an equilateral convex pentagon is $> 180^0$. Hence two nontriangular faces of $\mathcal{N}_f(k+1,F)$ cannot about, since the angles at one end of the common side would add up to more than 180^0, leaving $< 180^0 < 3\alpha$ for the remaining three angles at this vertex. Thus at most two non-triangular faces of $\mathcal{N}_f(k+1,F)$ can meet at a vertex of $\mathcal{N}_f(k,F)$ and if two do meet, then the triangles split up into two groups as desired.

LEMMA 2. *If* (1) *holds for* $k+1 = n+2$ ($n \geq 1$) *and if two non-triangular faces of* $\mathcal{T}(\mathcal{P})$ *meet at one vertex of* $\mathcal{N}_f(n,F)$, *then one of them has an angle* $\alpha_1 = 360^0 - 4\alpha$ (where α is the angle of a regular triangle of side $2r$) *at a vertex adjacent to the above vertex of* $\mathcal{N}_f(n,F)$.

PROOF. At the given vertex P of $\mathcal{N}_f(n,F)$ we have three equilateral triangles of side 2r spliting up into groups of one $\{T_1\}$, and two $\{T_2,T_3\}$ (Fig.2.). Either we can find a vertex of T_1 different from P (belonging to $\mathcal{N}_f(n+1,F)$) such that we have four equilateral triangles of side 2r there (which proves our lemma) or at any vertex of T_1 we have exactly two of the nontriangular faces F_1, F_2, F_3 of $\mathcal{N}_f(n+2,F)$, which are adjacent to T_1 (Fig. 2/a), or there is an equilateral triangle T_4 adjacent to T_1 and four nontriangular faces F_1, F_2, F_3, F_4 of $\mathcal{N}_f(n+2,F)$ which are adjacent to T_1, T_4 (Fig. 2/b).

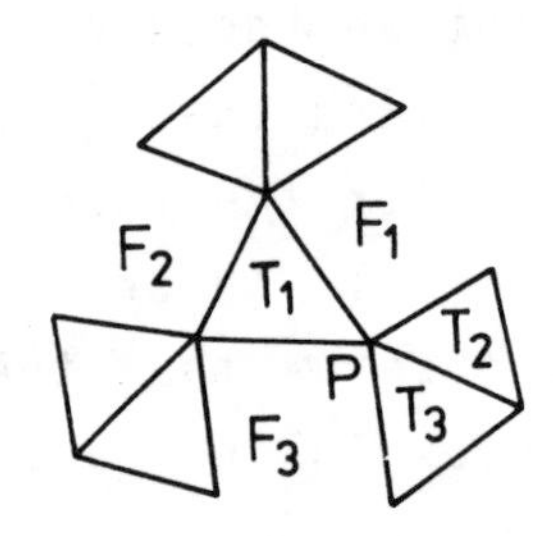

Figure 2/a

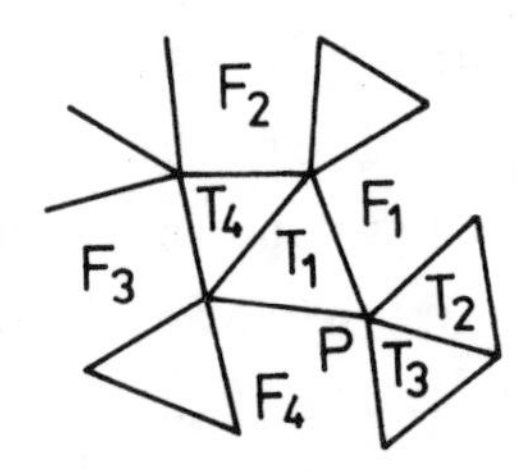

Figure 2/b

Since in the faces F_1, F_2, F_3 (F_1,F_2,F_3,F_4), the sum of two adjacent angles is larger than 180^0, there exists a vertex of T_1 (T_1 and T_4) where the sum of the angles of the nontriangular faces is larger than 180^0, which is a contradiction because the degree of that vertex is five, and so the above sum is $\le 360^0 - 3\alpha < 180^0$.

LEMMA 3. *Suppose that we have a convex equilateral pentagon of side* 2r *on the unit sphere such that the angles are* $\le 120^0$. *If* r *is small enough, then the angles are obtuse.*

PROOF. The lemma is correct without any restriction with respect to r, but we need only this weaker version of it. For the proof it is enough to observe that if r is small then we can check our lemma easily in the euclidean plane or we can refer to the article of R.M. Robinson [2] which contains a proof of a slightly stronger version of this lemma.

LEMMA 4. *If* (1) *holds for* $k+1=n+2$, $(n \geq 1)$ *and* r *is small enough, then we have at most two different shapes of rhombuses in* $\mathcal{N}_f(n,F)$.

PROOF. Suppose that we can find three different shapes of rhombuses in $\mathcal{N}_f(n,F)$ with angles $\alpha_i \geq \beta_i$, $\alpha_j \geq \beta_j$, $\alpha_k \geq \beta_k$. We may suppose also that $\alpha_i > \alpha_j > \alpha_k$. So $\alpha_1 (=360^0 - 4\alpha) > \alpha_j > \alpha_k$. Let $P_j, P_k \in \mathcal{N}_f(n,F)$ be vertices at which the angles α_j, α_k, occur. Obviously there is another nontriangular face of $\mathcal{N}_f(n+1,F)$ at P_j, P_k, respectively (Otherwise $\alpha_j = \alpha_i$ or $\alpha_k = \alpha_1$). This can only be a rhombus since $\alpha_j > \alpha_k > 90^0$ (see Lemma 3). Thus because of Lemma 2 this rhombus has angles $\alpha_1 \geq \beta_1$, which implies that $\beta_1 + 3\alpha + \alpha_j = 360^0$ and $\beta_1 + 3\alpha + \alpha_k = 360^0$ i.e. $\alpha_j = \alpha_k$, which is a contradiction.

LEMMA 5. *If* r *is small enough, then* $\deg_e(12,F) < 5$.

PROOF. Suppose that $\deg_e(12,F) = 5$. Lemma 1 implies that (1) holds for $k+1 = 6$ i.e. $k = 5$ and so by Lemma 4 we have at most two different shapes of rhombuses in $\mathcal{N}_f(4,F)$. At least one of our two rhombuses can be found in $\mathcal{N}_f(2,F)$. Otherwise there would be a pentagon in $\mathcal{N}_f(1,F)$ such that its neighbors would be regular triangles of side 2r lying in $\mathcal{N}_f(2,F)$ (here we used that r is small enough). This is a contradiction, because then all of the angles of our pentagon have to be close to 120^0 provided that r is small enough.

Thus there is a rhombus with angles $\bar{\alpha} \geq \bar{\beta}$ in $\mathcal{N}_f(2,F)$. Either $\bar{\alpha} = \alpha_1 = 360^0 - 4\alpha$ or there is an acute angle $\bar{\bar{\beta}}$ of another rhombus in $\mathcal{N}_f(3,F)$ such that $\bar{\bar{\beta}} + \bar{\alpha} + 3\alpha = 360^0$. In the second case the largest (i.e. the obtuse angle of our second rhombus) has to be $\alpha_1 = 360^0 - 4\alpha$, since we have only two different shapes of rhombuses in $\mathcal{N}_f(4,F)$. This means that in any case if r is small enough, then the acute angle (obtuse angle) of our rhombuses in $\mathcal{N}_f(4,F)$ have to be close to 60^0 (120^0). This implies that there is no pentagon in $\mathcal{N}_f(3,F)$, if r is small. To see this suppose that we have a pentagon in $\mathcal{N}_f(3,F)$, whose arbitrary angle is denoted by γ. As we know γ is obtuse so either $\gamma = 360^0 - 4\alpha$, or there is an acute angle β of a rhombus in $\mathcal{N}_f(4,F)$ such that $\gamma + \beta + 3\alpha = 360^0$. In both cases if r is small, then γ will be close to 120^0 because α and β are close to 60^0, which is a contradiction, as we saw before.

Finally we have to investigate the case when we have only regular triangles and rhombuses of side 2r in $\mathcal{N}_f(3,F)$. Obviously there is a rhombus $\bar{F}$ in $\mathcal{N}_f(1,F)$ if r is small. Hence in $\mathcal{N}_f(2,\bar{F})$ we have only regular triangles and at most two different shapes of rhombuses of side 2r. Starting with $\bar{F}$ we get the following two possible subgraphs in $\mathcal{N}_f(2,\bar{F})$ (provided that r is small) (Fig. 3).

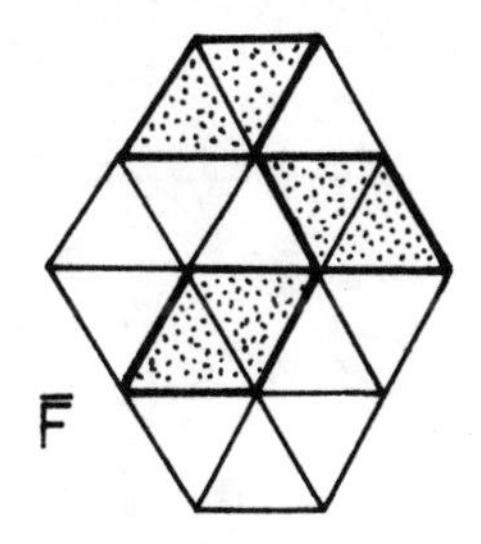

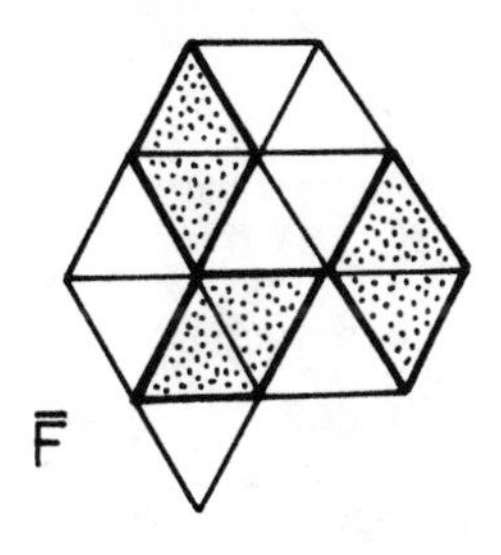

Figure 3

It is easy to check that both are impossible because the indicated 3 rhombuses of $\mathcal{N}_f(2,\bar{F})$ should be different, although we know that in this neighborhood we have at most two different shapes of rhombuses. This proves Lemma 5.

Now we are in a position to do the final steps of our investigation. Let us consider an arbitrary vertex of $G(\mathcal{P})$ and draw a circle of radius 28r centered at the given vertex. If r is small, then Lemma 5 implies that this circle contains at least one vertex of $G(\mathcal{P})$ with degree smaller than five. On the other hand simple volume estimats show that this "big" circle contains at most $\frac{29^2 \cdot r^2 \cdot \pi}{r^2 \cdot \pi} = 29^2 = 841$ vertices of $G(\mathcal{P})$. (r is small!) Hence for N vertices of $G(\mathcal{P})$ there are at least N/840 vertices of degree at most four. (Use the fact, that the circles of radius 28r (r is small) centred at the vertices of $G(\mathcal{P})$ with degree at most 4 cover the unit sphere.) Consequently

THEOREM. *There is an* $r_0 > 0$ *such that for any packing of congruent circles of radius* $r \le r_0$ *on the sphere of unit radius the average number of neighbors of the circles in the packing is at most*

$$\frac{5+\frac{4}{840}}{1+\frac{1}{840}} = 4{,}99881\ldots .$$

This answers our original question. I.e. we have proved (*).

We mention the following two facts. It is possible to get a smaller upper bound than 4,99881... but not much and our proof here has the advantage that it is relatively short. Also we could have concluded our theorem with the help of the following inequality $\frac{\Sigma i n_i}{\Sigma n_i} \le \frac{\Sigma i^2 n_i}{\Sigma i n_i}$, where n_i means the number of the vertices of $G(\mathscr{P})$ with degree i. The right side of this inequality makes it possible to restrict our problem to the faces, or rather to some collections of the faces of $G(\mathscr{P})$. We then compute the average degree of the vertices of $G(\mathscr{P})$ on each face and find an upper bound for certain groups of these average numbers.

3. CONSTRUCTIONS

L. Fejes Tóth found a packing of congruent circles on the unit sphere, where the radius was arbitrarily small, and the average number of neighbors of a circle in the packing was $4\frac{1}{3}$. We show how to improve this average number to $4\frac{2}{5}$ and we shall provide some justification for speculating that $4\frac{2}{5}$ might be the largest possible average degree for circles of radius approaching 0.

As before we will describe the packing by means of its graph, where the centers of the circles are the vertices, and two vertices are joined if the corresponding circles intersect.

First, it seems natural to only consider graphs that are rigid. If a graph is not rigid, we can "flex" the graph until it is rigid due to new contacts. We are trying to maximize the number of contacts for any fixed number of circles in the packing.

Second, if a graph is rigid it "probably" is infinitesimally rigid which implies, among other things, that $2v - 3 \leq e$, where v is the number of vertices of the graph (i.e. the number of circles in the packing) and e is the number of edges in the graph (i.e. the number of pairs of touching circles).

Third, it seems apparent that the only way to constract graphs with a large e, for a fixed v, is to "glue" several rigid subgraphs.

Fourth if the gluing is to be well-defined for the sphere for an interval of values for r, the radius of the circles, the only method that seems to work is to take two copies of the same graph and identifies them along corresponding vertices, after reflecting (about a line or a point) one to get the other. Note that for some graphs, for some special values of r, it may be possible to identify more than 2 points, but not for an entire interval on the sphere.

Note that if the graph has average degree d, then $dv = 2e$.

Thus we seem to be forced into starting with a graph G_0 with v_0 vertices, e_0 edges such that $2v_0 - 3 = e_0$. If $2v_0 - 3 < e_0$ we expect to find smaller pieces where equality holds. Thus the average degree of G_0 is

$$d_0 = \frac{2e_0}{v_0} = \frac{4v_0-3}{v_0} = 4 - \frac{3}{v_0}.$$

If we perform the operation described above of taking two copies of G_0 and identifying 2 of its vertices, to get a graph G_1, then

$$v_1 = 2v_0 - 2,\ e_1 = 2e_0,\ d_1 = \frac{2e_1}{v_1} = \frac{4e_0}{2v_0-2} = 4 - \frac{4}{2v_0-2},$$

computing the number of vertices, edges and average degree.

Doing the above k times we get a graph G_k with

$$v_k = 2v_{k-1} - 2 = 2^k v_0 - 2^{k+1} + 2,\ e_k = 2^k e_0,$$

$$d_k = \frac{2e_k}{v_k} = \frac{2^{k+1}e_0}{2^k v_0 - 2^{k+1} + 2} = \frac{2^{k+2}v_0 - 3\cdot 2^{k+1}}{2^k v_0 - 2^{k+1} + 2} =$$

$$= \frac{4v_0 - 6}{v_0 - 2 + \frac{1}{2^{k-1}}}.$$

Thus

$$\lim_{k\to\infty} d_k = \frac{4v_0 - 6}{v_0 - 2},$$

which means that in order to maximize d_∞ $(= \lim_{k\to\infty} d_k)$ we must minimize v_0.

L. Fejes Tóth found a graph G_0 coming from a packing as shown in Fig. 4 with $v_0 = 8$ yielding $d_\infty = 4\frac{1}{3}$.

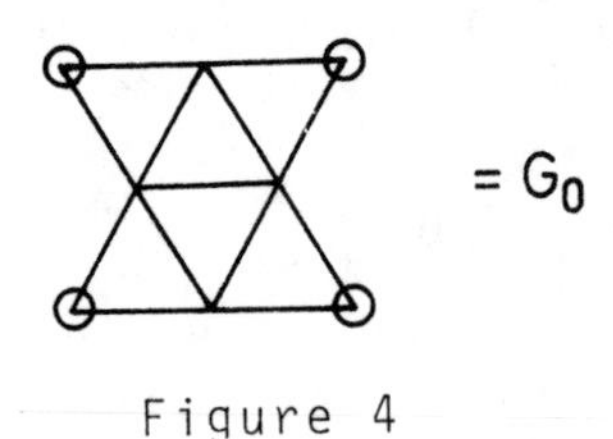

Figure 4

The circled vertices are those that can be identified to other copies of G_0 sucessively to get the limiting graph as shown in Fig. 5:

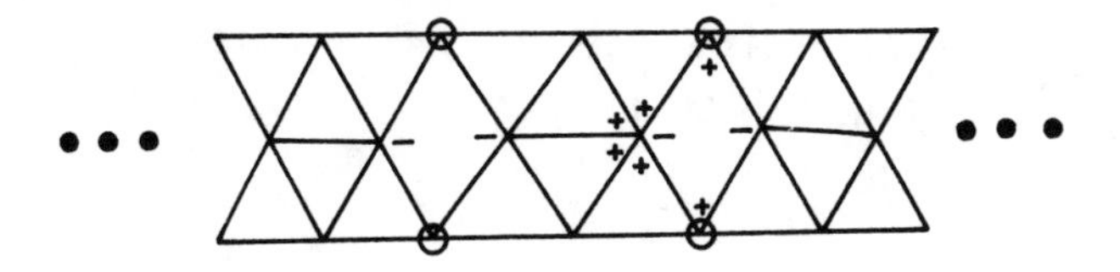

Figure 5

A plus or minus indicates that the angle is (slightly) larger than or smaller than (respectively) the limiting euclidean angle. When the limiting angle is 60^{0}, then we need that the angle be labelad + to insure that the graph comes from a packing of congruent circles.

The following (Fig. 6) is a graph G_0 with $v_0 = 7$, yielding $d_\infty = 4\frac{2}{5}$

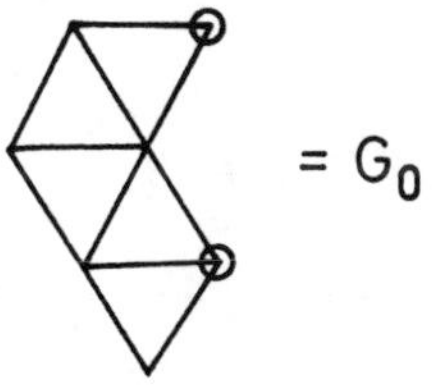

$= G_0$

Figure 6

The following (Fig. 7) describes the glueing process:

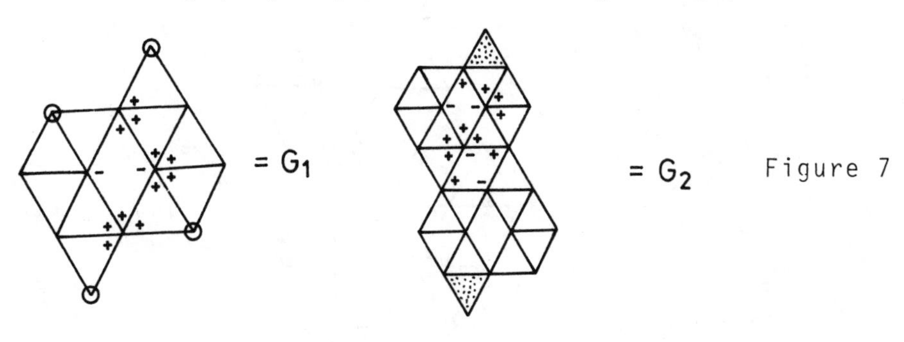

Figure 7

There are other variations on these constructions, by moving the shaded triangles to other positions before constructing G_3 for instance, but they all yield $d_\infty = 4\frac{2}{5}$.

For the hyperbolic case we can use the same general method as for the spherical case, but we need a different generator G_0 given below (Fig. 8).

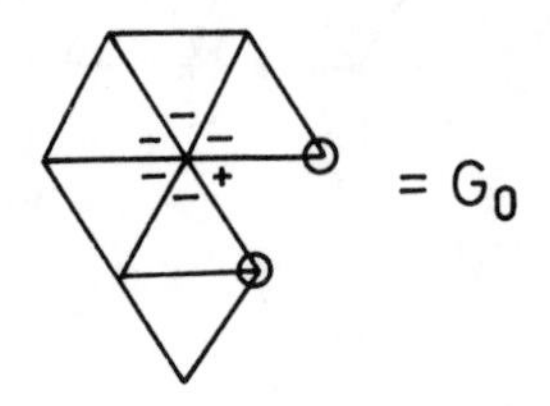

Figure 8

G_1 is obtained as follows (Fig. 9).

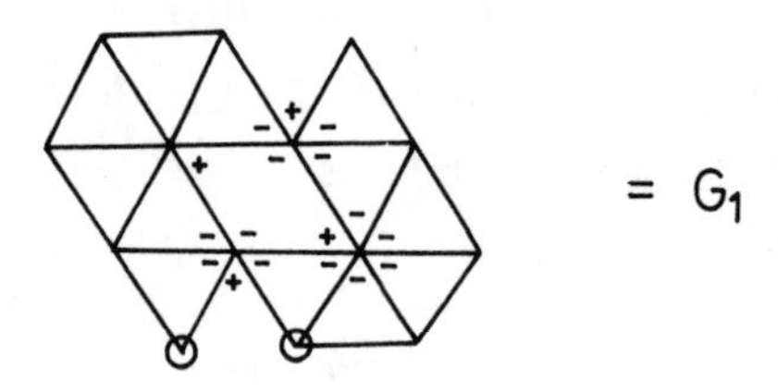

Figure 9

G_2, G_3 etc. are obtained by "translating" appropriate copies of the previous graph (Fig. 10).

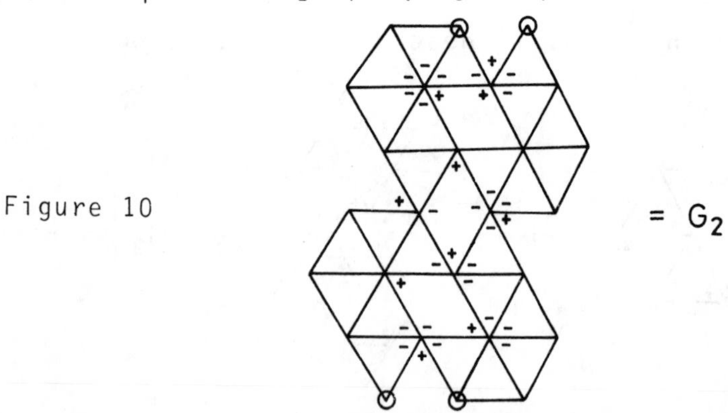

Figure 10

We have indicated with + or - signs which angles are larger or smaller than the Euclidean limit, as before. Every equilateral triangle must have all internal angles -, and as before when all but one angles around a vertex are labeled one sign, the last angle must have the other sign. Also if a rhombus has two opposite angles labeled +, then the other two angles must be labeled - .

Then with the above labeling facts, we can see that all angles whose Euclidean limit is 60^0 are either part of an equilateral triangle or are labeled + . Thus each G_i, $i = 0,1,2,\ldots$ corresponds to a packing in the hyperbolic plane.

We see that the number of vertices of G_0 is $v_0 = 8$, so the limiting average degree $d_\infty = 4 + \frac{2}{8-2} = 4\frac{1}{3}$.

A calculation showes that if we try a glueing as below (Fig. 11), then we do not get a packing at the circled vertices.

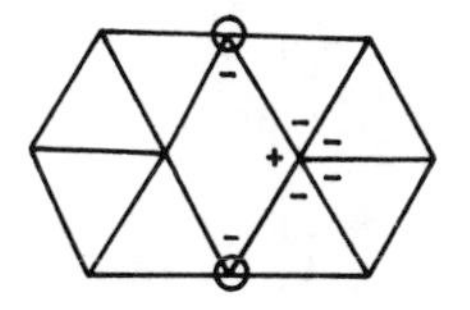

Figure 11

Thus it seems we need to start with a G_0 having at least 7 vertices and need to add the eighth to be able to continue. This glueing method should provide examples with average degree no larger than a $4\frac{2}{5}$ and $4\frac{1}{3}$ "*may*" be the largest average degree in the hyperbolic plane.

REFERENCES

[1] L. FEJES TÓTH, personal communication.

[2] R.M. ROBINSON, *Finite sets of points on a sphere with each nearest to five others*, Math. Ann. 179, 296-318 (1969).

K.BEZDEK
Eötvös Univ.
Dept. of Geometry
1088 Budapest
Rákóczi út 5
Hungary

R.CONNELLY
Cornell Univ.
Dept. of Math.
Ithaca, N.Y. 14853
U.S.A.

G. KERTÉSZ
Technical Univ.
Budapest
Faculty of Civil Eng.
Dept.of Mathematics
Budapest,
Stoczek u. 2.
1152
Hungary

COLLOQUIA MATHEMATICA SOCIETATIS JÁNOS BOLYAI
48. INTUITIVE GEOMETRY, SIÓFOK, 1985.

LINEARLY RELATED PLANE CONVEX SETS

T. BISZTRICZKY - J. SCHAER

Let A^2 denote the real affine plane with lines $p, q, r, \ldots$. Let F be a family of convex sets of A^2 and n be a positive integer. Then F may have the following properties:

G_n: For any n sets of F, there is a line meeting exactly these n sets of F.

H_n: The boundary of the convex hull of any subset F' of F meets at most n sets of F'; hence the convex hull of F' is the convex hull of at most n setst of F'.

I_n: Any line meets at most n sets of F.

I : Any line meets only a finite number of sets of F.

T_n: For any n sets of F, there is a line meeting these n sets.

T : There is a line meeting every set of F.

We observe that T_n is a weaker condition than G_n and that much attention has been directed towards determining when T_n ($n \geq 3$) implies T. To cite a few examples: Grünbaum [1] has shown that if F is a family of disjoint translates of a parallelogram then T_5 implies T; Klee [3] determined that if F is a family of parallel line segments then T_3 implies T; and finally Hadwiger [2] proved that if F is a family of infinitely many pairwise disjoint congruent compact convex sets in the Euclidean m-space E^m with nonempty interiors then T_{m+1} implies T.

As opposed to the preceding Helly type problems, we propose in this note to determine some consequences of the stronger condition G_n ($n \geq 3$) under the weaker assumption that F is any family of pairwise disjoint compact convex sets. The major consequence is that only finite F can satisfy G_n. More precisely, G_n implies H_{n+2} and I_{n+8}, and H_n and I imply that F is finite. The latter result is somewhat unexpected, as it follows by neither H_n nor I alone.

I. THE PROPERTY G_n

Let F be a family of pairwise disjoint compact convex sets of A^2. For any collection $\{A,B,C,\dots\}$ of subsets of A^2, we denote by $H(A \cup B \cup C \cup \dots)$ its convex hull. In this section we assume that F satisfies G_n, $n \geq 3$.

1. THEOREM. G_n *implies* H_{n+2}.

PROOF. Let $F' \subseteq F$ and $H = H(\bigcup_{A \in F'} A)$. Suppose that there are n+3 sets in F', say $A_1, A_2, \dots, A_{n+3}$, each of which meets the boundary ∂H of H.

We first note that $A_i \cap \partial H$ is connected for $i=1,2,\dots$ $\dots,n+3$. For if $A_i \cap \partial H$ is not connected for some i, then $\partial H \setminus A_i$ has at least two components. As A_i is a closed set and ∂H is a closed curve, it follows that $(\operatorname{int} H) \setminus A_i$ has at least two components. Then $H = H(\bigcup_{A \in F'} A)$ yields that each component of $H \setminus A_i$ contains a set of F'. Thus there are sets A and B in F' which lie in different components of $H\setminus A_i$. But then any line which meets A and B also meets A_i, and $|F'| = n$ by G_n. This contradicts our assumption that $|F'| \geq n+3$.

Since ∂H is a closed convex curve, we orient ∂H and label the sets A_i so that ∂H meets them in the cyclic order

$$A_1, A_2, \ldots, A_{n+3}, A_1.$$

By G_n, there is a line p meeting exactly the sets $A_1, A_3, A_5, A_7, \ldots, A_{n+3}$ and there is a line q meeting exactly the sets $A_2, A_4, A_6, \ldots, A_{n+3}$.

As p meets A_1 and A_3 and $A_1 \cap A_3 = \emptyset$, we obtain

$$\tilde{p} = H(p \cap (A_1 \cup A_3)) \setminus (A_1 \cup A_3) \neq \emptyset.$$

Next $A_2 \cap (p \cup A_1 \cup A_3) = \emptyset$ implies that A_2 is contained in a region R of H bounded by A_1, A_3 and $\tilde{p}$. Since ∂H meets the sets A_i in a cyclic order, it follows $R \cap \partial H$ meets only the sets A_1, A_2 and A_3. Thus A_1, A_2 and A_3 are the only labelled sets contained in R, and for $j = 4, \ldots, n+3$, $A_j \cap R \neq \emptyset$ implies $A_j \cap \tilde{p} \neq \emptyset$. Hence $p \cap (A_4 \cup A_6) = \emptyset$ yields that $A_4 \cup A_6 \subset H\setminus R$. Since H is convex, and $A_1 \cap \partial H$ are non-empty and connected, we obtain that any line that meets A_2 but not $\tilde{p} \cup A_1 \cup A_3$ also does not meet $H\setminus R$. As q meets $A_2 \subset (A_1 \cup A_3) = \emptyset$, we have

$$(1) \qquad q \cap \tilde{p} \neq \emptyset.$$

The preceding argument is symmetric with respect to p and q; that is,

$$q = H(q \cap (A_4 \cup A_6)) \setminus (A_4 \cup A_6) \neq \emptyset,$$

A_5 is contained in a region S of H bounded by A_4, A_6 and $\tilde{q}$, $A_1 \cup A_3 \subset H \backslash S$, and

$$(2) \qquad p \cap q \neq \emptyset.$$

Now (1) and (2) yield that $\tilde{p} \cap \tilde{q} \neq \emptyset$, and thus ∂H meets the sets A_1, A_3, A_4 and A_6 in the cyclic order A_1, A_4, A_3, A_6 (or its reverse A_1, A_6, A_3, A_4). As this contradicts the original ordering, there are no n+3 sets in F' which meet ∂H; thus H_{n+2}

2. Let $A_i \neq A_j$ in F and set $H_{ij} = H(A_i \cup A_j)$. If H_{ij} is a line segment, say $H_{ij} \subset t$, then G_n implies that the line t meets every set of F, and thus $|F| \leq n$. Hence we assume that H_{ij} is never a line segment (and accordingly, at most one set of F is a one-point set). Then there exist distinct common supporting lines t_{ij} and t_{ji} of A_i and A_j that are also supporting lines of H_{ij}. Let

$$t^*_{ij} = t_{ij} \cap H_{ij} \quad \text{and} \quad t^*_{ji} = t_{ji} \cap H_{ij}.$$

We may now also assume that no set of $F \backslash \{A_i, A_j\}$ meets both t^*_{ij} and t^*_{ji}; otherwise, $|F| \leq n$ by G_n.

3. Let p be a line which meets each of the sets $A_1, A_2, \ldots, A_{n+3}$ of F in such a manner that

$$(3) \qquad \tilde{p} \cap A_i \neq \emptyset \quad \text{for} \quad i=2,\ldots,n+2$$

where $\tilde{p} = H(p \cap (A_1 \cup A_{n+3}))$. Let $H = H(\bigcup_{i=1}^{n+3} A_i)$. By H_{n+2}, $A_j \subset \text{int } H$ for some j, $1 \leq j \leq n+3$.

LEMMA. *Let* q *be a line such that* $q \cap A_j \neq \emptyset$ *and* $q \cap A_i = \emptyset$ *for* $i \neq j$, $1 \leq i \leq n+3$. *Then* $q \cap \tilde{p} \neq \emptyset$.

PROOF. Case 1: $j=1$ or $j=n+3$. If $A_1 \subset \text{int } H$, then (3) implies that $A_1 \subset H_{k,\ell} = H(A_k \cup A_\ell)$ for some k and ℓ, $2 \le k < \ell \le n+3$. Since $q \cap (A_k \cup A_\ell) = \emptyset$ and $q \cap A_1 \subset H_{k,\ell}$, it follows that q separates A_k and A_ℓ in $H_{k,\ell}$, that is, A_k and A_ℓ are in distinct components of $H_{k,\ell} \backslash q$. But then any line which meets A_k and A_ℓ also meets q at a point in $H_{k,\ell}$ between A_k and A_ℓ. As p is such a line,

$$p \cap q \subset H(p \cap (A_k \cup A_\ell)) \subset \tilde{p}.$$

The proof is similar for $j = n+3$.

Case 2: $1<j<n+3$.

We note that $H_{1,n+3} = H(A_1 \cup A_{n+3}) \subseteq H$, and by (3) $p \cap A_i \subset H_{1,n+3}$ for $i = 2,\ldots,n+2$. If $A_j \subset H_{1,n+3}$, then by arguing as in Case 1 we obtain that $q \cap \tilde{p} \neq \emptyset$. Let $A_j \not\subset H_{1,n+3}$. From 2, A_j does not meet both $t^*_{1,n+3}$ and $t^*_{n+3,1}$, and we may assume that $A_j \cap t^*_{1,n+3} \neq \emptyset$. Then $A_j \subset \text{int } H$ implies that there exist A_i and A_k such that

(4) $\qquad A_i$ and A_k both meet $t^*_{1,n+3}$

and

(5) $\qquad A_j \backslash H_{1,n+3} \subset H_{i,k}$ for some $1 \le i \neq j \neq k \le n+3$.

Hence $A_j \subset H_{1,n+3} \cup H_{i,k}$, and it follows that $q \cap H_{1,n+3} \neq \emptyset$ or $q \cap H_{i,k} \neq \emptyset$. Since A_i and A_k both meet $\tilde{p}$, we obtain that $q \cap H_{1,n+3} \neq \emptyset$ in either case and thus again $q \cap \tilde{p} \neq \emptyset$.

4. THEOREM. G_n *and* $|F| > n+9$ *imply* I_{n+8}.

PROOF. Let F satisfy G_3 and suppose that there is a line p meeting twelve sets, say $A_1,\ldots,A_6$ and $B_1,\ldots,B_6$. We label these sets in such a manner that

(6) $\qquad H(p \cap (A_1 \cup A_6)) \cap H(p \cap (B_1 \cup B_6)) = \emptyset$

and for $i=2,3,4,5$;

(7) $\qquad A_i \cap H(p \cap (A_1 \cup A_6)) \neq \emptyset \neq B_i \cap H(p \cap (B_1 \cup B_6))$.

By (7), both collections $\{A_1,\ldots,A_6\}$ and $\{B_1,\ldots,B_6\}$ satisfy (3), and so there is

$$A_j \subset \text{int } H(\bigcup_{j=1}^{6} A_i) \text{ and a } B_k \subset \text{int } H(\bigcup_{i=1}^{6} B_i);$$

$$1 \leq j,\ k \leq 6.$$

If $|F| > 12$ then there is a set $C \in F$ distinct from the A_j's and the B_i's. By G_3, there is a line q meeting only A_j, B_k and C. Thus q satisfies the conditions in 3 with respect to both collections and so by the Lemma, $q \cap H(p \cap (A_1 \cup A_6)) \neq \emptyset$ and $q \cap H(p \cap (B_1 \cup B_6)) \neq \emptyset$. As $q \neq p$, this is a contradiction. Thus G_3 and $|F| > 12$ imply I_{11}.

We now assume $n \geq 4$, and for $3 \leq m \leq n-1$, G_m and $|F| > m+9$ imply I_{m+8}. Let F satisfy G_n and $|F| > n+9$. Let $A \in F$ and set $F' = F \setminus \{A\}$. Then G_n implies that for any n-1 sets of F', there is a line which meets A and only the chosen n-1 sets of F'. Hence F' satisfies G_{n-1} and $|F'| > (n-1)+9$. By the induction hypothesis, any line meets at most (n+1)+8 sets of F', and thus at most n+8 sets of F.

II. THE PROPERTIES H_n AND I.

In this section we assume that F satisfies I and H_n, $n \geq 2$.

5. Let $A \in F$, and denote by s(A) the set of supporting lines of A. We note that s(A) is a connected set in the

space of lines A^2. For $p \in s(A)$, let $Qp[Rp]$ denote the closed half-plane bounded by p which contains [does not contain] A. Clearly, Qp and Rp depend continuously on p. Finally, let $s^\circ(A) = s(A) \setminus s^*(A)$ where $s^*(A) = \{p \in s(A) | Rp$ contains an infinite number of sets of $F\}$.

LEMMA. *For* $A \in F$, *either* $s(A) = s^*(A)$ *or* $s(A) = s^\circ(A)$.

PROOF. For $p \in s^*(A)$, let

$$F_p = \{B \in F | B \subset \text{int } Rp\} \text{ and } H_p^* = H(\bigcup_{B \in F_p^*} B).$$

By I, H_p^* contains an infinite number of sets of F, and by H_n, H_p^* is the convex hull of at most n sets of F_p^*. Thus H_p^* is a closed and bounded convex set in int Rp. Let $q \in s(A)$ tend to p. Since H_p^* is closed and bounded, $p \cap H_p^* = \emptyset$ implies that $q \cap H_p^* = \emptyset$ if q is sufficiently close to p. As R_q tends to R_p as q tends to p, $H_p^* \subset \text{int } R_p$ yields that $H_p^* \subset R_q$ for q sufficiently close to p. Since H_q^* contains an infinite number of sets of F, all such q are in $s^*(A)$. Thus $s^*(A)$ is open in $s(A)$.

For $p \in s^\circ(A)$, let

$$F_p = \{B \in F | B \subset \text{int } Q_p\} \text{ and } H_p^\circ = H(\bigcup_{B \in F_p^\circ} B).$$

Since R_p contains a finite number of sets of F, I implies that H_p° contains all but a finite number of sets of F. We note again that H_p° is a closed and bounded convex set in int Q_p, and argue as in the preceding to obtain that $H_p^\circ \subset Q_t$ for any line $t \in s(A)$ sufficiently close to p. Since H_p° contains all but a finite number of sets of F, we obtain from I and $Q_t \cap R_t = t$ that R_t contains a finite number of sets of F whenever t is sufficiently close to p. Thus all such t are in $s^\circ(A)$, and $s^\circ(A)$ is open in $s(A)$. Since $s(A)$ is connected, the Lemma follows.

6. THEOREM. H_n *and* I *imply* $|F| < \infty$.

PROOF. Let $H = H(\bigcup_{A \in F} A)$. Assuming that $|F| \geq 2$, H is the convex hull of k sets of F, $2 \leq k \leq n$. Let A and B be two of these k sets. Since A and B are compact, convex and disjoint, there is a line t which strictly separates A and B. Let $P_A[P_B]$ be the closed half-plane bounded by t which contains A[B].

Since A is convex and $t \cap A = \emptyset$, there is a line $p \in s(A)$ that is parallel to t and separates t and A. Thus $t \subset R_p$ and $P_B \subset R_p$. Similarly, there is a line $q \in s(B)$ that is parallel to t such that $P_A \subset R_q$.

Since $A \subset H$ and $A \cap \partial H \neq \emptyset$, there is a line $s \in s(A)$ such that s supports H. Then $H \subset Q_s$ and I imply that R_s contains only a finite number of sets of F. Thus $s(A) = s^o(A)$ and we have that R_p, and in particular P_B, contain only a finite number of sets of F. Similarly, $B \subset H$ and $B \cap \partial H \neq \emptyset$ yield that R_q, and hence P_A, contain only a finite number of sets of F. As $A^2 = P_A \cup P_B$ and F has property I, F is a finite family.

COROLLARY. If F has the property G_n, $n \geq 3$, then $|F| < \infty$.

Finally, this study of linearly related plane convex sets raises some open questions. For example: Do these results generalize to higher dimension? What other relationships exist among G_n, H_n, I_n and I? For a fixed $n \geq 3$, we know that any family F of pairwise disjoint compact convex sets of A^2 with property G_n is finite but is there an upper bound for $|F|$?

Since G_n implies T_m for $n \geq m$, the latter question can be answered by T and 4 for two of the families mentioned in the introduction:

If F is a family of disjoint translates of a parallelogram with the property G_n, $n \geq 5$, then $|F| \leq n+9$.

If F is a family of parallel compact line segments with the property G_n, $n \geq 3$, then $|F| \leq n+9$.

REFERENCES

[1] GRÜNBAUM, B., *On common transversals*. Arch. Math. 9. (1958), 465-469.

[2] HADWIGER, H., *Über einen Satz Hellyscher Art*. Arch. Math. 7, (1956), 377-379.

[3] KLEE, V.L., *Common secants for plane convex sets*. Proc. A.M.S. 5, (1954), 639-641.

T. BISZTRICZKY, J. SCHAER
University of Calgary,
Dept. of Mathematics
Calgary, Alberta T2N 1N4
Canada

COLLOQUIA MATHEMATICA SOCIETATIS JÁNOS BOLYAI
48. INTUITIVE GEOMETRY. SIÓFOK, 1985.

ISOPERIMETRIC DIVISIONS INTO SEVERAL CELLS WITH NATURAL BOUNDARY

M.N. BLEICHER

ABSTRACT

This paper is an extension of [1] in which the problem of finding the figure of minimal perimeter with cells of preassigned area was considered. In this paper the problem in which certain fixed curves (natural boundaries) may be used without being counted as part of the perimeter is considered. Several necessary conditions are proved. Examples are given of divisions of the circle, equilateral triangle, and square into a small number of equal parts.

I. INTRODUCTION

In [1] the author considered the problem of the plane figure of fixed total perimeter L which surrounds n regions of fixed area ratio $a_1:a_2:\ldots:a_n$. In this paper analogous problems are considered in which, in addition to the fixed total perimeter L which may be used in any manner, a set, Γ, of fixed immovable curves are also given which may or may not be used in whole or part in forming the optimal configuration. Some theorems on necessary conditions are proven in Section II.

Section III gives examples which illustrate applications of the theorems as well as problems, such as non-unique solutions, the non-existence of solutions, and discontinuities of the solutions for fixed natural boundary as L increases.

These results are related to the work in [2], [3] and [4] in which the problem is restricted to the case in which the division is by polygonal arcs and Γ is a polygon the interior of which is being tesselated into convex parts the total area of which is the total area of the polygon Γ. This paper gives a theoretical background to the more computational work in [5].

II. MAIN RESULTS

Let Γ be the graph of the curve $\gamma(t)$ for t in some interval. In this paper we consider only curves Γ for which in any compact region of the plane, Γ has only finite length.

DEFINITION. Let $P \in \Gamma$. Γ has property A (for Area) at $P = \gamma(t_0)$ if there is a positive constant k and a $\delta > 0$ such that given a point $Q = \gamma(t_1)$ on Γ with $|t_0 - t_1| < \delta$, the line segment PQ and the graph Γ_0 of the curve $\gamma(t)$ for t between t_0 and t_1, bound an area of at most $k|PQ|^2$.

NOTE 1. By the area surrounded by Γ_0 and PQ, where PQ is a Jordan arc, is meant the measure of the union of all bounded components of the complement of $\Gamma_0 \cup PQ$ which were parts of unbounded components of the complement of Γ, (see Figure 1), i.e. the new area bounded by them.

The heavier curve is Γ_0. The shaded area is surrounded by PQ and Γ_0.

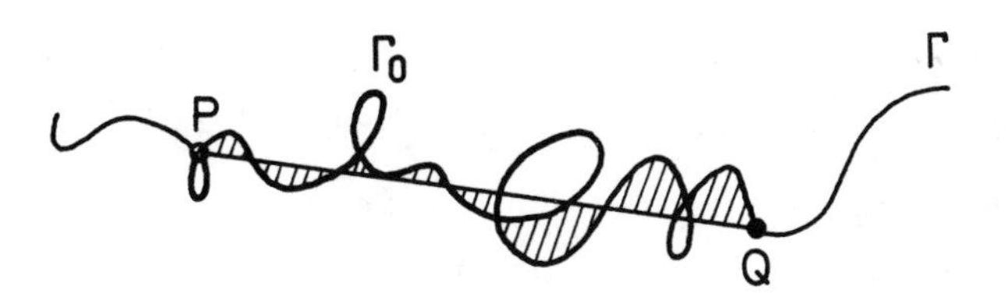

Figure 1

NOTE 2. The choices for t_0 and t_1 and the path connecting them may not be unique; in which case, let Γ_0^* be the union of all curves Γ_0 obtained from different choices. We then require that Γ_0^* and PQ surround an area at most $k|PQ|^2$.

NOTE 3. The curve Γ_0^* can effectively be replaced by at most two curves, one "above" and one "below" PQ.

DEFINITION. Let $P \in \Gamma$. Let PT be a line segment. PT is orthogonal to Γ (at P) if and only if for α defined by

$$\alpha = \lim \inf \{\measuredangle TPQ : Q \in \Gamma, \ |PQ| \to 0\}$$

the inequality $\alpha \geq \frac{\pi}{2}$ holds, where the angle measure $\measuredangle TPQ$ is always positive.

DEFINITION. A circular arc through P from one side is orthogonal to Γ if its tangent ray on that side is orthogonal to Γ.

THEOREM 1. *Let Λ be a fixed curve. If Λ is a curve of length* L, *which together with Γ bounds maximal area then, each component of $\Lambda\backslash(\Lambda\cap\Gamma)$ is a line segment or circular arc which is orthogonal to Γ at any end point at which Γ has property* A.

NOTE 4. If Γ is bounded, then by standard argument from the Blaschke selection theorem this optimal curve Λ will exist. In general Λ is not unique. See examples 4 and 5 in Section III.

NOTE 5. If Λ_1 is a circular arc of Λ then the area it bounds must lie on the same side of Λ_1 as does the center of the circle of which it is part. For, otherwise, replacing any part of the circular arc by its chord would increase the area, while using less length. This fact is used implicitly in the proofs of several theorems.

PROOF. Let Λ_0 be a component of $\Lambda\backslash(\Lambda\cap\Gamma)$. Let $P,Q \in \Lambda\cap\Gamma$ be the end points of Λ_0. If Λ_0 is not a line segment or circular arc then there must be a point with either no curvature or changing curvature. If a chord is drawn around that point, the part of Λ_0 which it cuts off can be replaced by a circular arc of the same length to enclose more area. (A detailed argument for this can be found in [1] or [6].)

Let P be an endpoint of Λ_0 at which Γ has property A. Then $P \in \Gamma \cap \Lambda$. If Λ_0 is not orthogonal to Γ at P then for every possible $\varepsilon > 0$ there is a ray Σ from P which makes an angle $\beta < \frac{\pi}{2}$ with the tangent ray, Ψ, to Λ_0 and such that there are points $Q \in \Sigma \cap \Gamma$ with $|PQ| < \varepsilon$.

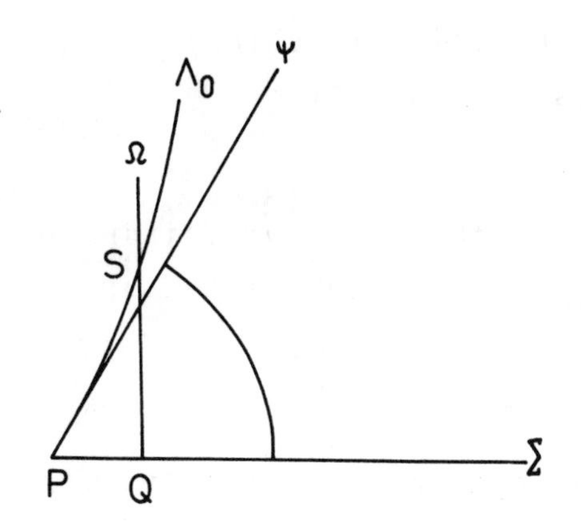

Figura 2a

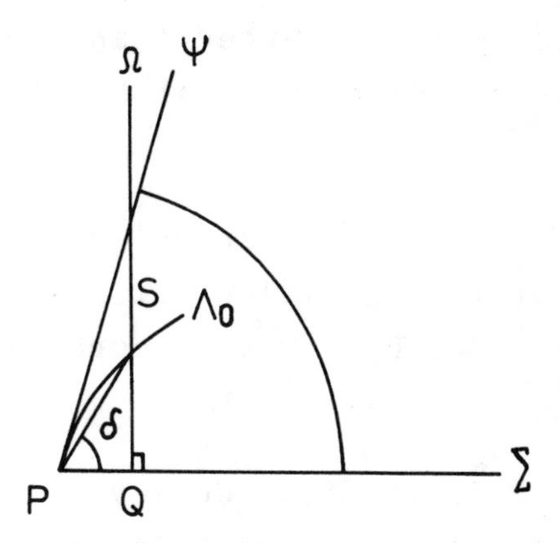

Figure 2b

Let Ω be the ray perpendicular to Σ through Q on the same side of Σ as Λ_0. Let S be a point on $\Lambda_0 \cap \Omega$ nearest Σ, such a point must exist for ε small enough. There are two cases (Figures 2a and 2b); S is outside the angle bounded by Σ and Ψ and S is inside the angle. Replace the arc PS by segment SQ. In the former case, Figure 2a, this surrounds more area and uses less length; thus, Λ could not have been optimal. In the latter case, Figure 2b, there is a length savings of at least $(1 - \sin \delta)\left|\frac{PQ}{\cos \delta}\right|$ where $\delta \leq \measuredangle QPS \leq \beta < \pi/2$. The area loss is at most $k^*|PQ|^2$ for some positive constant k^*. The existence of k^* follows from the fact that Γ has property A at P and because the area of the region PQS is less than $|PQ|^2 \tan \beta$, while for $\varepsilon > 0$ and small enough, $\frac{1-\sin \delta}{\cos \delta} \geq \frac{1-\sin \beta}{\cos \beta} > 0$. Thus we have a savings in length of order $|PQ|$ while the area loss is of order $|PQ|^2$. By picking a chord of the circular arc Λ_0 and replacing the segment it cuts off by a segment using the saved length, at least $(\frac{1-\sin \beta}{\cos \beta})|PQ|$, we increase the area by an amount of order $|PQ|$. For $|PQ|$ small enough this new area will be larger than the area loss at P. Thus if Λ_0 was not orthogonal at P, it was not optimal.

REMARK. A careful reading of the proof will indicate that the condition "at most $k|PQ|^2$" in property A can be replaced by a weaker condition.

DEFINITION. We say Λ is an optimal bounding set for natural boundary Γ if and only if

1. $\Gamma = \bigcup_1^n \Gamma_i$ where each Γ_i is a graph of a curve $\gamma_i(t)$, $t \in I_i$, where I_i is a finite interval.
2. Λ is a union of curves of total length L.
3. The total area of the bounded components of the complement of $\Lambda \cup \Gamma$ is greater than or equal to the total area of the bounded components for $\Lambda' \cup \Gamma$ for any Λ' near Λ. (Where Λ' near Λ means in some neighborhood of Λ in the Hausdorff metric).

COROLLARY. If Λ is an optimal bounding set for natural boundary Γ, then each component of $\Lambda \setminus (\Lambda \cup \Gamma)$ is a circular arc orthogonal to Γ at its end points, if Γ has property A at the end point.

PROOF. If Λ_1 is a component of $\Lambda\backslash(\Lambda\cap\Gamma)$, but is not needed as part of boundary of a component bounded by $\Lambda\cup\Gamma$, then Λ_1 is superfluous and Λ is not optimal. Thus, it may be assumed that Λ_1 is part of the boundary of a component, C. Theorem 1 then says Λ_1 must be a circular arc. One can replace Γ by the single curve Γ^* which forms the remainder of the boundary of C. This replacement can be done since each interval I_i parametrizing Γ^*_γ is finite, thus the total length in the parameter space is measurable and finite and thus can be parametrized on a finite interval. It is now assumed that P is an end point of Λ_1 which has property A.

The closed circular region $C(\delta)$ centered at P of radius δ may have finitely many or infinitely many components of $C(\delta)\cap\Gamma$. If there are only finitely many components then there is a distance $\beta > 0$ such that $C(\beta)$ has in it only parts of the component of $\Gamma\cap C(\delta)$ which contains P. If $C(\delta)$ has infinitely many components then only finitely many of those can also reach inside the circular region $C(\delta/2)$. This is true since any such component has a point on the circle of radius δ and one on the circle of radius $\delta/2$ and hence the component has length at least $\delta/2$. Thus there can be only finitely many such since the length of Γ in the compact set $C(\delta)$ is finite. Thus we obtain a circular region $C(\beta)$, as above.

It is possible to parametrize $\Gamma_2 = C(\beta)\cap\Gamma$ on a finite interval since it is connected and of finite length. If Γ has property A at P then so does Γ_2, since if bounded components of the complement of Γ are excluded then the area surrounded by the chord to Γ_2 is at most the area surrounded by the same chord to Γ. Thus Theorem 1 applied to $\Gamma^*\cup\Gamma_2$ implies that Λ_0 is orthogonal to Γ_2.

But for sufficiently small neighborhoods of P, Γ and Γ_2 are the same. Hence Λ_0 is orthogonal to Γ.

THEOREM 2. *Let Λ be an optimal boundary set for Γ, then if Λ consists of more than one circular arc, all the arcs have the same radius.*

PROOF. To obtain a contradiction, it is assumed that Λ_1 and Λ_2 are distinct circular arcs of differing radii. Without loss of generality it is assumed that Λ_1 is not a straight line segment. On each arc Λ_1 and Λ_2 there are points P_1 and P_2 respectively so that the circular region $C(\varepsilon_1)$ and $C(\varepsilon_2)$, respectively, meet no arcs of Γ or Λ except Λ_1 and Λ_2, respectively. Inside these regions equal length chords, R_1S_1 and R_2S_2, respectively centered at P_1 and P_2, are drawn, making them sufficiently short that each intercepts less than a semicircle. The intercepted arcs are denoted by Δ_1 and Δ_2, respectively. Placing these chords, together with Δ_1 and Δ_2 above the chords, end to end with $R_2 = S_1 = T$, say, on a line segment one notes that the circular arc (line segment) through R_1, T, and S_2 is shorter than the sum of the lengths of Δ_1 and Δ_2. On the other hand by bending the segment at T as a hinge and moving R_1 and S_2 downward and considering the circular arc from R_1 through T to S_2 one obtains in the limiting position when R_1 and S_2 coincide a circle with diameter $TR_1 = TS_2$. The perimeter of this circle is greater than the sum of the lengths of Δ_1 and Δ_2 since these were less than semicircles. By continuity there will be a position where the circular arc R_1TS_2 has the same length as the sum of Δ_1 and Δ_2. If the rest of this circle is drawn one obtains two figures of equal perimeter, the first the circle through R_1, T, and S_2 and the second consisting of Δ_1, Δ_2, and the circular arc from S_2 to R_1 which doesn't pass through T. The circle surrounds greater area by the classical isoperimetric theorem, but the only difference is in the area bounded by the chords R_1S_1 and R_2S_2 and the arcs Δ_1 and Δ_2, respectively, and the circular caps on these chords. Thus Δ_1 and Δ_2 can be replaced by these equal circular caps on the same chord to increase the total area bounded by Γ and Λ. This contradicts the optimality of Λ. It follows that all the pieces of Λ are circular arcs of equal radius.

THEOREM 3. *If* Λ *is optimal for natural boundary* Γ *and two arcs of* Λ, Λ_1 *and* Λ_2, *meet at a point* P *of* Γ *which has property* A, *then* Λ_1 *and* Λ_2 *have a common tangent line at* P *and are on opposite sides of* Γ.

PROOF. The tangent rays, R_1 and R_2, respectively to the arcs Λ_1 and Λ_2 at P form two angles which together contain the whole plane. There are two cases.

CASE 1: One angle contains neither arc Λ_1 nor Λ_2 and the other angle contains both Λ_1 and Λ_2.

CASE 2: Each angle contains one of the arcs.

Case 1 is considered first. In this case the angle containing no arcs can not measure less than π, for otherwise choose points Q_1 and Q_2, near enough to P on Λ_1 and Λ_2 respectively so that the chord Q_1Q_2 lies outside the circles of which Λ_1 and Λ_2 are part. Replace the curve Q_1PQ_2 by the shorter line segment Q_1Q_2. More area is bounded with less length; this violates the optimality of Λ.

If the two arcs Λ_1 and Λ_2 are part of the boundary of the same component of $\mathbb{R}^2\setminus(\Lambda\cup\Gamma)$ then $\Lambda_1 \cup \Lambda_2$ can be considered as one arc and hence is circular and the two tangent rays R_1 and R_2 form the tangent line to this circular arc at P.

If Λ_1 and Λ_2 are parts of the boundary of different regions then P must be on an arc of Γ which is part of the boundary of at least one of these regions. By Theorem 2 both Λ_1 and Λ_2 are orthogonal to Γ at P, hence they must make an angle of at least π with one another. Case 1 is completed.

In case 2, Λ_1 and Λ_2 can not be parts of the same circular arc, hence they must be parts of the boundaries of distinct regions.

Another part of the boundary of the region A_1 bounded by Λ_1 must have part of Γ which goes through P, say Γ_1 as its boundary. For, otherwise, Λ_1 would continue as a circular arc through P and by picking the right two circular arcs of the three which meet at P, the angle which contains neither would be less than π and as in case 1 a chord could be drawn to surround more area with less length. Similarly for Λ_2. It follows that P is in the relative interior of Γ and that R_1 and R_2 are each orthogonal to Γ at P. But this means that they are orthogonal to Γ at P from opposite sides and hence Γ has a tangent line at P and R_1 and R_2 are both orthogonal to it from opposite sides. Hence $R_1 \cup R_2$ is a line through P orthogonal to Γ. The theorem is proved.

Up to this point we have not needed to distinguish the dual problems of surrounding the maximal area with a given length and surrounding a given area with the minimal perimeter. This was true since if a given area could be surrounded by less perimeter, the saved perimeter could be used to increase the total area for the given length. Thus proving that a given condition was necessary for optimality for surrounding a given area also proved it was necessary for surrounding the maximal area with a given length*.

DEFINITION. Let a natural boundary Γ be given as well as a ratio $a_1:a_2:\ldots:a_n$, and a constant A. Suppose a collection of curves Λ of finite total length is such that $\Gamma \cup \Lambda$ divides the plane into components n of which have area ratio $a_1:a_2:\ldots:a_n$ and the total area of these n components is A. We call Λ_0 *optimal* for the given parameters if the total length of Λ_0 is minimal among all such collections of curves in some neightborhood of Λ_0.

NOTE 6. If Γ is a Jordan curve we can add the requirement that the area surrounded by $\Gamma \cup \Lambda$ is all inside Γ, and none of the proofs of the following theorems are effected. This assumption was part of the motivating examples of [5], in which case A was the area surrounded by Γ.

THEOREM 4. *If Λ is optimal for Γ with ratio $a_1:a_2:\ldots:a_n$ and area A, then:*

1) All arcs of Λ are line segments or circular arcs.

---

* The author wishes to thank the referee who pointed out the need for this clarification which the author had overlooked.

2) The curves of Λ meet Γ orthogonally, if in a neighbourhood of the common point the curves constituting Γ are C^1-curves (the common point included).

3) If arcs of Λ meet at points not on Γ, then three arcs meet at that point, and they meet at angles of $2\pi/3$.

4) If arcs of Λ meet at a point not on Γ, then the total curvature at the point is zero. The curvature of a line segment is zero. The curvature of a circular arc of radius r is $\pm$ 1/r, + if the center is on the right leaving the point, - otherwise.

PROOF. The proof of statements 1) and 2) are similar to the proof of the analogous statements in the previous theorems and are ommitted.

The proofs of statements 3) and 4) are analogous to the proofs of the main theorem in [1] and are omitted.

III. CONJECTURES AND SPECIAL CASES

Examples for the case of no natural boundary ($\Gamma = \Phi$) were given in [1].

We consider the special case of dividing the interior of a Jordan curve into N regions of given area, the sum of which is the interior of the region, in such a way as to minimize the total arc length. From the Blaschke selection theorem it is clear that there is a best way and that it satisfies the extremal conditions of Theorem 4. (See notes 4 and 6.) See [2,3] for the problem with the additional restriction that Λ is polygonal and Γ is a polygon, and all regions are convex.

We now give examples for low N for dividing the interior of a equilateral triangle, a square and a circle into N equal parts. An asterisk denotes a configuration which is conjectured best possible, but for which I have not seen adequate proof. The work in [5] is very helpful for eliminating many cases from consideration.

Also in the case of dividing the interior of a convex curve, it is clear that the boundary $\Lambda \cap \Gamma$ is connected. (Just slide a disconnected component until it touches some other part; at the point of contact, the angle conditions will be violated.) However the following conjecture is not yet proven.

CONJECTURE. For Γ connected, for an optimal Λ surrounding N cells in the plane with total area A such that the areas of the cells have ratio $a_1:a_2:\ldots:a_N$, $\Lambda \cup \Gamma$ is connected.

Even a proof for the case Γ of finite length would be useful.

EXAMPLE I. Γ is an equilateral triangle, $a_1 = \ldots = a_N = 1$.

Completely rigorous proofs of optimality have only been worked out for $N = 1, 2$, or 3; none-the less it is strongly conjectured that the ones shown in Fig. 3 for $N = 4, 6$ are optimal. The figure for $N = 5$ is merely schematic of the topological type of the optimal division.

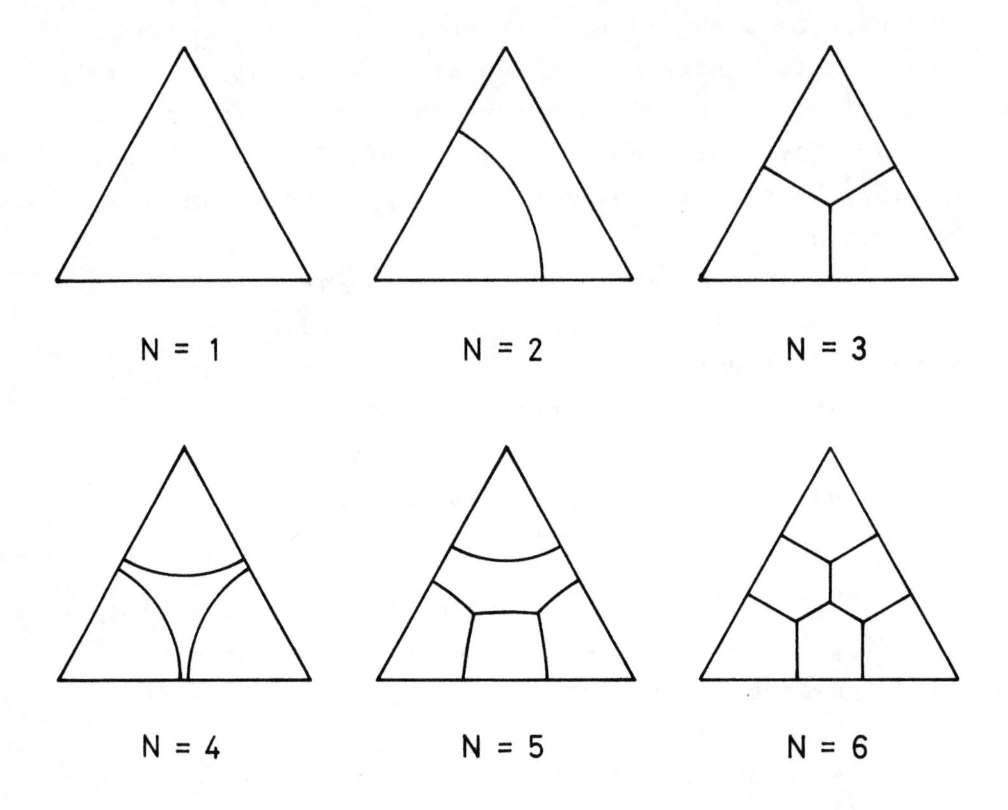

Figure 3

In Table 1, N is the number of cells, A the total area of the cells, T the total internal length used, T^* the sum of the internal length plus the perimeter of the triangle. The case of $N = \infty$ is the case of division of the plane into regular hexagonal tiles. This is thought to be the limiting value as $N \to \infty$, but the proof has been elusive.

TABLE 1

Division of Equilateral Triangle into N Equal Parts

N	A/N	T	T^2/NA	T^*-3+T	T^{*2}/NA
1	$\sqrt{3}/4$	0	0	3	$12\sqrt{3} =$ 20.78...
2	$\sqrt{3}/8$	$\frac{\sqrt{\pi}\ 3^{3/4}}{2\cdot 3}$ = 0.673...	$\frac{\pi}{6}$ = 0.523...	3.673...	15.581...
3	$\sqrt{3}/12$	$\frac{\sqrt{3}}{2}$ = 0.866...	$\frac{1}{\sqrt{3}}$ = 0.577...	3.866...	11.505...
4^*	$\sqrt{3}/16$	$\frac{3^{3/4}\sqrt{\pi}}{2\sqrt{2}}$ = 1.428...	$\frac{3\pi}{8}$ = 1.178...	4.428...	11.322...
5^*	$\sqrt{3}/20$	1.624...	1.218...	4.624...	9.875...
6^*	$\sqrt{3}/24$	$\sqrt{3}$ = 1.732...	$\frac{2}{\sqrt{3}}$ = 1.154...	4.732...	8.618...
∞^*			$2\sqrt{3}$ = 3.464...		3.464...

EXAMPLE II. Γ is a square, $a_1 = , \ldots , = a_N$

As before rigorous proofs are available only for $N = 1, 2, 3$.

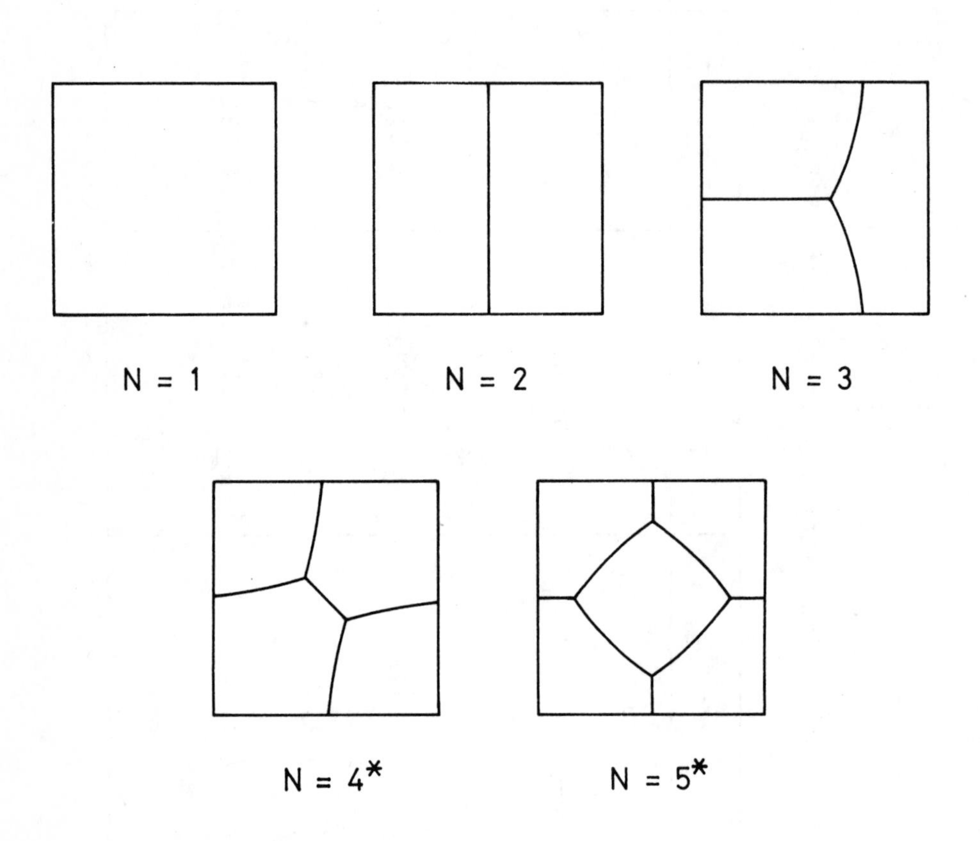

Figure 4

TABLE 2

N	A/N	T	T^2/NA	T^*	T^{*2}/NA
1	1	0	0	4	16
2	1/2	1	0.5	5	12.5
3	1/3	$\frac{\pi}{2}-\frac{1}{6}+\frac{\tan 15^{\circ}}{4}$	0.721...	5.471...	9.977...
4*	1/4	$\sqrt{2}+\frac{\pi}{3}-\sqrt{4-2\sqrt{3}}$	0.975...	5.975...	8.926...
5*	1/5	$2(1+\sqrt{\frac{1}{5}(\frac{\pi}{3}+1-\sqrt{3})}$	1.252...	6.502...	8.455...
∞*			$2\sqrt{3}=3.464...$		3.464...

In [5] it is shown that the solution is of the topological type of the example given for $N = 4$. It is also shown in [5, p. 54] that for $N = 5$ the figure must be one of seven topological types. Three of these types do not satisfy the necessary conditions of Theorem 4.

EXAMPLE III. Γ is a circle, $a_1 = \ldots = a_N = 1$

Fig. 5 shows divisions of a circle into $N \leq 6$ equal parts. For $N = 1$, 2 these are clearly optimal. For $N = 3$ it is known [5] that this is the right topological type. It is hard to imagine that this figure is not optimal, but proof is lacking. For $N = 4$ it is known [5] that this

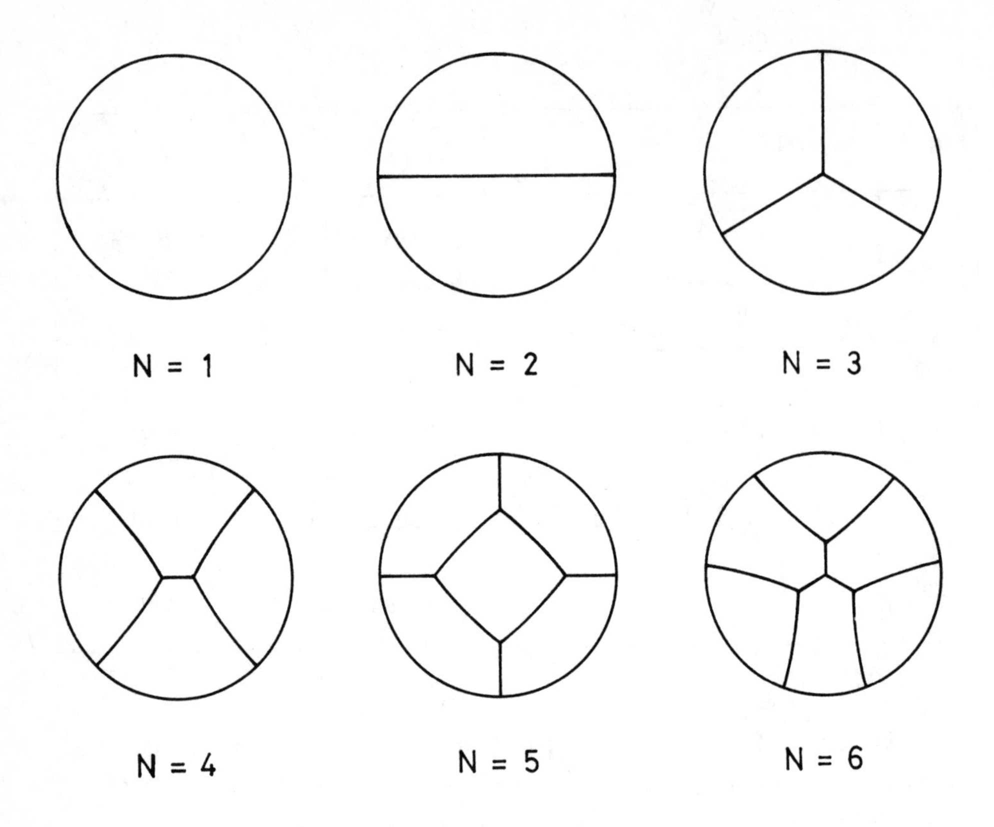

Figure 5

is the correct topological type and this figure satisfies all the necessary conditions of the theorem. For N = 5 Tomonaga [5] proves that the optimum is one of five topological types. Theorem 4 eliminates all but two of these. The author conjectures the figure shown is optimal, although he would not be shocked if the solution was of the other topological type (see Figure 6).

TABLE 3

N	A/N	T	T^2/NA	T^*	T^{*2}/NA
1	π	0	0	2π	4π
2	$\pi/2$	2	$2/\pi = 0.636...$	$2(\pi + 1)$	$\frac{2(\pi+1)^2}{\pi} = 10.919...$
3^*	$\pi/3$	3	$3/\pi = 0.954...$	$2\pi + 3$	$\frac{3(\frac{2\pi}{3} + 1)^2}{\pi} = 9.143...$
4^*	$\pi/4$	3.945...	1.238...	$2\pi + 3.945...$	8.326...
5^*	$\pi/5$	4.889...	1.522...	$2\pi + 4.889...$	7.947...
6^*	$\pi/6$	5.644...	1.690...	$2\pi + 5.644...$	7.547...

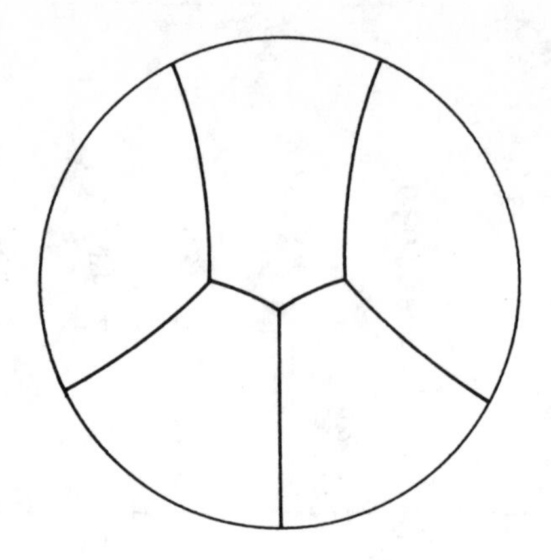

Figue 6

EXAMPLE IV. Non-existence of a solution

Let Γ consist of the intervals $[2n, 2n+1]$ on the x-axis, the line segments joining $(2n, 0)$ to $(2n, n)$, $n = 0, 1, 2, \ldots$ the line segments joining $(2n+1, 0)$ to $(2n+1, n+1)$ $n = 0, 1, 2, \ldots$ and the line segments joining $(2n-1, n)$ to $(2n, n)$ $n = 1, 2, \ldots$ (see Figure 7). Let $L = 1$. (Actually any finite value of $L \geq 1$ would do.) Then Λ can close any one rectangle, but there is clearly no largest one.

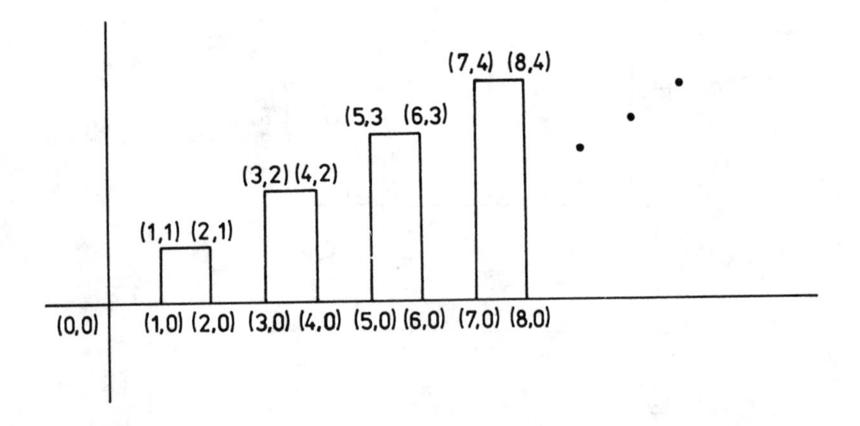

Figure 7

EXAMPLE V. Non-unique solutions.

Let Γ consist of two semicircles of radius 1, one the upper-half of the unit circle centered at the origin and the other the upper half of the unit circle centered at (3,0). Let $L = 2$. (Actually $2 \leq L < 4$ will work.) Then Λ can close either one of the semicircles, but not both, thus two distinct solutions.

Obviously one can construct examples with as many distinct solutions as one wishes, in fact infinitely many solutions for non-compact Γ.

EXAMPLE VI. Discontinuity of the solution.

We show here that for a fixed boundary Γ the maximum of the surrounded area is not a continuous function of the given length L. However it is conjectured that for Γ bounded this maximum area is an upper semicontinuous function of L.

Let Γ consist of the three line segments (0,1) to (0,0), (0,0) to (1,0), (1,0) to (1,1) (see Figure 8). Let $L < 1$, then the most area that can be surrounded by a figure of total length L is the quarter circle centered

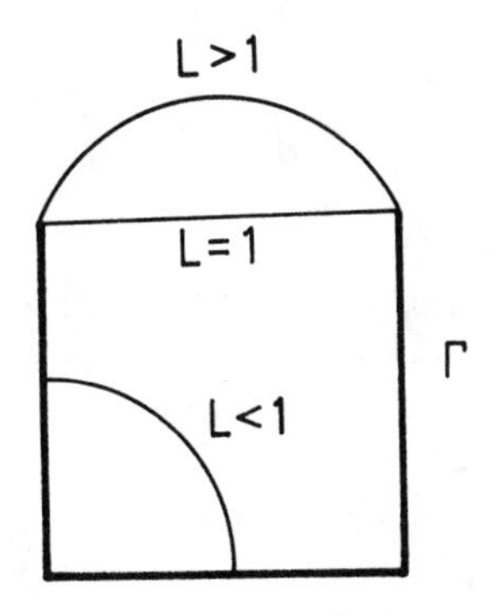

Figure 8

at the origin. The radius is $2L/\pi$ and the area is L^2/π. Thus if L tends to one the area tends to $\frac{1}{\pi} < 1$. However when L equals one it can close the gap from (0,1) to (1,1) and enclose a unit area.

REFERENCES

[1] BLEICHER, MICHAEL N., *Isoperimetric division into a finite number of parts in the plane*, to appear.

[2] FEJES TÓTH, G., *An isoperimetric problem for tessellations*, Studia Sci. Math. Hung. 10 pp. 171-173, 1975.

[3] FEJES TÓTH, L., *Isoperimetric problems for tilings*, Mathematika 32 (1985), 10-15.

[4] FEJES TÓTH, L., *Regular Figures*, Pergamon Press, Amsterdam, 1964.

[5] TOMONAGA, YASURO, *Geometry of Length and Area*, Dept. of Math., Utsunomiya University, Utsunomiya, 1974.

[6] YAGLOM, I.M. and BOLTYANSKII, V.G., *Convex Figures*, Holt. Rinehart and Winston, New York, 1961.

M.N. BLEICHER
University of Wisconsin, Math. Dept.,
Madison, Wisc. 53706, USA

COLLOQUIA MATHEMATICA SOCIETATIS JÁNOS BOLYAI
48. INTUITIVE GEOMETRY, SIÓFOK, 1985.

NOTE ON A THEOREM OF WIK

A. BLOKHUIS, K.A. POST and C.L.M. VAN PUL

ABSTRACT

A short proof is given for a theorem of Ingemar Wik concerning functions on the vertices of the Unit Cube satisfying a Lipschitz condition

In [1] I. Wik proves the following result:

THEOREM. *Let f be a function:* $\{0,1\}^n \to \mathbb{R}$ *satisfying* $|f(x) - f(y)| \leq d_H(x,y)$ *for all* $x,y \in \{0,1\}^n$.

Then

$$\sum_{x,y\in\{0,1\}^n} |f(x)-f(y)| \leq \sum_{x,y} |wt(x)-wt(y)|$$

$$\left(= n\binom{2n}{n}\right).$$

Here d_H *denotes the Hamming distance, i.e.* $d_H(x,y)$ *is the number of coordinates in which* x *and* y *differ, and* wt *is the weight function, i.e.* $wt(x) = d_H(x,0)$.

PROOF. Consider the set $\underline{F}$ of all functions: $\{0,1\}\to\mathbb{R}$, and $f(0) = 0$. Then $\underline{F}$ is a 2^n-1 dimensional vector space. Elements of $\{0,1\}^n$ will be identified with subsets of

$\{1,2,...,n\}$ (i.e. $x \leftrightarrow \{i : x(i) = 1\}$). Consider the convex polytope C in $\underline{F}$ consisting of all functions satisfying

$$|f(A) - f(B)| \leq |A \Delta B| \quad (\text{i.e. } |f(x)-f(y)| \leq d_H(x,y)) .$$

for all $A,B \subset \{1,2,...,n\}$. We want to maximize

$$S(f) := \sum_{A,B \subset \{1,2,...,n\}} |f(A)-f(B)| \text{ where f is in C.}$$

Note that S is a convex function, i.e.

$$(1) \qquad S(\lambda f + (1-\lambda)g) \leq \lambda S(f) + (1-\lambda)S(g) \quad (0<\lambda<1).$$

Moreover, equality in (1) occurs iff the following holds:

$$(2) \qquad \text{for all } A,B : f(A) \leq f(B) \Leftrightarrow g(A) \leq g(B).$$

Since C is a convex set the maximum of S is attained at some vertex of C.

LEMMA 1. *All vertices of* C *have integral coordinates.*

PROOF. A vertex is determined by a set $\underline{H}$ of 2^n-1 linearly independent bounding hyperplanes. Form a graph with as points all subsets of $\{1,2,...,n\}$ and $A \sim B$ if $f(A) - f(B) = \pm|A \Delta B|$ is in $\underline{H}$.

Since the hyperplanes in $\underline{H}$ are independent, this graph does not contain circuits. But a graph on 2^n points with 2^n-1 edges and no circuits is a tree. Since $f(\emptyset) = 0$ all $f(A)$ can be computed and are integral (since $|A \Delta B|$ is always integral).

COROLLARY. $f(A) - f(B) \equiv |A \Delta B| \pmod 2$ if f is a vertex of C.

PROOF. Let $A = A_0, A_1, \ldots, A_m = B$ be a path in the graph corresponding to the vertex f. Then

$$f(A) - f(B) = \sum_{i=0}^{m-1} f(A_i) - f(A_{i+1}) \equiv \sum_{i=0}^{m-1} |A_i \Delta A_{i+1}| \equiv$$

$$\equiv |A \Delta B| \pmod 2.$$

Note in particular that

$$(3) \qquad |f(A) - f(A \cup \{i\})| = 1 \quad \text{for all} \quad A, i \notin A.$$

REMARK. The maximum of S cannot be realized by a non-vertex of C, for let $f = \sum_i a_i f_i$ with $S(f)$ maximal, $a_i > 0$ and $\Sigma a_i = 1$, f_i vertices of C.

Then $S(f) = \Sigma a_i S(f_i)$ and by (2): $f_i(A) \leq f_i(B)$ iff $f_j(A) \leq f_j(B)$ (for all i, j, A, B).

But together with (3) this yields $f_i = f_j$, hence f is a vertex.

LEMMA 2. *If* $S(f)$ *is maximal then for all* i: $f(A) - f(A \cup i)$ *is independent of* A ($i \notin A$).

PROOF. Let $\underline{A}_i = \{A : i \notin A \text{ and } f(A) - f(A \cup i) = -1\}$,

$\underline{B}_i = \{B : i \notin B \text{ and } f(B) - f(B \cup i) = +1\}$

(note that $|f(A) - f(A \cup i)| = 1$ since f is a vertex).

Define $f' : f'(A) = f(A)$ if A or $A \setminus \{i\} \in \underline{A}_i$,

$f' : f'(B) = f(B \Delta \{i\})$ if B or $B \setminus \{i\} \in \underline{B}_i$.

and let $f^\wedge(A) := f'(A) - f'(\emptyset)$ for <u>all</u> A.

Claim 1. $f^\wedge \in C$. This is easily verified by distinguishing the several possibilities for $f^\wedge(A) - f^\wedge(B)$.

Claim 2. $S(f^\wedge) = S(f)$. This follows from the fact that the "multisets" $\{f(A): A \subset \{1,2,\dots,n\}$ and $\{f'(A): A \subset \{1,2,\dots,n\}$ are equal, and $S(f') = S(f^\wedge)$.

Claim 3. $f^\wedge$ is not a vertex of C if $\underline{A}_i$ and $\underline{B}_i$ are both nonempty, for let $A \in \underline{A}_i$, $B \in \underline{B}_i$, then

$$|f^\wedge(A) - f^\wedge(B)| \equiv |f(A) - f(B \cup i)| \equiv |A \Delta B| + 1 \pmod 2$$

and use the corollary to lemma 1.

Hence if $S(f)$ is maximal then either $\underline{A}_i$ or $\underline{B}_i$ is empty.

Finally fix f such that $S(f)$ is maximal, and let $D := i : \underline{B} \neq \emptyset$.

LEMMA 3. $f(A) = |D \Delta A| - |D|$ *for* $A \subset \{1,2,\dots,n\}$.

PROOF. $f(\emptyset) = 0$ and $f(A \cup i) - f(A) = 1$ if $i \notin C$, and 0 if $i \in C$. This completely determines f.

FINAL REMARK. The first half of this note (until lemma 2) applies to arbitrary (bipartite) graphs G and functions f satisfying $|f(u) - f(v)| \leq d_G(u,v)$, where d_G denotes the graph distance, and $f(u) = 0$ for some point in G.

REFERENCE

[1] I. WIK, *A Lipschitz Condition for Functions on the vertices of the Unit Cube*, Univ. of UMEÅ, dept. of Math. (S-90187 UMEÅ) nr 4, 1984 (ISSN 0345-3928).

A. BLOKHUIS, K.A. POST, C.L.M. van PUL
Dept. of Math. Techn. Univ. Eindhoven
Den Dolech 1, Eindhoven, Netherlands

COLLOQUIA MATHEMATICA SOCIETATIS JÁNOS BOLYAI
48. INTUITIVE GEOMETRY, SIÓFOK, 1985.

SCHLAEFLI'S DIFFERENTIAL FORM IN THE NON-EUCLIDEAN GEOMETRY OF POLYTOPES AND SOME OF ITS APPLICATIONS

J. BÖHM

1. SCHLAEFLI'S DIFFERENTIAL FORM

Schlaefli's differential form deals with the differential of the volume of an r-simplex $S^{(r+1)}$ in a non-Euclidean space of dimension r and constant curvature $\kappa \neq 0$. Without lack of generality the curvature may be normalized to $\kappa = +1$ (elliptic space) or $\kappa = -1$ (hyperbolic space). We denote the vertices of $S^{(n)}$ $(n := r+1)$ by $1, 2, \ldots, n$, the face opposite to the vertex j by $S_j^{(n-1)}$ and the dihedral angle between $S_j^{(n-1)}$ and $S_k^{(n-1)}$ and on $S_{jk}^{(n-2)} := S_j^{(n-1)} \cap S_k^{(n-1)}$ by its measure v_{jk} for $j, k \in \{1, 2, \ldots, n\}$. Then Schlaefli's formula reads

$$d(\mathrm{vol}_{n-1}(S^{(n)})) = \frac{\kappa}{n-2} \sum_{\substack{j,k=1 \\ j<k}}^{n} \mathrm{vol}_{n-3}(S_{jk}^{(n-2)}) dv_{jk}, \quad n \geq 3$$

$$\text{with } \mathrm{vol}_0(S_{jk}^{(1)}) := 1.$$

L. Schlaefli [12] established this formula for $\kappa = 1$ and for arbitrary $r = n-1 \geq 2$ in 1852. N.I. Lobachevski already knew this form for the 3-dimensional hyperbolic orthogonal tetrahedron in 1830, and also C.F. Gauss used it in the same case in 1832, but he did not publish it during his lifetime. Gauss' thoughts were first printed in Gauss' collected papers in 1900 and commented by P. Staeckel. In 1936 H. Kneser [8] proved the validity of this differential form for any $r \geq 2$ and $\kappa = \pm 1$ by algebraic methods. Kneser used the half of the surface of the $(r+1)$-dimensional unit sphere $(o,1)$ in an $(r+1)$-Euclidean space as a model of an elliptic space and for the hyperbolic space he took the surface of a sheet of the hypersphere $(o,\sqrt{-1})$ in an $(r+1)$-pseudo-Euclidean space with radius $\sqrt{-1}$. By the metric $ds^2 = \sum_{j=1}^{n-1} x_j^2 + \kappa dx_n^2$ the modified distance from o is given by $\rho^2 = \kappa(\sum_{j=1}^{n-1} x_j^2 + \kappa x_n^2)$ (≥ 0). On the surface of such a half hypersphere $(o,\sqrt{\kappa})$, $x_n \geq 0$, the simplex $S^{(n)}$ is placed. Let $C :=$ cone $(o,S^{(n)})$ denote the cone with apex o and rays through $S^{(n)}$ then we have

$$H(S^{(n)}) := \int_{(C)} e^{-\rho_2^2} dx_1 \ldots dx_n = 2^{\frac{n}{2}-1} \Gamma(\tfrac{n}{2}) \mathrm{vol}_{n-1}(S^{(n)}).$$

Calculating dH by introducing spherical coordinates one gets Schlaefli's differential form in both cases $\kappa = \pm 1$ after skilful algebraic transformations. For the hyperbolic space another modification of the model is possible: The hypersphere with radius $\sqrt{-1} = i$ in an Euclidean $(r+1)$-space can be used. Locally the surface of this hypersphere represents the hyperbolic space. Then Schlaefli's differential form is formally the same for $\kappa = 1$ and $\kappa = -1$, namely

$$d(\mathrm{vol}_{n-1}(S^{(n)})) = \frac{1}{n-2} \sum_{\substack{j,k=1 \\ j<k}}^{n} \mathrm{vol}_{n-3}(S_{jk}^{(n-2)}) dv_{jk} \qquad (\kappa = \pm 1;\ n \geq 3).$$

For $\kappa = -1$ we interpret

$$(*) \qquad \mathrm{vol}_{n-1}(S^{(n)}) := \mathrm{i}^{n-1}\mathrm{vol}_{n-1}(S^{(n)}).$$

With the help of Schlaefli's determinant

$$\Delta_h(i_1,i_2,\ldots,i_h) = \begin{vmatrix} 1 & -\cos v_{i_1i_2} & -\cos v_{i_1i_3} & \ldots & -\cos v_{i_1i_n} \\ -\cos v_{i_2i_1} & 1 & -\cos v_{i_2i_3} & \ldots & -\cos v_{i_2i_n} \\ \vdots & & & & \\ -\cos v_{i_ni_1} & -\cos v_{i_ni_2} & -\cos v_{i_ni_3} & \ldots & 1 \end{vmatrix}$$

one can see whether a simplex exists and in which space: To realise a simplex we must have $\Delta_h > 0$ for all $h=1,2,\ldots,r$ and all combinations $(i_1,i_2,\ldots,i_h)$. The sign of Δ_{r+1} then decides in the cases:

$$\kappa = 1 \iff \Delta_{r+1} > 0$$

$$(\kappa = 0 \iff \Delta_{r+1} = 0)$$

$$\kappa = -1 \iff \Delta_{r+1} < 0.$$

2. ORTHOSCHEMES

An r-dimensional simplex $R^{(n)}$, $n = r+1$, is called an orthoscheme iff the vertices of the $(n-1)$-simplex $R^{(n)}$ can be labelled by $1,2,\ldots,n$ so that $(1,2,\ldots,k)\perp(k,k+1)$ $(k=2,3,\ldots,n-1)$. Then all dihedral angles are right angles apart from the $n-1$ essential dihedral angles. The measure of these essential dihedral angles may be $v_k := v_{k,k+1}$. The apex of that dihedral angle with the measure v_j is the $(r-2)$-dimensional suborthoscheme with the set of vertices $\{1,2,\ldots,n\}\setminus\{j,j+1\}$ $(1 \le j \le n-1)$. Hence Schlaefli's differential form for an orthoscheme $R^{(n)}$ yields

$$d(\mathrm{vol}_{n-1}(R^{(n)})) = \frac{1}{n-2} \sum_{j=1}^{n-1} \mathrm{vol}_{n-3}(R_j^{(n-2)}) dv_j \quad (n \ge 3)$$

for $\kappa = +1$, and in the case $\kappa = -1$ with volume as in (*).

Every polytope can be dissected after completing into orthoschemes (Ergänzung durch Orthoscheme) by constructing suitable orthogonals. The sharper conjecture by H. Hadwiger states that every polytope can be dissected into orthoschemes. Up to now this has been proved only for $r \le 4$. But the possibility of generating a polytope by orthoschemes makes clear that it is sufficient to deal with the calculation of the orthoscheme volume to get a polytope volume. For example a regular simplex $S_{\mathrm{reg}}^{(n)}$ in a space of constant curvature $\kappa(\kappa=\pm 1)$ has equal dihedral angles of measure $v_{jk} = 2\alpha$ (for all j,k). All the edges have the same length $2a$ with

$$\cos 2a = \frac{\cos 2\alpha}{1-(n-2)\cos 2\alpha} .$$

A regular simplex $S_{\mathrm{reg}}^{(n)}$ exists

$$\text{for } \kappa = 1 \text{ iff } -1 < \cos 2\alpha < \frac{1}{n-1}$$

$$(\text{for } \kappa = 0 \text{ iff } \quad \cos 2\alpha = \frac{1}{n-1})$$

$$\text{for } \kappa = -1 \text{ iff } \frac{1}{n-1} < \cos 2\alpha \le \frac{1}{n-2} .$$

A dissection of $S_{\mathrm{reg}}^{(n)}$ into $n!$ orthoschemes $R^{(n)}(\alpha,\frac{\pi}{3},\ldots,\frac{\pi}{3})$ with $\alpha,\frac{\pi}{3},\ldots,\frac{\pi}{3}$ as the measures of the essential dihedral angles (fundamental orthoscheme of the regular simplex) is always possible.

3. SIMPLEX VOLUME

The knowledge of the volume of the orthoscheme suffices to calculate the volume of a simplex or generally of a polyhedron. For an orthoscheme $R^{(n)}(v_1,v_2,\dots,v_{n-1})$ a formal calculation via Schlaefli's differential form is possible in the following way. Let the measures $v_2,v_3,\dots,v_{n-1}$ of the dihedral angles be constant and let v_1 be variable. Integrating Schlaefli's differential form $d(\mathrm{vol}_{n-1}(R^{(n)})) = \frac{1}{n-2}\mathrm{vol}_{n-3}(R_1^{(n-2)})dv_1$ over v_1 from the point shaped case to the value v_1, one gets

$$(1) \qquad \mathrm{vol}_{n-1}(R^{(n)}) = \frac{1}{n-2}\int_{\Delta_n=0}^{v_1} \mathrm{vol}_{n-3}(R_1^{(n-2)})dv_1 .$$

In particular, dissecting a regular simplex $S_{\mathrm{reg}}^{(n)}(2\alpha)$ into orthoschemes Schlaefli's differential form gives

$$\begin{aligned}\mathrm{vol}_{n-1}(S_{\mathrm{reg}}^{(n)}(2\alpha)) &= n!\ \mathrm{vol}_{n-1}(R^{(n)}(\alpha,\tfrac{\pi}{3},\dots,\tfrac{\pi}{3}))\\ &= \frac{n!}{n-2}\int_{\Delta_n=0}^{\alpha} \mathrm{vol}_{n-3}(R_1^{(n-2)})dv_1 .\end{aligned}$$

For dimension $r \geq 3$ elementary values of the volume of a regular elliptic simplex are known, namely $\alpha = \frac{\pi}{3}$ and $\alpha = \frac{\pi}{4}$:

$$\mathrm{vol}_{n-1}(S_{\mathrm{reg}}^{(n)})_{\alpha=\frac{\pi}{3}} = \frac{1}{n+1}\ \mathrm{vol}_{n-1}(S_{n-1})$$

$$\mathrm{vol}_{n-1}(S_{\mathrm{reg}}^{(n)})_{\alpha=\frac{\pi}{4}} = \frac{1}{2^n}\ \mathrm{vol}_{n-1}(S_{n-1}),$$

$\mathrm{vol}_{n-1}(S_{n-1})$ is the volume of the surface of the n-dimensional unit hypersphere in the Euclidean space.

Hadwiger's conjecture was that there are no more elementary values for regular non-Euclidean simplices in dimension $r \geq 5$. Stimulated by a result of H. Ruben [11] from the year 1960 Hadwiger proved in 1978 an explicit formula for a regular elliptic simplex in the case $-\frac{1}{n-1} < \cos 2a \leq 0$ as a consequence of Schlaefli's differential form as follows:

$$\mathrm{vol}_{n-1}(S^{(n)}_{\mathrm{reg}}(2\alpha)) = c_n \int_{-\infty}^{\infty} e^{-p^2} \left(\int_{-\infty}^{A_n(2a)p} e^{-s^2}\, ds \right)^n dp$$

$$c_n = \frac{2}{\sqrt{\pi}\,\Gamma(\frac{n}{2})}, \quad A_n(2a) = \sqrt{\frac{-\cos 2a}{1+(n-1)\cos 2a}} .$$

Changing the lower bound of the integral in (1) to a suitable $\tilde{v}_1$, one gets the volume of a truncated orthoscheme.

4. SCHLAEFLI'S REDUCTION FORMULA

For even dimension the volume of a simplex can be reduced to the volumes of suitable simplices of lower dimensions. The formula was already known for the case $r = 2$ as the volume of an elliptic or hyperbolic triangle. Schlaefli (1852) [12] and later H. Poincare (1905) [10] showed the reducibility of the volume of an even dimensional simplex in the elliptic case in different proofs. In 1925 H. Hopf [7] proved the validity of reducing the volume in both cases $\kappa = \pm 1$ and E. Peschl [9] proved the equivalence of the results of Schlaefli and Poincarè. Schlaefli's reduction formula specializes to orthoschemes as follows. Let $f^{(2m+1)}$ be the normed volume of an $2m$-orthoscheme $R^{(2m+1)}$ (Schlaefli function) and let $\bar{F}_{2m-2h}$ be the sum of Schlaefli functions $f^{(2m+2h)}$ of all spherical vertex figures of order $2h+1$, which can be represented as a product of Schlaefli functions of non orthogonal (elliptic) orthoschemes of odd dimension. Then Schlaefli's reduction formula gives

$$f^{(2m+1)}(v_1,v_2,\ldots,v_{2m}) = \sum_{h=0}^{n} (-1)^h b_h \bar{F}_{2m-2h}$$

$$b_h = \frac{1}{h+1}\binom{2h}{h} \quad (m \geq 1).$$

For any dimension $2m = r \geq 3$ the structure of this formula for an orthoscheme $S^{(2m+1)}$ may be written in terms of Schlaefli functions

$$f^{(2m+1)}(v_1,v_2,\ldots,v_{2m}) = f^{(2m)}(v_1,v_2,\ldots,v_{2m-1}) +$$

$$+ f^{(2m)}(v_2,v_3,\ldots,v_{2m}) + o(f^{(2m-2)}).$$

For example, in the dimension $r = 4$ the formula yields

$$f^{(5)}(v_1,v_2,v_3,v_4) = f^{(4)}(v_1,v_2,v_3) + f^{(4)}(v_2,v_3,v_4) +$$

$$+ f^{(2)}(v_1)\, f^{(2)}(v_4) - (f^{(2)}(v_1) + f^{(2)}(v_2) + f^{(2)}(v_3) +$$

$$+ f^{(2)}(v_4)) + 2.$$

The proof (by induction) uses Schlaefli's differential form.

Two applications of this formula are now given here.

1) Calculating the volume of a regular asymptotic simplex in the $(n-1)$-dimensional hyperbolic space, that is, of a regular simplex with ideal vertices, one obtains

$$2\alpha = \arccos\frac{1}{n-2} \quad \text{and} \quad W_n = \alpha = \frac{1}{2}\arccos\frac{1}{n-2} \; .$$

On the other hand, for odd dimension, i.e. $r = 2m-1$, then one gets from (1) for the fundamental orthoschemes

$$i^{2m-1}\mathrm{vol}_{2m-2}(R^{(2m)}(W_{2m},\tfrac{\pi}{3},\ldots,\tfrac{\pi}{3})),$$

and for even dimension $r = 2m$ by Schlaefli's formula we write for the normed Schlaefli function of the fundamental orthoscheme of the regular simplex

$$f^{(2m+1)} = f^{(2m)}(W_{2m+1},\tfrac{\pi}{3},\tfrac{\pi}{3},\ldots,\tfrac{\pi}{3}) + f^{(2m)}(\tfrac{\pi}{3},\tfrac{\pi}{3},\ldots,\tfrac{\pi}{3}) +$$
$$+ o(f^{2m-2})).$$

The first summand is zero because it is the volume of a point shaped Euclidean simplex. The second summand is elementary because it is one of the special volumes mentioned above. Therefore the first terms to calculate will be the Schlaefli functions $f^{(2m-2)}$, $f^{(2m-4)}$ and so on and this yields then

$$\kappa^{m}\cdot\mathrm{vol}_{2m}(R^{(2m-2)}_{\mathrm{asymp}}) =$$

$$= a_{m,m-1}\pi\;\mathrm{vol}_{2m-3}(R^{(2m-2)}(\tfrac{1}{2}\arccos\tfrac{1}{2m-2},\tfrac{\pi}{3},\ldots,\tfrac{\pi}{3})) + \ldots + a_{m0}\pi^{m} =$$

$$= \sum_{h=0}^{m-1} a_{mh}\pi^{m-h}\mathrm{vol}_{2h-1}(R^{(2h)}(\tfrac{1}{2}\arccos\tfrac{1}{2m-2},\tfrac{\pi}{3},\ldots,\tfrac{\pi}{3}))$$

$$(\mathrm{vol}_{-1}(R^{(0)}) := 1).$$

Special values are

$$r = 2:\ \mathrm{vol}_{2}R^{(3)} = \frac{1}{3}!\,\pi$$

$r = 3$: $\mathrm{vol}_3 R^{(4)} = \frac{1}{4}(\Lambda_2(\frac{\pi}{3}) - 2\Lambda_2(\frac{\pi}{6})) = \frac{1}{4!}(\frac{\pi}{2}\log 2 - 3\Lambda_2(\frac{\pi}{6}))$

($\Lambda_2(z) := -\int_0^z \log\cos t\,dt$ with the identity

$2\Lambda_2(\frac{\pi}{3}) - 3\Lambda_2(\frac{\pi}{6}) = \frac{\pi}{6}\log 2$; published first by

H.S.M. Coxeter in 1935 [4]).

$r = 4$: $\mathrm{vol}_4 R^{(5)} = \frac{1}{5!}(\frac{4\pi^2}{3} - \frac{10}{3}\pi\arccos\frac{1}{3})$

$r = 5$: $\mathrm{vol}_5 R^{(6)} = \mathrm{vol}_5(R^{(6)}(\frac{1}{2}\arccos\frac{1}{4}, \frac{\pi}{3}, \frac{\pi}{3}, \frac{\pi}{3}, \frac{\pi}{3}))$

$r = 6$: $\mathrm{vol}_6 R^{(7)} = \bar{a}_{30}\pi^3 + \bar{a}_{31}\pi^2\arccos\frac{1}{5} +$

$+ \bar{a}_{32}\pi\,\mathrm{vol}_3(R^{(4)}(\frac{1}{2}\arccos\frac{1}{5}, \frac{\pi}{3}, \frac{\pi}{3}))$

with $\bar{a}_{30} = \frac{17}{37800}$, $\bar{a}_{31} = -\frac{1}{900}$, $\bar{a}_{32} = \frac{1}{45}$; $R^{(4)}$ is an elliptic orthoscheme.

2) If $R^{(2m+1)}$ is degenerated to a point in the Euclidean case then it is

$$\begin{aligned}\mathrm{vol}_{2m}(\mathrm{R}^{(2m+1)}) &= 0 \\ &= \mathrm{vol}_{2m-1}(R^{(2m)}(v_1, v_2, \ldots, v_{2m-1})) \\ &\quad + \mathrm{vol}_{2m-1}(R^{(2m)}(v_2, \ldots, v_{2m})) + \\ &\quad + o(R^{(2m-2)})\end{aligned}$$

and this yields

$$\mathrm{vol}_{2m-1}(R^{(2m)}(v_1, \ldots, v_{2m-1})) + \mathrm{vol}_{2m-1}(R^{(2m)}(v_2, v_3, \ldots, v_{2m})) = o(R^{(2m-2)}).$$

The geometric interpretation of this last formula is that the two $(2m-1)$-dimensional orthoschemes are neighbour orthoschemes in the sense of a generalisation of the figure of Gauss's pentagramma mirificum in the dimension $r = 2$.

5. ORTHOGONAL DEGENERATION AND SUPPLEMENTARY SIMPLEX

These notions will be considered only in the elliptic case.

1) If in a simplex S a face S_k and its complement S_{n-k} are mutually orthogonal, then Schlaefli's differential form yields

$$\mathrm{vol}\ S = c \cdot \mathrm{vol}\ S_k \cdot \mathrm{vol}\ S_{n-k}$$

where c depends only on the dimension r of the simplex S and the dimension of the face S_k.

2) Supplementing a simplex $S^{(n)}$ to a two-angle, called S, by lengthening all the edges with a common vertex a_0, and S_0 is the polar face of this vertex a_0, then becaus of 1)

$$\mathrm{vol}_r S = c \cdot \mathrm{vol}_{r-1} S_0$$

holds. S can be elementarily dissected into the two simplices $S^{(n)}$ and the supplementary simplex to $S^{(n)}$ denoted by $S^{(n)}_{\mathrm{sup}}$. Then for the sum of the volume of these two simplices there is

$$\mathrm{vol}_r S^{(n)} + \mathrm{vol}_r (S^{(n)}_{\mathrm{sup}}{}') = c \cdot \mathrm{vol}_{r-1} S_0 .$$

6. STRUCTURE OF THE VOLUME FUNCTION

For even dimension the volume of an orthoscheme $R^{(2m+1)}$, $\kappa = \pm 1$, can always be reduced to the volume of odd-dimensional elliptic simplices. Hence it is sufficient to consider $R^{(2m)}$ with odd dimension $r = 2m-1$. By a method applied by Coxeter [4] in the dimension three and generalised by the author [2] to higher dimensions, one obtains an extended Schlaefli's differential form with $(r-1)^2$ variables in generalisation. Integrating this form one obtains after specialisation (cf. [3], p. 232) $\mathrm{vol}_{2m-1}(R^{(2m)})$. It can be represented by generalised polylogarithm functions of order m and less. A polylogarithm function is recursively given by

$$\mathrm{Li}_1(z) = -\log(1-z) \quad \text{and} \quad \mathrm{Li}_m(z) = \int_0^z \frac{\mathrm{Li}_{m-1}(t)}{t} dt$$

and the generalised polylogarithm function can be represented by

$$\tilde{\mathrm{Li}}_m(z, a_1, a_2, \ldots, a_{m-2}, a_{m-1}) = \int_0^z \frac{\tilde{\mathrm{Li}}_{m-1}(t, a_1, a_2, \ldots, a_{m-2})}{t - a_{m-1}} dt$$

$$(m \geq 1).$$

7. THE ISOPERIMETRIC PROBLEM

The last application of Schlaefli's differential form discussed here is the isoperimetric problem for a non-Euclidean simplex S in a space of dimension r with constant curvature $\kappa = 1$ or $\kappa = -1$. Let the dihedral angles of S have the measures v_{jk} $(j,k = 1,2,\ldots,r+1)$ and let the surface of S be denoted by

$$F = \sum_{j=1}^{r+1} S_j .$$

The isoperimetric problem under consideration now is the following one:

Which simplices S have maximum volume, and constant surface, symbolically

$$\mathrm{vol}_r S = \max \quad \text{and} \quad \mathrm{vol}_{r-1} F = \text{const.}$$

In the case $r = 3$, $\kappa = -1$ L. Fejes Tóth [5] showed by a nice elementary geometric proof that the regular simplex is the unique solution. But these ideas do not work in three dimensional elliptic case. Therefore one needs another idea, and this is the use of Schlaefli's differential form.

The dual problem of the isoperimetric problem formulated above is to solve

$$\mathrm{vol}_r S = \text{const} \quad \text{and} \quad \mathrm{vol}_{r-1} F = \min.$$

Hence it is necessary in the first case that

$$\mathrm{vol}_r S + \lambda_1 \mathrm{vol}_{r-1} F = \text{extremal}$$

and in the second case that

$$\mathrm{vol}_{r-1} F + \lambda_2 \mathrm{vol}_r S = \text{extremal}.$$

In both cases this implies

$$\frac{\partial \mathrm{vol}_{r-1} F}{\partial v_{jk}} = \lambda_0 \cdot \frac{\partial \mathrm{vol}_r S}{\partial v_{jk}} \quad (\lambda_1 \cdot \lambda_2 \neq 0;\ j,k = 1,\ldots,r+1;\ j<k).$$

Schlaefli's differential form yields

$$\lambda_0 \mathrm{vol}_{r-2}(S_{jk}) = \sum_{\substack{l=1 \\ p,q=1 \\ p<q;\,p,q\neq l}}^{r+1} \mathrm{vol}_{r-3}(S_{l;p,q}) \frac{\partial v_{l;pq}}{\partial v_{jk}} .$$

One solution is

$$v_{jk} = v \text{ for all } j,k \text{ with a suitable measure } v,$$

and this is represented by a regular simplex with dihedral angles of measure v.

For dimension $r = 2$ and $\kappa = \pm 1$ this is known to be the unique solution. For dimension $r = 3$ one has to consider more conditions.

For the dual problem ($r = 3$, $\kappa = \pm 1$) one can write

$$\mathrm{vol}_3 S(v_1, v_2, v_3, v_4, v_5, v_6) = \mathrm{const}$$

with v_j as the measure of the six dihedral angles. Then it follows

$$v_6 = \mathrm{v}_6(v_1, v_2, v_3, v_4, v_5)$$

and hence one gets from

$$A = 4\pi + \mathrm{vol}_2 F = \min$$

the condition

$$A(v_1, v_2, v_3, v_4, v_5, v_6(v_1, v_2, v_3, v_4, v_5)) = \min .$$

Necessary condition is

$$\frac{\partial A}{\partial v_j} = 0 \quad (j=1,2,\ldots,5).$$

Again one solution is $v_j = v$ $(j=1,2,\ldots,5)$ for a suitable v and

$$v_6 = v_6(v,v,v,v,v) = v.$$

Then it is sufficient for a solution that the quadratic form of derivations of second order is positive definite. That means one has to consider the matrix of the form

$$\boldsymbol{B} = (b_{jk})_{j,k=1,\ldots,5} = \begin{pmatrix} b_{11} & \frac{1}{2}b_{11} & \frac{1}{2}b_{33} & b_{14} & \frac{1}{2}b_{11} \\ \frac{1}{2}b_{11} & b_{11} & \frac{1}{2}b_{33} & \frac{1}{2}b_{11} & b_{14} \\ \frac{1}{2}b_{33} & \frac{1}{2}b_{33} & b_{33} & \frac{1}{2}b_{33} & \frac{1}{2}b_{33} \\ b_{14} & \frac{1}{2}b_{12} & \frac{1}{2}b_{33} & b_{11} & \frac{1}{2}b_{11} \\ \frac{1}{2}b_{11} & b_{14} & \frac{1}{2}b_{33} & \frac{1}{2}b_{11} & b_{11} \end{pmatrix}$$

with positive definite principal minors. These are

$$\boldsymbol{B}_1 = b_{11},\ \boldsymbol{B}_2 = \frac{3}{4}b_{11}^2,\ \boldsymbol{B}_3 = \frac{1}{4}b_{11}b_{33}(b_{11}+2b_{14}),$$
$$\boldsymbol{B}_4 = \frac{1}{4}b_{14}b_{33}^2(b_{33}+2b_{14}),\ \boldsymbol{B}_5 = \frac{3}{4}b_{33}^2 b_{14}^2$$

with

$$b_{11} = (1-\cos a) + \cos a\left(2\frac{\sin a}{a} - (1+\cos a)\right)$$

$$b_{14} = 1 - \frac{\sin a}{a}$$

$$b_{33} = 2\left\{\left(\frac{\sin a}{a} - \cos a\right) + \cos a\left(2\frac{\sin a}{a} - (1+\cos a)\right)\right\}$$

and $\cos a = \frac{\cos v}{1-2\cos v}$.

Then $\frac{1}{\kappa}b_{jk} > 0$ and $\boldsymbol{B}_j > 0$ for $\kappa = 1$, $j = 1,2,\ldots,5$. For $\kappa = -1$ read $-A$ for A and $-b_{jk}$ for b_{jk}, then $\boldsymbol{B}_j > 0$ for $j=1,2,\ldots,5$. Therefore, the surface area of a 3-simplex in a space of constant curvature $\kappa = \pm 1$ with given constant volume of the simplex will attain a relative minimum in the case of a regular simplex. The dual assertion also holds:

THEOREM. *For all tetrahedra in elliptic (or hyperbolic) space with equal surface volume the tetrahedron volume for the regular ones is a (relative) maximum.*

REFERENCES

[1] AOMOTO, K., *Analytic structure of Schläfli function.* Nagoya Math. J. 68 (1977), 1-16.

[2] BÖHM, J., *Zu Coxeters Integrationsmethode in gekrümmten Räumen.* Math. Nachr. 27 (1964), 179-214.

[3] BÖHM, J. and E. HERTEL, *Polyedergeometrie in* n-*dimensionalen Räumen konstanter Krümmung.* DVW Berlin 1980 und Birkhäuser Basel-Boston-Stuttgart 1981.

[4] COXETER, H.S.M., *The functions of Schläfli and Lobatschefsky.* Quart. J. Math. Oxford 6 (1935), 13-29.

[5] FEJES TÓTH, L., *On the isoperimetric property of the regular hyperbolic tetrahedra.* Publ. Math. Hung. Acad. Sci. 8 (A) (1963), 53-57.

[6] HAAGERUP, U. and H.J. MUNKHOLM, *Simplices of maximal volume in hyperbolic n-space*. Acta Math. 147 (1981), 1-11.

[7] HOPF, H., *Curvatura integra Clifford-Kleinscher Raumformen*. Nachr. Ges. Wiss. Göttingen, Math.-phys. Kl. aus dem Jahre 1925 (1926), 131-141.

[8] KNESER, H., *Der Simplexinhalt in der nichteuklidischen Geometrie*. Deutsche Math. 1 (1936), 337-340.

[9] PESCHL, E., *Winkelrelationen am Simplex und die Eulersche Charakteristik*. Bayr. Akad. Wiss. Math.-nat. Kl. (1956), 319-345.

[10] POINCARÉ, H., *Sur la généralisation d'un théoreme élémentaire de géométrie*. C.R. Acad. Sci. Paris (1), 140 (1905), 113-117.

[11] RUBEN, H., *On the geometrical moments of skew-regular simplices in hyperspherical space, with some applications in geometry and mathematical statistics*. Acta Math. 103 (1960), 1-23

[12] SCHLÄFLI, L., Ges. math. Abh. 1 (*Theorie der vielfachen Kontinuität; aus dem Jahra 1852*) Birkhäuser Basel 1950, 227 ff.

[13] SEIDEL, J.J., *On the volume of a hyperbolic simplex*. Stud. Sci. Math. Hung. 21 (will appear in 1986).

[14] WITT, E., *Rekursionsformel für Volumina sphärischer Polyeder*. Arch. Math. 1 (1949), 317-318.

J. BÖHM
Friedrich-Schiller-Universität Jena
Sektion Mathematik
DDR-6900 Jena
UHH, 17. OG

COLLOQUIA MATHEMATICA SOCIETATIS JÁNOS BOLYAI
48. INTUITIVE GEOMETRY, SIÓFOK, 1985.

A NEW POLYHEDRON OF GENUS 3 WITH 10 VERTICES

J. BOKOWSKI - U. BREHM

1. INTRODUCTION

It would be very desirable to have a fast algorithm to decide for a combinatorially given d-dimensional simplicial complex whether it can be embedded in R^d. In this paper we give an approach to this aim. Our approach yielded a new 3-dimensional polyhedron of genus 3 with 10 vertices. Our construction in general and in this particular case is done in two steps. First we determine all chirotopes which are compatible with the given complex. Chirotopes have been considered by different authors and it is known that this combinatorial structure is equivalent to oriented matroids, see [2] and the references given there. Compatible in this context means that in case of realizability of these chirotopes selfintersections are excluded.

In a second step we have to decide whether at least one of these chirotopes is realizable. The problem of finding a fast algorithm to decide the geometric realizability of chirotopes is one of the central problems in the theory of oriented matroids, see [3]. Using the idea of solvability-sequence such decisions became possible in many cases, see [3] and the references given there.

We would like to thank Frederik Anheuser for his excellent help in writing the corresponding programs. The result is thus due to him as well.

2. CONSTRUCTION OF COMPATIBLE CHIROTOPES

For any set of n points $p_1,\ldots,p_n$ spanning R^{d-1} (or, viewed projectively, n oriented lines $p_1,\ldots,p_n$ through 0 in R^d) we define in the following a "*complete set of hyperline-configurations*". For each sequence $p_{i_1},\ldots,p_{i_{d-2}}$ of d - 2 affinely independent points we consider their oriented affine hull $L = \langle p_{i_1},\ldots,p_{i_{d-2}}\rangle_{or}$ and the corresponding oriented orthogonal 2-dimensional linear subspace $L^{\perp}$. For any other point p_k we regard the intersection $\langle p_k, p_{i_1},\ldots,p_{i_{d-2}}\rangle_{or} \cap L^{\perp}$ which gives either 0 or an oriented line through 0.
Drawing these lines equidistributed through 0 whilst keeping their order we get a hyperline-configuration (cf. Fig. 1a). We label the tip of the corresponding arrow by the point p_k. The hyperline-configuration determined by L can be described as a cyclically ordered sequence of the points (or sets of points) which are not contained in L and their "formal inverses" $\bar{p}_k$, by writing down the points at the tips of the arrows or the formal inverse of the points at the other end of the arrows in the counter-

clockwise order. In case that several points determine the same or the inverse hyperline, we write the set of these points in our sequence. For an example cf. Fig. 1b. Of course the second half of the sequence is the same as the first half with all points replaced by their inverse ($\bar{\bar{p}}_k = p_k$).

It is very easy to read off all orientations of ordered d-simplices from these hyperline-configurations.

On the other hand, let be given a complete set of the d-simplices which we get by immediate reading off are consistent (i.e. for every d-simplex its orientation read off from different hyperline-configurations coincides).

In this case we define a map $\chi : \Lambda(n,d) \to \{-1,0,1\}$ by assigning to every ordered d-simplex $\lambda \in \Lambda(n,d) :=$ $:= \{(\lambda_1,\ldots,\lambda_d) \in N^d \mid 1 \le \lambda_1 < \ldots < \lambda_d \le n\}$ its orientation

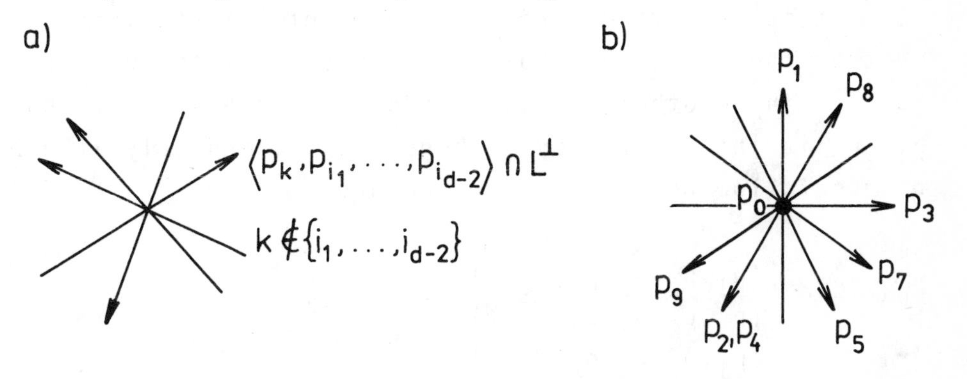

corresponding sequence:

$p_0-(p_1,\bar{p}_5,\bar{p}_7,\bar{p}_3,p_9,\{p_2,p_4,\bar{p}_8\},\bar{p}_1,p_5,p_7,p_3,\bar{p}_9,\{\bar{p}_2,\bar{p}_4,p_8\})$

Figure 1

Hyperline configurations

according to the hyperline-configurations. We call such a map (extended to all d-tupels by determinant-rules) an (oriented) d-*chirotope*.

It can be shown that this definition is equivalent to chirotopes defined by Graßmann-Plücker-relations and this is turn is another way of defining oriented matroids which were studied independently by several authors, see [2] for details and references.

In the following we describe the fundamental idea to determine a corresponding chirotope (or all corresponding chirotopes) for a combinatorially given complex. In our case we have to consider only a given simplicial 2-manifold.

We determine by induction all 4-chirotopes with the same number of vertices according to a further restriction. This is done by successively filling in new points in already existing hyperline-configurations and using thus determined orientations for the following hyperline-configurations to fill in only in a consistent way.

It is easy to describe in terms of orientations whether the convex hull of two points meets the convex hull of three other points (→Radon-partition). A triangle $p_1p_2p_3$ is intersected by an edge p_4p_5 if and only if one of the following combinations occurs

$$(*)\qquad \begin{array}{ll} (p_1,p_2,p_3,p_5) & +1 \\ (p_1,p_2,p_3,p_4) & -1 \\ (p_1,p_2,p_5,p_4) & -1 \\ (p_2,p_3,p_5,p_4) & -1 \\ (p_3,p_1,p_5,p_4) & -1 \end{array}$$

or with all signs reversed (-*).

During our induction process we make sure that no edge of the given 2-manifold M meets a triangle of M (assuming realizability), i.e. if $p_1p_2p_3$ is a triangle of M and p_4p_5 any edge of M then we make sure that the combinations (*) and (-*) of orientations do not occur.

Furthermore we make sure that the following combinations of orientations never occur (affine condition).

$$
(**) \qquad \begin{array}{lr}
(p_1,p_2,p_3,p_4) & -1 \\
(p_1,p_2,p_3,p_5) & +1 \\
(p_1,p_2,p_5,p_4) & +1 \\
(p_1,p_5,p_3,p_4) & +1 \\
(p_5,p_2,p_3,p_4) & +1
\end{array}
$$

or with all signs reversed (-**), since (**)(or(-**)) obviously cannot occur as a combination of orientations for points in R^3.

Using a computer we found (among others) the following chirotope which is compatible with the 2-manifold M described in the next section.

$1,2-(0,3,8,6,5,4,\bar{9},7)$ $1,3-(7,4,6,\bar{9},5,8,2,0)$ $1,4-(2,0,8,5,\bar{7},6,3,9)$
$1,5-(2,0,8,\bar{4},9,\bar{7},3,6)$ $1,6-(7,4,\bar{3},\bar{9},5,2,0,8)$ $1,7-(2,0,8,5,\bar{9},4,6,3)$
$1,8-(6,5,4,\bar{9},7,\bar{3},2,0)$ $1,9-(4,6,3,\bar{7},5,8,0,2)$ $1,0-(3,8,6,5,4,\bar{9},7,2)$
$2,3-(9,1,8,7,\bar{0},5,4,6)$ $2,4-(0,6,3,9,5,8,1,7)$ $2,5-(0,6,3,9,\bar{4},8,1,7)$
$2,6-(3,9,8,1,7,\bar{0},5,4)$ $2,7-(4,5,6,3,8,0,1,9)$ $2,8-(4,5,6,\bar{1},3,\bar{7},0,9)$
$2,9-(7,1,8,5,4,6,3,0)$ $2,0-(9,1,7,8,3,6,5,4)$ $3,4-(9,7,1,8,5,2,6,0)$
$3,5-(7,1,9,\bar{6},8,\bar{4},2,0)$ $3,6-(4,\bar{0},9,7,1,5,8,2)$ $3,7-(1,0,8,2,5,\bar{9},6,4)$
$3,8-(6,5,4,\bar{0},7,\bar{2},1,9)$ $3,9-(0,2,8,5,1,7,6,4)$ $3,0-(4,6,5,2,8,7,1,9)$
$4,5-(7,8,1,\bar{0},\bar{2},9,3,6)$ $4,6-(2,5,8,1,7,\bar{0},9,3)$ $4,7-(9,3,6,1,8,5,0,2)$
$4,8-(9,3,6,\bar{7},5,\bar{1},0,2)$ $4,9-(3,0,6,2,5,8,1,7)$ $4,0-(2,7,1,8,5,6,9,3)$
$5,6-(0,2,\bar{4},8,3,9,1,7)$ $5,7-(6,3,1,9,8,\bar{4},0,2)$ $5,8-(7,4,2,\bar{0},\bar{2},9,3,6)$
$5,9-(0,2,\bar{4},8,\bar{6},\bar{3},7,1)$ $5,0-(2,7,1,8,\bar{4},9,3,6)$ $6,7-(9,3,\bar{4},1,8,0,2,5)$
$6,8-(0,2,\bar{9},\bar{3},5,4,7,1)$ $6,9-(3,0,\bar{4},2,8,5,1,7)$ $6,0-(5,2,7,1,8,\bar{4},9,3)$
$7,8-(9,1,\bar{2},3,\bar{0},6,4,5)$ $7,9-(4,6,3,1,5,8,0,2)$ $7,0-(9,1,\bar{2},3,8,6,5,4)$
$8,9-(7,1,\bar{0},\bar{2},4,5,6,3)$ $8,0-(6,5,4,\bar{9},2,\bar{3},7,1)$ $9,0-(3,4,6,5,8,1,7,2)$

3. COMBINATORIAL DESCRIPTION OF THE NEW POLYHEDRON

In fig. 2 we give the orientable manifold M of genus 3 with 10 vertices, which we embed into R^3 as a polyhedron. M has A_4 as its automorphism group. It is generated by the permutations (1,2,3)(5,9,8)(6,7,0) and (2,4,3)(5,0,8)(6,7,9) of the vertices.

M is combinatorially different from the polyhedron described in [5], since 1,2,3,4 are the only 9-valent vertices and none of 123, 124, 134, 234 is a triangle of M, whereas the polyhedron described in [5] contains a triangle having only 9-valent vertices, namely cdd'. The alternative polyhedron described in [5] has even a different graph of edges.

REMARK. If we remove from M either the three edges 12, 13 and 14 (marked by a pair of dots in Fig. 2) or the three edges 23, 24 and 42 (marked by six dots in Fig. 2) (thus replacing each of the three pairs of triangles adjacent to the removed edges by a quadrangle), we get a weakly neighborly polyhedral map (i.e. every pair of vertices is joined by an edge or a diagonal). For the exact definition and motivation of weakly neighborly polyhedral map cf. [6]. The two weakly neighborly polyhedral maps which we derived from M are not isomorphic. They are the first examples of orientable w.n.p. maps of genus 3. This is especially interesting since there exists no orientable w.n.p. map of genus 2, cf. [1]. Each of the two w.n.p. maps has Z_3 as its automorphism group, generated by the permutation (2,4,3)(5,0,8)(6,7,9) of the vertices. It would be interesting to decide whether the two w.n.p. maps are geometrically realizable (as a polyhedron containing three convex quadrangles).

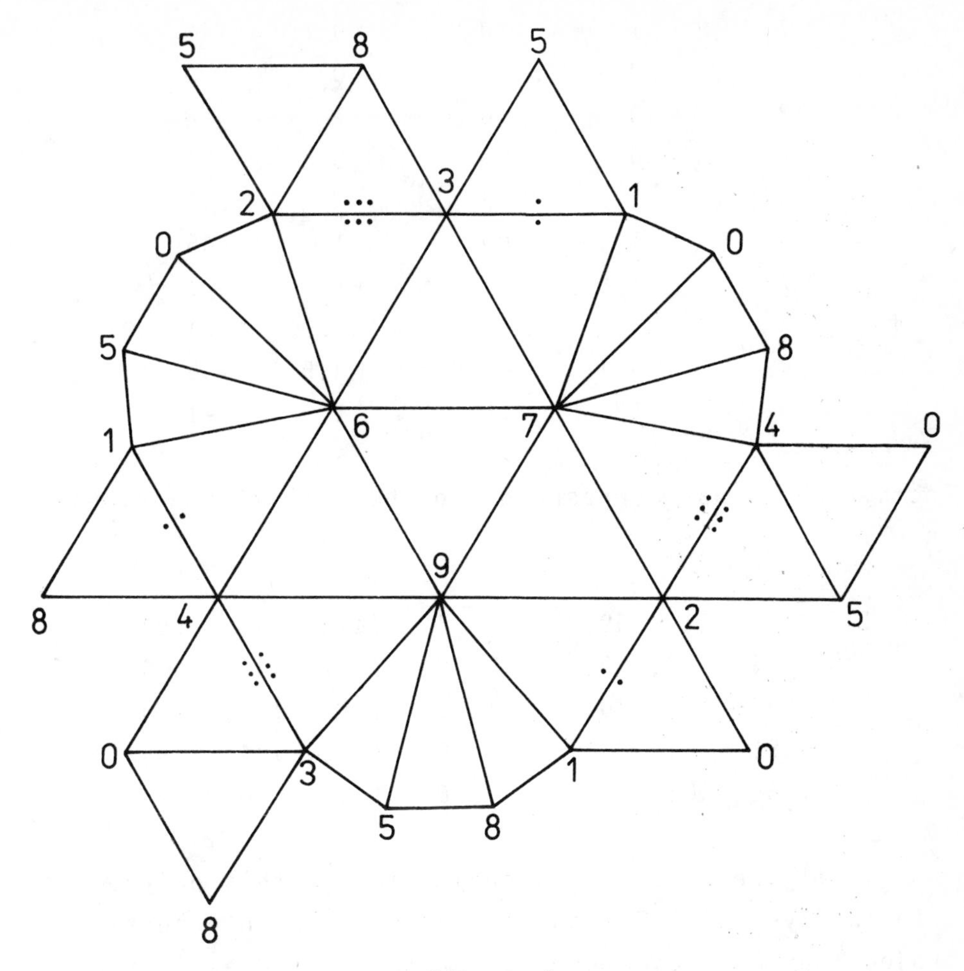

Figure 2

4. COORDINATIZATION OF THE MANIFOLD

To determine coordinates for our manifold we used a modification of the method in which a solvability-sequence is determined. This method was successful in similar cases and will be described in detail in [3].

Using this method we found the following homogeneous coordinates for the vertices. Any hyperplane in R^4 which intersects all corresponding halflines with endpoint (0,0,0,0) determines a realization of our manifold.

1	1	0	0	0
2	0	1	0	0
3	0	0	1	0
4	0	0	0	1
5	1	9	1	113
6	-1	300	301	11865
7	1,5	-1	-1	1
8	1,8	11	1,07	1
9	0,1	-1	0,5	-30
0	-1	16	1,5	-1

Another chirotope corresponds to the following coordinates in R^3:

1	(28, 114, 59)	6	(113, 147, 59)
2	(72, 7, 0)	7	(0, 114, 55)
3	(145, 153, 70)	8	(42, 100, 43)
4	(98, 191, 31)	9	(65, 184, 216)
5	(56, 106, 38)	0	(193, 0, 2)

In figure 3 we show a projection of this polyhedron onto the xy-plane. For any two edges with intersecting projections we indicate which of the two edges is above the other one. The facets of the convex hull of the vertices are the 8 triangles 029, 039, 034, 042, 247, 279, 349 and 479. We get the filled polyhedron (i.e. the compact point set which has the polyhedron as its boundary) as the (topological) closure of the difference set of the convex hull of the vertices minus the union of the following (closed) tetrahedra:

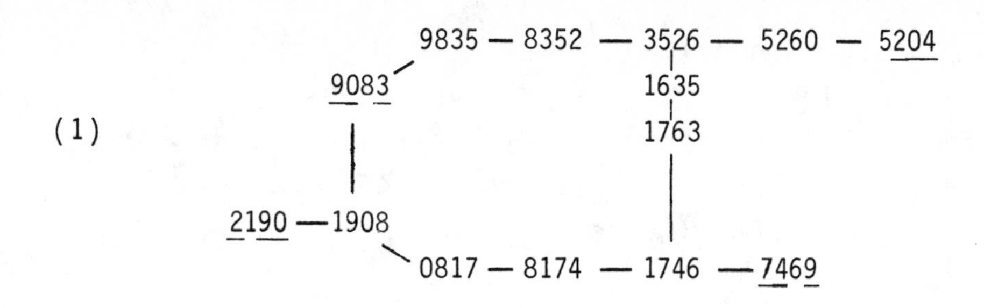
(1)
9835 — 8352 — 3526 — 5260 — 5204
9083
1635
1763
2190 —1908
0817 — 8174 — 1746 —7469

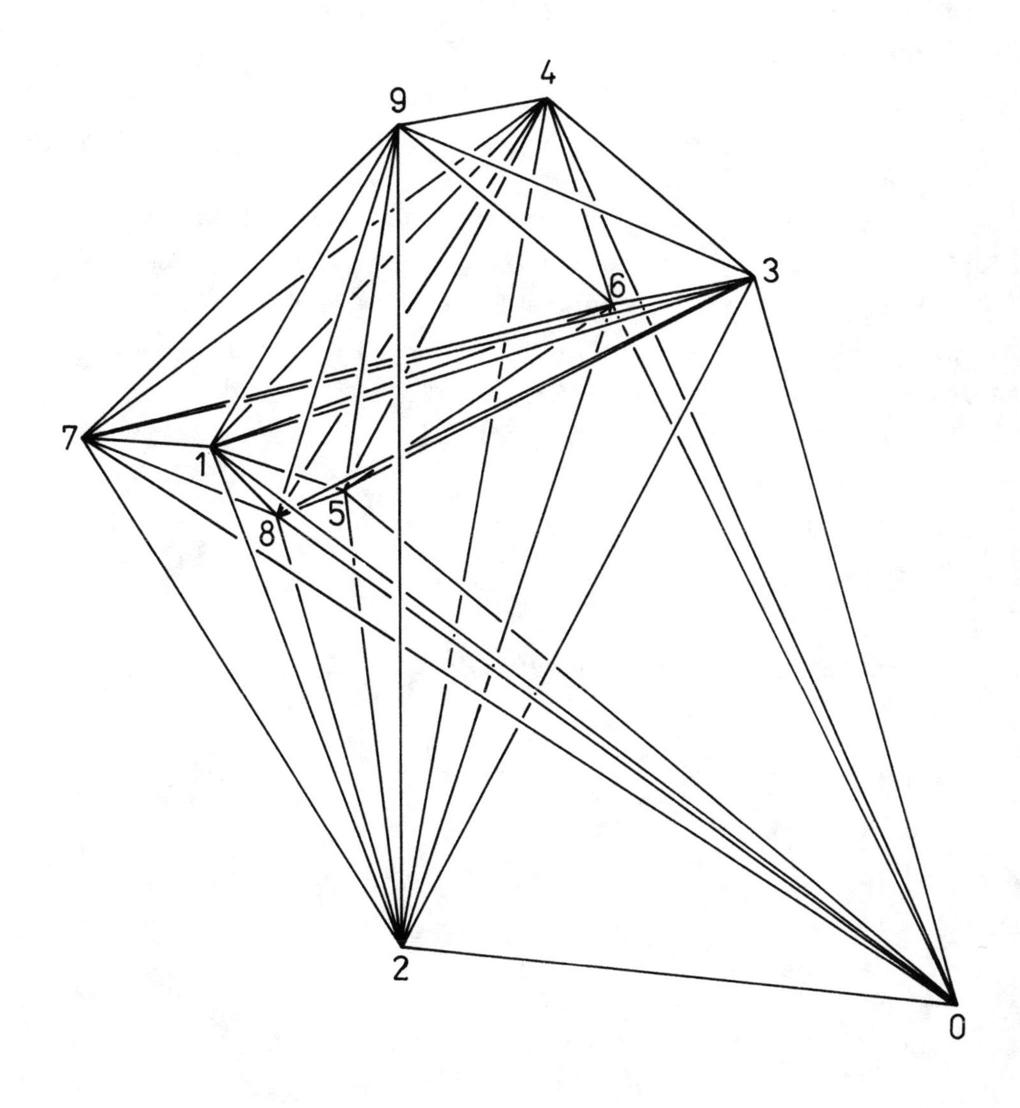
4
9
3
6
7
1
5
8
2
0

Figure 3

Two tetrahedra are neighbors in (1) if and only if they have a triangle in common. The triangles which lie on the convex hull are indicated by underline.
The meet of any two of the tetrahedra of (1) is either a common face (triangle, edge or vertex) or empty. In order to check this regard Fig. 3 and check that all edges are drawn in the right way (especially the overcrossing in the projection).

It is easy to check that the polyhedron described by (1) and the facets of the convex hull is (combinatorially) the same as that one given by Fig. 2.

To faciliate the imagination how the polyhedron looks like we show the polyhedron in three parts (Fig. 4a, b, c). In Fig. 4a) is shown the convex hull minus the union of the tetrahedra 2190, 1908, 9083, in Fig. 4b) is shown the union of the tetrahedra 9835, 8352, 3526, 5260, 5204, in Fig. 4c) is shown the union of the tetrahedra 0817, 8174, 1746, 7469, 1763, 1635 (cf. (1)).

The triangles along which the three parts are glued together are 398, 024, 479, 018, 356.

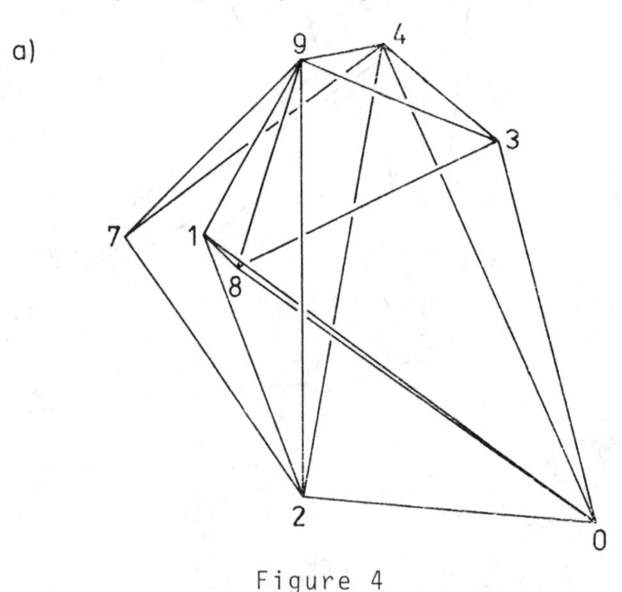

Figure 4

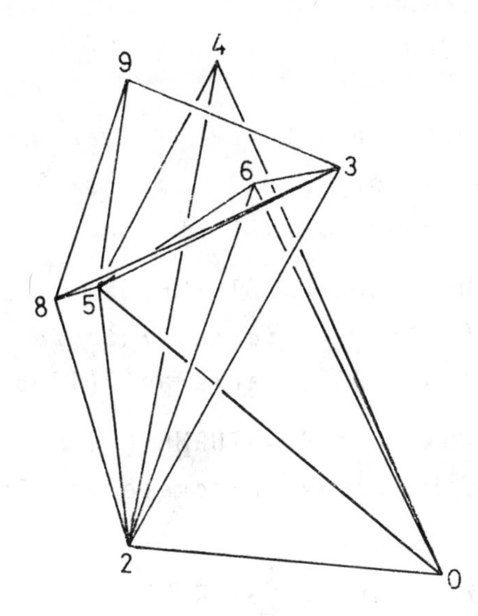

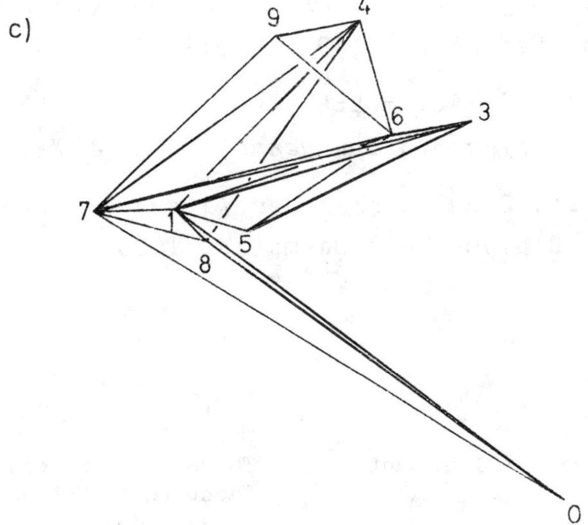

Figure 4

REFERENCES

[1] A. ALTSHULER and U. BREHM, *Non-existence of weakly neighborly polyhedral maps on the orientable 2-manifold of genus 2.* J. Comb. Theory, Series A. (to appear).

[2] J. BOKOWSKI and B. STURMFELS, *Problems of geometrical realizability - Oriented matroids and chirotopes.* Preprint Nr. 901, Darmstadt (1985).

[3] J. BOKOWSKI and B. STURMFELS, *Coordinatization of oriented matroids.* Discrete and Computational Geometry. (to appear)

[4] J. BOKOWSKI and B. STURMFELS, *Polytopal and nonpolytopal spheres - An algorithmic approach.* Israel J. Math. (to appear)

[5] U. BREHM, *Polyeder mit zehn Ecken vom Geschlecht drei.* Geometriae Dedicata 11(1981), 119-124.

[6] U. BREHM and A. ALTSHULER, *On weakly neighborly polyhedral maps of arbitrary genus.* Israel J. Math. (to appear)

[7] B. STURMFELS, *Zur linearen Realisierbarkeit orientierter Matroide.* Diplomarbeit Darmstadt 1985.

J. BOKOWSKI
Technische Hochschule Darmstadt
Fachbereich Mathematik
Schloßgartenstr. 7.
6100 Darmstadt
West-Germany

U. BREHM
Technische Universität Berlin
Fachbereich Mathematik
Strasse des 17. Juni 135.
1000 Berlin 12
West-Germany

COLLOQUIA MATHEMATICA SOCIETATIS JÁNOS BOLYAI
48. INTUITIVE GEOMETRY, SIÓFOK, 1985.

ON AN EXTREMUM PROPERTY OF THE REGULAR SIMPLEX IN S^d

K. BÖRÖCZKY*

Dedicated to László Fejes Tóth

It is well known in the two-dimensional case that even on the spherical surface in a given circle the area of a triangle is the largest for the inscribed regular triangle. This can be easily obtained with the help of the Lexell-circle [1]. In this paper we will consider the cases when $d \geq 3$.

Let B be a ball in S^d (the d-dimensional spherical space) with radius r, where $r < \pi/2$.

THEOREM. *If we have a simplex* S *in a ball* B *in* S^d *and* S_0 *denotes an inscribed regular simplex of* B, *then the volumes* $V(S)$, $V(S_0)$ *of the simplexes satisfy*

$$V(S) \leq V(S_0),$$

and equality is attained iff S *and* S_0 *are congruent.*

*Research supported by Hungarian National Foundation for Scientific Research grant no; 1238.

PROOF. By continuity considerations there exists a simplex with largest volume. As a result, it suffices to prove, that the volume of a non-regular simplex cannot be maximal.

We will show that if a simplex has a vertex inside the ball B, that it cannot have a largest volume. In fact choose a point from the opposite (d-1)-dimensional face and move the vertex from the choosen point a bit away along the line containing these two points. Then the volume of the simplex increases.

So in the following we will only consider inscribed simplexes. A non-regular simplex $S = \mathrm{conv}(A_1, A_2, \ldots, A_{d+1})$ has at least two neighbouring edges with different lengths. Let for example $A_1A_2 > A_1A_3$.

Let h be the (d-1)-dimensional spherical space which is the perpendicular bisector of A_2A_3 with poles D and E. Let us consider the symmetrization of Steiner with poles D and E and with hyperplane h. This means that each chord of a line orthogonal to h translates in its own line into a position symmetrical with respect to h. While the length of a chord remains unchanged, in a spherical space this is obviously a volume-increasing transformation. The simplex S can be obtained as the union of triangles $T = A_2A_3N$, where N runs over each point of the (d-2)-dimensional simplex $(A_1A_4A_5\ldots A_{d+1})$.

Our symmetrization transforms these triangles A_2A_3N in their own two-dimensional plane. Performing this symmetrization the (d-2)-dimensional simplex $(A_1A_4A_5\ldots$ $\ldots A_{d+1})$ will be projected onto a (d-2)-dimensional simplex $(A_1'A_4'A_5'\ldots A_{d+1}')$. The image of a point N is N'. Let us denote the d-dimensional simplex $(A_1'A_2A_3A_4'A_5'\ldots$ $\ldots A_{d+1}')$ by S', the triangle A_2A_3N' by T' and the image of S and T by the symmetrization by S^* and T^*, respectively.

PROPOSITION. T' *contains* T^* *and* S' *contains* S^*.

In order to prove our statement it suffices to show that T' contains T^*, since the second half of the statement follows from this.

Let S^2 be a unit sphere in the Euclidean 3-space. Furthermore let the triangle ABC of S^2 be congruent to the triangle A_2A_3N (Fig. 1).

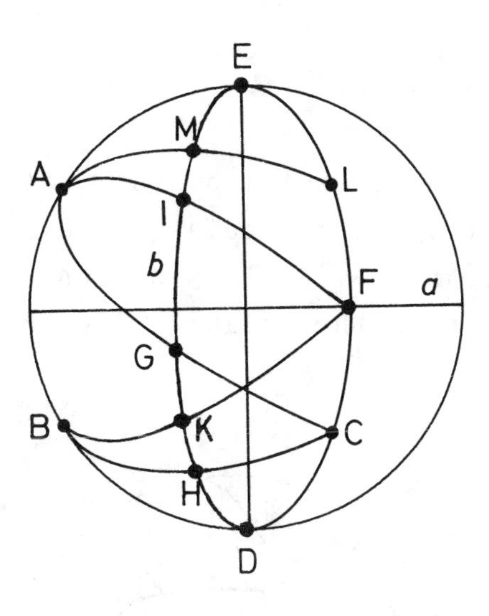

Figure 1

Let a be the great circle on S^2 which is the perpendicular bisector of AB with poles D and E, where

$$\widehat{EA} + \widehat{AB} + \widehat{BD} = \widehat{ED}.$$

Let F be the intersection of the great circles a and the arc $\widehat{DCE}$, and let L be the reflection of C in a. Let us consider any great circle $b = \widehat{DE}$, which intersects the arcs $\widehat{AC}$, $\widehat{BC}$, $\widehat{AF}$, $\widehat{BF}$, $\widehat{AL}$ in the points G, H, I, K, M, respectively.

We have to prove that

$$\widehat{IK} > \widehat{GH},$$

which is equivalent to

$$\widehat{IG} > \widehat{KH} = \widehat{IM}.$$

Let O denote the centre of the unit sphere in the Euclidean 3-space. Let the points $\bar{F}$, $\bar{G}$, $\bar{I}$, $\bar{M}$ be the images of the points F, G, I, M by central projection from the centre O to the triangle ACL (Fig. 2).

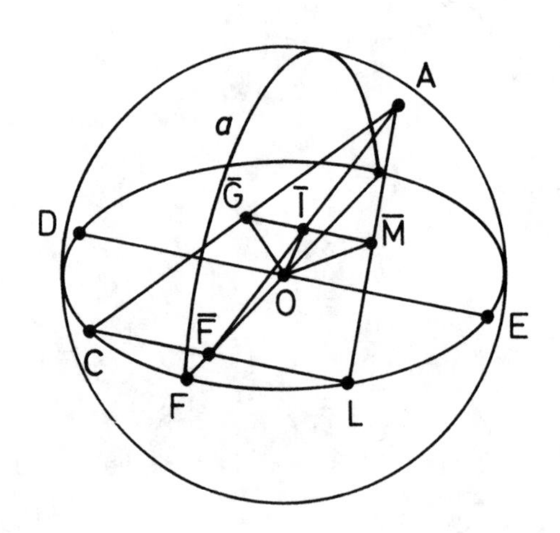

Figure 2

We have $\widehat{IG} = \sphericalangle \bar{I}O\bar{G}$ and $\widehat{IM} = \sphericalangle \bar{I}O\bar{M}$.

We can also obtain the points $\bar{G}$, $\bar{I}$, $\bar{M}$ as the intersections of the plane of the great circle b with the segments AC, A$\bar{F}$, AL, respectively. So the segment $\bar{G}\bar{M}$ is parallel to DE and CL, and $\bar{I}$ is the midpoint of the segment $\bar{G}\bar{M}$ (see Fig. 2).

Since EA < AD so E$\bar{I}$ < $\bar{I}$D. Now it follows that $\bar{G}O < \bar{M}O$. Therefore from the triangle $O\bar{G}\bar{M}$ we have

$$\widehat{IG} = \sphericalangle \bar{I}O\bar{G} > \sphericalangle \bar{I}O\bar{M} = \widehat{IM}.$$

As a result we have proved that $\widehat{IK} > \widehat{GH}$, hence our original proposition follows.

Hence we have proved that for a non-regular inscribed simplex S for the volumes V(S), V(S*), V(S') the following inequalities are true:

$$V(S) < V(S^*) < V(S') .$$

Obviously the simplex S' lies inside the ball B. As a result the simplex S could not have the maximal volume, therefore the original theorem is proved.

REFERENCE

[1] FEJES TÓTH, L., *Lagerungen in der Ebene, auf der Kugel und im Raum*, Berlin-Göttingen-Heidelberg, 1953, 22-23.

KÁROLY BÖRÖCZKY
Eötvös Loránd University, Institute of Mathematics,
Department of Geometry, Budapest, VIII. Rákóczi út 5.
H-1088

COLLOQUIA MATHEMATICA SOCIETATIS JÁNOS BOLYAI
48. INTUITIVE GEOMETRY, SIÓFOK, 1985.

THE RICH LINE PROBLEM OF P. ERDŐS

CHIH-HAN SAH

Let n points be given in the plane with $n > 1$. A line L is called *rich* if it passes at least $n^{1/2}$ of these points. Let $f(n)$ denote the maximum number of rich lines as the points are varied. P. Erdős raised the question: What is a good asymptotic estimate of $f(n)$? In general, $f(n)/n^{1/2}$ is bounded by some universal constant (to be called the *rich line constant*) from above, (cf. e.g. the paper of P. Erdős in these proceedings). The more precise problem calls for the determination of the rich line constant (with some sort of estimate on the error term).

By using a square lattice with $[n^{1/2}] + 1$ points on each side and by removing small square corners as needed, it is easy to see that:

$$\liminf_n f(n)/n^{1/2} \geq 2.$$

It has been suggested that perhaps the rich line constant is 2. In fact, a greedier guess is that $f(n) \leq 2(n^{1/2} + 1)$.

It is not difficult to see that $f(2) = 1$, $f(3) = 3$, $f(4) = 6$, $f(5) = 2$. However, the Pappus configuration yields 9 points and 10 lines with each line passing through exactly 3 points. Thus $f(9) \geq 10$ and the greedy guess is wrong. We will use some rather simple constructions to show the following result:

THEOREM. *Let* $n = p^2 \geq 4$. $f(p^2)$ *is at least* $3p$ *for odd* p *and at least* $3p+3$ *for even* p. *We have* $\liminf_n f(n)/n^{1/2} \geq 3$.

PROOF. We begin with the case $p = 2k+1 > 1$. Let C_1 denote a regular p-gon with edges labelled in order as E_i, $-k \leq i \leq k$. Since p is odd, the lines L_i extending E_i are never parallel and lead to $p(p-1)/2 = p.k$ points of intersections. These points give us the vertices of k concentric regular p-gons C_j, $1 \leq j \leq k$, where the vertices of C_j are the intersections of L_s abd L_t with $s-t \equiv \pm j \bmod p$.

We next let C_0 be the regular p-gon formed by the midpoints of E_i. Each L_i clearly passes through exactly p points--one vertex of C_0 and 2 vertices of C_j, $1 \leq j \leq k$. Let σ be a similarity fixing the center of symmetry and carrying C_k onto C_0. Denote $\sigma(C_j)$ by C_{j-k}, $0 \leq j \leq k$. The vertices of the p regular p-gons C_i, $-k \leq i \leq k$, therefore give us p^2 points. Since σ involves a nontrivial contraction, L_i, $\sigma(L_i)$, $-k \leq i \leq k$, give us 2p rich lines not passing through the center of symmetry. It is easy to see that C_{odd} is similar to C_1 and C_{even} is similar to C_0 under central homotheties and that each of the p axes of symmetries passes through exactly one vertex of each C_j, $-k \leq j \leq k$. We therefore have at least 3p rich lines through the p^2 vertices.

We go to the case $p = 2k \geq 2$. We construct regular (p+1)-gons C_j, $1 \leq i \leq k$, as in the preceding case. Let ρ be a central similarity carrying C_k onto C_1 and define $\rho(C_j)$ to be C_{j-k+1}, $1 \leq j \leq k$. The vertices of C_j, $-k+1 < j \leq k$, therefore give us $(p+1)(p-1)$ points. To these we add the center of symmetry to get p^2 points. Each L_i passes through

$2k = p$ points-- 2 vertices from each C_j, $1 \le j \le k$. Since ρ also involves a nontrivial contraction, L_i, $\rho(L_i)$, $-k \le i \le k$, account for 2p+2 rich lines. The introduction of the center of symmetry gives us p+1 more rich lines in the way of p+1 axes of symmetry. We therefore have at least 3p+3 rich lines.

Lastly, for $n \ne p^2$, we may remove some vertices of some outer regular polygons lying in some small angular domain with apex at the center of symmetry and conclude that $\liminf_n f(n)/n^{1/2} \ge 3$.

CHIH-HAN SAH
Department of Mathematics
SUNY at Stony Brook
N.Y. 11794, U.S.A.

COLLOQUIA MATHEMATICA SOCIETATIS JÁNOS BOLYAI
48. INTUITIVE GEOMETRY, SIÓFOK, 1985.

A PACKING OF 840 BALLS OF RADIUS 9°0'19" ON THE 3-SPHERE

H.S.M. COXETER

Dedicated to László Fejes Tóth

1. INTRODUCTION

When the dodecahedral facets of the 120-cell {5,3,3} are centrally projected onto a concentric 3-sphere of unit radius, their inspheres project into a packing of 120 balls of angular radius 18° [Fejes Tóth 1965, pp. 298-299]. (The density 0.77412... of this packing exceeds the density 0.74048... of the familiar cubic close packing in Euclidean 3-space.) The centres of the 120 balls are the vertices of the 600-cell {3,3,5}. By reducing the radius to 9° and inserting 720 new balls whose centres arise from the midpoints of the edges, we obtain a packing of 120 + 720 balls each touching or nearly touching 12 others. Since the 720 new balls have no mutual contacts, this packing, of density 0.68747..., is unstable. However, by slightly displacing each of these 720 in a suitable direction and very slightly increasing the radius of all the 840, we can increase the number of contacts and thus achieve a stable packing of density 0.68865... . The chief difficulty in carrying out this procedure lies in choosing the 'suitable direction' for all the 720 balls simultaneously.

2. THE COMPOUND OF FIVE 600-CELLS

The edges of the spherical honeycomb {3,3,5} [Coxeter 1974, p. 52] consist of 720 arcs of length 36° whose midpoints are the centres of the 720 'new balls'. All contacts can be destroyed by slightly displacing each of these balls in *any* direction orthogonal to the relevant arcs. We are thus confronted by a whole pencil of possible directions for each displacement. Can we consistently make a 'best' choice of direction in each of the 720 pencils? As one step towards an answer, we observe that each edge of {3,3,5} is orthogonally bisected by a pentagonal face of the dual honeycomb {5,3,3}, thus specializing ten of the infinitely many directions in the pencil: towards, or away from, each vertex of the pentagon. But can we consistently choose, in all the 720 pentagonal faces of the 120-cell {5,3,3}, one vertex of each?

A neat solution to this puzzle is provided by the *vertex-regular compound*

$$\{5,3,3\}[5\{3,3,5\}]$$

[Coxeter 1973, pp. 269-270; Du Val 1964, p. 76], in which the 600 vertices of {5,3,3} are distributed among 5 inscribed {3,3,5}'s. Each {3,3,5} has 600 tetrahedral facets, 120 of which are inscribed in the 120 dodecahedral facets of {5,3,3} (one in each), thus distinguishing *one* vertex of each pentagonal face, which is just what we were seeking.

Figure 1 shows one dodecahedral facet of {5,3,3}, surrounding a vertex A (not shown) of the reciprocal {3,3,5}. B, C, D are the centres of three pentagonal faces of this dodecahedron; E is one end of the common edge of the first two faces, while F is a vertex of the third, opposite to its common edge with the first. More precisely, E and F are two adjacent vertices of one of the {3,3,5}'s inscribed in {5,3,3}; thus EF is an edge of one of the regular tetrahedra inscribed in the dodecahedron.

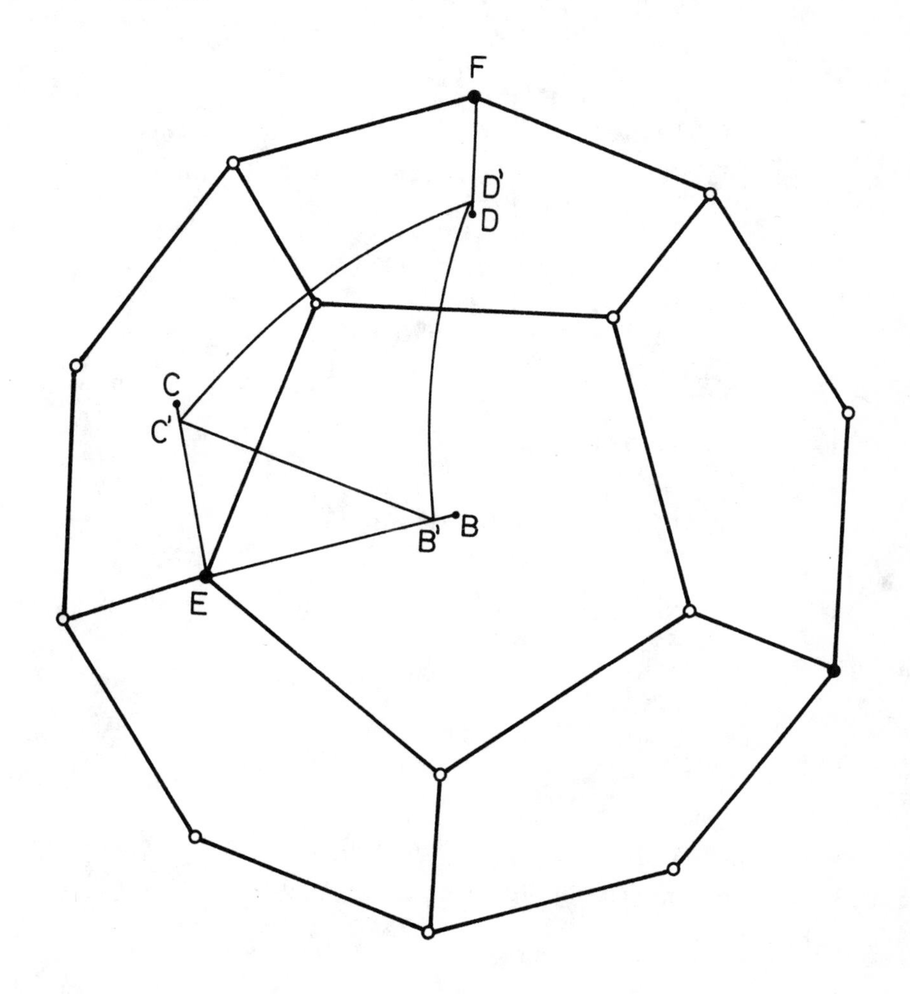

Figure 1

In the {3,3,5} reciprocal to {5,3,3}, three of the edges through A have B, C, D as their midpoints. Thus A, B, C, D are the centres of four of the 840 balls in the unstable packing. The proposed displacement leaves A fixed while moving B, C, D to certain positions B', C', D' on the lines (i.e., great circles) EB, EC, FD.

3. THE USE OF COORDINATES

One possible set of coordinates for the 120 vertices of {3,3,5} (found by P.H. Schoute in 1893) consists of the permutations of:

$$(\pm 2, \pm 2, 0, 0) ,$$

$(\tau,\tau,\tau,\tau^{-2})$ and $(\tau^{2},\tau^{-1},\tau^{-1},\tau^{-1})$ with an even number of - signs, and

$(\sqrt{5}, 1, 1, 1)$ with an odd number of - signs

(24 + 32 + 32 + 32 = 120; [Coxeter 1973, p. 239]),

where

$$\tau = (1 + \sqrt{5})/2 .$$

These 120 points occur among the 600 vertices of a {5,3,3} whose remaining 480 vertices consist of the permutations of

$$(\tau,\tau,\tau,\tau^{-2}),\ (\tau^2,\tau^{-1},\tau^{-1},\tau^{-1}),\ (\sqrt{5},\ 1,\ 1,\ 1)$$

with the remaining distributions of signs, and the *even* permutations of

$$(\pm\tau^2,\pm\tau^{-2},\pm 1,\ 0),\ (\pm\sqrt{5},\pm\tau^{-1},\pm\tau,0),\ (\pm 2,\pm 1,\pm\tau,\pm\tau^{-1})$$

($96 + 96 + 96 + 192 = 480$; [Coxeter 1973, p. 157]).

These 600 points all lie on a 3-sphere whose radius will be reduced from $2\sqrt{2}$ to 1 if we divide all the coordinates by $2\sqrt{2}$; or we may simply regard them as *homogeneous* coordinates, so that the (angular) distance between two points (x) and (y) is

$$\operatorname{arc\,cos} \frac{\Sigma xy}{(\Sigma x^2 \cdot \Sigma y^2)^{1/2}} \quad .$$

Homogeneous coordinates for all the points named in §2 may conveniently be chosen as follows:

$$A = (0,\ 0,\ 0,\ 1)\ ,$$

$$B = (\tau,0,1,\sqrt{5}\tau^2),\ C = (0,1,\tau,\sqrt{5}\tau^2),\ D = (0,-1,\tau,\sqrt{5}\tau^2),$$

$$E = (\tau^{-1},\tau^{-1},\tau^{-1},\tau^2),\ F = (-\tau^{-1},-\tau^{-1},\tau^{-1},\tau^2),$$

$$B' = B + \varepsilon E = (\tau + \tau^{-1}\varepsilon,\tau^{-1}\varepsilon,\ 1 + \tau^{-1}\varepsilon,\tau^2(\sqrt{5} + \varepsilon)),$$

$$C' = C + \varepsilon E = (\tau^{-1}\varepsilon,\ 1 + \tau^{-1}\varepsilon,\tau + \tau^{-1}\varepsilon,\ \tau^2(\sqrt{5} + \varepsilon)),$$

$$D' = D + \varepsilon F = (-\tau^{-1}\varepsilon,-1 - \tau^{-1}\varepsilon,\tau + \tau^{-1}\varepsilon,\ \tau^2(\sqrt{5} + \varepsilon)),$$

where ε is positive or negative according as the displacement BB' (for instance) is towards E or away from E.

If ε is sufficiently small, the simultaneous displacement of the 720 balls of radius ρ destroys their contact with the 120, without allowing them to intersect (or touch) one another. We can imagine $|\varepsilon|$ to increase steadly until one of the three distance B'C', B'D', C'D' becomes equal to AB' (= AC' = AD'). Then the 120 + 720 balls, suitably dilated, will have many contacts with one another. To determine such an optimal ε, we observe first that

$$\cos AB' = \tau^2(\sqrt{5} + \varepsilon)/\lambda \tag{3.1}$$

$$\begin{aligned}\lambda^2 &= (\tau+\tau^{-1}\varepsilon)^2 + (\tau^{-1}\varepsilon)^2 + (1+\tau^{-1}\varepsilon)^2 + \tau^4(\sqrt{5}+\varepsilon)^2 = \\ &= 4(\sqrt{5}\tau^3 + 2\tau^3\varepsilon + 2\varepsilon^2) ,\end{aligned} \tag{3.2}$$

$$\begin{aligned}\cos B'C' &= \{(\tau+\tau^{-1}\varepsilon)\tau^{-1}\varepsilon+\tau^{-1}\varepsilon(1+\tau^{-1}\varepsilon) + (1+\tau^{-1}\varepsilon)(\tau+\tau^{-1}\varepsilon) + \\ &+ \tau^4(\sqrt{5}+\varepsilon)^2\}/\lambda^2 = 2(\tau^6 + 4\tau^3\varepsilon + 4\varepsilon^2)/\lambda^2 = 1 - 2/\lambda^2,\end{aligned} \tag{3.3}$$

$$\begin{aligned}\cos B'D' &= \{-(\tau+\tau^{-1}\varepsilon)\tau^{-1}\varepsilon-\tau^{-1}\varepsilon(1+\tau^{-1}\varepsilon)+(1+\tau^{-1}\varepsilon)(\tau+\tau^{-1}\varepsilon) + \\ &+ \tau^4(\sqrt{5}+\varepsilon)^2\}/\lambda^2 = 2(\tau^6 +\sqrt{5}\tau^4\varepsilon+2\tau\varepsilon^2)/\lambda^2 ,\end{aligned} \tag{3.4}$$

$$\begin{aligned}\cos C'D' &= \{-(\tau^{-1}\varepsilon)^2-(1+\tau^{-1}\varepsilon)^2+(\tau+\tau^{-1}\varepsilon)^2+\tau^4(\sqrt{5}+\varepsilon)^2\}/\lambda^2 = \\ &= 2(\tau^6+6\tau^2\varepsilon + 2\tau\varepsilon^2)/\lambda^2 ,\end{aligned} \tag{3.5}$$

Since the coefficients of ε in (3.3), (3.5), (3.4) satisfy the inequalities

$$4\tau^3 > 6\tau^2 > \sqrt{5}\tau^4$$

and $|\varepsilon| < \tau$, C'D' always lies between B'C' and B'D':

$$B'C' < C'D' < B'D' \text{ if } \varepsilon > 0$$

and

$$B'D' < C'D' < B'C' \text{ if } \varepsilon < 0 .$$

Thus, while ε increases steadily from zero, the first opportunity for a new contact occurs when $AB' = B'C'$; and while $|\varepsilon|$ (with $\varepsilon < 0$) increases steadily from zero, the first opportunity occurs when $AB' = B'D'$. In either case $C'D'$ is irrelevant, and there is no need to make further use of that distance.

4. TWO QUARTIC EQUATIONS

For a stable packing with $\varepsilon > 0$ the equation $AB' = B'C'$ yields

$$\tau^2(\sqrt{5} + \varepsilon)\lambda = 2(\tau^6 + 4\tau^3\varepsilon + 4\varepsilon^2) .$$

Squaring and collecting terms, we obtain (for ε) the quartic equation

$$(4.1) \qquad 6\varepsilon^4 + 10\tau^3\varepsilon^3 + (49 + 22\sqrt{5})\varepsilon^2 + 4\tau^6\varepsilon - \tau^4 = 0.$$

The relevant root (kindly computed by John Leech) is

$$(4.2) \qquad \varepsilon \approx 0.085193165 \quad ,$$

the common radius of the dilated balls is

$$(4.3) \qquad \rho = \frac{1}{2} B'C' = \arcsin 1/\lambda \approx 0.1571374 \approx 9^\circ 0'12''$$

and the density of the packing is

$$(4.4) \qquad 840(\rho - \sin\rho\cos\rho)/\pi \approx 0.688226919 \;.$$

Similarly, for a stable packing with $\varepsilon < 0$, the equation $AB' = B'D'$ yields

$$\tau^2(\sqrt{5} + \varepsilon)\lambda = 2(\tau^6 + \sqrt{5}\tau^4\varepsilon + 2\tau\varepsilon^2)$$

and another quartic equation

$$(4.5) \qquad 4\varepsilon^4 + 4(4+\sqrt{5})\varepsilon^3 + (23+13\sqrt{5})\varepsilon^2 + 4\sqrt{5}\tau^3\varepsilon + 2\tau = 0 \;.$$

The relevant root is

$$(4.6) \qquad \varepsilon \approx -0.097998303 \;,$$

the radius is

$$(4.7) \qquad \rho = \frac{1}{2} AB' = \frac{1}{2} \operatorname{arc\,cos} \{\tau^2(\sqrt{5}+\varepsilon)/\lambda\} \approx 0.1571697 \approx 9^\circ 0' 19''$$

and the density is

$$(4.8) \qquad 840(\rho - \sin\rho\cos\rho)/\pi \approx 0.688650073 \;.$$

This is almost certainly the greatest possible density for a packing of 840 congruent balls on the 3-sphere.

5. THE SYMMETRY GROUP

We recall that a dodecahedral facet of the 120-cell can serve as a fundamental region for the binary icosahedral group of order 120 [Coxeter 1970, p. 41]. In each dodecahedron we are using a tetrahedral subgroup of its icosahedral symmetry group. Therefore the symmetry group of the whole packing is the non-crystallographic rotation group

$$(\langle 3,3,2\rangle/\langle 3,3,2\rangle;\ \langle 5,3,2\rangle/\langle 5,3,2\rangle)$$

of order $120 \times 12 = 1440$, the subdirect product of the binary tetrahedral group $\langle 3,3,2\rangle \cong SL(2,3)$ and the binary icosahedral group $\langle 5,3,2\rangle \cong SL(2,5)$ [Coxeter 1974, pp. 80, 105]. In the notation of Du Val [1964, p. 57] it is

$$24 \cdot (T/T;\ I/I) \quad .$$

Since this group of order 1440 is transitive on the 720 'displaced' balls, each ball is invariant for a subgroup of order 2. In fact, such a subgroup is generated by the half-turn about the diameter EB of the ball with centre B'. (From the four-dimensional standpoint this is the half-turn about the plane of the great circle EB.)

6. THE NUMBER OF CONTACTS

The pentagon with centre B is the interface between the dodecahedron with centre A and another dodecahedron, derivable by means of the half-turn about BE. Figure 2 names some further points:

$$G' = G + \varepsilon E \qquad \text{with } \varepsilon \approx 0.085 \quad ,$$

$$B'' = B + \varepsilon E,\ D'' = D + \varepsilon F,\ H'' = H + \varepsilon J \qquad \text{with } \varepsilon \approx -0.098 \ .$$

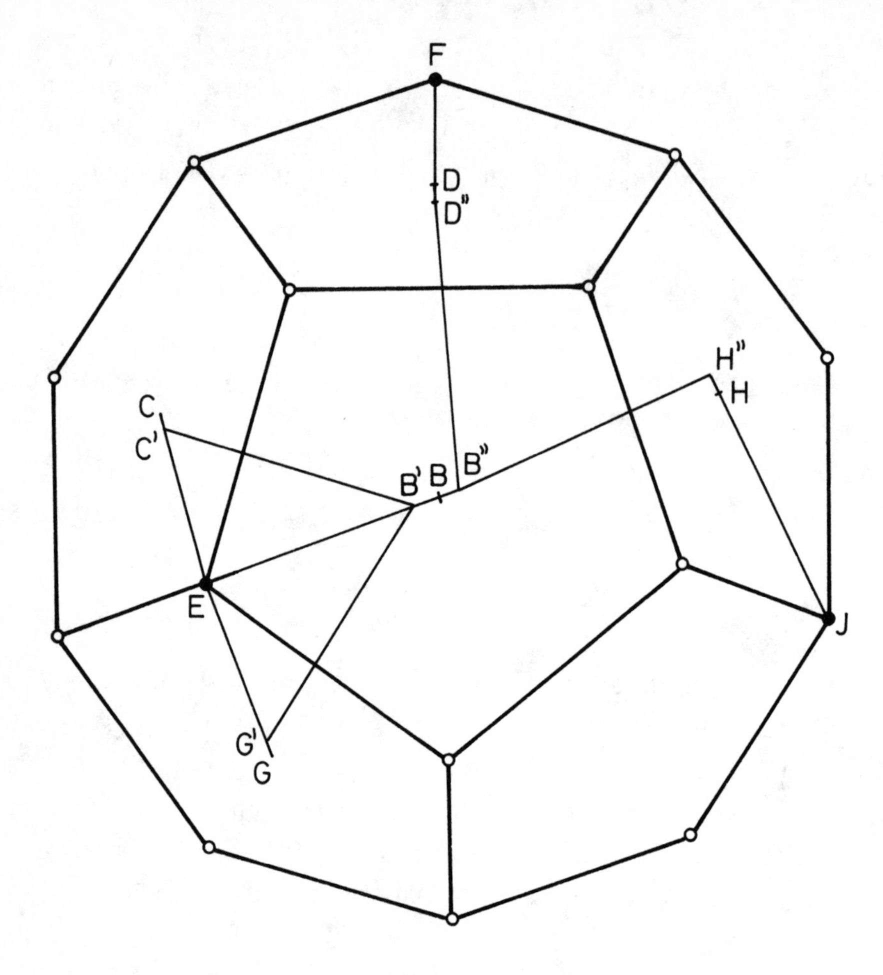

Figure 2

The relationship between various pairs of pentagons shows that

$$G'B' = B'C' \quad \text{and} \quad H''B'' = B''D'' \; ;$$

in fact, B'C'G' and B"D"H" are equilateral triangles, yielding triads of mutually tangent balls (with $\varepsilon > 0$ and $\varepsilon < 0$, respectively).

We conclude that, in either kind of stable packing, the complete list of contacts (among the 120 + 720 balls) is as follows: each of the 120 touches twelve of the 720; each of the 720 touches two of the 120 and four others of the 720.

REFERENCES

[1] H.S.M. COXETER, *Twisted Honeycombs*. Regional Conference Series in Mathematics, No. 4, American Mathematical Society. 1970.

[2] H.S.M. COXETER, *Regular Polytopes*, 3rd ed., Dover, New York. 1973.

[3] H.S.M. COXETER, *Regular Complex Polytopes*. Cambridge University Press. 1974.

[4] P. DU VAL, *Homographies, Quaternions and Rotations*. Oxford University Press. 1964.

[5] L. FEJES TÓTH, *Reguläre Figuren*. Akadémiai Kiadó, Budapest, 1965.

H.S.M. COXETER
University of Toronto
Dept. of Mathematics
Toronto, Ont. Canada
M5S 1A1

COLLOQUIA MATHEMATICA SOCIETATIS JÁNOS BOLYAI
48. INTUITIVE GEOMETRY, SIÓFOK, 1985.

VARYING TWO TANGENT HYPERBOLAS

J. van de CRAATS

1. INTRODUCTION

In explaining mathematical ideas to 'the man in the street', or 'the intelligent schoolgirl', we nowadays have a powerful tool: the computer. In particular, computer graphics can be used to illustrate geometry in a spectacular way. Geometric objects can be shown on a screen, rotating in space, in various projections. Conics appear as soon as circles are involved.
Varying the parameters of a problem can be done instantaneously; loci of points and envelopes of lines are no longer abstract concepts, but visible objects.
The underlying mathematics is analytic affine and projective geometry, and it is from this field that our subject is chosen.

2. TWO HYPERBOLAS

Consider in the real affine plane four parallel lines d_1, d_2, d_3 and h (say, in vertical direction), and a triangle with vertices $A_i \in d_i$ and sides a_i ($A_i \notin a_i$). The intersection of the lines a_i and d_i is called B_i. Take two hyperbolas H_1 and H_2, both with h as one of their asymptotes, H_1 tangent to a_1, a_2 and a_3, and H_2 through B_1, B_2 and B_3 (H_1 and H_2 are uniquely determined by these conditions). It may be a surprise to discover that the hyperbolas H_1 and H_2 always will thouch each other.

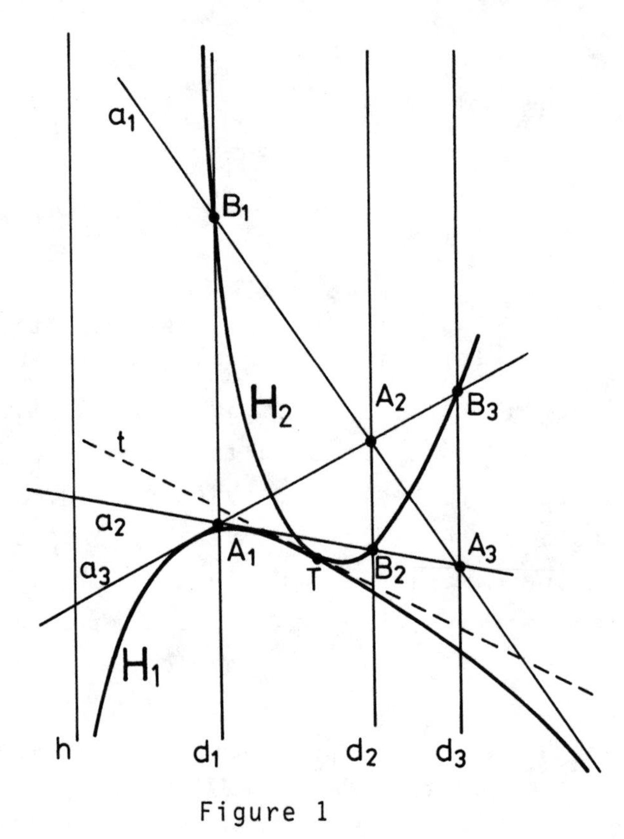

Figure 1

3. WHY ARE H_1 AND H_2 TANGENT ?

By projecting the asymptote h to the line at infinity, the hyperbolas H_1 and H_2 become parabolas with parallel axes, and the tangency of H_1 and H_2 follows from a theorem of De Cicco [2]. This theorem may be considered as a

limiting case of Baker's projective version of Feuerbach's theorem on the tangency of the nine-point-circle of a triangle and its incircles (cf. Coxeter [1], p. 219 and also Van de Craats [3], p. 409 ff.).
Returning to the case where H_1 and H_2 are hyperbolas, we vary the asymptote h parallel to itself, while keeping fixed the lines d_i, a_i and the points A_i and B_i.

Then the point of tangency T of H_1 and H_2 will describe a certain locus, while the common tangent line t envelops a certain curve. We want to determine this locus and this envelope. The proper tool is the use of homogeneous (projective) coordinates.

4. HOMOGENEOUS COORDINATES

Introduce homogeneous point coordinates (x_1,x_2,x_3) and line coordinates $[X_1, X_2, X_3]$ with $(x_1,x_2,x_3) \in [X_1,X_2,X_3] \Leftrightarrow x_1X_1 + x_2X_2 + x_3X_3 = 0$, taking $A_1 = (1,0,0)$, $A_2 = (0,1,0)$, $A_3 = (0,0,1)$ and $(1,1,1)$ as the point at infinity in the vertical direction. Then

$$B_1 = (0,1,1)\ ,\ B_2 = (1,0,1)\ ,\ B_3 = (1,1,0)\ ,$$
$$a_1 = [1,0,0]\ ,\ a_2 = [0,1,0]\ ,\ a_3 = [0,0,1]\ ,$$
$$d_1 = [0,1,-1]\ ,\ d_2 = [-1,0,1]\ ,\ d_3 = [1,-1,0]\ .$$

The line h has line coordinates $[\alpha_1, \alpha_2, \alpha_3]$ for some α_1, α_2, α_3 with (since $(1,1,1) \in h$):

$$(4.1) \qquad \alpha_1 + \alpha_2 + \alpha_3 = 0.$$

It is immediate to verify that the hyperbola H_1 is given in line coordinates by

(4.2) $$H_1 : \alpha_1^2 X_2 X_3 + \alpha_2^2 X_3 X_1 + \alpha_3^2 X_1 X_2 = 0$$

and that H_2 is given in point coordinates by

(4.3) $$H_2 : \alpha_1(x_1{}^2 + x_2 x_3) + \alpha_2(x_2{}^2 + x_3 x_1) + \alpha_3(x_3{}^2 + x_1 x_2) = 0.$$

By using methods similar to the ones employed in [4], one finds that the coordinates of the common point T of H_1 and H_2 and their common tangent line t are

(4.4) $$T = ((\alpha_2 - \alpha_3)^2, (\alpha_3 - \alpha_1)^2, (\alpha_1 - \alpha_2)^2)$$

and

$$t = [2\alpha_1{}^3 + \alpha_1\alpha_2\alpha_3, 2\alpha_2{}^3 + \alpha_1\alpha_2\alpha_3, 2\alpha_3{}^3 + \alpha_1\alpha_2\alpha_3]$$

(these results can also be verified directly from (4.2) and (4.3), using (4.1). This provides an alternative proof that H_1 and H_2 are tangent.)

5. THE LOCUS OF T

Since the common point T of H_1 and H_2 is given by

$$T : (x_1, x_2, x_3) = ((\alpha_2 - \alpha_3)^2, (\alpha_3 - \alpha_1)^2, (\alpha_1 - \alpha_2)^2)$$

we have (with appropriate choice of signs)

$$\pm x_1{}^{\frac{1}{2}} = \alpha_2 - \alpha_3$$

$$\pm x_2{}^{\frac{1}{2}} = \alpha_3 - \alpha_1$$

$$\pm x_3{}^{\frac{1}{2}} = \alpha_1 - \alpha_2 \quad .$$

It follows that

$$(5.1)\qquad C : x_1^{\frac{1}{2}} \pm x_2^{\frac{1}{2}} \pm x_3^{\frac{1}{2}} = 0$$

is the equation of the locus of T. Squaring two times shows that C is the conic

$$(5.2)\qquad C : x_1^2 + x_2^2 + x_3^2 = 2(x_2x_3 + x_3x_1 + x_1x_2) .$$

In line coordinates C has the simple form

$$(5.3)\qquad C : X_2X_3 + X_3X_1 + X_1X_2 = 0.$$

From (5.2) and (5.3) it is evident that C passes through the points B_i, and that the lines a_i are the tangent lines in these points. E.g., $T = B_1 = (0,1,1)$ occurs when $h : [\alpha_1, \alpha_2, \alpha_3] = [-2,1,1]$. Then h, d_1, d_2, d_3 is a harmonic quadruple of lines through the infinite point (1,1,1).

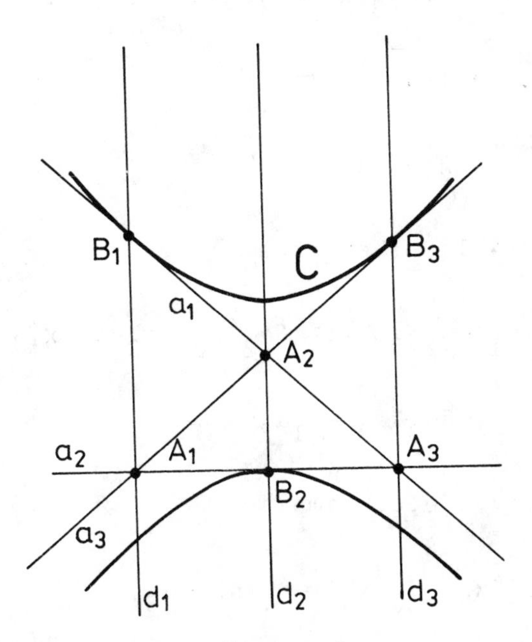

Figure 2

6. THE ENVELOPE OF t

From (4.5) it follows that the equation of t in point coordinates is

$$(2\alpha_1^3 + \alpha_1\alpha_2\alpha_3)\, x_1 + (2\alpha_2^3 + \alpha_1\alpha_2\alpha_3)\, x_2 + \tag{6.1}$$
$$+ (2\alpha_3^3 + \alpha_1\alpha_2\alpha_3)\, x_3 = 0.$$

Following Coxeter (cf. [4], section 4) we observe that the point (x_1, x_2, x_3) where the line t is in contact with its envelope also satisfies the differential equation obtained by differentiating (6.1) with respect to the parameters:

$$(6\alpha_1^2 x_1 + \alpha_2\alpha_3 \Sigma x_i)\, d\alpha_1 + (6\alpha_2^2 x_2 + \alpha_3\alpha_1 \Sigma x_i) d\alpha_2 +$$
$$+ (6\alpha_3^2 x_3 + \alpha_1\alpha_2 \Sigma x_i) d\alpha_3 = 0.$$

This should be equivalent to

$$d\alpha_1 + d\alpha_2 + d\alpha_3 = 0$$

on account of (4.1). Therefore

$$6\alpha_1^2 x_1 + \alpha_2\alpha_3 \Sigma x_i = 6\alpha_2^2 x_2 + \alpha_3\alpha_1 \Sigma x_i =$$
$$= 6\alpha_3^3 x_3 + \alpha_1\alpha_2 \Sigma x_i.$$

The last equality, e.g., implies

$$6(\alpha_2^2 x_2 - \alpha_3^3 x_3) = \alpha_1(\alpha_2 - \alpha_3) \Sigma x_i =$$
$$= -(\alpha_2+\alpha_3)(\alpha_2-\alpha_3) \Sigma x_i = (\alpha_3^2 - \alpha_2^2) \Sigma x_i,$$

i.e.,

$$\alpha_2^{\,2}(6x_2 + \Sigma x_i) = \alpha_3^{\,2}(6x_3 + \Sigma x_i).$$

It follows that

$$\alpha_1^{\,2} : \alpha_2^{\,2} : \alpha_3^{\,2} = y_1^{-1} : y_2^{-1} : y_3^{-1}$$

where

$$(6.2)\qquad \left.\begin{aligned} y_1 &= 6\ x_1 + \Sigma x_i = 7\ x_1 + x_2 + x_3 \\ y_2 &= 6\ x_2 + \Sigma x_i = x_1 + 7\ x_2 + x_3 \\ y_3 &= 6\ x_3 + \Sigma x_i = x_1 + x_2 + 7\ x_3 \end{aligned}\right\}$$

Therefore, the envelope is given by

$$(6.3)\qquad Q : y_1^{-\frac{1}{2}} \pm y_2^{-\frac{1}{2}} \pm y_3^{-\frac{1}{2}} = 0.$$

Squaring two times shows that Q is a *quartic curve*:

$$(6.4)\qquad Q : y_2^{\,2}y_3^{\,2} + y_3^{\,2}y_1^{\,2} + y_1^{\,2}y_2^{\,2} = $$

$$= 2(y_2y_3 + y_3y_1 + y_1y_2).$$

7. GEOMETRIC PROPERTIES OF THE ENVELOPE

Intersecting Q with the line $y_2 = y_3$ yields

$$y_2^{\,4} = 4\ y_1y_2^{\,3}$$

so $(y_1,y_2,y_3) = (1,0,0)$ is a *cusp* with cuspoidal tangent line $y_2 = y_3$. The other, 'fourth' point of intersection of Q with $y_2 = y_3$ is $(y_1,y_2,y_3) = (1,4,4)$, where the tangent line to Q is given by

$$8\, y_1 + y_2 + y_3 = 0.$$

For the lines $y_3 = y_1$ and $y_1 = y_2$ we get similar results. In (x_1,x_2,x_3) coordinates the cusps are $(-8,1,1)$, $(1,-8,1)$ and $(1,1,-8)$. These are points on the lines d_i, and exactly these lines are the cuspoidal tangent lines. The remaining points of intersection of Q with the lines d_i are the points $(x_1,x_2,x_3) = (0,1,1)$ etc., i.e., the points B_i, the tangent lines to Q in these points being the lines a_i.

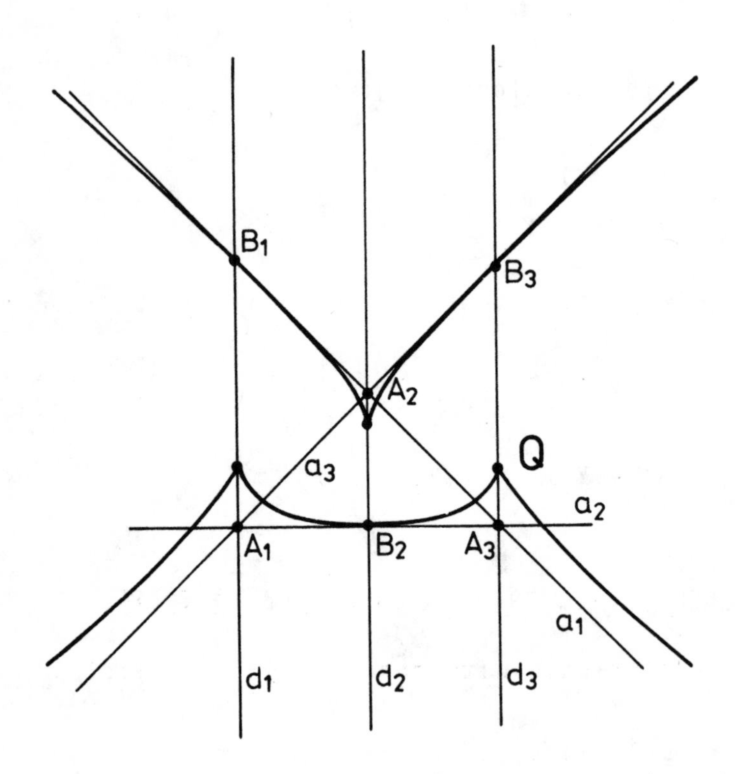

Figure 3

It may be shown that Q is projectively equivalent to Steiner's well-known hypocycloid (deltoid) (cf [4]). This is illustrated in fig. 4, where the triangle $A_1A_2A_3$ is transformed into an equilateral triangle with the point at infinity in the vertical direction (1,1,1) transformed into its center. The locus C of the points T then simply is the incircle of the triangle.

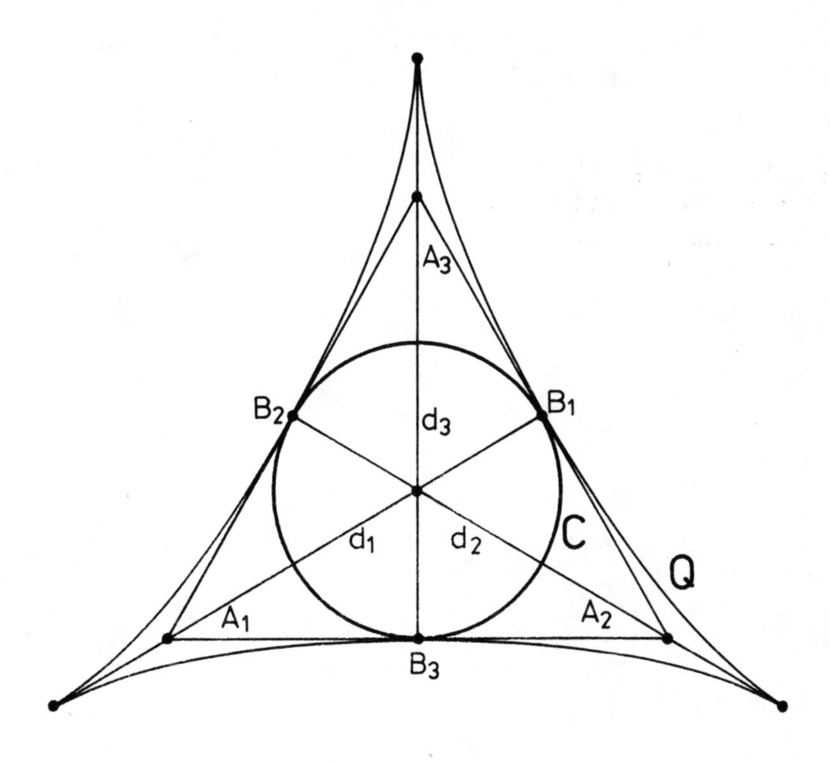

Figure 4

REFERENCES

[1] COXETER, H.S.M. (1983), *The affine aspect of Yaglom's Galilean Feuerbach*, Nieuw Archief voor Wiskunde (4), vol. 1, 212-223.

[2] DE CICCO, J. (1939), *An Analog of the Nine-point Circle in the Kasner plane*, Am.Math. Monthly 46, 627-634.

[3] VAN DE CRAATS, J. (1983), *On Galilean Geometry*, Nieuw Archief voor Wiskunde (4), vol. 2, 403-419.

[4] VAN DE CRAATS, J. (1986), *An unexpected appearance of Steiner's hypocycloid*, Aequations Mathematicae 30, 239-251.

J. van de CRAATS
Marinus de Jongstraat 12.
4904 PL Oosterhout - NB
The Netherlands

COLLOQUIA MATHEMATICA SOCIETATIS JÁNOS BOLYAI
48. INTUITIVE GEOMETRY, SIÓFOK, 1985.

ON THE REGULARITY CONDITION OF A DISCRETE POINT SYSTEM

P. ENGEL

Dedicated to László Fejes Tóth on occasion of his 70th birthday

1. REGULAR POINT SYSTEMS

We consider a point set X in the three-dimensional Euclidean space E^3 which, following Hilbert [1], fulfils three conditions:

i. The point set X is discrete, that is, around each point of X an open ball of fixed radius $r > 0$ can be drawn which contains no other point of X.

ii. Every interstitial ball, that is, every open ball which can be embedded into E^3 such that it avoids all points of X, has a radius less or equal to a fixed finite R.

iii. The point set X looks the same if seen from every point of X.

A point set X which fulfils the first two conditions is called a discontinuum or, following Delone [2,3], an (r,R)-system. It has important applications in the theory of quasi-crystals and amorphous matter.

A more rigourous characterization of the third condition was given by Sohncke [4]:

iii. A discrete point set X is regular if from any two points of it straight line segments are drawn to all the other points of X and these line systems are directly or symmetrically congruent.

A point set X which fulfils all three conditions is called, following Sohncke, a regular point system. The third condition implies an infinite discrete group of isometries which acts transitively on the points of X. Generally, a regular point system corresponds to an orbit of a space group G. It has applications in the theory of ideal crystals.

The third condition can be weakened if we consider the local properties of a discrete point set. It will be shown that the regularity of a point set X already is asserted if these line systems for every point $x_i \in X$ are directly or symmetrically congruent within a given "shpere of regularity" centered at x_i. Delone and his coleagues [3] obtained the first results concerning local properties of point sets. The main result of this paper is that we give examples of regular point systems for which a radius of 6R is needed as a sphere of regularity and thus disprove a conjecture of Galiulin [6] which states that 4R should be sufficient.

These investigations are of fundamental importance in crystallography because no far reaching forces have to be assumed in order to explain the crystalline state of matter.

2. DIRICHLET DOMAIN PARTITIONS

In what follows we assume the point set X to be a (r,R)-system in E^3. The metric and topologic properties of X can best be seen from the Dirichlet domain partition.

DEFINITION 1. For $x_0 \in X$ the Dirichlet dimain $D(x_0)$ is a subset of the space E^3 containing all points which are closer to x_0 than to any other point $x_j \in X$.

Given x_0 we take any other point $x_j \in X$ and determine the bisecting plane H_j^o which is normal to the straight line segment $\overline{x_0 x_j}$. By construction all points $p \in H_j^o$ have equal distance from x_0 and x_j. The plane H_j^o separates the space into two open half-spaces H_j^+ and H_j^-. We assume H_j^+ to contain x_0. It follows that all points $p \in H_j^+$ lie closer to x_0 than to x_j. Thus the Dirichlet domain $D(x_0)$ is obtained by intersecting all open half-spaces H_j^+,

$$D(x_0) := \bigcap_{x_j \in X \setminus x_0} H_j^+ .$$

The half-spaces H_j^+ are convex hence, $D(x_0)$ is convex. For any point $p \in D(x_0)$ we consider the open ball $B(p, |\overline{x_0 p}|)$ with center in p and radius $|\overline{x_0 p}|$. We take such a ball for each point $p \in D(x_0)$ and obtain the region [7]

$$Q_0 := \bigcup_{p \in D(x_0)} B(p, |\overline{x_0 p}|) .$$

LEMMA 1. *All points* $x_j \in X$ *which generate faces of* $D(x_0)$ *lie on the boundary of* Q_0.

PROOF: Suppose there exists a point $x_j \in X$ inside some ball $B(p, |\overline{x_0 p}|)$, $p \in D(x_0)$. Then the point p would be closer to x_j than to x_0. Thus $p \notin D(x_0)$ which is a contradiction. A point $x_j \in X$ outside of Q_0 does not contribute to $D(x_0)$ because the bisecting plane H_j^0 does not intersect $D(x_0)$.

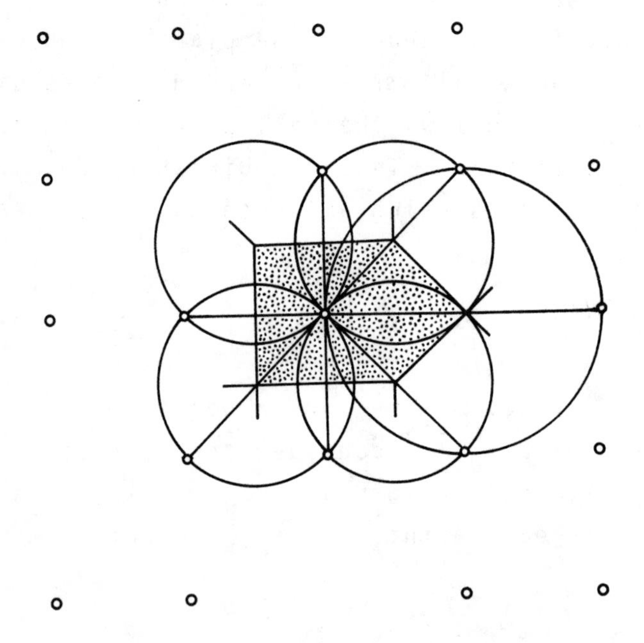

Figure 1

The Dirichlet domain of the point $x_0 \in X$ together with the region Q_0 (X is generated from x_0 by the plane group G=p4)

LEMMA 2. *In a* (r,R)*-system* X *the region* Q_i *of every point* $x_i \in X$ *is contained in a ball of radius* $2R$ *and center* x_i.

PROOF. By lemma 1 all points of X generating faces of $D(x_i)$ lie on the boundary of Q_i. By condition ii) the radius R is the supreme of radii of all interstitial balls. Therefore, Q_i is contained within a ball $B(x_i, 2R)$.

LEMMA 3. *In a* (r,R)*-system* X *the Dirichlet domain* $D(x_i)$ *of every point* $x_i \in X$ *is a bounded polyhedron with a finite number of boundary faces.*

PROOF. From lemma 2 it follows that $Q_i(x_i) \subset B(x_i, 2R)$. By condition i) there are only a finite number of points of X within $B(x_i, 2R)$.

LEMMA 4. *In a* (r,R)*-system* X *the Dirichlet domain* $D(x_i)$ *of any point* $x_i \in X$ *is contained within a ball of radius* R *and center* x_i.

PROOF. This immediately follows from lemma 2.

We denote by $cl(D(x_i)$ the closure of $D(x_i)$. We recall that the vertices v_j are the extreme points of $cl(Dx_i)$.

LEMMA 5. *In a* (r,R)*-system* X *the region* Q_i *of any Dirichlet domain* $D(x_i)$, $x_i \in X$, *is determined by the union of all open balls at the vertices* v_j *of* $clD(x_i)$,

$$Q_i = \bigcup_{v_j \in clD(x_i)} B(v_j, |\overline{x_i v_j}|).$$

PROOF. For any point $p \in D(x_i)$ there exists a closed straight line segment $[p_1, p_2]$, p_1, $p_2 \in clD(x_i)$, such that $p \in int[p_1, p_2]$. We have

$$B(p, |\overline{x_i p}|) \subset [B(p_1, |\overline{x_i p_1}|) \cup B(p_2, |\overline{x_i p_2}|)] \ .$$

Every point $p \in D(x_i)$ can be represented as a linear combination of the vertices $v_j \in clD(x_i)$. From this the above theorem follows.

LEMMA 6. *If we attach at each point* x_i *of a* (r,R)-*system* X *a closed ball of radius* R *then*

$$\bigcup_{x_i \in X} clB(x_i, R) = E^3.$$

PROOF. By construction we have $\bigcup_{x_i \in X} clD(x_i) = E^3$. By lemma 4 now $D(x_i) \subset B(x_i, R)$ holds which proves lemma 6.

To every (r,R)-system corresponds a unique Dirichlet domain partition. On the other hand different (r,R)-systems X, X' may have the same Dirichlet domain partition. An (r,R)-system together with its Dirichlet domain partition is called a marked Dirichlet domain partition. With the marked Dirichlet domain partition we have a complementary representation of the (r,R)-system and most topologic and metric properties become very illustrative.

3. THE REGULARITY CONDITION

We consider a family of three-dimensional closed tiles $\{T_i \mid i \in I\}$, where each tile T_i is a subset of the

Euclidean space, $T_i \subset E^3$, and I is an infinite set of indices.

The family $\{T_i | i \in I\}$ is called a space partition if $\bigcup_{i \in I} T_i = E^3$ and $\mathrm{int}T_i \cap \mathrm{int}T_j = \emptyset$, for all $i,j \in I$, with $i \neq j$.

Following Grünbaum and Shephard [8], a space partition is called normal if there exists a ball B of fixed finite radius R and a ball B' of fixed radius $r > 0$ such that $B' \subset T_i \subset B$, for every $i \in I$.

We say that T_i is congruent to T_j if there exists an isometry $\varphi_{ij} \in \mathbb{E}(3)$, $\varphi_{ij}: T_i \to T_j$. Following Grünbaum and Shephard [9], a space partition is called monohedral if the tiles T_i are congruent to a prototile T, for every $i \in I$.

A space partition is called face-to-face if the intersection of every pair of tiles is either empty, or a vertex, or an edge, or a face of each.

In what follows we consider a normal space partition P. The cell complex $C^1(T_i) := \{T_j | T_j \cap T_i \neq \emptyset\}$ is called the first corona of T_i. Similarly, for $k \in \mathbb{N}$, $C^k(T_i) := \{T_j | T_j \cap C^{k-1}(T_i) \neq \emptyset\}$ is the k-corona of T_i. We define $C^0(T_i) := T_i$. We say $C^k(T_i)$ is congruent to $C^k(T_j)$ if there exists an isometry $\varphi_{ij}: C^\nu(T_i) \to C^\nu(T_j)$, $\nu = k, k-1, \ldots, 0$.

Let $K[T_i] := \{\sigma_v | \sigma_v: T_i \to T_i\}$ be the stabilizer of T_i in $\mathbb{E}(3)$. Similarly let $K[C^1(T_i)] := \{\sigma_v | \sigma_v: C^1(T_i) \to C^1(T_i), T_i \to T_i\}$ be the stabilizer of $C^1(T_i)$. It follows that

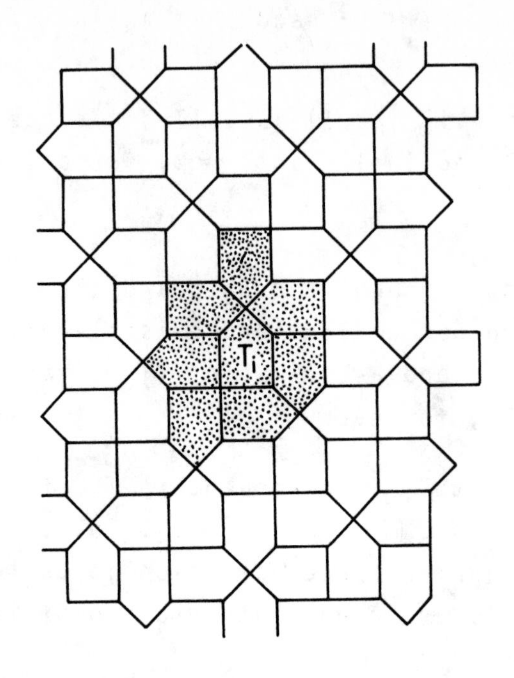

Figure 2

The first corona of a tile T_i. Here $K[T_i] = \{1,m\}$ (1 is the identity, m is the reflection in a mirror line ,
$K[(C^1(T_i)] = \{1\}$

$K[C^1(T_i)] \leq K[T_i]$ because every $\sigma \in K[C^1(T_i)]$ maps T_i onto itself. We write $K[T_i]$ as a union of cosets with respect to $K[C^1(T_i)]$,

$$K[T_i] = \hat{\sigma}_1 K[C^1(T_i)] \cup \ldots \cup \hat{\sigma}_m K[C^1(T_i)],$$

where $\hat{\sigma}_1$ is the identity.

Let now P_0 be a fixed normal space partition such that all tiles $T_i \in P_0$ have congruent first coronas. By this P_0 is monohedral. Schoenflies [5] stated, without proof, that a tiling is regular if all tiles are surrounded by its "neighbouring tiles" in a congruent way. If we assume that the neighbouring tiles correspond to the first corona then this statement is not true in general. In order to show this we take the tiles of P_0 and try to assemble them in a different way, under the condition that each tile obtains a congruent first corona. This is like putting together a puzzle. In the following we have to formalize this game introducing the concept of geometric extension.

DEFINITION 2. For $T_j \in C^1(T_i)$, $i \neq j$, an isometry $\gamma_{ij} \in \mathbb{E}(3)$, $\gamma_{ij}: T_i \to T_j$, $C^1(T_i) \to C^1(T_j)$, is called a geometric extension if the overlap domain $O_{ij} := C^1(T_i) \cap C^1(T_j)$ is a subcomplex of $C^1(T_i)$.

A geometric extension γ_{ij} requires that for $T_i \cap T_j \neq \emptyset$ it holds that $\gamma_{ij}^{-1}(T_j) = T_i$, $\gamma_{ij}^{-1}(T_i) = T_k$. Since $\gamma_{ij}^{-1}(T_i \cap T_j) = T_k \cap T_i \neq \emptyset$ it follows that $T_k \in C^1(T_i)$. Hence $\gamma_{ij}^{-1}(O_{ij}) = O_{ik} \subset C^1(T_i)$, where O_{ik} is a subcomplex of $C^1(T_i)$. The overlap domain O_{ik} may be identical to O_{ij} but, in general, this is not required.

We call an ordered sequence of tiles $T_1,\ldots,T_s$, where $T_i \cap T_{i+1} 1 \neq \emptyset$, $i=1,\ldots,s-1$, a chain of tiles. If $T_s = T_i$ we say that the chain is closed. We call the successive geometric extensions $\gamma_{12},\gamma_{23},\ldots,\gamma_{s-1,s}$ a geometric extension along the chain $T_1,\ldots,T_s$.

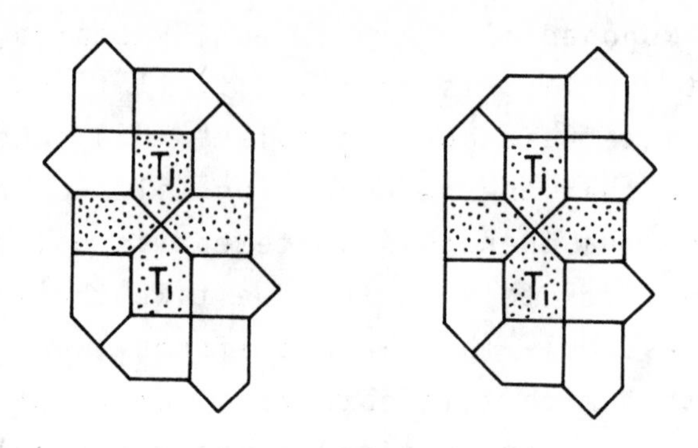

Figure 3

The two alternative settings for a multivaried geometric extension with overlap domain O_{ij}. The second one cannot be continued to be a tiling of the plane

In what follows we assume that $K[C^1(T_i)]$ is a proper subgroup of $K[T_i]$. If for some λ and $\rho_v \in \hat{\sigma}_\lambda\, K[C^1(T_j)]$, $\rho_v : T_j \to T_j$, there exists an overlap domain $O_{jh} \subset C^1(T_j)$ such that $\rho_v(O_{jh}) = O_{ij}$, then for $\lambda \neq 1$, $\rho_v \gamma_{ij} : C^1(T_i) \overset{\gamma_{ij}}{\to} C^1(T_j) \overset{\rho_\nu}{\to} C^1_\nu(T_j)$ is an alternative geometric extension, where $C^1_\nu(T_j)$ is not assumed to be a subcomplex of P_0. Such an example is shown in Figure 3. In general, O_{jh} is not required to be identical to O_{ij}. For all such λ let $S_{ij} := \{\hat{\sigma}_1 K[C^1(T_j)], \ldots, \hat{\sigma}_\ell K[C^1(T_j)]\}$ be the union of the corresponding cosets of $K[T_j]$. We call $\rho_v \gamma_{ij}$, $\rho_v \in S_{ij}$, a multivaried S_{ij}-extension, if S_{ij} contains more than one coset. Clearly multivaried γ_{ij}-extensions only occur it $K[C^1(T_i)] < K[T_i]$.

We next consider two geometric extensions γ_{ij} and γ_{ik}. We choose $T_k \subset C^1(T_i)$, $k \neq j$, such that $T_k \cap T_j \neq \emptyset$ and we say that γ_{ij} and γ_{ik} are contiguous geometric extensions. For a given $\rho_\nu\gamma_{ij}$, $\rho_\nu \in S_{ij}$, we say $\rho_\mu\gamma_{ik}$, $\rho_\mu \in S_{ik}$, can be realized if $O_{jk} := C^1_\nu(T_j) \cap C^1_\mu(T_k)$ is a subcomplex of $C^1_\nu(T_j)$. Otherwise $\rho_\mu\gamma_{ik}$ cannot be realized. In general two contiguous multivaried extensions can only be realized for some of the cosets say $\bar{S}_{ij} \subset S_{ij}$ and $\bar{S}_{ik} \subset S_{ik}$ respectively. We note that $\bar{S}_{ij}$ and $\bar{S}_{ik}$ include the trivial cosets $K[C^1(T_j)]$ and $K[C^1(T_k)]$ respectively.

We say that two contiguous multivaried geometric extensions $\rho_\nu\gamma_{ij}$ and $\rho_\mu\gamma_{ik}$ are not coupled if for every $\rho_\nu \in \bar{S}_{ij}$ the geometric extensions $\rho_\mu\gamma_{ik}$ can be realized for all $\rho_\mu \in \bar{S}_{ik}$. Otherwise they are called coupled. We have to consider this two cases separately.

Let us assume that all pairs of contiguous multivaried extensions are not coupled.

LEMMA 7. *If in a normal space partition* P_0 *all tiles* $T_i \in P_0$ *have congruent first coronas and only if no coupled extensions occur then the assemblage of* P_0 *is unique.*

PROOF. Starting from the tile $T_0 \in P_0$ we assemble the space partition under the condition that every tile T_i, $i \in I$, has a congruent first corona. We perform geometric extensions along all possible chains $T_0, T_j, \ldots$. Suppose we thereby encounter a multivaried S_{sv}-extension. By the congruence of the first coronas there exists a multivaried S_{0k}-extension. We prove that we can uniquely build up chains T_0, T_1, ..., T_s such that $T_i \cap T_k \neq \emptyset$, $i \neq k$, for $i=0,1,\ldots,s$. The overlap domain $O_{0k} \subset C^1(T_0)$ is given. By the monohedral condition a tile may not be surround another tile therefore, there exists at least another tile $T_1 \in O_{0k}$ such that $T_1 \neq T_0$, T_k. By assumption $\rho_v\gamma_{01}$ and $\rho_\mu\gamma_{0k}$ are not coupled. Thus for every $\rho_\mu \in \bar{S}_{0k}$ we can realize all extensions $\rho_v\gamma_{01}$, $\rho_v \in \bar{S}_{01}$. Since $O_{1k} = C^1_\nu(T_1) \cap C^1_\mu(T_k)$ this means that O_{1k} remains conserved and hence is uniquely determined. As above there exists another tile $T_2 \in O_{1k}$ such that $T_2 \neq T_0$, T_1, T_k. As before we can show that O_{2k} is uniquely determined. Similarly, we can show that O_{ik}, i=3,...,s, are uniquely determined. In this way we construct all such chains and finally $C^1(T_k)$ becomes uniquely determined. This proves lemma 7.

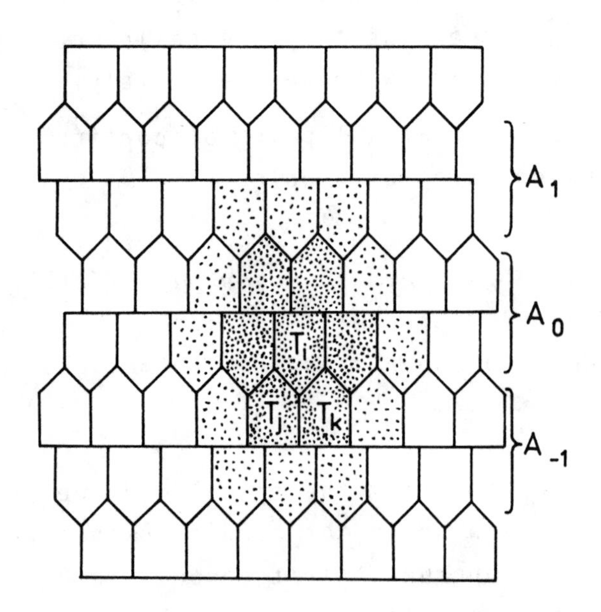

Figure 4

Layered structure with coupled extension

LEMMA 8. *A monohedral space partition* P_0 *is regular if it can be assembled in a unique way starting from any tile of it.*

PROOF. We have to prove that there exists an infinite group of isometries which acts transitively on the tiles of P_0. Let the tiling P_0 be assembled in a unique way starting from the tile $T_i \in P_0$. By assumption, for any other tile $T_j \in P_0$ as starting tile, we can assemble the same tiling P_0 in a unique way. By the monohedral condition and by the unique assemblage it follows that T_i and T_j have congruent k-coronas, for k=0,1,... . We show that there exists an isometry φ which maps T_i onto T_j and thereby maps P_0 onto itself. Indeed, for any $T_s \in P_0$ we can find a chain from T_i to T_s. Because of the unique assemblage at least one congruent chain exists from T_j to some $T_v \in P_0$. Hence φ maps T_s onto T_v. This is true for every pair of tiles T_i and T_j and all tiles T_s. This proves lemma 8.

In the case that coupled extension occurs lemma 7 does not hold. An example of coupled extension is shown in Figure 4. By the congruence of the first coronas only the double layer A_0 can be uniquely assembled starting from the tile T_i. For the adjacent double layers A_{-1} and A_1 respectively two different settings are possible. In this case the ambiguity is resolved if we require the congruence of the second corona. Considering this example and the other examples given in Figure 6 we propose the following conjecture.

CONJECTURE 9. The assemblage of a normal space partition P_0 is uniquely determined if all tiles $T_i \in P_0$ have congruent second coronas.

The (r,R)-system determines a unique marked Dirichlet domain partition. By lemma 4 and condition i) it is normal. By conjecture 9 the congruence of the second coronas should be sufficient that the Dirichlet domain partition is uniquely assembled. It follows by lemma 8 that under this condition the (r,R)-system is regular. By lemmas 1 and 2 the Dirichlet domain of the point $x_i \in X$ is determined through all points $x_j \in X$ within a ball of radius 2R and center x_i. It follows that the first corona is determined through all points within a ball of radius 4R and a ball of radius 6R is sufficient to determine the second corona.

By conjecture 9 the ball of radius 6R should be an upper limit. In most cases a ball of radius 4R or even 2R is sufficient to ensure regularity. Since the Dirichlet domains are convex polyhedra and the Dirichlet domain partition is a very special face-to-face tiling we have to show that the ball of radius 6R is essential. In the plane E^2 we can show that no face-to-face tiling exists which requires the congruence of the second corona in order to establish the regularity. However, in threeee-dimensional space E^3 such face-to-face tilings exist. The first known example of a Dirichlet domain partition where the congruence of all the first coronas is not sufficient to guarantee the regularity, is shown in Figure 5. The Dirichlet domain is a square prism with a slanting roof. Translates of the Dirichlet domain are assembled to form a layer. Two congruent layers are put together head-to-head to form a double layer. Congruent double layers are stacked one upon the other. Each double layer can be rotated by $\pm 90^0$ with respect to the previous one. Thereby the congruence of the first coronas is conserved. If we rotate always by $+90^0$ then a regular Dirichlet domain partition results with space group $P4_122$ ($P4_322$ respectively if we rotate always by -90°). If we rotate alternately by $+90^o$ and -90° then again a regular Dirichlet domain partition results which has directly and symmetrically congruent first coronas but different second coronas with respect to the first partition. The space group is C2/c. Besides these two regular partitions an infinite number of different irregular partitions exist where the sense of rotation is arbitrarily changed. All these partitions have congruent first coronas.

A list of Dirichlet domains which show these exeptional properties is shown in Figure 6. In the first column is given the space group when the sense of rotation is always the same. Where two space groups are given in Figure 6, the rotation angle is $\pm 120^o$ for $P3_112$ and $P3_121$, and $\pm 60^o$ for $P6_122$. By the rotation of 60° or 120° respectively non-congruent first coronas result. The second column shows the Dirichlet domain and the third column gives a view from top of it.

These are the only known examples of point sets in E^3 which have a very high local regularity but which may be globaly irregular. They all have a layered structure similar to the one shown in Figure 4. These kinds of point sets are fundamental in order to understand the crystalline state of matter. Layered crystal structures which show irregularities have been known for a long time. These structures are called polytypes. Some examples are SiC, ZnS, and CdI_2. For these compunds many polytypes were discovered. This clearly indicates that atoms in a crystal only recognize their nearest neighbours.

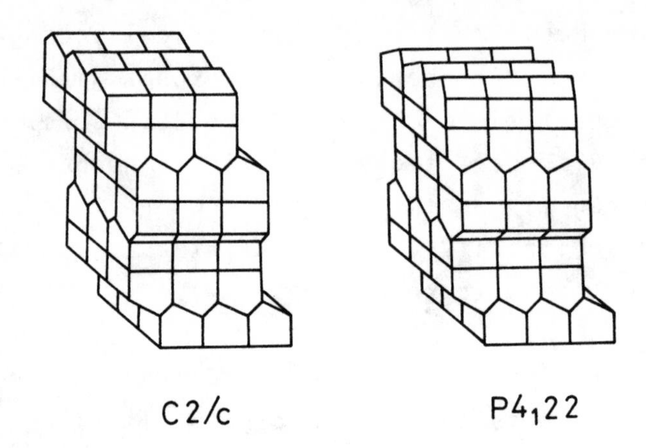

Figure 5

Two different regular Dirichlet domain partitions which have congruent first coronas

ACKNOWLEDGEMENT. I am indebted to Hans Debrunner and Peter Mani for many stimulating conversations. I am also grateful to the referee for very helpful comments.

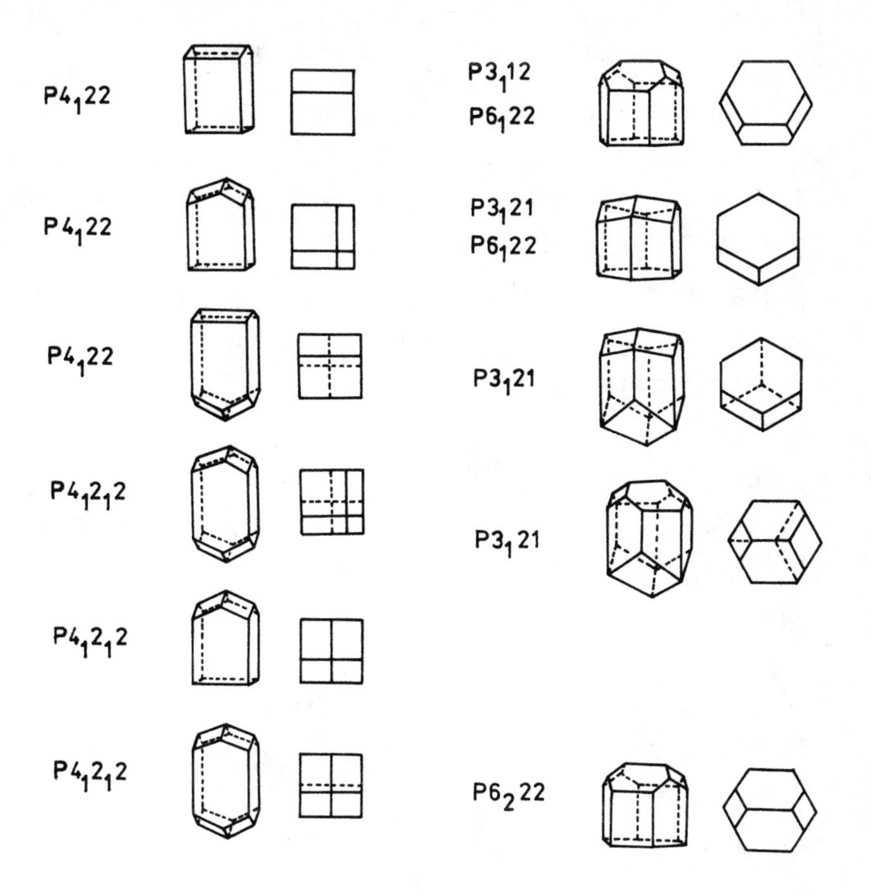

Figure 6

List of known Dirichlet domains which allow irregular space partitions each having congruent first coronas

REFERENCES

[1] HILBERT, D., COHN-VOSSEN, S., *Anschauliche Geometrie*. Wiss. Buchgesellschaft, Darmstadt (1973).

[2] DELONE, B.N., PADUROV, N. and ALEKSANDROV, A.D., *Matematičeskie osnovy structurnogo analiza kristallov*. ONTI gosudarstvennoe techniko-teoretičeskoe izdatel'-stvo. Leningrad, Moskva (1934).

[3] DELONE, B.N., DOBILIN, N.P., ŠTOGRIN, M.I. and GALIULIN, R.V., *A local criterion for regularity of a system of points*. Sov. Math. Dokl. 17 (1976), 319-322.

[4] SOHNCKE, L., *Die regelmässigen ebenen Punktsysteme von unbegrenzter Ausdehnung*. J. reine angew. Math. 77 (1874), 47-101.

[5] SCHOENFLIES, A., *Über reguläre Gebietstheilungen des Raumes*. Nachrichten der kgl. Gesell. der Wiss. und der Georg-Augusts-Univ. zu Göttingen, 9 (1888), 223-237.

[6] GALIULIN, R.V., *Delaunay systems*. Sov. Phys. Crystallogr. 25 (1980), 517-520.

[7] ENGEL, P., *Über Wirkungsbereichsteilungen von kubischer Symmetrie*. Z. Kristallogr. 154 (1981), 199-215.

[8] GRÜNBAUM, B. and SHEPHARD, G.C., *Isotoxal tilings*. Pacific J. of Math. 76 (1978), 407-430.

[9] GRÜNBAUM, B. and SHEPHARD, G.C., *Tilings with congruent tiles*. Bull. Amer. Math. Soc. 3 (1980), 951-973.

P. ENGEL
Univ. of Bern, Lab. for Crystallography, Freiestr. 3, CH-3012

COLLOQUIA MATHEMATICA SOCIETATIS JÁNOS BOLYAI
48. INTUITIVE GEOMETRY, SIÓFOK, 1985.

SOME COMBINATORIAL AND METRIC PROBLEMS IN GEOMETRY

P. ERDŐS

Some of these problems were discussed at our recent meeting at Siófok.

1. Let there be given n points in the plane in general position, i.e. no three on a line and no four on a circle. Let f(n) denote the largest integer so that these points determine at least f(n) distinct distances. Determine or estimate f(n) as well as possible. I have no example to show that

$$(1) \qquad f(n)/n^2 \to 0$$

and, on the other hand, I cannot prove

$$(2) \qquad f(n)/n \to \infty.$$

I feel that (1) holds, but I am less sure about (2). An old problem of mine states that if n points are in general position and $n > n_0$, then it cannot happen that the points determine n-1 distinct distances so that the ith distance occurs i times (in some order). I. Palásti and Liu have an example which shows that 7 such points are possible. $f(n) \geq n$ for $n > n_0$ would of course show that my conjecture is true.

A related problem states as follows. Let $x_1, \ldots, x_n$ be in general position. Denote by $d(x_i)$ the number of distinct distances from x_i. Trivially, $d(x_i) \geq (n-1)/3$ for every i. I am sure that there is an absolute constant $c > 0$ (i.e. independent of n and the position of the points) so that

$$(3) \qquad D(n) = \max_i d(x_i) > (1 + c)n/3.$$

Is it true that there is a set $x_1, \ldots, x_n$ (in general position) for which

$$(4) \qquad D(n) < (1 - c)n \ ?$$

It is rather frustrating that I got nowhere with (3) and (4). Perhaps (3) remains true if we only assume that no four of our points are on a circle or even if no circle whose center is one of the x_i's goes through more than three of the other x_i's. It would also be of interest to prove or disprove

$$(5) \qquad \sum_{i=1}^{n} d(x_i) > (1 + c)n^2/3.$$

An old and no doubt very difficult problem of mine states as follows. Let $x_1, \ldots, x_n$ be n points in the plane (not necessarily in general position). Is it true that

$$(6) \qquad \max_i d(x_i) > cn/(\log n)^{1/2},$$

and perhaps even

$$(7) \qquad \sum_{i=1}^{n} d(x_i) > cn^2/(\log n)^{1/2} \ ?$$

It is very easy to show that for some $c > 0$, $\max d(x_i) > cn^{1/2}$. The only non-trivial result which points in the direction of (6) and (7) is an unpublished result of J. Beck, who proved

$$(8) \qquad \max_i d(x_i)/n^{1/2} \to \infty.$$

The proof of (8) is quite complicated.

2. Croft, Purdy and I conjectured that if n points in the plane are given then for $k \le n^{1/2}$ the number of distinct lines which contain $\ge k$ of them is less than cn^2/k^3. This conjecture was proved by Szemerédi and Trotter [1], but the best value of c is not known. Their value is almost certainly very far from being best possible. In particular, if $k = \sqrt{n}$ we obtain that the number of distinct lines which contain at least $\sqrt{n}$ of our points is less than $c\sqrt{n}$. This result is interesting since it shows the difference between finite geometries and points in the Euclidean plane. In a finite geometry of $n = p^2 + p + 1$ points one has n lines containing $p + 1 > \sqrt{n}$ points. The best value of c is not known. It is trivial and shown by the lattice points in the plane that one can give n points so that there should be $2\sqrt{n} + 2$ lines which contain $\sqrt{n}$ of our points and I thought that perhaps this is best possible, but Sah showed that one can find $(3 + o(n))\sqrt{n}$ such lines. This construction appears for the first time in the proceedings of this meeting. Perhaps it gives the best possible value of c.

3. Let there be n points in the plane, no five on a line. Denote by $g(n)$ the maximum number of distinct lines each containing four of our points. Is it true that $g(n)/n^2 \to 0$? This is an old conjecture of mine and I offer 100 dollars for a proof or disproof. Kárteszi proved that $g(n) > c\, n \log n$ is possible and Grünbaum [2] proved that $g(n) > c\, n^{3/2}$ is possible. Perhaps $g(n) < C\, n^{3/2}$, but this may be too optimistic.

4. Let $x_1, \ldots, x_n$ be n distinct points in the plane. Denote by $D(x_1,\ldots,x_n)$ the number of distinct distances determined by our points. Put

$$g(n) = \min_{x_1,\ldots,x_n} D(x_1,\ldots,x_n).$$

An old and no doubt very difficult conjecture of mine states [3]

$$c_1 \cdot n/(\log n)^{1/2} < g(n) < c_2 \cdot n/(\log n)^{1/2}. \tag{9}$$

The upper bound is easy and is shown by the lattice points in the plane, but I offer 500 dollars for a proof or disproof of the lower bound. Of course, (9) would follow immediately from (6).

Here we are not concerned about the value of $g(n)$. Let $x_1,\ldots,x_n$ be a set of points which determines $g(n)$ distinct distances. For which n is it true that $x_1,\ldots,x_n$ is uniquely determined up to similarity? Clearly this holds for $n = 3$, the triangle must be equilateral. There is no uniqueness for $n = 4$ since $g(4) = 2$ and this can be implemented by a square or by two equilateral triangles having an edge in common. Also, $g(5) = 2$ and it seems that the regular pentagon is the only solution. A detailed proof was given by a colleague from Zagreb. (Unfortunately, I do not have his letter.) Now $g(6) = g(7) = 3$, $g(8) = 4$ and it is easy to see that there is no uniqueness here either. I thought that for $n > 5$, $g(n)$ can always be implemented in more than one way. But a colleague remarked in conversation: $g(9) = 4$ and is implemented by the regular nonagon. Is there any other way? Recently this has been decided in the positive by Gy. Hegyi, who found the following example: the six vertices of a regular hexagon, its center and the mirror images of the center with respect to two neighbouring sides. Is it true that for $n > n_0$, $g(n)$ can always be implemented in more than one way? At the moment I do not see how to attack this problem.

5. Let there be given n points in the plane. Denote by $f(n)$ the largest integer so that for every choice of the n points there should be $f(n)$ of them no two of which have distance 1. A simple example of L. and W. Moser shows that $f(n) \le 2n/7$. L. Székely [4] proved that $f(n) > n/5$, and in fact a somewhat sharper result. Determine $f(n)$ as accurately as possible. In particular, is it true that $f(n) \ge n/4$?

A related problem states as follows: Let there be given n points in the plane and assume that 1 is the shortest distance between any two of them. Join two of them if their distance is 1. This graph is clearly planar. Denote by $g(n)$ the largest integer such that this graph always has an independent set of size $g(n)$. I thought that perhaps $g(n) = n/3$ but F. Chung and R. L. Graham, and independently J. Pach, gave a construction which shows $g(n) \le 6n/19$. Their construction appears in Fig.1.

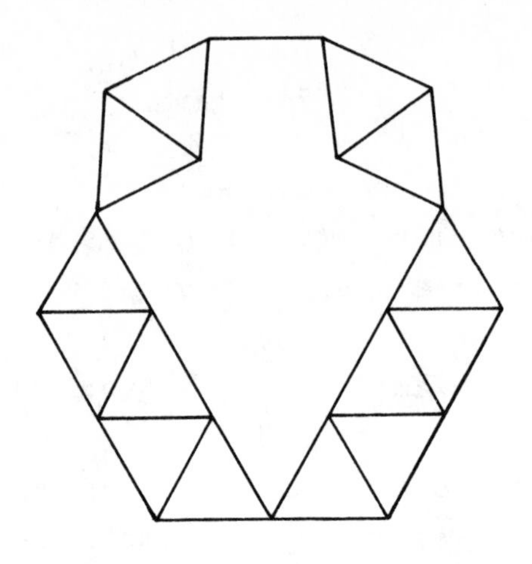

Figure 1

By the way R. Pollack [5] has a simple proof of $g(n) \geq n/4$. The determination of $g(n)$, or even $\lim g(n)/n$, is perhaps not quite easy.

6. An old theorem of Anning and myself [6] states that if S is an infinite set in the plane so that the distance between any two points of S is an integer, then S must be linear. If we only know that the distances must all be rational, S does not have to be linear but probably must have very special structure. Ulam conjectured 40 years ago that S cannot be everywhere dense and Besicovitch (independently) conjectured that the set of limit points of S cannot contain some convex n-gon for $n > n_0$.

An old problem whose origin I cannot trace states: Are there for any n, n points in general position, i.e. no three on a line and no four on a circle, so that all the $\binom{n}{2}$ distances are integers? J. Lagrange [8] found six such points, see Fig.2. Harborth just wrote me that he and Kemnitz have shown this was the example with minimal diameter, and in fact the only one with diameter at most 220.

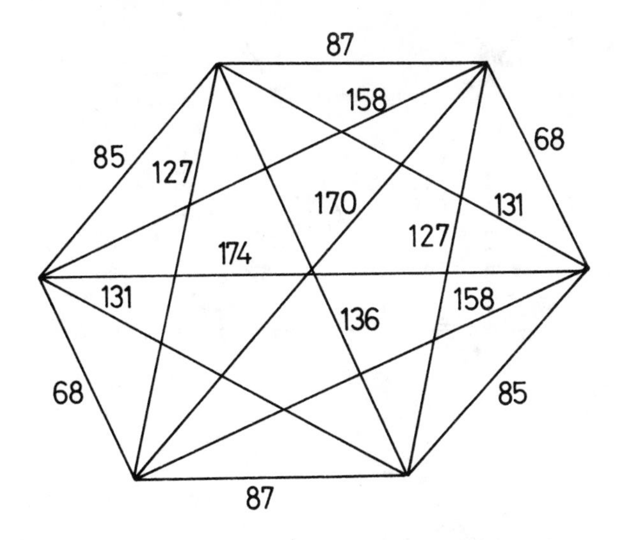

Figure 2

7. During our meeting, G. Fejes Tóth and I raised the following problem: Can one find a finite set of unit intervals in the unit square, no two of which intersect and which are maximal with respect to this property? To my surprise, Danzer found a simple example (Fig.3). This costed me 10 dollars.
Another example was found by another participant of our meeting (Fig.4 where, say, the elongation of the upper side of the lower left quadrangle passes through the lower right vertex of the square). Evidently in both examples the position of the segments can be varied. It is not clear what happens if the unit square is replaced by other regions. Also it is not clear what happens in the unit square if we insist that the only common point

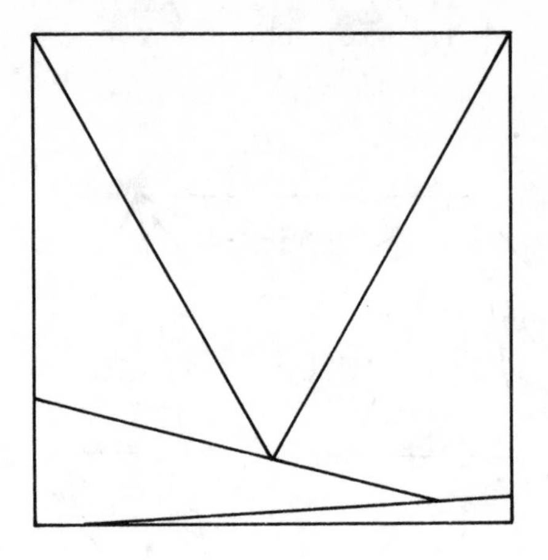

Figure 3

two of our intervals can have is their endpoint. Let R be any region and let there be given in R a maximal set of disjoint unit intervals. Can such a set ever be denumerable?

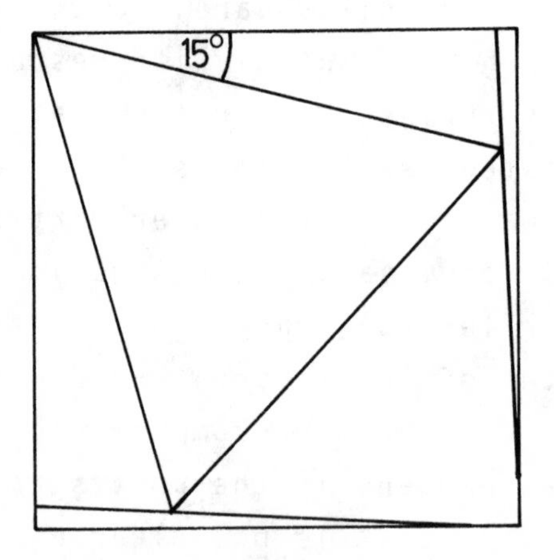

Figure 4

8. Another old problem of mine states: The vertices of a convex n-gon determine at least $[\frac{n}{2}]$ distinct distances. This conjecture was proved by Altman [7]. I further conjectured that in a convex n-gon there always is a vertex so that the number of distinct distances from this vertex is at least $[\frac{n}{2}]$. As far as I know this conjecture is still open. I also conjectured that in a convex n-gon there always is a vertex which has no three other vertices equidistant from it. This conjecture was disproved by Danzer, his example appears in Fig.5. This

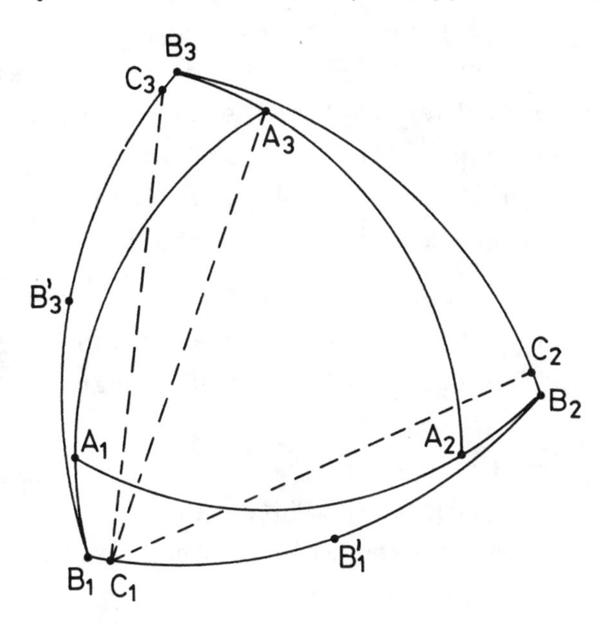

Figure 5

is a convex nonagon $A_1B_1C_1A_2B_2C_2A_3B_3C_3$ of threefold rotational symmetry, satisfying $A_1A_2 = A_1A_3 = A_1B_3$, $B_1B_2 = B_1C_2 = B_1B_3$, $C_1C_2 = C_1A_3 = C_1C_3$. It is constructed in the following way. Take a Reuleaux triangle $A_1A_2A_3$. Elongate the arc A_3A_1 beyond A_1 and choose a point B_1 on this elongation, close to A_1. Analogously we define B_2, B_3 (taking into account the threefold rotational

symmetry of the figure) and we draw the Reuleaux triangle $B_1B_2B_3$. Denote B_i' the midpoint of the side B_iB_{i+1} of this Reuleaux triangle ($B_4 = B_1$). Choose a point C_1 on the arc B_1B_1' of the side B_1B_2 and analogously choose points C_2, C_3, by taking into account the rotational symmetry of the figure. For $C_1 = B_1$ we have $C_1C_3 = B_1B_3 > B_1A_3 = C_1A_3$, while for $C_1 = B_1'$ we have $C_1C_3 = B_1'B_3' < B_1'A_3 = C_1A_3$ (provided B_1A_1 is sufficiently small). Hence for some intermediate position of C_1 we will have $(C_1C_2 =) C_1C_3 = C_1A_3$. The nonagon constructed with this C_1 will satisfy all the requirements.
Perhaps in every convex polygon there is a vertex which does not have four other vertices equidistant from it.
Finally Szemerédi conjectured that if $x_1,\dots,x_n$ are n points no three on a line then they determine at least $[\frac{n}{2}]$ distinct distances, but he can only prove this with $[\frac{n}{3}]$.

G. Purdy and I plan to publish a survey article soon on this and related problems and perhaps later a book.

For many related problems see W. Moser, Problems in Discrete Geometry, Mimeograph Notes (1981). A new edition will soon appear in collaboration with J. Pach.

REFERENCES

[1] E. SZEMERÉDI and W.T. TROTTER, *Extremal problems in Discrete Geometry*, Combinatorica 3 (1983), 381-392.

[2] B. GRÜNBAUM, *New views on some old questions of combinatorical geometry*, Coll. Internaz. Teorie Comb. Rome (1973), Tom. 1, 451-468.

[3] P. ERDŐS, *On sets of distances of n points*, Amer. Math. Monthly 53 (1946), 248-250. See also On some problems of elementary and combinatorial geometry, Ann. Mat. Pura ed Applicata, (4) 103 (1975), 99-108.

[4] L. SZÉKELY, *Measurable chromatic number of geometric graphs and sets without some distances in Euclidean space*, Combinatorica 4 (1984), 213-218.

[5] R. POLLACK, *Increasing the minimum distance of a set of points*, J. Comb. Th., Ser. A, 40 (1985), 450.

[6] W.H. ANNING and P. ERDÖS, *Integral Distances*, Bull. Amer. Math. Soc. 51 (1945), 598-600. See also P. Erdős, Integral Distances, ibid. 996.

[7] E. ALTMAN, *Some theorems on convex polygons*, Canad. Math. Bull. 15 (1972), 329-340, see also: On a problem of P. Erdös, Amer. Math. Monthly 70 (1963), 148-157.

[8] J. LAGRANGE, *Points du plan dont les distances mutuelles sont rationnelles*, Séminaire de Théorie des Nombres de Bordeaux, Année 1982-83, exp. 27, 29. 4. 1983.

P. ERDŐS
Mathematical Institute of the
Hungarian Academy of Sciences
H-1364 Budapest
P.O.B. 127
Hungary

COLLOQUIA MATHEMATICA SOCIETATIS JÁNOS BOLYAI
48. INTUITIVE GEOMETRY, SIÓFOK, 1985.

CHARACTERIZATION OF CENTRALLY SYMMETRIC CONVEX DOMAINS IN PLANES OF CONSTANT CURVATURE

G. FEJES TÓTH and A. KEMNITZ

Confirming a conjecture of McMullen [1], Rogers [3] gave the following characterization of centrally symmetric plane convex sets:

In the euclidean plane let C, C' be distinct inner points of a convex domain K. If each chord of K through C has the same length as the parallel chord of K through C', then K is centrally symmetric about the midpoint of the line segment CC'.

Larman and Tamvakis [2] showed that this statement remains true if we drop the condition that C and C' are interior to K and, consider chords in parallel lines through C and C'. They also proved an analogous result in n-dimensional euclidean space. In the present note we show that Rogers' result can be extended to the hyperbolic and the spherical geometry if we consider, instead of parallel chords, chords contained in lines symmetric about the midpoint of the segment CC'. We note without proof that one can extend analogously the result of Larman and Tamvakis to the hyperbolic plane, but such an extension to the sphere seems to involve difficulties.

Unifying the terminology we shall refer to the hyperbolic plane, the euclidean plane and the sphere as planes of constant curvature. We shall suppose that the curvature of the plane is -1, 0, and 1, respectively. By a convex domain we mean a closed set with non-empty interior, which is the intersection of some open half-planes. We note that in the case of the sphere, this means that a convex domain is contained in an open hemisphere.

THEOREM. *In a plane of constant curvature, let* K *be a convex domain and* C, C' *be distinct inner points of* K. *Let* M *be the midpoint of the segment* CC'. *Further, let* $f(t)$ *be a strictly increasing convex function defined for* $t \geq 0$. *Let* ℓ *be a line through* C *and* ℓ' *be the line obtained from* ℓ *by reflecting in* M. *Let* ℓ *and* ℓ' *meet the boundary of* K *in* P *and* Q, *and* P' *and* Q', *respectively. If for any lines* ℓ *and* ℓ' *through* C *and* C', *which are symmetric about* M, *we have*

$$f(CP) + f(CQ) = f(C'P') + f(C'Q'),$$

then K *is centrally symmetric about* M.

PROOF OF THE THEOREM. Let V and V' be the points of intersection of the line CC' with the boundary of K, and suppose that V, C, C' and V' are situated in this order on the chord VV' of K. Let P_0 and P_0' be two points on the boundary of K such that the lines CP_0 and $C'P_0'$ are symmetric about M, and P_0 and P_0' lie on different sides of the chord VV'. We suppose that $CP_0 \geq C'P_0'$. Starting with the points P_0 and P_0', we construct inductively two sequences of points $P_0, P_1, \ldots$ and $P_0', P_1', \ldots$, which lie on the boundary of K and on different sides of the chord VV' of K, such that C lies on the segment $P_i'P_{i+1}$ and the lines CP_i and $C'P_i'$ are symmetric about M. We claim that

$$(1) \qquad CP_i \geq C'P_i'$$

for $i = 1,2,\ldots$, with strict inequality if $CP_0 \neq C'P_0'$.

The proof of (1) follows by induction. By supposition, (1) is true for $i = 0$. Now suppose that $CP_{i-1} \geq C'P_{i-1}'$. Then the line $P_i'C'$ intersects the segment CP_{i-1} in a point, say R_i, such that R_i and P_{i-1}' are symmetric about M. Let S_i be the second point of intersection of the boundary of K with the line $P_i'C'$. Then we have

$$f(CP_{i-1}') = f(C'R_i) \leq f(C'S_i).$$

On the other hand, since the lines $P_{i-1}'P_i$ and $P_i'S_i$ are symmetric about M, we have by supposition

$$f(CP_i) + f(CP_{i-1}') = f(C'S_i) + f(C_i'P_i').$$

The last two relations imply that

$$f(CP_i) \geq f(C'P_i').$$

Hence (1) follows immediately by the strict monotonity of $f(t)$. The case of equality is obvious.

Next we shall show that

$$(2) \qquad \lim_{i\to\infty} P_i = V \quad \text{and} \quad \lim_{i\to\infty} P_i' = V'.$$

We consider the triangle $CC'P_i'$. In hyperbolic and euclidean geometry, the sum of the angles of a triangle is at most π, so that in these cases we have

(3) $\quad \measuredangle CC'P_i' + \measuredangle C'CP_i' < \pi$

However (3) holds also in the case of spherical geometry. For, assuming that $\measuredangle CC'P_i' + \measuredangle C'CP_i' \geq \pi$, the line (great circle) CP_i intersects the closed segment $C'P_i'$ in a point, say T_i. We have

$$\frac{\sin CT_i}{\sin C'T_i} = \frac{\sin \measuredangle CC'T_i}{\sin \measuredangle C'CT_i} = 1,$$

which, together with $\measuredangle CC'T_i + \measuredangle C'CT_i = \pi$, implies that $CT_i + C'T_i = \pi$. It follows by (1) that

$$P_iT_i = P_iC + CT_i > C'P_i' + CT_i > C'T_i + CT_i = \pi,$$

which is impossible by the convexity of K. This proves (3) for the case of spherical geometry.

Using (3), we see that

$$\measuredangle VCP_i = \measuredangle V'C'P_i' = \pi - \measuredangle CC'P_i' > \measuredangle C'CP_i' = \measuredangle VCP_{i+1}.$$

The monotonity of the angles $\measuredangle VCP_i$ implies that the sequences $P_0, P_1, \ldots$ and $P_0', P_1', \ldots$ converge, say to $\bar{P}$ and $\bar{P}'$, respectively. Furthermore we have

(4) $\quad \measuredangle C'C\bar{P}' + \measuredangle CC'\bar{P}' = \pi$

and

(5) $\quad C\bar{P} \geq C'\bar{P}'$.

In the case of hyperbolic and euclidean geometry, (2) follows immediately by (4). In the case of spherical geometry, we obtain from (4) by a calculation similar to that used in the proof of (3) that

$$C\bar{P}' + C'\bar{P}' = \pi.$$

Combining this with (5) we see that $C\bar{P} + C\bar{P}' \geq \pi$, which is impossible by the convexity of K. This completes the proof of (2).

The next step in the proof is to show that

$$CV = C'V'. \tag{6}$$

Suppose that $CV \neq C'V'$. Then the quantities $CV - C'V'$ and $CV' - C'V$ have different signs. It follows by continuity that there is a pair of points P_0, P_0' on the boundary of K, on different sides of the line VV' such that the lines CP_0 and $C'P_0'$ are symmetric about M and

$$CP_0 = C'P_0'.$$

Constructing the sequences $P_0, P_1, \ldots$ and $P_0', P_1', \ldots$ with this special pair of points P_0 and P_0', we have $CP_i = C'P_i'$ for $i = 1,2,\ldots$. Hence we obtain $CV = C'V'$ by (2).

In order to finish the proof of the Theorem, we have to show that $CP_0 = C'P_0'$ for an arbitrary pair of points P_0, P_0', which are on the boundary of K and on different sides of VV', such that the lines CP_0 and $C'P_0'$ are symmetric about M. We assume that $CP_0 \neq C'P_0'$ and seek a contradiction. Without loss of generality we suppose that

$$CP_0 > C'P_0'.$$

We consider the sequences $P_0, P_1, \ldots$ and $P_0', P_1', \ldots$ associated with P_0 and P_0'. In accordance with our previous notations, let R_i be the point of intersection of the line $P'_{i+1}C'$ with the segment CP_i and let S_i be the second point of intersection of $P'_{i+1}C'$ with the boundary of K. We write

$$\alpha_i = \measuredangle CC'P'_i, \quad \beta_i = \measuredangle C'CP'_i,$$

$$\gamma_i = \measuredangle CP'_iC', \quad \delta_i = \measuredangle P_iS_iR_i.$$

Further let δ be the angle, enclosed by the half-line VC with the half-tangent of K at V, pointing in that half-plane bounded by the line VV' which contains P_0. Obviously we have

$$(7) \qquad 0 < \delta < \pi.$$

It follows by (2) that

$$(8) \qquad \lim_{i\to\infty} \alpha_i = \pi, \quad \lim_{i\to\infty} \delta_i = \delta, \quad \lim_{i\to\infty} \beta_i = \lim_{i\to\infty} \gamma_i = 0,$$

$$(9) \qquad \lim_{i\to\infty} CP'_i = CV' > C'V' = \lim_{i\to\infty} C'P'_i$$

and

$$(10) \qquad \lim_{i\to\infty} CP_i > CC'.$$

Further, in the case of spherical geometry, we have

$$(11) \qquad \lim_{i\to\infty}(CP'_i + C'P'_i) = CV' + C'V' < \pi.$$

We also observe that (2) and (6) imply that

(12) $$\lim_{i\to\infty} P_iS_i = \lim_{i\to\infty} P_iR_i = 0.$$

Using the relations (7)-(12), one easily sees that there are positive constants $k_1 < 1$ and $k_2 < 1$ and a natural number r such that for $i \geq r$ the following inequalities hold:

(13) $$CP_i' > C'P_{i+1}',$$

(14) $$\begin{cases} \dfrac{\text{sh } C'P_i'}{\text{sh } CP_i'} < k_1, \ \dfrac{\text{sh } CC'}{\text{sh } CP_i'} < k_1 & \text{in the hyperbolic plane,} \\[2ex] \dfrac{C'P_i'}{CP_i'} < k_1, \ \dfrac{CC'}{CP_i'} < k_1 & \text{in the euclidean plane,} \\[2ex] \dfrac{\sin C'P_i'}{\sin CP_i'} < k_1, \ \dfrac{\sin CC'}{\sin CP_i'} < k_1 & \text{on the sphere,} \end{cases}$$

(15) $$\frac{\text{sh } P_iS_i}{\text{sh } P_iR_i} > k_1 \cdot \frac{P_iS_i}{P_iR_i}, \quad \frac{\sin P_iS_i}{\sin P_iR_i} > k_1 \cdot \frac{P_iS_i}{P_iR_i}$$

and

(16) $$\sin \delta_i > k_2 > \sin \alpha_i.$$

We have by construction

(17) $$CP_i - C'P_i' = R_iP_i.$$

Further, since the chords P'_iP_{i+1} and $P'_{i+1}S_i$ of K go through C and C', respectively, and since they are symmetric about M, we have

$$f(CP_{i+1}) + f(CP'_i) = f(C'P'_{i+1}) + f(C'S_i),$$

or equivalently,

$$f(CP_{i+1}) - f(C'P'_{i+1}) = f(C'S_i) - f(CP'_i).$$

It follows by (13) and the convexity of the function f(t) that

$$C'S_i - CP'_i \leq CP_{i+1} - C'P'_{i+1} \quad (i \geq r).$$

Observing that

$$C'S_i - CP'_i = C'S_i - C'R_i = R_iS_i,$$

we see that

$$CP_{i+1} - C'P'_{i+1} \geq R_iS_i \tag{18}$$

for $i \geq r$.

The sine law applied to the triangle $P_iS_iR_i$, combined with the inequalities (15), yields

$$\frac{\sin \gamma_i}{\sin \delta_i} > k_1 \frac{P_iS_i}{P_iR_i} .$$

Therefore we have

$$(19) \qquad S_iR_i \geq P_iR_i - P_iS_i > P_iR_i\left(1 - \frac{\sin \gamma_i}{k_1 \sin \delta_i}\right).$$

Applying now the sine law to the triangle $CC'P_i$, we obtain for the hyperbolic, euclidean and spherical geometry

$$\sin \alpha_{i+1} = \sin \beta_i = \sin \alpha_i \frac{\text{sh } C'P_i'}{\text{sh } CP_i'}, \ \sin \gamma_i = \sin \alpha_i \frac{\text{sh } CC'}{\text{sh } CP_i'},$$

$$\sin \alpha_{i+1} = \sin \beta_i = \sin \alpha_i \frac{C'P_i'}{CP_i'}, \ \sin \gamma_i = \sin \alpha_i \frac{CC'}{CP_i'},$$

$$\sin \alpha_{i+1} = \sin \beta_i = \sin \alpha_i \frac{\sin C'P_i'}{\sin CP_i'}, \ \sin \gamma_i = \sin \alpha_i \frac{\sin CC'}{\sin CP_i'},$$

respectively. This and (14) imply that

$$(20) \qquad \sin \alpha_{i+1} < k_1 \sin \alpha_i \quad \text{and} \quad \sin \gamma_i < k_1 \sin \alpha_i$$

for $i \geq r$.

The combination of the relations (16)-(20) yields

$$\frac{CP_{i+1} - C'P_{i+1}'}{CP_i - C'P_i'} > \left(1 - \frac{\sin \alpha_i}{k_2}\right) > 0 \qquad (i \geq r).$$

Hence we obtain by induction

$$CP_i - C'P_i' > \prod_{j=r}^{i-1}\left(1 - \frac{\sin \alpha_j}{k_2}\right)(CP_r - C'P_r').$$

However (20) implies that the sum

$$\sum_{j=r}^{\infty} \sin \alpha_j$$

is convergent, and therefore so is the product

$$\prod_{j=1}^{\infty} \left(1 - \frac{\sin \alpha_j}{k_2}\right).$$

This means that in addition to the inequality $CP_i - C'P_i' > 0$, which follows immediately by (1), and the assumption $CP_0 - C'P_0' > 0$, we have

$$\lim_{i\to\infty} (CP_i - C'P_i') > 0.$$

But this contradicts (2) and (6). Therefore we have $CP_0 = C'P_0'$. This completes the proof of the Theorem.

ACKNOWLEDGEMENT

The present work was done during a visit of the first named author to the Technical University of Braunschweig. We thank the DFG (Deutsche Forschungsgemeinschaft) for the financial support. The research was supported also by the Hungarian National Foundation for Scientific Research under grant no. 1238.

REFERENCES

[1] P.M. GRUBER and R. SCHNEIDER: *Problems in geometric convexity, Problem 59*. Contributions to Geometry, ed. by W. Tölke and J.M. Wills, Birkhäuser Verlag Basel, Boston, Stuttgart (1979).

[2] D.G. LARMAN and N.K. TAMVAKIS: *A characterization of centrally symmetric convex bodies in* E^n. Geometriae Dedicata 10 (1981), 161-176.

[3] C.A. ROGERS: *An equichordal problem*. Geometriae Dedicata 10 (1981), 73-78.

G. FEJES TÓTH
Mathematical Institute of the
Hungarian Academy of Sciences
H-1364 Budapest
P.O.B. 127
Hungary

A. KEMNITZ
Wümmeweg 10
D-3300 Braunschweig
West Germany

COLLOQUIA MATHEMATICA SOCIETATIS JÁNOS BOLYAI
48. INTUITIVE GEOMETRY, SIÓFOK, 1985.

PACKING AND COVERING WITH CONVEX DISCS

A. FLORIAN

1. INTRODUCTION

We all know how we have to arrange equal coins on a large table if we want to place the greatest possible number of them so that no two overlap. We have to place the coins in such a way that each touches six others (Fig. 1.). In 1892, the Norwegian mathematician A. Thue [17] gave an account of a result showing that the upper density (to be defined later) of any packing of equal circles in the plane satisfies

$$d \le \frac{\pi}{\sqrt{12}} = 0,90689...$$

which is attained for the hexagonal arrangement. $\pi/\sqrt{12}$ is the ratio of the area of a circle to the area of the regular hexagon circumscribed about the circle. Somewhat later, in 1910, Thue [18] gave another proof using very different ideas.

The dual counterpart of the above question is the following: How have we to arrange equal circular discs if we want them to cover a large table without gaps such

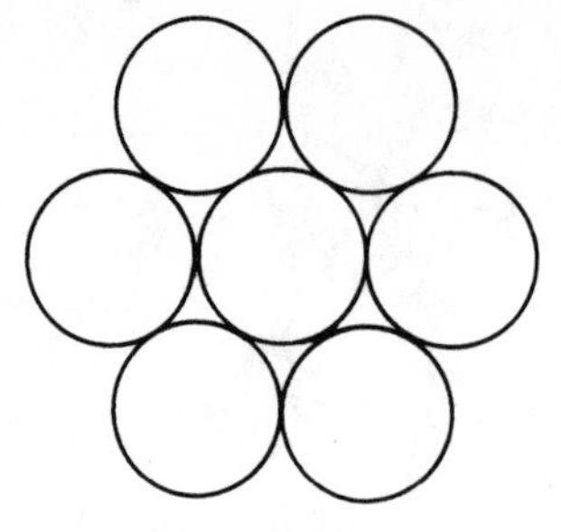

Figure 1

Figure 2

that the number of the discs involved is as small as possible? The earliest result on the subject seems to be the proof of Kershner [15] in 1939 that the lower density of any covering of the plane by equal circles satisfies

$$D \geq \frac{2\pi}{\sqrt{27}} = 1,20919\ldots .$$

Equality is attained for a system of circles, with their centres at the points of a hexagonal lattice, and their common radius chosen so that they just cover the plane (Fig. 2). $2\pi/\sqrt{27}$ is the ratio of the area of a circle to the area of the regular hexagon inscribed in the circle.

A great many refinements, generalizations and variants of these results are contained in the excellent books by L. Fejes Tóth [6], [7] and C.A. Rogers [16]. In [6], [7] the author confines his attention mainly to problems in two and three dimensions, while Rogers' book concentrates principally on problems in n-dimensional space, with n large. For a comprehensive review including more recent results in both fields we refer to [2], [9].

2. NOTATIONS AND NOTIONS

In this talk I shall try to give a short account of some problems of packing and covering. For this purpose we shall need some basic notions. By a *convex disc* we mean a convex compact subset of the Euclidean plane with non-empty interior. A sequence of convex discs $C_1, C_2, \ldots$ is said to form a *packing into a set* R if no two of them have an inner point in common and each disc C_i is contained in R. The sequence $C_1, C_2, \ldots$ is said to *cover* R if each point of R belongs to at least one of the discs $C_1, C_2, \ldots$. If R is the whole plane we shall simply speak of packing or covering. We shall denote the area of a set M (i.e. the Lebesgue measure of M) by a(M). Let $C_1, C_2, \ldots, C_n$ be convex discs and let R be a bounded set of positive area. If $a(C_i) = a_i$ we shall call the ratio

$$\frac{a_1 + \ldots + a_n}{a(R)}$$

the *density of* $C_1, \ldots, C_n$ *relative to* R.

Let $C_1, C_2, \ldots$ be a sequence of convex discs. Let C be a circle with fixed centre O and radius r. The *upper density* of the sequence is defined to be

$$\limsup_{r\to\infty} \frac{1}{\pi r^2} \sum_i a(C_i), \qquad C_i \cap C \neq \emptyset ,$$

where the sum is taken over all those discs C_i which have a point in common with the circle C. The *lower density* of the sequence is defined to be

$$\liminf_{r\to\infty} \frac{1}{\pi r^2} \sum_i a(C_i), \qquad C_i \subset C ,$$

where the sum is taken over all those discs C_i which are contained in C. It can be shown that the values of these densities do not depend on the choice of O.

It has been proved convenient to consider packings in or coverings of a convex polygon with at most six sides, shortly a *hexagon*. The corresponding statements regarding the whole plane then follow by a limiting process.

3. PACKING AND COVERING; ORDINARY PERIMETER

Many problems of packing and covering have the following form: *Let $\mathscr{C}$ be a given class of convex discs. What is the densest packing and what is the most economical covering of the plane by discs from $\mathscr{C}$?*

Let a and p be positive constants satisfying the isoperimetric inequality

$$\frac{p^2}{a} \geq 4\pi .$$

In particular, let $\mathscr{C}(a,p)$ be the class of all convex discs with area not less than a and perimeter not greater than p. Both problems are interesting only in the case when $p^2/a < 8\sqrt{3}$, that means that p is less than the perimeter of a regular hexagon of area a. For otherwise we can tile the plane by discs from $\mathscr{C}(a,p)$.

Let P be a regular k-gon of in-radius r. We consider a *smooth regular k-gon* $(P_{-\rho})_\rho$, that is the outer parallel domain of the inner parallel domain of P at some distance $\rho \leq r$ (Fig. 3). We call P the *case* of $(P_{-\rho})_\rho$.

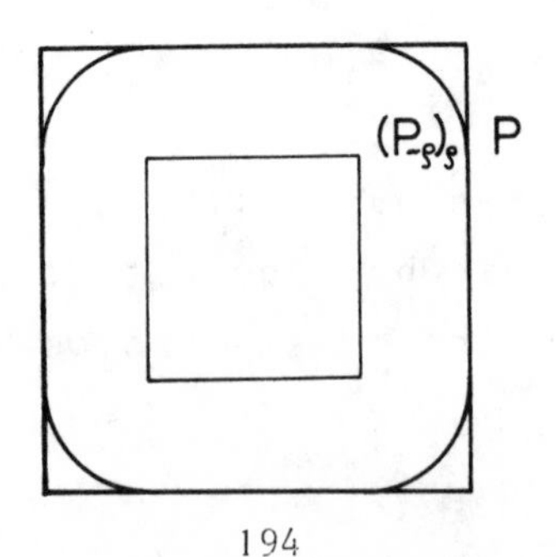

Figure 3

We define a function f(a,p,k) by

$$f(a,p,k) = \begin{cases} \frac{k}{4\pi^2}\tan\frac{\pi}{k}\left(p-\sqrt{(p^2-4a\pi)(1-\frac{\pi}{k}\cot\frac{\pi}{k})}\right)^2, & \text{if } \frac{p^2}{a} < 4k\tan\frac{\pi}{k}, \\ a, & \text{if } \frac{p^2}{a} \geq 4k\tan\frac{\pi}{k}, \end{cases}$$

where the upper term on the right-hand side is the area of the case of a smooth regular k-gon of area a and perimeter p. The solution to the above packing problem follows from

THEOREM 1. *Let the convex discs* $C_1,\ldots,C_n$ *from* $\mathscr{C}(a,p)$ *be packed in a convex hexagon* H, *and write* $a(C_i) = a_i$. *Then*

$$\frac{a_1 + \ldots + a_n}{a(H)} \leq \frac{a}{f(a,p,6)}$$

In the case $p^2/a < 8\sqrt{3}$ the ratio a/f(a,p,6) is the density of a smooth regular hexagon of area a and perimeter p relative to its case. A densest packing of the plane with discs from $\mathscr{C}(a,p)$ arises by placing smooth regular hexagons of area a and perimeter p so that their cases form a hexagonal tiling (Fig. 4).

We give an outline of the proof of Theorem 1 making, however, the assumption that all discs have area a.

Let $C_1,\ldots,C_n$ be packed in a hexagon H. By use of the Blaschke selection theorem and Euler's formula it can be shown that there are convex polygons $P_1,\ldots,P_n$ with the following properties:

(i) $C_i \subset P_i \subset H$ for $i = 1,\ldots,n$;

(ii) no two of the polygons overlap

$\text{int } P_i \cap \text{int } P_j = \emptyset \quad (i \neq j)$;

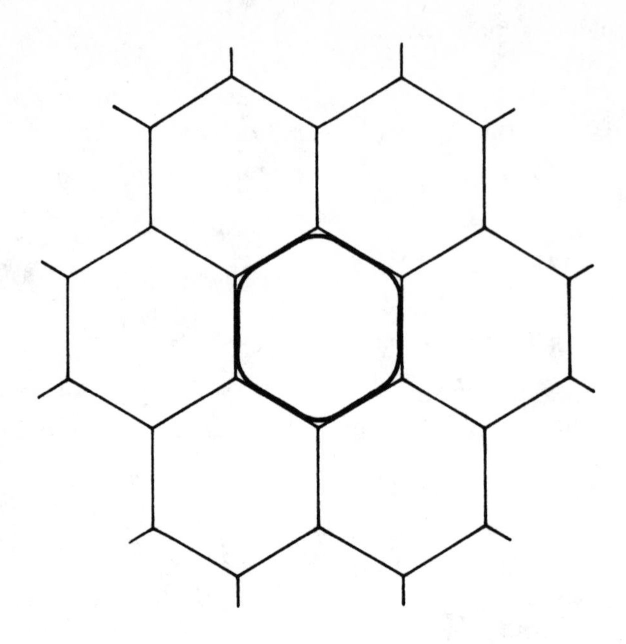

Figure 4

(iii) the average number of sides of the polygons $(k_1 + \ldots + k_n)/n = \bar{k} \le 6$.

For the proof of Theorem 1 we need two lemmas:

LEMMA 1. *If* C *is a disc from* $\mathscr{C}(a,p)$ *and* P *a convex* k-*gon with* $C \subset P$, *then*

$$a(P) \ge f(a,p,k)$$

with equality if C *is a (possibly degenerate) smooth regular* k-*gon of area* a *and perimeter* p, *and* P *is the case of* C.

LEMMA 2. f *is a convex and decreasing function of* k *for* $k \ge 3$.

By using the lemmas and Jensen's inequality we have

$$a(H) \geq \sum_{j=1}^{n} a(P_j) \geq \sum_{j=1}^{n} f(a,p,k_j) \geq nf(a,p,\bar{k}) \geq nf(a,p,6) ,$$

which is equivalent with the statement to be proved.

It should be remarked that even this simplified version of Theorem 1 implies Thues result.

Theorem 1 can be deduced from a more general theorem by L. Fejes Tōth [5], [6, p. 174].

THEOREM 2. *Let the convex discs* $C_1,\ldots,C_n$ *with average area* $\bar{a}$ *and perimeter not greater than* p *be packed in a convex hexagon* H. *Then the density of* $C_1,\ldots,C_n$ *relative to* H *satisfies*

$$\frac{a_1 + \ldots + a_n}{a(H)} \leq \frac{\bar{a}}{f(\bar{a},p,6)} .$$

Theorem 2 can be proved by showing that f is a convex function of the two variables a and k, and using Lemma 1 and Jensen's inequality.

According to A. Heppes [14], Theorem 2 continues to hold for a packing of not necessarily convex discs.

L. Fejes Tōth and A. Heppes [11], [6, p. 175] proved a similar inequality involving the average perimeter of the discs instead of the average area.

THEOREM 3. *Let the convex discs* $C_1,\ldots,C_n$ *with area* a *and average perimeter* $\bar{p}$ *be packed in a convex hexagon* H. *Then the density of* $C_1,\ldots,C_n$ *relative to* H *satisfies*

$$\frac{n\,a}{a(H)} \leq \frac{a}{f(a,\bar{p},6)} .$$

In both theorems the bound for the density can be attained for $n = 1$ and can be approximated for large values of n.

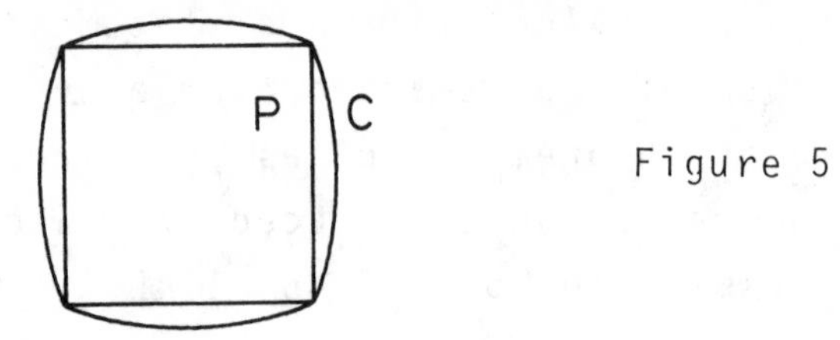

Figure 5

Let us now turn to the problem of *covering* the plane by discs from $\mathscr{C}(a,p)$. We define a *regular arc-sided k-gon* to be a convex disc C obtained from a regular *k-gon* P by joining each two consecutive vertices of P by congruent circular arcs. P is called the *kernel* of C (Fig. 5). We conjecture that a thinnest covering of the plane by discs from $\mathscr{C}(a,p)$ arises by placing regular arc-sided hexagons of area a and perimeter p so that their kernels form a hexagonal tiling (Fig. 6).

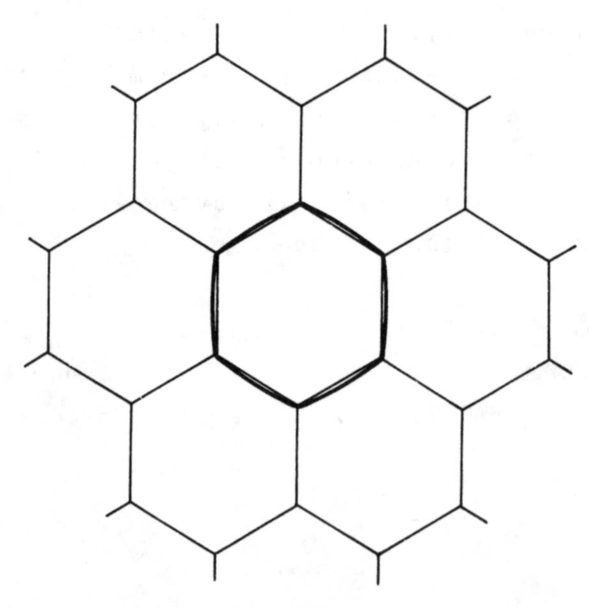

Figure 6

In a joint paper by G. Fejes Tóth and myself [3] we proved this conjecture under a certain restriction. We shall say that two convex discs C_1 and C_2 *intersect simply* if they satisfy one of the following conditions [4], [1]:

(i) C_1 and C_2 have no inner point in common,

(ii) one of them is completely contained in the other, or

(iii) C_1 and C_2 overlap, and the boundary of their intersection can be split up into two non-overlapping connected arcs, one belonging to the boundary of C_1 and the other to that of C_2.

We proved the conjecture for such coverings, where any two discs intersect simply. It seems to be difficult to get rid of this restriction.

In the particular case when $p^2/a = 4\pi$ we obtain Kershner's theorem mentioned at the beginning of this talk.

The result is a corollary of a more general theorem.

Let a,p,k be given such that

$$\frac{p^2}{a} < 4k \tan \frac{\pi}{k} .$$

It can be shown that there is exactly one regular arc-sided k-gon of area a and perimeter p, say C. Let P be the kernel of C. We define the function $\bar{f}(a,p,k)$ by

$$\bar{f}(a,p,k) = \begin{cases} a(P) & \text{if } \frac{p^2}{a} < 4k \tan \frac{\pi}{k} , \\ a & \text{if } \frac{p^2}{a} \geq 4k \tan \frac{\pi}{k} . \end{cases}$$

Then we have

THEOREM 4. *Let* $\{C_1,\ldots,C_n\}$ *be a collection of convex discs of average area* $\bar{a}$ *and perimeter not greater than* p. *If* $\{C_1,\ldots,C_n\}$ *covers a hexagon* H *such that any two discs intersect simply, then*

$$\frac{a_1 + \ldots + a_n}{a(H)} \geq \frac{\bar{a}}{\bar{f}(\bar{a},p,6)} .$$

A variant of the problem solved by Theorem 4 arises if the hexagon H is covered by n convex discs of equal area a and average perimeter $\bar{p}$. We conjecture that

$$\frac{na}{a(H)} \geq \frac{a}{\bar{f}(a,\bar{p},6)} .$$

The proof of this inequality, however, seems to be rather difficult.

The proofs of Theorem 2 and Theorem 3 are based on the solution of the following problems: Find

$$\min_{P \supset C} a(P) \qquad \text{and} \qquad \max_{P \subset C} a(P),$$

where the minimum and the maximum is to be taken over all discs C from $\mathcal{C}(a,p)$ and all k-gons P which contain C or are contained in C respectively. The minimum is attained if C is a smooth regular k-gon of area a and perimeter p, and P is the case of C. The maximum is attained if C is a regular arc-sided k-gon of area a and perimeter p, and P is the kernel of C.

In a recent paper [13] I considered the following more general problem: *Find a member of* $\mathscr{C}(a,p)$ *and a* k-*gon of given area* a_0 *such that their intersection has the greatest possible area.* We denote this maximal area by $M(a,p;k,a_0)$.

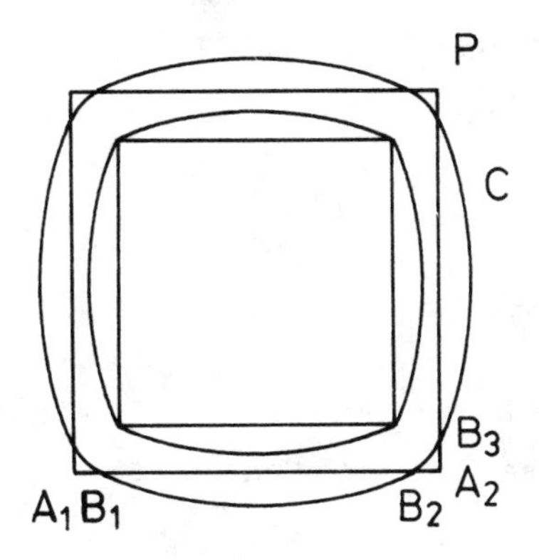

Figure 7

Let us assume that

$$\frac{p^2}{a} < 4k \tan \frac{\pi}{k}, \tag{i}$$

and that

$$\bar{f}(a,p,k) < a_0 < f(a,p,k). \tag{ii}$$

Outside this interval the solution to the problem is given by a regular arc-sided k-gon and its kernel, and by a smooth regular k-gon and its case. To describe the solution if condition (ii) is satisfied, we consider an outer parallel domain C of a regular arc-sided k-gon (Fig. 7). C is bounded by k circular arcs congruent with $\widehat{B_1B_2}$, and k circular arcs congruent with $\widehat{B_2B_3}$, where the radius of $\widehat{B_2B_3}$ is less than that of $\widehat{B_1B_2}$. The lines joining the endpoints of every arc of type $\widehat{B_1B_2}$ enclose a regular k-gon P which we call the *central* k-*gon of* C. By the central k-gon of a circle C we mean any regular k-gon concentric with C. Then we have

THEOREM 5. *Let* C *be a disc from* $\mathscr{C}(a,p)$ *and* P *a* k-*gon of given area* a_0, *and let us assume that (i) and (ii) are satisfied. If* C *and* P *are such that* $a(C \cap P)$ *is maximal, then* C *is a parallel domain of a regular arc-sided* k-*gon, and* P *is the central* k-*gon of* C. *Furthermore,* C *has area* a *and perimeter* p.

We observe that the extremal configuration occuring in any other case may be regarded as a degenerate parallel domain of a regular arc-sided k-gon and its central k-gon.

We point out a consequence of Theorem 5 which is of interest by itself. One of the most usual ways of measuring the deviation between two convex discs, X and Y, is given by their area deviation

$$\delta^A(X,Y) = a(X \cup Y) - a(X \cap Y).$$

Which member of $\mathscr{C}(a,p)$ has the least possible area deviation from any convex k-gon? The following corollary of Theorem 5 answers this question, where $t = t(k) =$ $= (k/\pi)\tan(\pi/k)$.

COROLLARY. *Suppose that* $p^2/4a\pi < t$. *There is exactly one disc* C *from* $\mathscr{C}(a,p)$ *and one* k-*gon* P *such that* $\delta^A(C,P)$ *is minimal.* C *and* P *are characterized by the following properties:*

(*i*) $a(C) = a$, $p(C) = p$;

(*ii*) *If* $p^2/4a\pi < (1+t)^2/(1+3t)$, C *is a parallel domain of a regular arc-sided* k-*gon, and* P *is the central* k-*gon of* C. *Any side of* P, *say* A_1A_2, *meets the boundary of* C *at points* B_1, B_2 *such that*

$$A_1B_1 = B_2A_2 = \frac{1}{4} A_1A_2 .$$

(*iii*) *If* $P^2/4a\pi \geq (1+t)^2/(1+3t)$, C *is a smooth regular* k-*gon, and* P *is the case of* C.

A problem which is not yet solved so far is connected with the function $M(a,p;k,a_0)$ mentioned above. Let $C_1,\dots,C_n$ be discs from $\mathscr{C}(a,p)$, and let H be a hexagon. Let S denote the part of H covered by the union of $C_1,\dots,C_n$. We conjecture that

$$a(S) \leq n\, M(a,p;\, 6, a(H)/n).$$

In the case when $p^2 = 4a\pi$ this inequality has been proved by L. Fejes Tōth [7, p. 80].

4. PACKING AND COVERING; AFFINE PERIMETER

In his review [9] L. Fejes Tōth writes: "From a certain point of view it is more natural, and also more interesting, to consider, instead of the class $\mathscr{C}(a,p)$, a class in which the part of the ordinary perimeter is played by the affine perimeter." The affine length of a curve, introduced by G. Pick and W. Blaschke, is an additive measure of arc which is invariant under affine area-preserving transformations. If the curve is rectifiable and the curvature κ of the curve at a point P is a continuous function of the ordinary length s, then the affine length of the curve is defined by $\int|\kappa|^{1/3}ds$. A more intuitive definition makes use of chains of triangles covering the arc. The affine length of the whole boundary of a convex disc C is called the *affine perimeter* of C.

If a is the area and Λ the affine perimeter of C, we have the "isoperimetric inequality" discovered by Blaschke

$$\Lambda^3 \leq 8\pi^2 a$$

with equality only for the ellipses. This inequality suggests considering problems of packing and covering with convex discs of area not greater than a and affine perimeter not less than Λ.

Let A be an affinely regular k-gon (Fig. 8). Let B be the k-gon spanned by the mid-points of the sides of A. We define a *regular conic*-k-*gon* to be a convex disc bounded by k affinely equivalent arcs of a conic joining consecutive vertices of B and touching the respective sides of A. We call A the *case* and B the *kernel* of the conic-k-gon. The conic-k-gons with case A and kernel B form an array joining A and B. The array contains an

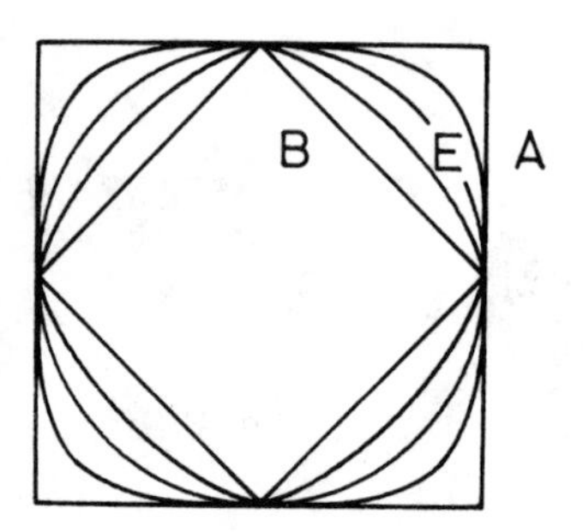

Figure 8

ellipse E. The members of the array between B and E are said to be *elliptic*, and those between E and A *hyperbolic*. The ellipse E is considered to be both elliptic and hyperbolic.

Let $A(a,\Lambda,k)$ be the area of the case of a regular hyperbolic conic-k-gon of area a and affine perimeter Λ. Similarly, let $B(a,\Lambda,k)$ be the area of the kernel of a regular elliptic conic-k-gon of area a and affine perimeter Λ.

In a joint paper with L. Fejes Tőth [10] we proved the following theorems, where a and Λ are positive numbers such that $\Lambda^3 \leq 8\pi^2 a$. The proofs are based on two results by L. Fejes Tőth [8].

THEOREM 6. *Let the convex discs* $C_1,\ldots,C_n$ *with area* a *and affine perimeter not less than* Λ *form a packing in a hexagon* H. *Then the density of the discs relative to* H *satisfies*

$$\frac{n\,a}{a(H)} \leq \frac{a}{A(a,\Lambda,6)} .$$

THEOREM 7. *Let the convex discs* $C_1,\ldots,C_n$ *with area* a *and affine perimeter not less than* Λ, *any two of them intersecting simply, form a covering of a hexagon* H. *Then the density of the discs relative to* H *satisfies*

$$\frac{n\,a}{a(H)} \geq \frac{a}{B(a,\Lambda,6)} .$$

In a subsequent paper [12] I proved two stronger theorems. I state only the theorem concerning the packing.

THEOREM 8. *If the convex discs* $C_1,\ldots,C_n$ *with average area* $\bar{a}$ *and average affine perimeter* $\bar{\Lambda}$ *form a packing in a hexagon* H, *then the density of the discs relative to* H *satisfies*

$$\frac{n\,\bar{a}}{a(H)} \leq \frac{\bar{a}}{A(\bar{a},\bar{\Lambda},6)} .$$

A similar inequality holds for the density of covering, if we restrict ourselves to coverings by simply intersecting discs.

Let $\mathscr{C}(a,\Lambda)$ be the class of all convex discs with area not greater than a and affine perimeter not less than Λ, where $\Lambda^3 \leq 8\pi^2 a$. Let $k \geq 3$ be a given integer. We conclude this review with the following open problem: *Find that member* C *of* $\mathscr{C}(a,\Lambda)$ *and that* k-*gon* P *for which* $\delta^A(C,P)$ *is minimal.*

I hope, I could give you the impression that the problems considered here belong to Intuitive Geometry and that they offer a range of possibilities for future work.

REFERENCES

[1] BAMBAH, R.P. and ROGERS, C.A., *Covering the plane with convex sets.* J. London Math. Soc. 27(1952), 304-314.

[2] FEJES TÓTH, G., *New results in the theory of packing and covering.* Convexity and its applications (Edited by P.M. Gruber and J.M. Wills), Birkhäuser Basel-Boston-Stuttgart (1983), 318-359.

[3] FEJES TÓTH, G. and FLORIAN, A., *Covering of the plane by discs.* Geom. Dedicata 16(1984), 315-333.

[4] FEJES TÓTH, L., *Some packing and covering theorems.* Acta Univ. Szeged, Acta Sci. Math. 12/A (1950), 62-67.

[5] FEJES TÓTH, L., *Filling of a domain by isoperimetric discs.* Publ. Math. Debrecen 5(1957), 119-127.

[6] FEJES TÓTH, L., *Regular Figures.* Pergamon Press Oxford-London-New York (1964).

[7] FEJES TÓTH, L., *Lagerungen in der Ebene, auf der Kugel und im Raum,* 2nd ed. Springer Berlin-Heidelberg-New York (1972).

[8] FEJES TÓTH, L., *Approximation of convex domains by polygons*. Studia Sci. Math. Hung. 15(1980), 133-138.

[9] FEJES TÓTH, L., *Density bounds for packing and covering with convex discs*. Expositiones Math. 2(1984), 131-153.

[10] FEJES TÓTH, L. and FLORIAN, A., *Packing and covering with convex discs*. Mathematika 29(1982), 181-193.

[11] FEJES TÓTH, L. and HEPPES, A., *Filling of a domain by equiareal discs*. Publ. Math. Debrecen 7(1960), 198-203.

[12] FLORIAN, A., *Packing and covering with convex discs*. Studia Sci. Math. Hungar. (to appear).

[13] FLORIAN, A., *Approximation of convex discs by polygons*. Discrete and computational geometry (to appear).

[14] HEPPES, A., *Filling of a domain by discs*. Magyar Tud. Akad. Mat. Kutató Int. Közl. 8(1963), 363-371.

[15] KERSHNER, R., *The number of circles covering a set*. Amer. J. Math. 61(1939), 665-671.

[16] ROGERS, C.A., *Packing and Covering*. Cambridge University Press (1964).

[17] THUE, A., *Om nogle geometrisk taltheoretiske Theoremer*. Forhdl. Skand. Naturforsk. 14(1892), 352-353.

[18] THUE, A., *Über die dichteste Zusammenstellung von Kreisen in einer Ebene*. Christiana Vidensk. Selsk. Skr. 1(1910), 3-9.

A. FLORIAN
Universität Salzburg
Institut für Mathematik
Hellbrunnerstrasse 34.
A-5020 Salzburg

COLLOQUIA MATHEMATICA SOCIETATIS JÁNOS BOLYAI
48. INTUITIVE GEOMETRY, SIÓFOK, 1985.

SIX PHASES FOR THE EIGHT-LAMBDAS AND EIGHT-DELTAS CONFIGURATIONS

RICHARD K. GUY

0. MOTIVATION

Our interest came from a desire to tile an integer-sided square with integer-sided triangles, as few in number as possible [1]. We soon found it convenient to recover such tilings from those of the *unit square* with *rational triangles*. The name rational triangle is often given to those with rational area, as well as rational sides, i.e. *Heron triangles*. It turned out that only such triangles were suited to our purpose.

It was also convenient to work with real variables, rather than rational ones, although the geometry was everywhere suggested by our quest for rational solutions. We found that we needed at least four triangular tiles. Two tiles (Fig. 1) clearly do not give a rational solution. With three tiles (Figs. 2,3,4) only Fig.4 gives an essentially new tiling, and this, we were able to prove, does not occur rationally.

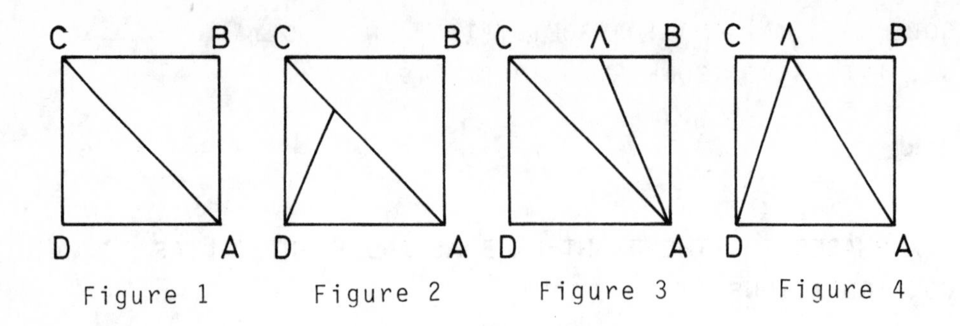

Figure 1 Figure 2 Figure 3 Figure 4

There are four essentially new candidates for tilings with four triangles: the *nu-configuration* (Fig.5), the *delta-configuration* (Fig.6), the *chi-configuration* (Fig.7) and the *kappa configuration* (Fig.8).

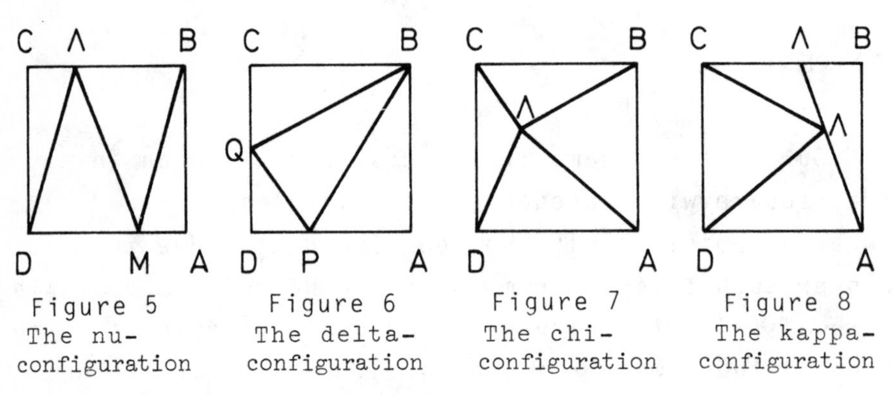

Figure 5 The nu-configuration

Figure 6 The delta-configuration

Figure 7 The chi-configuration

Figure 8 The kappa-configuration

We conjecture [2] that the chi-configuration does not exist with all four distances rational, but we keep the possibility in our analysis by noting that each of

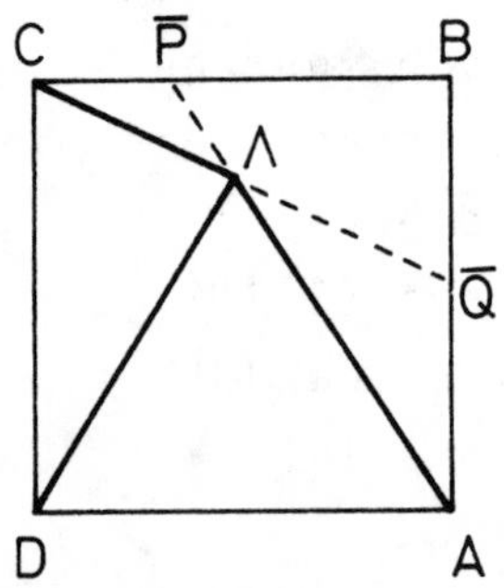

Figure 9 A lambda-configuration and its two associated kappa-configurations

Figs. 7 & 8 (as well as Figs. 2 & 4) contains a *lambda-configuration* (Fig.9) which has two related kappa-configurations. The lambda-configuration turns out to have a dual relation with the delta-configuration: only the nu-configuration needed separate treatment.

1. THE LAMBDA-DELTA DUALITY

Take a *solution-point*, S, in the plane of the unit square ABCD, with coordinates (X,Y) relative to DA and DC as axes; $P\bar{P}$, $Q\bar{Q}$ are the ordinate and abscissa through S. The *delta* is the triangle PBQ, whose sides are diagonals of the rectangles DPSQ, $AB\bar{P}P$ & $BCQ\bar{Q}$ (Fig.10Δ). Now switch attention to the other diagonals, DS, $A\bar{P}$ & $C\bar{Q}$, of these rectangles. These concur, by Pappus's theorem, and form a *lambda* (Fig.10λ)

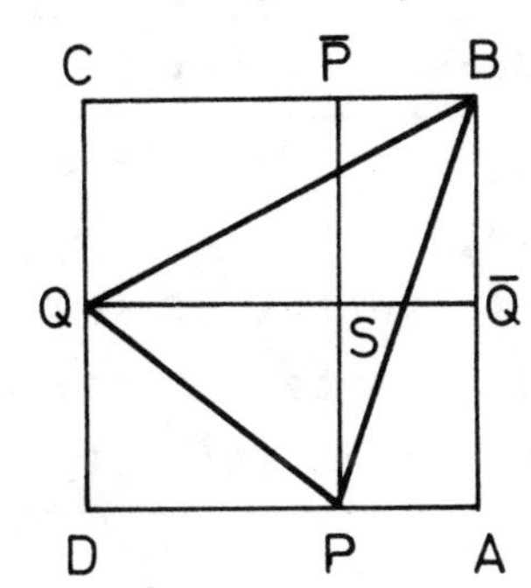

Figure 10Δ The delta-construction

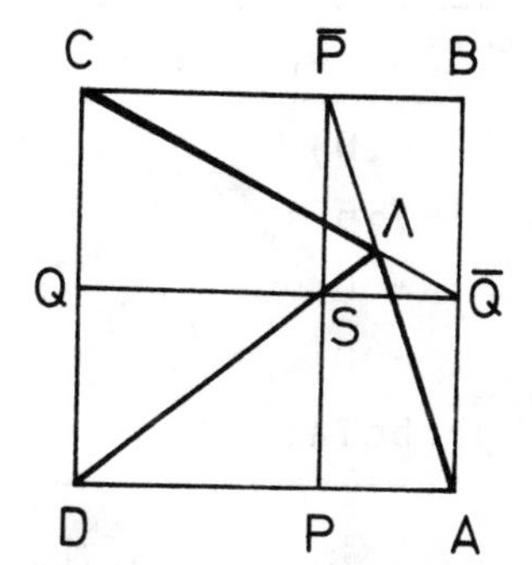

Figure 10λ The lambda-construction

The point of concurrence is the *lambda-point*, Λ with coordinates (x,y). The relations between the lambda-point and the solution-point, (X,Y), are

$$(1) \qquad x = \frac{X}{X+Y-XY}, \qquad y = \frac{Y}{X+Y-XY},$$

$$(2) \qquad X = \frac{x+y-1}{y}, \qquad Y = \frac{x+y-1}{x}.$$

It is important not to confuse the lambda-point with the solution-point. The lambda-point, (x,y), lives in the *inversive plane*, whose single point at infinity corresponds to the *unit hyperbola*, $XY = X+Y$ or $(X-1)(Y-1) = 1$ in the (X,Y)-plane. The solution-point, $S = (X,Y)$, lives in the *projective plane*, whose line at infinity corresponds to the origin, $(0,0)$, in the (x,y)-plane. The lambda-point and the solution-point lie on the same *slope-line* through the origin, D.

The fixed points of the transformations (1) & (2) lie on two sides of the square, $X = 1(x=1)$ & $Y = 1(y=1)$, except for the points A(1,0) & C(0,1) in the (x,y)-plane, which are singularities. We list the singularities in Table 1.

Solution-space, (X,Y)	Lambda-space, (x,y)
Origin, D(0,0)	$x+y=1$, negative diagonal
Line at infinity	Origin, D(0,0)
X-axis, $Y = 0$	A(1,0), vertex of square
Y-axis, $X = 0$	C(0,1), vertex of square
unit hyperbola, $XY = X+Y$	The point at infinity

Table 1 The singularities of the transformations (1) & (2).

2. THREE INVOLUTIONS

We used a computer to search for integer solutions and found many more numerical coincidences than we expected. This is because such solutions occur in orbits of eight. The relations between these sets of eight solutions are best described geometrically, in terms of three involutions.

Reflexion in the main diagonal of the square, X = Y, x = y, is obvious. We thought we were halving our work by ignoring it. In fact we were blinding ourselves to the existence of the third involution.

Inversion was already known [2]. Figure 11 shows a lambda-point, Λ, at distances $\Lambda A, \Lambda D, \Lambda C$ from three corners of the square ABCD. Draw a second square, $A'B'\Lambda D$, with $D\Lambda$ as side. Triangle DAA' is congruent to $DC\Lambda$, being a rotation through a right-angle about D. So $AA' = C\Lambda$ and A is a lambda-point for the new square. The three distances to three corners of this square are the same as for the original lambda-point in the original square, except that the middle distance, ΛD, and the side of the square, AD, have swapped roles. Rotate and reflect the new square and normalize it to have unit side, then the point corresponding to A is Λ_1, say, the inverse of Λ in the *unit circle*, centre D.

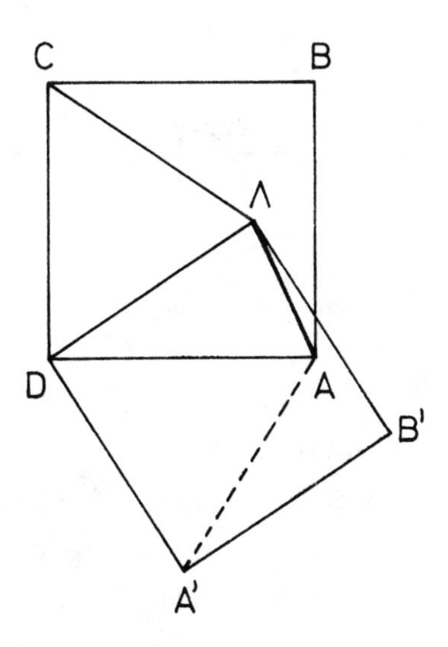

Figure 11 Two lambdas, related by inversion

The third involution, *alternation*, is most easily visualized where it was discovered, in the solution-space. We shall see that it also has a simple and pleasing geometric interpretation in the lambda-space.

Let S be a solution-point, and S_1 the solution-point corresponding to the lambda-point, Λ_1. This is *not* the inverse of S with respect to the unit-circle; although it is often convenient to refer to it as the inverse solution. Let us calculate its coordinates. If $S = (X,Y)$, then the corresponding Λ has coordinates $(x,y) = (X/(X+Y-XY), Y/(X+Y-XY))$, by (1). The inverse, Λ_1, of Λ has coordinates

$$(x_1,y_1) = (x/(x^2+y^2),\ y/(x^2+y^2)) \tag{3}$$

The solution-point, S_1, corresponding to Λ_1, has coordinates

$$\begin{aligned}(X_1,Y_1) &= ((x_1+y_1-1)/y_1,\ (x_1+y_1-1)/x_1), \text{ by (2)}\\ &= ((x+y-x^2-y^2)/y,\ (x+y-x^2-y^2)/x), \text{ by (3)}\\ &= \left(\frac{(X+Y)(X+Y-XY)-X^2-Y^2}{Y(X+Y-XY)},\ \frac{(X+Y)(X+Y-XY)-X^2-Y^2}{X(X+Y-XY)}\right), \text{ by (1)}\end{aligned} \tag{4}$$

$$(X_1,Y_1) = \left(\frac{X(2-X-Y)}{X+Y-XY},\ \frac{Y(2-X-Y)}{X+Y-XY}\right)$$

We now have four solution-points, $S = (X,Y)$, $S_1 = (X_1,Y_1)$ and their reflexions $S_2 = (X_2,Y_2) = (Y,X)$ and $S_3 = (X_3,Y_3) = (Y_1,X_1)$. If we take a solution, with its reflected inverse, for example (X,Y) with (X_3,Y_3) and *alternate* the coordinates, we obtain two more solutions,

$$(X_4,Y_4) = (X_3,Y) = (Y_1,Y) \quad \text{and}$$
$$(X_7,Y_7) = (X,Y_3) = (X,X_1).$$

Similarly (X_1,Y_1) and (X_2,Y_2) generate the solutions

$$(X_5,Y_5) = (X_1,Y_2) = (X_1,X) \quad \text{and} \quad (X_6,Y_6) = (X_2,Y_1) = (Y,Y_1)$$

These are depicted in Fig. 12. Note that (X_4,Y_4) is the reflexion of (X_6,Y_6) and (X_5,Y_5) is the reflexion of (X_7,Y_7).
Moreover (see (4) above),

$$(X_4,Y_4) = \left(\frac{Y(2-X-Y)}{X+Y-XY}, Y\right) \quad \text{and} \quad (X_5,Y_5) = \left(\frac{X(2-X-Y)}{X+Y-XY}, X\right)$$

are inverse solutions (i.e. correspond to lambda-points which are inverse with respect to the unit circle) since the inverse of (X_4,Y_4) is, by (4),

$$\left(\frac{\frac{Y(2-X-Y)}{X+Y-XY}\left\{2 - \frac{Y(2-X-Y)}{X+Y-XY} - Y\right\},\ Y\left\{2 - \frac{X(2-X-Y)}{X+Y-XY} - Y\right\}}{\frac{Y(2-X-Y)}{X+Y-XY} + Y - \frac{Y^2(2-X-Y)}{X+Y-XY}}\right) =$$

$$= \left(\frac{X(2-X-Y)}{X+Y-XY}, X\right) = (X_5,Y_5)$$

Similarly, (X_6,Y_6) and (X_7,Y_7) are inverse solutions in this sense.

In Table 2 we have calculated the coordinates of the eight solution-points and of the eight lambda-points, both in terms of (X,Y), the coordinates of the corresponding lambda-point.

The subscripts for the solution-points, the lambda-points and their coordinates have been chosen so that an inverse is obtained by nim-adding 1 ($\ddagger 2$). Alternation is an involution between *pairs* of points, whose numbers are obtained by $\ddagger 4$ and $\ddagger 7$. Table 3 is a nim-addition table from 0 to 7.

i	Solution-points (X_i, Y_i)	Lambda-points (x_i, y_i)
0	$(X,Y) = \left(\frac{x+y-1}{y}, \frac{x+y-1}{x}\right)$	$(x,y) = \left(\frac{X}{X+Y-XY}, \frac{Y}{X+Y-XY}\right)$
1	$\left(\frac{X(2-X-Y)}{X+Y-XY}, \frac{Y(2-X-Y)}{X+Y-XY}\right) = \left(\frac{x+y-x^2-y^2}{y}, \frac{x+y-x^2-y^2}{x}\right)$	$\left(\frac{x}{x^2+y^2}, \frac{y}{x^2+y^2}\right) = \left(\frac{X(X+Y)-XY)}{X^2+Y^2}, \frac{Y(X+Y-XY)}{X^2+Y^2}\right)$
2	$(Y,X) = \left(\frac{x+y-1}{x}, \frac{x+y-1}{y}\right)$	$(y,x) = \left(\frac{Y}{X+Y-XY}, \frac{X}{X+Y-XY}\right)$
3	$\left(\frac{Y(2-X-Y)}{X+Y-XY}, \frac{X(2-X-Y)}{X+Y-XY}\right) = \left(\frac{x+y-x^2-y^2}{x}, \frac{x+y-x^2-y^2}{y}\right)$	$\left(\frac{y}{x^2+y^2}, \frac{x}{x^2+y^2}\right) = \left(\frac{Y(X+Y-XY)}{X^2+Y^2}, \frac{X(X+Y-XY)}{X^2+Y^2}\right)$
4	$\left(\frac{Y(2-X-Y)}{X+Y-XY}, Y\right) = \left(\frac{x+y-x^2-y^2}{x}, \frac{x+y-1}{x}\right)$	$\left(\frac{x(x+y-x^2-y^2)}{y(x^2+(1-y)^2)}, \frac{x(x+y-1)}{y(x^2+(1-y)^2)}\right) = \left(\frac{2-X-Y}{1+(1-Y)^2}, \frac{X+Y-XY}{1+(1-Y)^2}\right)$
5	$\left(\frac{X(2-X-Y)}{X+Y-XY}, X\right) = \left(\frac{x+y-x^2-y^2}{y}, \frac{x+y-1}{y}\right)$	$\left(\frac{y(x+y-x^2-y^2)}{x((1-x)^2+y^2)}, \frac{y(x+y-1)}{x((1-x)^2+y^2)}\right) = \left(\frac{2-X-Y}{(1-X)^2+1}, \frac{X+Y-XY}{(1-X)^2+1}\right)$
6	$\left(Y, \frac{Y(2-X-Y)}{X+Y-XY}\right) = \left(\frac{x+y-1}{x}, \frac{x+y-x^2-y^2}{x}\right)$	$\left(\frac{x(x+y-1)}{y(x^2+(1-y)^2)}, \frac{x(x+y-x^2-y^2)}{y(x^2+(1-y)^2)}\right) = \left(\frac{X+Y-XY}{1+(1-Y)^2}, \frac{2-X-Y}{1+(1-Y)^2}\right)$
7	$\left(X, \frac{(X(1+X-Y))}{X+Y-XY}\right) = \left(\frac{x+y-1}{y}, \frac{x+y-x^2-y^2}{y}\right)$	$\left(\frac{y(x+y-1)}{x((x-1)^2+y^2)}, \frac{y(x+y-x^2-y^2)}{x((1-x)^2+y^2)}\right) = \left(\frac{X+Y-XY}{(1+X)^2+1}, \frac{2-X-Y}{(1-X)^2+1}\right)$

Table 2 Orbits of eight solution-points and eight lambda-points.

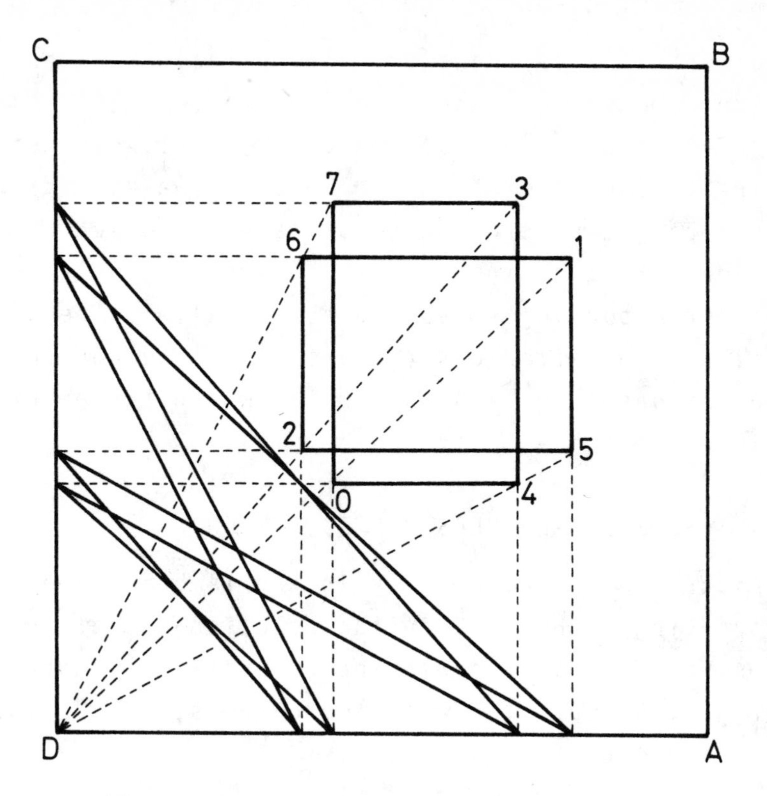

Figure 12 An orbit of eight solution-points and the side opposite to B of each of the eight associated deltas.

0	1	2	3	4	5	6	7
1	0	3	2	5	4	7	6
2	3	0	1	6	7	4	5
3	2	1	0	7	6	5	4
4	5	6	7	0	1	2	3
5	4	7	6	1	0	3	2
6	7	4	5	2	3	0	1
7	6	5	4	3	2	1	0

Table 3 A nim-addition table

3. THE EIGHT-DELTAS CONFIGURATION

Figure 12 also shows one side of each of the eight solution-points. To avoid confusion, we have not drawn the other sixteen sides of the deltas, which coincide in pairs through the point B. The sides we have drawn form two congruent crossed quadrilaterals which are reflexions of each other. They also have parallel sides, though corresponding sides are not parallel. The four directions are reflexions (in a side of the square) of the four slope lines (through D) on which the eight solution-points lie.

4. THE EIGHT-LAMBDAS CONFIGURATION

Figure 13 is the diagram in the lambda-space corresponding to Fig. 12 in the solution-space. Notice that pairs of lambda-points are not only collinear with the origin, D, but also with the corners, A & C. The sets of three points

A07 A15 A26 A34
C04 C26 C25 C37

are each collinear. Here and in the rest of this section we denote lambda-points by their subscripts.

Table 4 shows which points are collinear. *Odious* numbers are those with an odd number of digits 1 in their binary representation:

1 = 001 2 = 010 4 = 100 7 = 111

and *evil* numbers have an even number of such digits:

0 = 000 3 = 011 5 = 101 6 = 110.

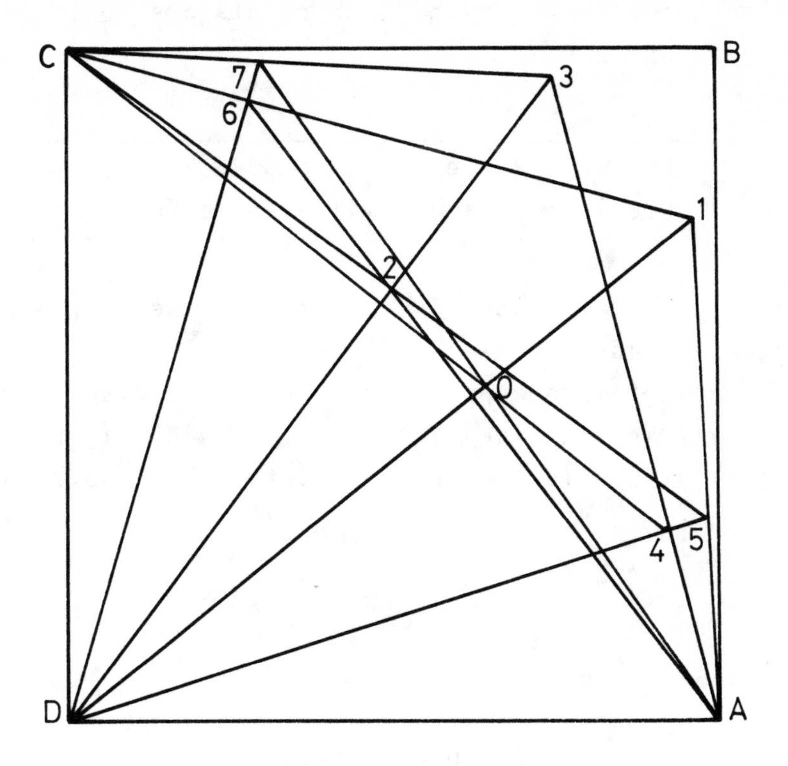

Figure 13 The eight-lambdas configuration

Larger numbers, 4 to 7	Smaller numbers, 0 to 3	Solution-points have the same X-coordinate. Lambda-points are collinear with A(1,0).	Solution-points have the same Y-coordinate. Lambda-points are collinear with C(0,1)
4,7 (odious)	0,3 (evil)	$\overset{*}{+}7$ (odd)	$\overset{*}{+}4$ (even)
5,6 (evil)	1,2 (odious)	$\overset{*}{+}4$ (odd)	$\overset{*}{+}7$ (odd)

Table 4 Collinearities of lambda-points with A or C.

A mnemonic for remembering which points are collinear with A and which with C is

Larger: Alike (odious-odd, evil-even)
Smaller: Cimilar (evil-even, odious-odd)

where odd or even respectively refer to the 7 or 4 which is to be nim-added.

The eight lambda-points of Fig.13 are discussed in some detail in [3]. They form a *Clifford* configuration, 8_4, of eight circles and eight points [4], denoted by

$$\begin{pmatrix} 0 & 1 & 4 & 5 \\ 3 & 2 & 7 & 6 \end{pmatrix}$$

i.e. the set of eight circles

0145	0176	0246	0275
3276	3245	3175	3146

with four circles through each point, four points on each circle. If we include the pairs of points (A,C), (D,∞), then there are two more Clifford configurations,

$$\begin{pmatrix} A & D & 0 & 4 \\ C & \infty & 3 & 7 \end{pmatrix} \text{ and } \begin{pmatrix} A & D & 1 & 6 \\ C & \infty & 2 & 5 \end{pmatrix}$$

which combine with the original one to form a $24_4 12_8$ configuration: 24 circles and 12 points, 4 points on each circle, 8 circles through each point. It has an automorphism group of order $2^7 3^2 = 1152$.

5. A NUMERICAL EXAMPLE

Table 5 exhibits a member of each of the four reflected pairs of solution-points and lambda-points, together with the lengths of the delta-sides, PQ,BP,BQ, and the lambda-distances, ΛD,ΛA,ΛC. To facilitate recall of the related integer solutions, these are expressed as fractions with a common denominator, and are not always in their lowest terms.

Solution number	Solution-point X	Y	Delta-sides PQ	BP	BQ	Lambda-point x	y	Lambda-distances ΛD	ΛA	ΛC
0	$\frac{329}{780}$	$\frac{17}{45}$	$\frac{1325}{2340}$	$\frac{2703}{2340}$	$\frac{2756}{2340}$	$\frac{14802}{22472}$	$\frac{3315}{5618}$	$\frac{375}{424}$	$\frac{289}{424}$	$\frac{329}{424}$
1	$\frac{329}{416}$	$\frac{17}{24}$	$\frac{1325}{1248}$	$\frac{1275}{1248}$	$\frac{1300}{1248}$	$\frac{2632}{3125}$	$\frac{7072}{9375}$	$\frac{424}{375}$	$\frac{289}{375}$	$\frac{329}{375}$
4	$\frac{17}{24}$	$\frac{17}{45}$	$\frac{289}{360}$	$\frac{375}{360}$	$\frac{424}{360}$	$\frac{45}{52}$	$\frac{6}{13}$	$\frac{51}{52}$	$\frac{25}{52}$	$\frac{53}{52}$
5	$\frac{329}{416}$	$\frac{329}{780}$	$\frac{5593}{6240}$	$\frac{6375}{6240}$	$\frac{7208}{6240}$	$\frac{260}{289}$	$\frac{416}{867}$	$\frac{52}{51}$	$\frac{25}{51}$	$\frac{53}{51}$

Table 5 A numerical example

6. WHERE ARE THE SOLUTION-POINTS AND WHERE ARE THE LAMBDA-POINTS?

Our concern is to locate each of the eight points in either set, in a variety of situations. We do this by partitioning the solution-space and the lambda-space: first, so that each of the eight points lies in a different region; additionally, according to various geometrical desiderata. The first classification is clearly defined. The boundaries of the regions are the double-points of the three involutions. The second is more arbitrary, and will be discussed in section 8. From many possibilities we have selected one of the simplest.

7. DOUBLE-POINTS

The double-points of reflexion are given by $x = y$, i.e. $X/(X+Y-XY) = Y/(X+Y-XY)$ or $X = Y$. This in turn gives $(x+y-1)/y = (x+y-1)/x$. i.e. $x = y$ or $x+y = 1$. The negative diagonal, $x+y = 1$, corresponds to the origin $(X,Y) = (0,0)$.

The double-points of inversion lie on the unit circle, $x^2+y^2 = 1$, which gives $X^2+Y^2 = (X+Y-XY)^2$, i.e. $X = 0$ or $Y = 0$ or the *inverting hyperbola,*

$$(X-2)(Y-2) = 2.$$

This corresponds to $x^2+y^2 = 1$, while $X = 0$, $Y = 0$ correspond respectively the points $C(0,1)$, $A(1,0)$ in the lambda-space (compare Table 1 in section 1). The inverse of $x+y = 1$ is the *circumcircle*,

$$x^2 + y^2 = x + y.$$

This corresponds to $X^2+Y^2 = (X+Y)(X+Y-XY)$, i.e. $X = 0$, $Y = 0$ and

$$X + Y = 2.$$

The double-points of alternation are given by

$$X = X_4 = \frac{Y(2-X-Y)}{X+Y-XY} \quad \text{and} \quad Y = Y_7 = \frac{X(2-X-Y)}{X+Y-XY}$$

These are the *alternating cubics*. The Xa-*cubic* has equation

$$X^2Y - (X+Y)^2 + 2Y = 0$$

and the Ya-*cubic* is

$$XY^2 - (X+Y)^2 + 2X = 0.$$

Each is the inverse, and the reflexion, of the other. The parabola $Y = (X-1)^2$ is asymptotic to the Xa-cubic; $X = (Y-1)^2$ is asymptotic to the Ya-cubic, while the hyperbola $(X-1)(Y-1) = 2$ serves as asymptote to both curves.

The Xa-cubic corresponds to

$$\frac{(x+y-1)^3}{y^2x} - (x+y-1)^2\left(\frac{1}{y}+\frac{1}{x}\right)^2 + \frac{2(x+y-1)}{x} = 0$$

i.e. $x+y = 1$ and the xa-*cubic*

$$y(x^2+y^2) + x^2-y^2-x = 0.$$

The only real asymptote to the xa-cubic is $y+1 = 0$ (more precisely, $y = -1+\frac{1}{x}+\frac{2}{x^2}$), which the curve crosses at the point $(-2,-1)$. The ya-*cubic* has equation

$$x(x^2+y^2) - x^2+y^2-y = 0.$$

The tangent at the origin to each of the four cubics is a coordinate axis. The tangents from the origin have equation $y = mx$ ($Y = mX$), where

$$x^3(1+m^2)-x^2(1-m^2)-xm = 0$$

has equal roots, i.e. $(m^2-1)^2+4m(1+m^2) = 0$, $(m+1)^4 = 8m^2$, whose real roots are given by $(m+1)^2 = -2\sqrt{2m}$,

$$m = -(1+\sqrt{2}) \pm \quad 2(1+\sqrt{2}) \tag{5}$$

$$m = -4{\cdot}611581789 \quad \text{or} \quad m = -0{\cdot}216845335$$

In Table 6 we collect the results of this section, together with the singularities of Table 1.

locus in lambda-space	*‡	locus in solution-space
main diagonal, $x=y$	2	$X=Y$
negative diagonal, $x+y=1$	$0^{\dagger}$ or 2	$D=(0,0)$ the origin
corner of square, $A(1,0)$	$0^{\dagger}$ or 1	$Y=0$, X-axis, side of square
corner of square, $C(0,1)$	$0^{\dagger}$ or 1	$X=0$, Y-axis, side of square
unit circle, $x^2+y^2=1$	1	$(X-2)(Y-2)=2$, inverting hyperbola
circumcircle, $x^2+y^2=x+y$	$0^{\dagger}$ or 2	$X+Y=2$
xa-cubic, $y(x^2+y^2)+x^2-y^2=x$	$4^{\dagger}$ or 7	$X^2Y-(X+Y)+2Y=0$ Xa-cubic
ya-cubic, $x(x^2+y^2)-x^2+y^2=y$	$4^{\dagger}$ or 7	$XY^2-(X+Y)^2+2X=0$ Ya-cubic
point at infinity	0	$(X-1)(Y-1)=1$ unit hyperbola
the origin $D=(0,0)$	0	the line at infinity
use the value marked thus:	†	inside, or crossing the boundary of
either the *inner semicircle* on AC as diameter, passing through B.		**or** the *inner triangle*, with vertices $(X,Y)=(0,0),(0,2)$ & $(2,0)$.

Table 6 Double-point loci and singularities.

No point on any of these curves corresponds to a rational solution for the original problem of tiling the square. These curves partition the lambda-space into 24 regions (Fig.14), and the solution-space into 32 regions (Fig.15). We label with zero the following four regions of the solution-space (Fig.15):

(i) the 45° sector bounded by the negative X-axis and the negative part of the main diagonal, $X = Y \leq 0$ (and the line at infinity);

(ii) the infinite region between the Ya-cubic and the X-axis with $X \geq 2$;

(iii) the region enclosed by the Xa-cubic and the main diagonal between $0 \leq X = Y \leq 2 - \sqrt{2}$;

(iv) the sector bounded by part of the main diagonal, $X = Y \geq 2 + \sqrt{2}$, part of the ya-cubic (and the line at infinity).

The corresponding regions of the lambda-space (Fig.14) are:

(i), (ii), (iii) the roughly triangular region bounded by the Ya-cubic between D & A, the xa-cubic between A & $(1/\sqrt{2}, 2/\sqrt{2})$, and the main diagonal from there to D. The three parts of this region are separated by the x-axis and by the negative diagonal, $x + y = 1$.

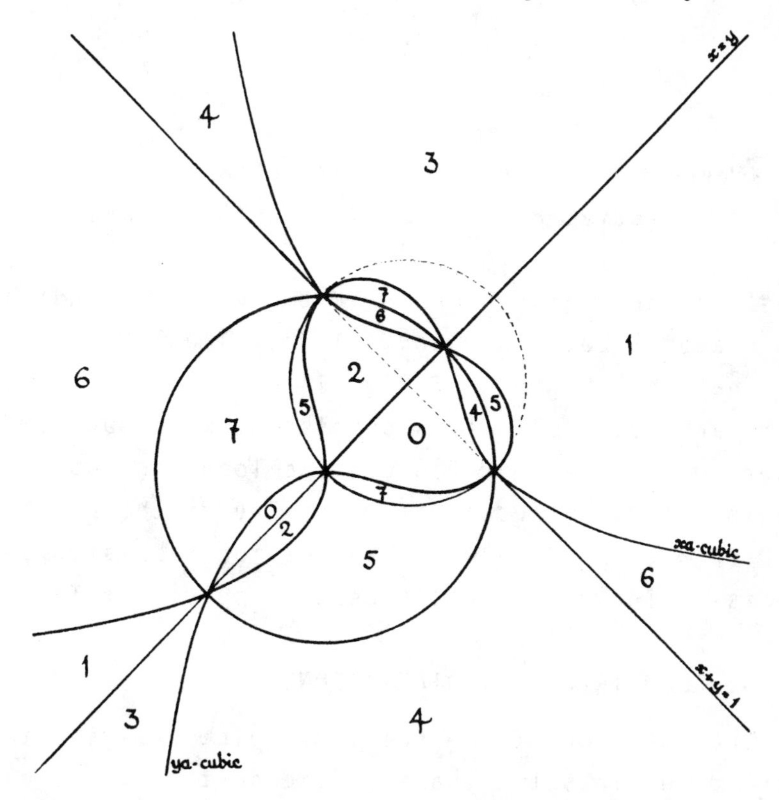

Figure 14 Regions of the lambda-space showing relative positions of lambda-points

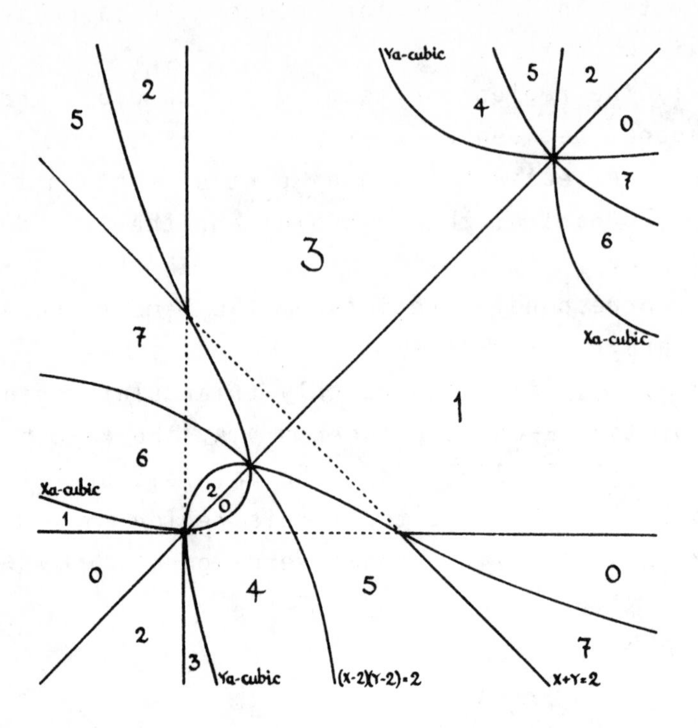

Figure 15 Regions of the solution-space showing relative positions of solution points

(iv) the region bounded by the ya-cubic and the main diagonal between D and $(-1/\sqrt{2}, -1/\sqrt{2})$.

The other regions of Figs.14 & 15 contain lambda-points and solution-points consistent with our earlier numbering if, whenever you cross a locus in Table 6 (or pass in a given direction through any of the points A,C,D or ∞), the nim-sum of the region labels on either side is as indicated in the central column of Table 6.

8. GEOMETRICAL CONSIDERATIONS

The position of the solution-point relative to the square determines the shape of the delta. It may also be of interest to know if the lambda-point is inside or outside the square. For example, the lambda forms "genuine" kappas (Fig.9) just if the lambda point is inside the triangle ABC.

We add the following loci to our list. The sides of the square, $x = 0$, $y = 0$ correspond to points at infinity in the solution space. The sides $x = 1$ ($X=1$), $y = 1$ ($Y=1$) are the fixed points of the transformation. The inverses of these sides are the x-*circle*, $x^2 + y^2 = x$ and the y-*circle*, $x^2 + y^2 = y$. The corresponding curves in the solution-space are ($Y = 0$ and) the X-*parabola*, $Y = X(1-X)$, and ($X = 0$ and) the Y-*parabola*, $X = Y(1-Y)$. These loci are collected in Table 7.

lambda-sapce		solution-sapce	
sides of square	$x = 0$		points at infinity
	$y = 0$		
	$x = 1$	$X = 1$	fixed points of the transformation
	$y = 1$	$Y = 1$	
x-circle	$x^2 + y^2 = x$	$Y = X(1-X)$	X-parabola
y-circle	$x^2 + y^2 = y$	$X = Y(1-Y)$	Y-parabola

Table 7 Additional geometrical boundaries

These add 24 regions to the lambda-space and 16 regions to the solution-space; a total of 48 regions in each. So we have sets of eight solutions in each of six *phases*.

9. WHAT IS A PHASE?

We have tried several criteria for distinguishing phases.

Double-points. We have already seen how the desire to have each of the eight solution-points or lambda-points in separate regions induces many more than eight regions. At least four phases occur "naturally".

Slopes. The eight solution-points and eight lambda-points lie in pairs on each of four slope lines. Two slopes are reciprocals of the others, so we need look only at the two slopes between ± 1. These are distinct unless the point is on an alternating cubic; such points do not give rational solutions. Here we have distinguished only between positive and negative slopes. Column 2 of Table 8 contains + if all slopes are positive; - if all are negative; $\pm$ if there are two of each. If more phases were to be distinguished, a finer spectrum of slopes could be considered. For example, the four intervals, two negative, two positive, between -1,m,0,1/2,1, where m is the slope (5) of a tangent from the origin to the alternating cubics. We have chosen our 'zero' solution to have slope between m and 1.

Deltas. These have one vertex at B, and the others, P & Q, on the X- & Y-axes. The vertex P may be between D and A, or on the positive side of A, or on the negative side of D. Denote these possibilities by 0,+ & - respectively. Similarly, Q may be on DC, on DC produced, or on CD produced, so there are nine types of delta, six essentially different. The delta-types are shown in column 3 of Table 8.

Lambdas. The sides of the square partition the lambda-space into nine regions, which we classify as 00,0+,... much as we have just classified the deltas. Column 4 of Table 8 lists the lambda-types. We have also noted that a lambda-point yields a "genuine" pair of kappas just if it lies inside the half-square ABC.

Figures 16 & 17 each show the 48 regions mentioned at the end of section 8. These are labelled with the phase letter, P to V and the solution or lambda-number,

Phase label	Slopes	Deltas	Lambdas
P	+	all eight, 00.	all eight, 00. 8 kappa-pairs.
Q	+	four 00 and two each of 0+, +0.	four 00 and two each of 0+, +0. 4 kappa pairs
R	+	all eight, ++.	all eight, --.
T	-	two each of 0-, -0, +-, -+.	two each of 0-, -0, +-, -+.
U	±	one each of 0+,+0,0-, -0,+-,-+; two of --.	one each of 0+,+0,0-, -0,+-,-+; two of --.
V	±	two each of ++, +-, -+, --.	two each of 00, ++, 0-, -0.

Table 8 Distinguishing features of phases

0 to 7. The numerical example of section 5 illustrates phase P, as do Figs. 12 & 13. The eight deltas and eight lambdas for each of phases Q, R, T, U & V are illustrated in Figs. 18Q to 18V. We have also experimented with 4, 5, 8, 12, 17, 20 and 22 for the number of different phases; the exact number is a matter of taste.

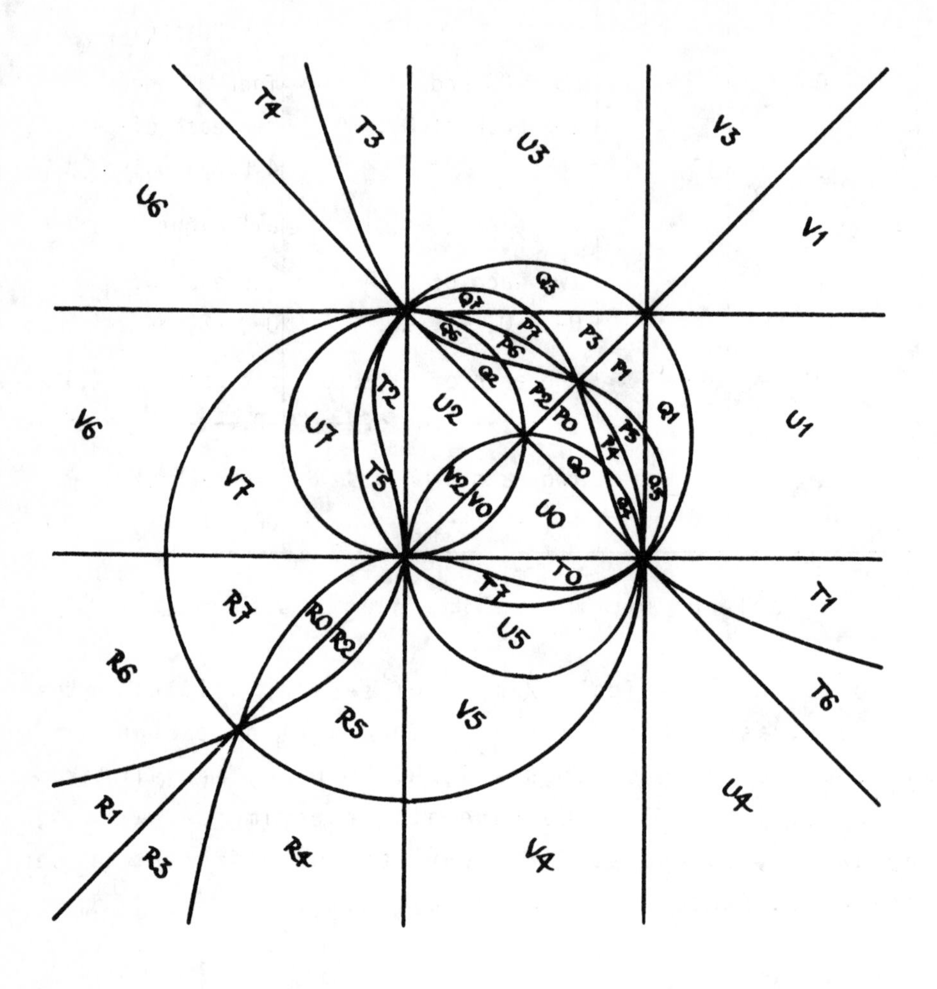

Figure 16 Lambda-space partitioned into 48 regions: eight lambda-points in each of six phases

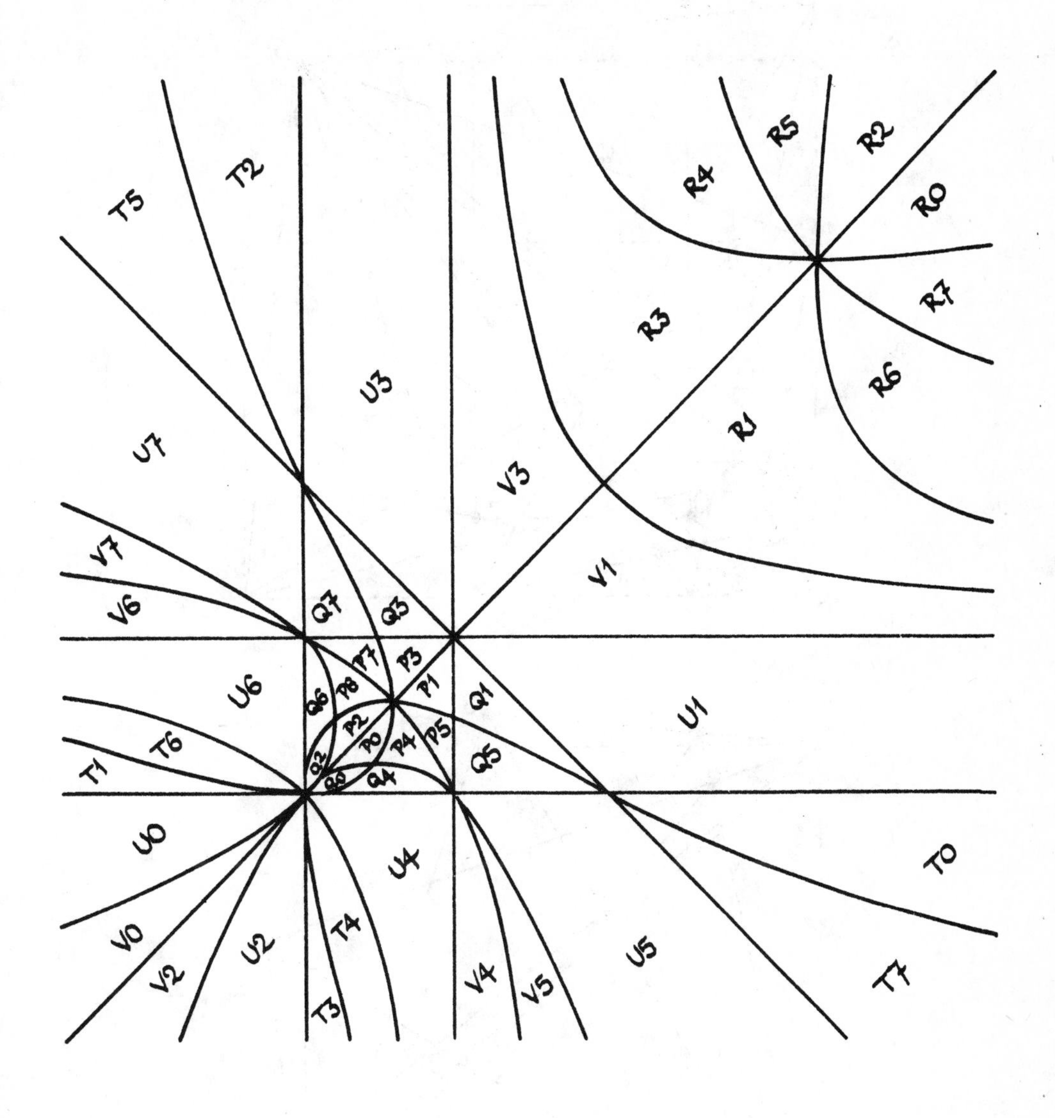

Figure 17 Solution-space partitioned into 48 regions: eight solution-points in each of six phases.

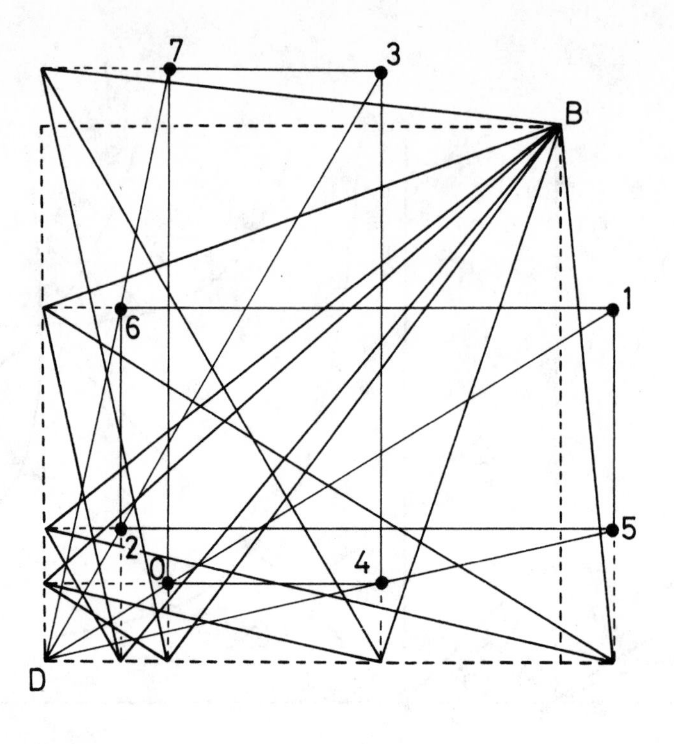

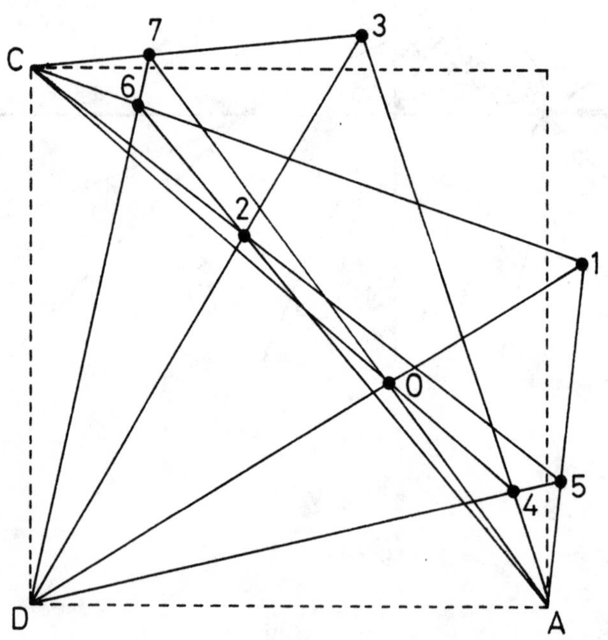

Figure 18Q

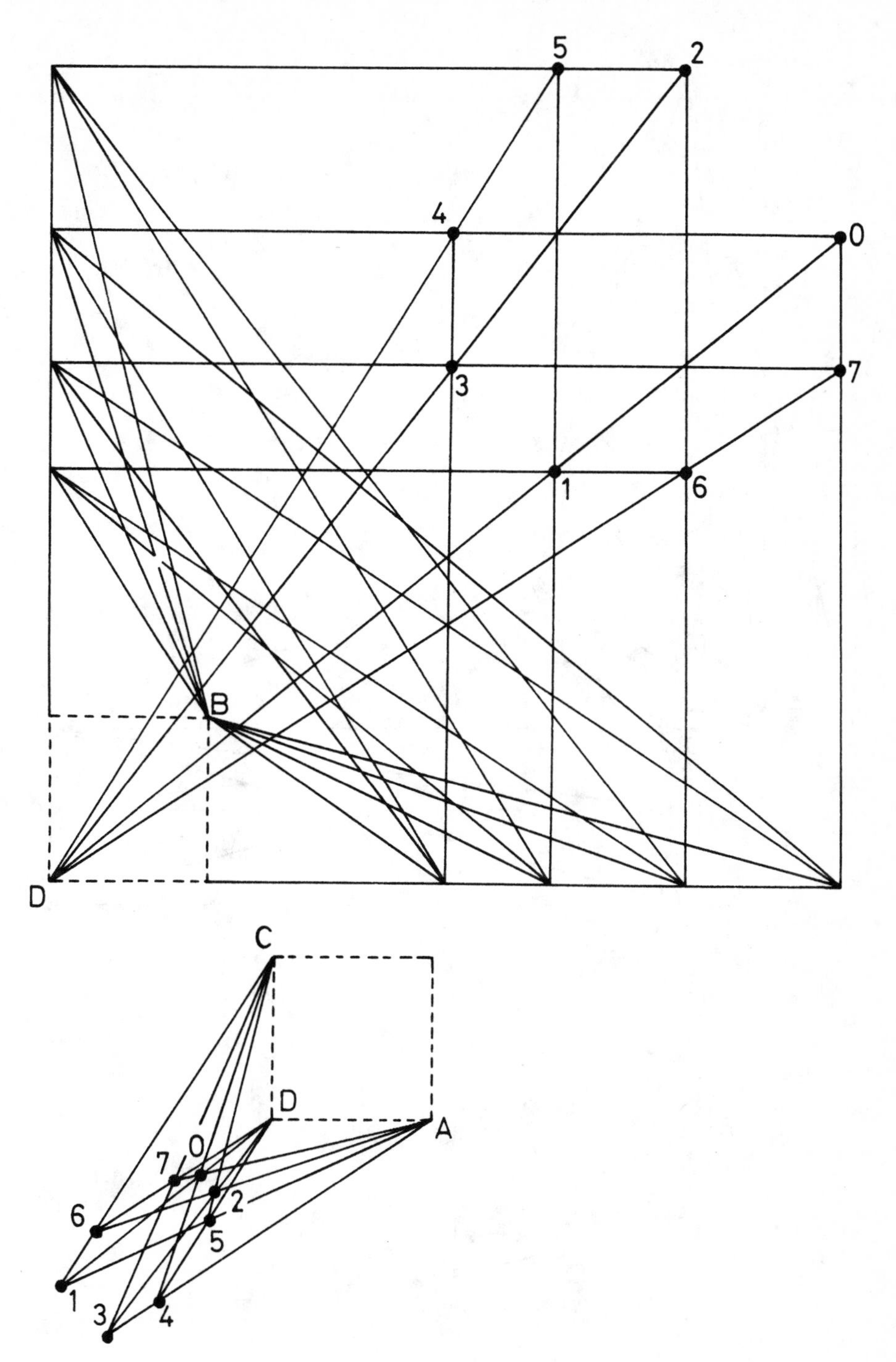

Figure 18R

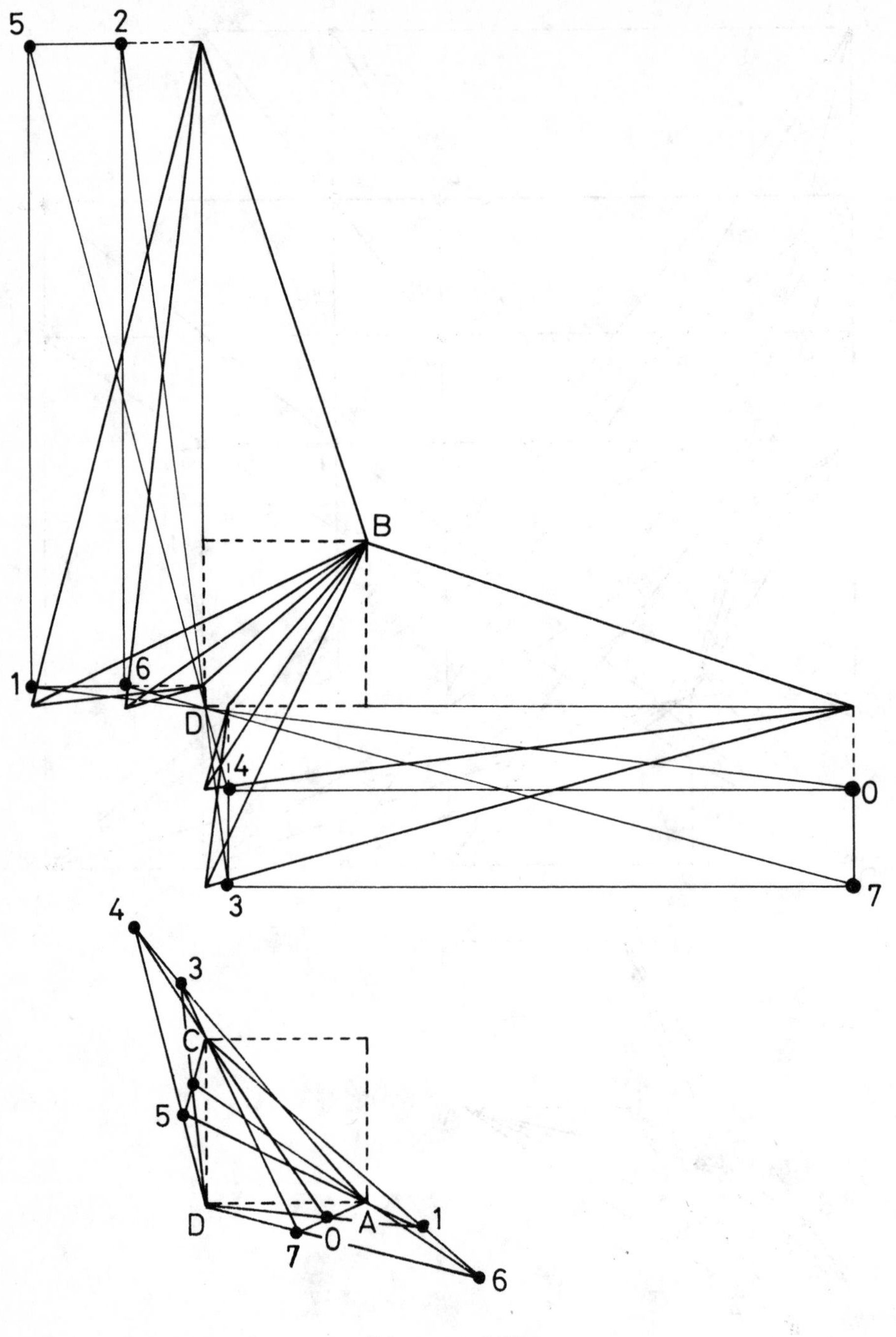

Figure 18T

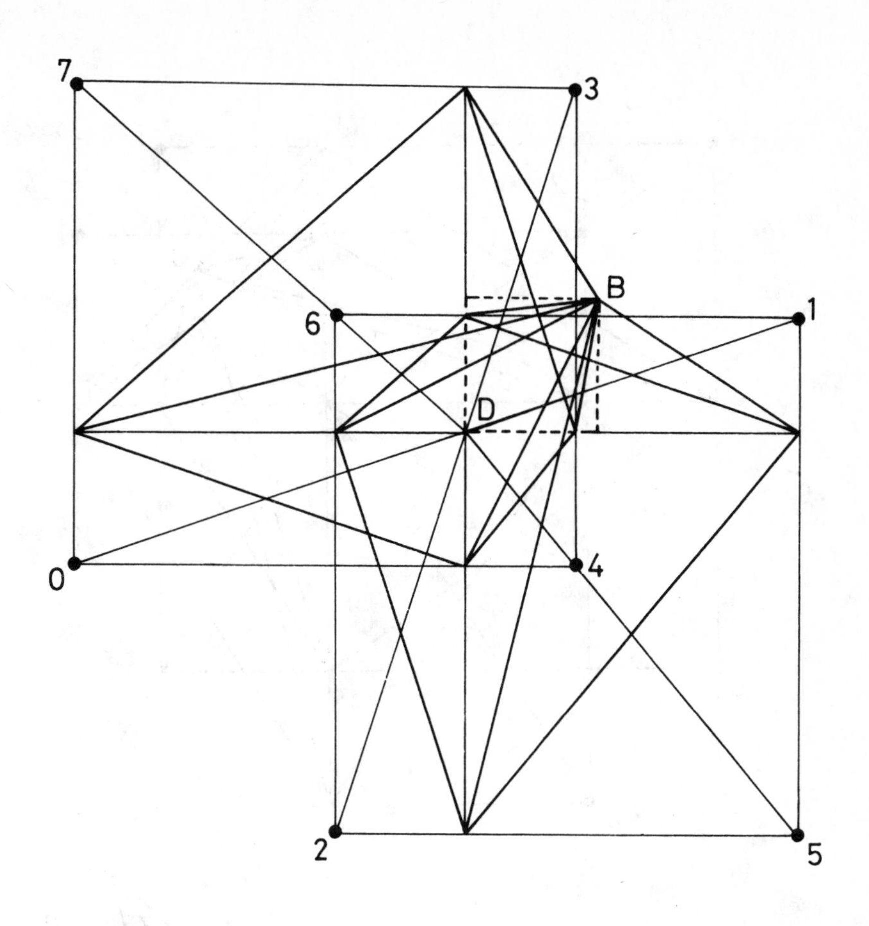

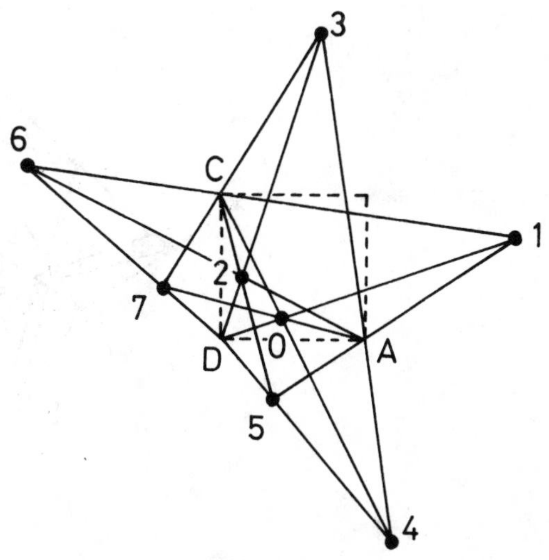

Figure 18U

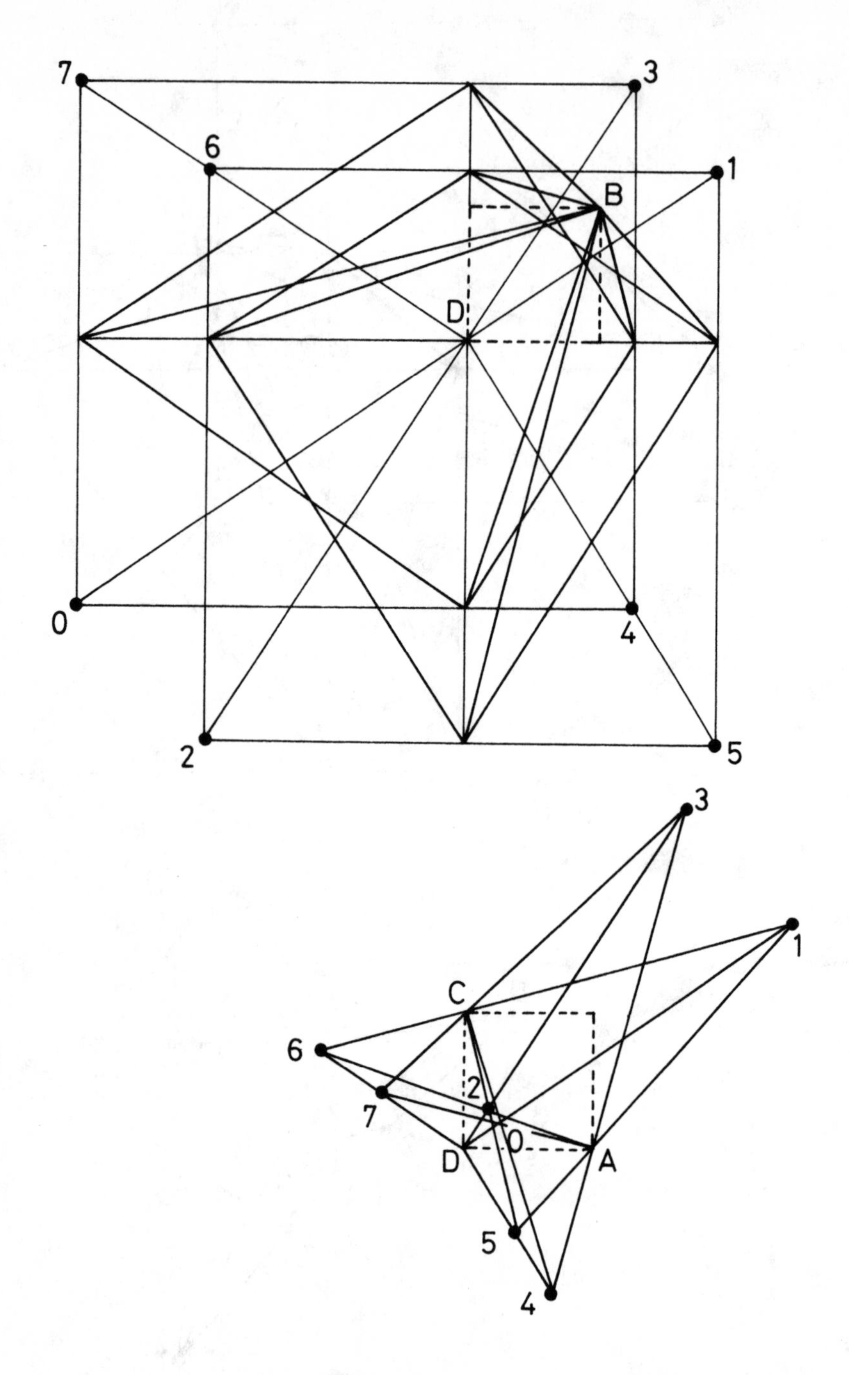

Figure 18V

REFERENCES

[1] R. B. EGGLETON, A. S. FRAENKEL, RICHARD K. GUY and J. L. SELFRIDGE, *Tiling the square with rational triangles*, Acta Arith. (submitted).

[2] RICHARD K. GUY, *Unsolved Problems in Number Theory*, Springer, 1981, Problem D19.

[3] RICHARD K. GUY, *The eight-lambdas configuration arising from some Pythagorean diophantine equations*, Colloq. Discrete Geom., Salzburg, 1985.

[4] J. F. RIGBY, *Half-turns and Clifford configurations in the inversive plane*, J. London Math. Soc. (2), 15(1977) 521-533; MR 57 13687.

R. K. GUY
Department of Mathematics & Statistics,
The University of Calgary
Calgary, Alberta T2N 1N4
Canada

COLLOQUIA MATHEMATICA SOCIETATIS JÁNOS BOLYAI
48. INTUITIVE GEOMETRY, SIÓFOK, 1985

REGULAR POINT SETS WITH UNIT DISTANCES

H. HARBORTH

1. INTRODUCTION

It is an unpublished problem of P. Erdős and G. B. Purdy to ask for the minimum number $p(n)$ of points in the plane such that each point has distance 1 to exactly n other of these $p(n)$ points. In this paper the following exact values are proved.

THEOREM. *The first values of the above defined function* $p(n)$ *are*

$$p(1) = 2, \ p(2) = 3, \ p(3) = 6, \ p(4) = 9, \ p(5) = 18. \tag{1}$$

The first two values of (1) are trivial. In general $p(n)$ exists. This can be seen from projections of the n-cube in the plane where only the images of the edges are of length 1. Since there are only a finite number of distances, projections are possible, which avoid further distances 1. Thus

(2) $$p(n) \leq 2^n$$

is a first upper bound.

If each point of an optimal point set with exactly r distances is substituted by an optimal point set with exactly n-r distances, then this composition gives

(3) $$p(n) \leq p(r)p(n-r).$$

With (1) follows a better upper bound

(4) $$p(n) \leq \begin{cases} 3^{n/2}, & \text{if } n \equiv 0 \pmod 2, \\ 2\cdot 3^{(n-1)/2}, & \text{if } n \equiv 1 \pmod 2. \end{cases}$$

See Figures 1,2,3 for n = 3,4,5.

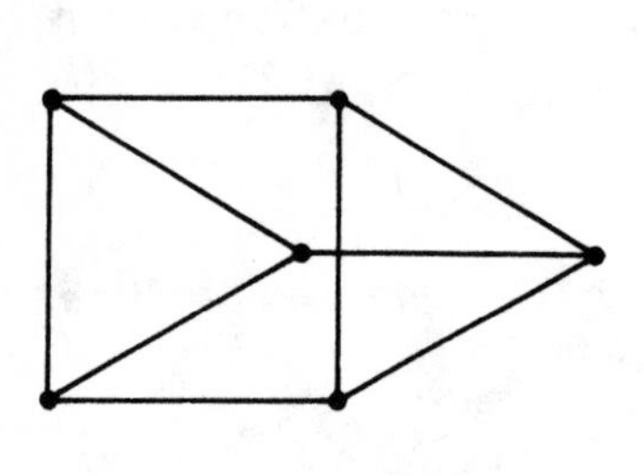

Figure 1

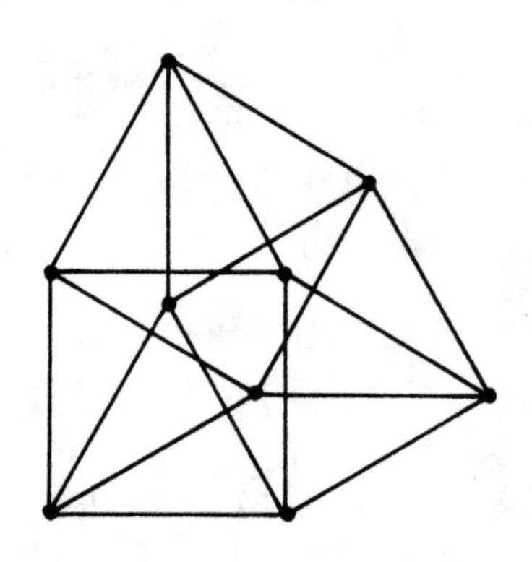

Figure 2

2. PROOFS FOR n=3 AND n=4

For n=3 there are $3p(3)/2$ distances 1, so that $p(3) \equiv 0 \pmod 2$. Since 4 points in the plane cannot have pairwise distances 1, the third value of (1) follows together with (4).

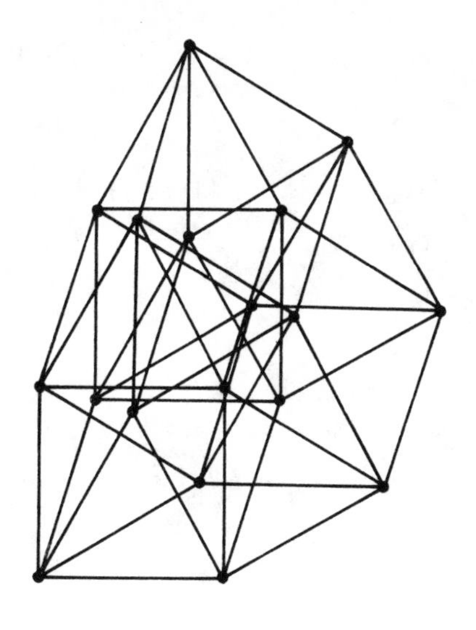

Figure 3

For $n = 4$ inequality (4) yields $p(4) \leq 9$ (see Figure 2). In case $p(4) \leq 8$ consider any point Q, and those 4 points with distance 1 to Q. Then there remain at most 3 points each with at most 2 distances 1 to these 4 points, so that 3 distances 1 have to occur between these 4 points of a unit circle. However, only an infinite set of points can have this property of Q for all points, and $p(4) = 9$ is proved.

3. PROOF FOR n=5

Again (4) yields $p(5) \leq 18$ (see Figure 3). The number of distances 1 is $5p(5)/2$, so that $p(5) \equiv 0 \pmod 2$. Thus $p(5) \leq 16$ can be assumed to give an indirect proof of $p(5) \geq 18$.

Let S be a set of points where each point has distance 1 to exactly 5 other points of S, $|S| \leq 16$, and $|S| \equiv 0 \pmod 2$.

Let the origin be one point P of S, and different unit vectors z_1, z_2, z_3, z_4, z_5 be those 5 points of S with distance 1 to P.

Case 1. No pair of the five points z_i has distance 1. - Then all 10 points $z_i + z_j$, $i \neq j$, belong to S, since otherwise $|S| > 16$. In general it holds for different unit vectors z_r, z_s, z_t

$$(5) \qquad |z_r - z_t| = 1 \text{ if and only if } |z_s + z_r - (z_s + z_t)| = 1,$$

and

$$(6) \qquad |Q-(z_r+z_s)| = 1,\ |Q-(z_r+z_t)| = 1 \text{ implies } |Q-(z_s+z_t)| = 1$$

$$\text{and } Q = z_r+z_s+z_t.$$

Thus the 3 remaining distances 1 from any point z_i+z_j of S have to go to z_r+z_s, z_r+z_t, and z_s+z_t, that means $z_i+z_j = z_r+z_s+z_t$. Then, for example, $z_1+z_2 = z_3+z_4+z_5$, and $z_1+z_3 = z_2+z_4+z_5$, and by subtraction the contradiction $z_2 = z_3$ follows.

Case 2. Only one pair, say z_1 and z_2, of the five points z_i has distance 1. - Then the pairs z_1+z_i and z_2+z_i for $i = 3,4,5$ also have distances 1. Five subcases are discussed. All not mentioned sums z_i+z_j belong to S.

2.1. $z_1+z_2 \notin S$. - There are 15 points, and an additional point Q must exist, which does not have distance 1 to the points z_i. Because of (5) from z_3+z_4 distances 1 can exist only to Q, z_1+z_5, and z_2+z_5, so that (6) yields $z_3+z_4 = z_1+z_2+z_5$. By the same arguments follows $z_3+z_5 = z_1+z_2+z_4$, and subtraction implies $z_4 = z_5$, a contradiction.

2.2. z_1+z_2, $z_1+z_3 \notin S$. - Two unit vectors x_1 and x_2 determine 2 points z_1+x_1, and z_3+x_2 of S, so that there are 16 points. From z_4+z_5 distances 1 can exist only to z_2+z_3, z_1+x_1, and z_3+x_2. Then only one of the points z_2+z_3, z_3+z_4, and z_3+z_5 can have distances 1 also to both points z_1+x_1, z_3+x_2, and by (5) and (6) two of the three identities $z_2+z_3 = z_1+z_4+z_5$, $z_3+z_4 = z_1+z_2+z_5$, and $z_3+z_5 = z_1+z_2+z_4$ hold, which give by subtraction one of the contradictions $z_2 = z_4$, $z_2 = z_5$, or $z_4 = z_5$.

2.3. z_1+z_2, $z_3+z_4 \notin S$. - There are 16 points together with z_3+x_1, and z_4+x_2, where x_1, x_2 are unit vectors. If never $z_i+z_j=z_r+z_s+z_t$ from (6), then z_3+z_5, and z_4+z_5 have distances 1 to both points z_3+x_1, and z_4+x_2, and the remaining 6 sums of vectors z_i have one distance 1 to z_3+x_1, or to z_4+x_2. Since only 8 of these 10 distances 1 are possible to z_3+x_1, and z_4+x_2. Thus for 2 points (6) can be used, and $z_i+z_j = z_r+z_s+z_t$ either with $z_i+z_r = z_j+z_s+z_t$ gives $z_r = z_j$ by subtraction, or with $z_r+z_s = z_i+z_j+z_t$ gives $z_t = 0$ by addition, which both are contradictions.

2.4. $z_1+z_2 \in S$, and z_1+z_3, $z_2+z_3 \notin S$. - Together with z_3+x_1, z_3+x_2, where x_1, x_2 are unit vectors, 16 points are chosen. Because of (5) it follows $z_4+z_5 = z_3+x_1+x_2$. Then z_1+z_2, and z_3+z_5 each cannot have also distance 1 to both points z_3+x_1, and z_3+x_2, so that with (5) and (6) from $z_1+z_2 = z_3+z_4+z_5$, and $z_3+z_5 = z_1+z_2+z_4$ by addition the contradiction $z_4 = 0$ is received.

2.5. $z_1+z_2 \in S$, and z_1+z_3, $z_2+z_4 \notin S$. - Together with z_3+x_1, and z_4+x_2 there are 16 points. At most 2 of the points z_1+z_2, z_1+z_4, z_2+z_3, and z_3+z_4 can by (5) have distances 1 to both points z_3+x_1, and z_4+x_2. Then for the remaining 2 points $z_i+z_j = z_r+z_s+z_t$ is fulfilled, and the same contradiction follows as in 2.3.

It is now proved that for fewer than 18 points each point must be a vertex point of at least two unit triangles.

Case 3. Each of the p(5) points is vertex point of exactly two unit triangles. - If t(5) denotes the number of all unit triangles, then $3t(5) = 2p(5)$, so that $p(5) \equiv 0 (\mathrm{mod}\ 3)$, and together with $p(5) \equiv 0 (\mathrm{mod}\ 2)$ only $p(5) = 12$ remains. If 2 triangles with vertex P have an edge in common, then P, its 5 neighbours, and $4+4+3+3+2 = 16$ neighbours of that 5 points are together 22 points. The 16 points can coincide in at most 9 pairs, so that more than 12 points remain. If all triangles are edge disjoint, again P, its 5 neighbours, and $4+3+3+3+3 = 16$ neighbours of that 5 points, diminished by at most 8 points because of coinciding pairs, gives more than 12 points.

Case 4. Four pairs of the five points z_i have distance 1. - Then the points z_i are consecutive vertex points of a regular hexagon with center point P. Together with 3+3+2+2+2 neighbour points of the points z_i there would be 18 points. Every coinciding pair of these neighbour points has 3 new neighbours. At least one of them cannot coincide with one of the 18 points, so that fewer than 18 points are impossible.

Case 5. Three pairs of the five points z_i have distance 1, such that these are consecutive sides of a regular hexagon. - Let z_1, z_2, z_3, z_4 in this sequence be the consecutive vertex points of a regular hexagon with center point P, that is, $z_4 = -z_1$, $z_2 = z_1 + z_3$, $z_3 = z_2 + z_4$.

5.1. $z_2 + z_3 \in S$. - The 6 points $P, z_1, z_2, z_3, z_4, z_2 + z_3$ have $1+3+1+1+3+3 = 12$ neighbour points. As in Case 4 this number of 18 points cannot be diminished by coincidences.

5.2. $z_1 + z_2 \in S$ (or symmetrically $z_3 + z_4 \in S$). - The 6 points $P, z_1, z_1 + z_2, z_2, z_3, z_4$ have $1+2+3+1+2+3 = 12$ neighbour points. It follows as in Case 4, that there are at least 18 points.

5.3. $z_1 + z_2$, $z_2 + z_3$, $z_3 + z_4 \notin S$. - If the set $S_1 = \{z_1 + z_5, z_2 + z_5, z_3 + z_5, z_4 + z_5\}$ belongs to S, and also the set $S_2 = \{z_1 + x_1, z_1 + x_2, z_2 + x_3, z_3 + x_4, z_4 + x_5, z_4 + x_6\}$ with x_i unit vectors, then there are already 16 points in S.

The points of S_1 also form 4 consecutive vertex points of a regular hexagon with center point z_5. Thus z_5 can be assumed to lie in the same half plane of the line through P and z_1 as z_2 and z_3, since otherwise z_5 and P can be interchanged.

From S_1 to S_2 there are $2+1+1+2 = 6$ distances 1, so that within S_2 there remain 9 distances 1. It is easily checked from [3, p. 222], that only 4 graphs with 6 points and 9 lines are possible in the plane with unit lines: Among them are 3 graphs, which are discussed already in Case 4, 5.1, and 5.2, and the remaining one is that of Figure 1. So 2 points in S_2 always have a distance less than 2, and each point of S_2 has exactly one distance 1 to S_1.

It is impossible that z_1+x_1 and z_1+x_2 both have distance 1 to z_1+z_5, since then z_4+x_5, or z_4+x_6 with distance 1 to z_4+z_5 has a distance at least 2 to z_1+x_1 or to z_1+x_2, z_4+x_5 and z_4+x_6 with distance 1 to z_3+z_5 would coincide with z_3 or z_4+z_5, so that only one distance 1 from z_2+z_5 remains for the 2 distances 1 from z_4+x_5 and z_4+x_6 to S_1. If z_1+x_1 or z_1+x_2 has distance 1 to z_2+z_5, then it equals z_2 or z_1+z_5. Therefore either z_1+x_1 or z_1+x_2 has distance 1 to z_3+z_5 or to z_4+z_5. Because of symmetry the same arguments yield, that either z_4+x_5 or z_4+x_6 has distance 1 to z_1+z_5, or to z_2+z_5. In contrary to the last two propositions it is easily seen, that either
$|z_3+z_5-z_1| = |z_3+z_4+z_5| > 2$ and $|z_4+z_5-z_1| = |2z_4+z_5|>2$, or
$|z_1+z_5-z_4| = |2z_1+z_5| > 2$ and $|z_2+z_5-z_4| = |z_1+z_2+z_5|>2$.

Case 6. Three pairs of the five points z_i have distance 1, such that two distances have a point in common, and one distance is separate. - Let $z_2 = z_1+z_3$ have distances 1 to z_1, and to z_3, and z_4 has distance 1 to z_5. Then 18 subcases have to be discussed to receive contradictions for at most 16 points. The sums z_1+z_2, and z_2+z_3 do not belong to S, since otherwise 5.2 can be used. All other not in the subcase mentioned sums of z_i belong to S. Always x_i are unit vectors, and z_1+x_1, z_3+x_2 are always points of S. Let S_1 denote the set of all points z_i+z_j, which must have distances 1 to points of S_2, which denotes the set of all points z_r+x_s.

The distance 1 between the sums of z_i is possible only from z_1+z_4 to z_1+z_5, and to z_2+z_4, from z_1+z_5 to z_2+z_5, from z_2+z_4 to z_2+z_5, and to z_3+z_4, from z_2+z_5 to z_3+z_5, and from z_3+z_4 to z_3+z_5. Distances 1 from z_4+z_5 to z_1+z_4, to z_1+z_5, to z_3+z_4, and to z_3+z_5 yield 3 unit triangles in line as in 5.2. From z_1+z_4 to z_3+z_4, and from z_1+z_5

to z_3+z_5 the distance $\sqrt{3}$ occurs. If z_2+z_4 to z_4+z_5, or z_2+z_5 to z_4+z_5 has distance 1, then z_5 or z_4 coincides with z_1 or z_3. For the remaining 6 pairs of sums distances 1 imply always coincidences with points z_i or z_i+z_j, for example, if z_1+z_4 has distance 1 to z_2+z_5, then z_1+z_4 coincides either with z_5, or with z_2+z_4.

6.1. $z_4+z_5 \notin S$. - There are 14 points. At least 6 distances 1 have to occur from S_2 to S_1. However, there are only 4 distances 1 from S_1 to S_2, and thus 14 points are impossible. Two additional points (only with distances 1 to S_1 or S_2) together with S_2 have at most 6 distances 1 to the remaining points, so that 7 distances 1 should exist between 4 points, and therefore also 16 points are impossible.

6.2. z_4+z_5, $z_1+z_4 \notin S$. - With z_1+x_3, and z_4+x_4 there are 15 points in S, and an additional point Q must exist (distances 1 only to S_1 or S_2). From S_1 exist 5 distances 1 to Q and S_2. At most 5 distances 1 occur within S_2. These distances form a unit rhomb with a unit diagonal. Then at most 2 distances 1 are possible from Q to S_2, and therefore at most 4 distances 1 are possible from Q and S_1 to S_2, in contradiction to 6 distances 1, which should come from S_2. If 4 distances 1 occur within S_2, then 8 distances exist to Q and S_1, and this implies 4 distances to S_1, and 4 to Q. The 4 distances in S_2 form a rhomb, or contain a triangle, so that 4 distances 1 from Q to S_2 are impossible. Fewer than 4 distances 1 within S_2 would imply at least 10 distances to Q and S_1, and 5 distances 1 from Q to S_2 are impossible.

6.3. z_4+z_5, $z_2+z_4 \notin S$. - With z_1+x_3, and z_2+x_4 there are 15 points, and again Q exists. The 5 points of S_1 have 7 distances 1 to Q and S_2. If within the 4 points of S_2 there occur 5,4, or 3 distances 1, then 6,8, or 10 distances have to go from S_2 to Q and S_1, which forces, that 2,3, or 4 distances 1 exist from Q to S_2. These distances in all cases are impossible, or 4 unit triangles occur as in 5.2.

6.4. z_4+z_5, z_1+z_4, $z_2+z_4 \notin S$. - With z_1+x_3, z_2+x_4, z_4+x_5, and z_4+x_6 there are 16 points. From S_1 exist 6 distances 1 to S_2, so that within the 6 points of S_2 exactly 9 distances 1 must occur. As in 5.3 it is checked from [3, p. 222], that only Figure 1 remains for S_2, so that each point of S_2 has exactly one distance 1 to S_1. However, neither z_4+z_5 nor z_4+x_6 has distance 1 to z_1+z_5, to z_3+z_5, or to z_2+z_5, since otherwise coincidence with z_5 or z_1+z_4, z_5 or z_3+z_4, or z_5 or z_2+z_4 is induced. Thus z_4+x_5 and z_4+x_6 both have distance 1 to z_3+z_4, and then one of them coincides with z_2+z_4.

6.5. z_4+z_5, z_1+z_4, $z_3+z_4 \notin S$. - With z_1+x_3, z_3+x_4, z_4+x_5, z_4+x_6 there are 16 points. If z_2+z_4 has distance 1 to z_1+x_1 or z_1+x_3, then these coincide with z_2 or z_1+z_4. If z_2+z_4 has distance 1 to z_3+x_2 or z_3+x_4, then these coincide with z_2 or z_3+z_4. Thus both distances 1 from z_2+z_4 have to go to z_4+x_5, and to z_4+x_6, however, then one of these points coincides with z_3+z_4.

6.6. z_4+z_5, z_1+z_4, $z_1+z_5 \notin S$. - With z_1+x_3, z_1+x_4, z_4+x_5, z_5+x_6 there are 16 points. If z_2+z_4 has distance 1 to z_1+x_1, z_1+x_3, or z_1+x_4, to z_3+x_2, to z_4+x_5, or to z_5+x_6, then z_1+x_1, z_1+x_3, or z_1+x_4 coincides with z_2 or

z_1+z_4, z_3+x_2 with z_2 or z_3+z_4, z_4+x_5 with z_1+z_4 or z_3+z_4, or z_5+x_6 with z_4 or z_2+z_5. However, one distance 1 should exist from z_2+z_4 to S_2.

6.8. z_4+z_5, z_1+z_4, $z_3+z_5 \notin S$. - With z_1+x_3, z_3+x_4, z_4+x_5, z_5+x_6 there are 16 points. As in 6.6 no distance 1 occurs from z_2+z_4 to S_2, whereas one distance 1 should exist.

6.9. z_4+z_5, z_2+z_4, $z_2+z_5 \notin S$. - With z_2+x_3, z_2+x_4, z_4+x_5, z_5+x_6 there are 16 points. Neither z_2+x_3, nor z_2+x_4 has distance 1 to z_1+z_4, to z_1+z_5, to z_3+z_4, or to z_3+z_5, since otherwise z_2+x_3 or z_2+x_4 coincides with z_1 or z_2+z_4, with z_1 or z_2+z_5, with z_3 or z_2+z_4, or with z_3 or z_2+z_5. If z_4+x_5 has distance 1 to z_1+z_5, or to z_3+z_5, then z_4+x_5 coincides with z_5 or z_1+z_4, or with z_5 or z_3+z_4. If z_5+x_6 has distance 1 to z_1+z_4, or to z_3+z_4, then z_5+x_6 coincides with z_4 or z_1+z_5, or with z_4 or z_3+z_5. No point of S_2 has distance 1 to both points z_1+z_4 and z_3+z_4, or to both points z_1+z_5 and z_3+z_5, since otherwise this point coincides with z_4 or $z_1+z_3+z_4 = z_2+z_4$, or with z_5 or $z_1+z_3+z_5 = z_2+z_5$. Thus at most 6 distances 1 can go from S_2 to S_1, whereas 8 distances remain from S_1 to S_2.

6.10. z_1+z_4, $z_1+z_5 \notin S$. - With z_1+x_3, z_1+x_4 there are 15 points, and Q is an additional point. Since there are 7 distances 1 from S_1 to Q and S_2, and S_2 has 4 points, the same situation as in 6.3 is given.

6.11. z_1+z_4, $z_2+z_5 \notin S$. - With z_1+x_3, z_2+x_4, and an additional point Q there are 16 points. No distance 1 can occur from z_2+x_4 to z_1+z_5, or to z_3+z_5, since otherwise z_2+x_4 coincides with z_1 or z_2+z_5, or with z_3 or z_2+z_5.

From the 4 points Q, z_3+x_2, z_1+x_1, z_1+x_3 only one distance 1 is possible to z_1+z_5, or to z_3+z_5, since otherwise they coincide with z_5 or $z_1+z_3+z_5 = z_2+z_5$. However, from z_1+z_5 and z_3+z_5 should $3+2 = 5$ distances 1 exist to S_2.

6.12. $z_1+z_4, z_3+z_5 \notin S$. - With z_1+x_3, z_3+x_4, and an additional point Q there are 16 points. Neither z_2+z_4 nor z_2+z_5 has distances 1 to S_2, since otherwise coincidences with z_2, z_1+z_5, z_2+z_5, z_3+z_4, or z_3+z_5 occur. Thus Q has distances 1 to z_2+z_4, and to z_2+z_5, and $Q = z_2+z_4+z_5$ follows. Then $|Q-(z_1+z_5)| = |z_2-z_1+z_4| = |z_3+z_4| \neq 1$, and $|Q-(z_3+z_4)| = |z_2-z_3+z_5| = |z_1+z_5| \neq 1$, if coincidences are avoided. Therefore 8 distances 1 have to occur from S_1 and Q to S_2, 2 distances from each of the points Q, z_1+z_5, z_3+z_4, z_4+z_5. Then 2 points of S_2 must be $z_1+z_4+z_5$ and $z_3+z_4+z_5$ if coincidences are avoided. These 2 points then each have 3 distances 1 to Q and S_1, and thus there is only one distance 1 within S_2, which is impossible, since 4 distances 1 occur within the 4 points of S_2.

6.13. $z_2+z_4, z_2+z_5 \notin S$. - With z_2+x_3, z_2+x_4, and Q there are 16 points. Let S_3 be S_1 without z_4+z_5. No distance 1 is possible from z_2+x_3, and z_2+x_4 to S_3, if coincidences are avoided. The points Q, z_1+x_1, z_3+x_2 each have at most 2 distances 1 to S_3, since distances 1 to z_1+z_4, and z_3+z_4, or to z_1+z_5, and z_3+z_5 yield coincidences with z_4 or z_2+z_4, or with z_5 or z_2+z_5. Thus at most 6 distances 1 can go from Q and S_2 to S_3, whereas 8 should exist.

6.14. $z_1+z_4, z_2+z_4, z_1+z_5 \notin S$. - With z_1+x_3, z_1+x_4, z_2+x_5, z_4+x_6 there are 16 points. Because of coincidences at most one distance 1 is possible from z_2+z_5 to S_2, namely to z_2+x_5, however, 2 distances 1 should exist.

6.15. z_1+z_4, z_2+z_4, $z_2+z_5 \notin S$. - With z_1+x_3, z_2+x_4, z_2+x_5, z_4+x_6 there are 16 points. Because of coincidences no distances 1 exist from z_1+z_5 to z_2+x_4, z_2+x_5, z_4+x_6. At most one distance 1 can occur from z_1+z_5 to z_1+x_1 or z_1+x_3, since otherwise there is coincidence with z_1+z_4. So at most 2 distances 1 can occur from z_1+z_5 to S_2, whereas 3 should exist.

6.16. z_1+z_4, z_2+z_4, $z_3+z_5 \notin S$. - With z_1+x_3, z_2+x_4, z_3+x_5, z_4+x_6 there are 16 points. Avoiding coincidences a distance 1 from z_2+z_5 is possible only to the point z_2+x_4, however, 2 distances 1 should exist from z_2+z_5 to S_2.

6.17. z_1+z_4, z_3+z_4, $z_1+z_5 \notin S$. - With z_1+x_3, z_1+x_4, z_3+x_5, z_4+x_6 there are 16 points. No distances 1 from z_2+z_5 to S_2 can occur, if coincidences are avoided, however, one distance should exist.

6.18. z_1+z_4, z_3+z_4, $z_2+z_5 \notin S$. - With z_1+x_3, z_2+x_4, z_3+x_5, z_4+x_6 there are 16 points. No distances 1 occur from z_2+z_4 to z_1+x_1, z_1+x_3, z_3+x_2, x_3+x_5, if coincidences are avoided, however, 3 distances 1 should exist from z_2+z_4 to S_2.

Now all cases of the assumption $p(5) \leq 16$ are discussed, and $p(5) = 18$ is proved. It remains open, whether $p(6)$ is smaller than 27.

4. VARIATIONS

At the end three variations of the above problem are mentioned.

I. Let $p_1(n)$ be the corresponding minimum number of points, if the distances 1 are not allowed to intersect one another, that means, if two distances 1 intersect, then only one of them counts. Then $p_1(n)$ does not exist for $n \geq 6$ by Euler's formula. Also $p_1(5)$ does not exist [1]. The exact values $p_1(1) = 2$, $p_1(2) = 3$, $p_1(3) = 8$ are known, and $p_1(4)$ is bounded by $p_1(4) \leq 52$ [4].

II. Let $p_2(n)$ be the corresponding minimum number, if moreover the distance 1 is the smallest distance in the point set. Then $p_2(1) = 2$, $p_2(2) = 3$, $p_2(3) = 16$, and $p_2(n)$ does not exist for $n \geq 4$ [5].

III. Nearly nothing is known for higher dimensions. Only for dimension 3, if $p_3(n)$ corresponds to $p_2(n)$ in the plane, then $p_3(1) = 2$, $p_3(2) = 3$, $p_3(3) = 4$, $p_3(4) = 6$, $p_3(5) = 12$, $p_3(6) \leq 240$, and $p_3(n)$ does not exist for $n \geq 10$ [2].

REFERENCES

[1] A. BLOKHUIS (unpublished).

[2] G. FEJES TÓTH and H. HARBORTH, *Kugelpackungen mit vorgegebenen Nachbarnzahlen*, Studia Sci. Math. Hungar. (to appear).

[3] F. HARARY, *Graph Theory*, Addison-Wesley, Reading 1969.

[4] H. HARBORTH, *Äquidistante, reguläre Punktmengen*, 2. Kolloquium Geometrie und Kombinatorik 1983, Technische Hochschule Karl-Marx-Stadt 1985, 81-86.

[5] F. ÖSTERREICHER and W. ROHM, *Die minimale 3-Nachbarn-packung kongruenter Kreise*. Math. Semesterberichte 30 (1983), 49-60.

H. HARBORTH
Bienroder Weg 47,
D-3300 BRAUNSCHWEIG
West Germany

COLLOQUIA MATHEMATICA SOCIETATIS JÁNOS BOLYAI
48. INTUITIVE GEOMETRY, SIÓFOK, 1985.

DIAMETERS OF INTEGRAL POINT SETS

H. HARBORTH - A. KEMNITZ

A point set in Euclidean d-space E^d is said to be rational (integral) if the points of the set have pairwise rational (integral) distances. Point sets in E^2 with cardinality n are also called n-gons. The maximum distance of two points of a point set is called diameter of the set.

Obviously, the questions for existence of rational and for existence of integral point sets are equivalent in the case of finite cardinality.

For any positive integer $n > d$ one can find integral point sets of cardinality n in E^d such that not all points are in the same hyperplane (Steiger [9]).

On the other hand, there does not exist in E^d an infinite set of non-collinear points with all pairwise distances integral (Anning, Erdős [1],[3]).

For $d = 2$ there are rational point sets which are dense in subsets of E^2, for example subsets of the unit circle [1],[8].

For any fixed n Müller [6] has constructed integral non-degenerate (no three points collinear) n-gons with all points also on a circle. Harborth [4], [5] proved that for any $n \geq 5$ one can find integral n-gons with no three points collinear and, moreover, no n-1 points on a circle. As a generalization, we may ask:

Do there exist n points in the plane with pairwise integral distances, no three points collinear, and no $n-\ell$ on a circle?

For $n \geq 6$ and $\ell \geq 2$ such point sets are not known.

At the end of the fifties, Schoenberg asked whether each polygon in the plane may be approximated arbitrary closely by a rational polygon. Besicovitch [2] proposed an affirmative answer. However, until now the proposition is only proved for triangles, quadrangles, and some special classes of n-gons for $n \geq 5$.

More generally, Ulam [10, p. 40] asked: Does there exist a dense subset of the plane such that all distances between its points are rational?

A positive answer to one of the above questions would obviously imply the existence of integral n-gons with the cited properties.

Here we will ask for the minimum diameters of such integral point sets. First we give some results in E^2.

The minimum diameter of an integral triangle is 1 (equilateral unit triangle). For any given k one can find $\frac{(k+1)^2-\varepsilon}{4}$ pairwise distinct integral triangles with diameter k with $\varepsilon = 0$ for k odd and $\varepsilon = 1$ for k even.

The minimum diameter of four noncollinear points with pairwise integral distances is 4. Non-degenerate integral quadrangles also have the diameter at least 4. The minimum diameter of integral quadrangles without three collinear points and without all points on a circle is 8 (see Fig. 1).

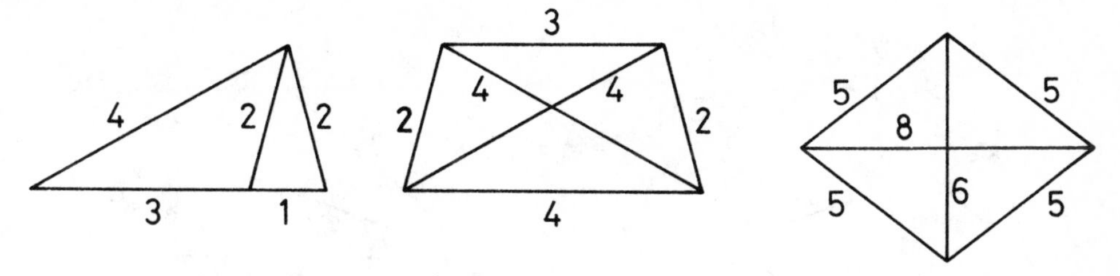

Figure 1
Integral quadrangles.

To determine systematically all integral quadrangles one can proceed as follows. The integral triangles $\Delta_1 = (k,a,b)$ and $\Delta_2 = (k,c,d)$ with the common diameter k can be combined in four different ways to a quadrangle (Fig. 2). Five of the six distances in the quadrangle obtained are integers: k,a,b,c,d. Writing

$$f_{\pm}(k,a,b,c,d) = a^2+d^2 - \frac{1}{2k^2}\Big\{(k^2+a^2-b^2)(k^2+d^2-c^2) \pm \\ \pm \sqrt{[(k^2-d^2)^2-2(k^2+d^2)c^2+c^4]\cdot[(k^2-a^2)^2-2(k^2+a^2)b^2+b^4]}\Big\},$$

the missing distance e is determined by

$$(1) \qquad e^2 = f_{\pm}(k,a,b,c,d)$$

and

$$(2) \qquad e^2 = f_{\pm}(k,a,b,d,c),$$

respectively. Here the plus and minus signs correspond to the cases when the two triangles lie on the same side and on different sides of the line of the common side, respectively. Every positive integer solution $e \leq k$ of (1) or (2) determines an integral quadrangle with diameter k.

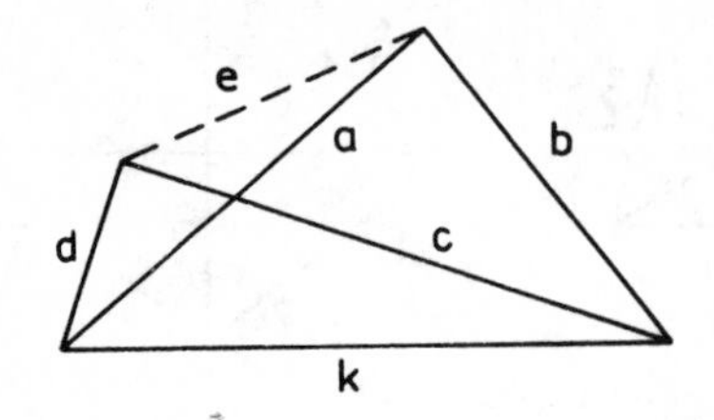

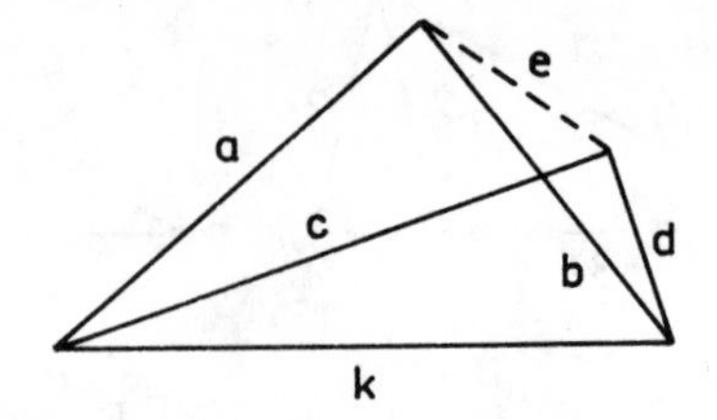

Figure 2

Possibilities of combining triangles $\Delta_1 = (k,a,b)$, $\Delta_2 = (k,c,d)$ to a quadrangle.

The number of possible combinations of triangles can be considerably decreased using the following observation. The area A_Δ of a non-degenerate rational triangle Δ can be written uniquely as $A_\Delta = q\sqrt{r}$ with q rational and $r = 1$ or squarefree integer. The integer r will be called characteristic of Δ. By using the cosine rule several times the following Lemma is easily proved.

LEMMA. *In any configuration of* n *points in the plane with pairwise rational distances all non-degenerate triangles have the same characteristic.*

Thus, it suffices to combine only triangles having the same characteristic. With the aid of the computer Amdahl 470 of the Technical University Braunschweig all possibilities of combining two integral triangles with the same characteristic and diameter not exceeding 160 have been checked. For example, there are 27, 258, 763, 1819, 3240, 5418, 8123, 11652 integral quadrangles for diameters not exceeding 20, 40, 60, 80, 100, 120, 140, 160, respectively, where four points on a circle are not counted.

Similarly, integral pentagons can be constructed from integral quadrangles of the same characteristic (the characteristic of each of its triangles) and of the same diameter. Using this systematic method we have got 7 as the minimum diameter for non-trivial integral pentagons (n-gons are called trivial if all points are collinear), 8 for non-degenerate integral pentagons, and 73 for integral pentagons without three points on a line and without four points on a circle (see Fig. 3).

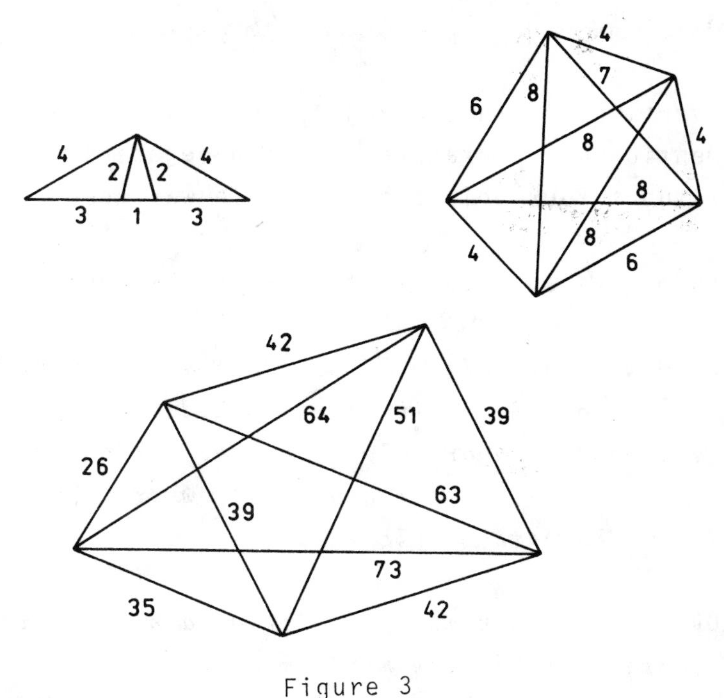

Figure 3
Integral pentagons.

Integral hexagons can be found in a corresponding way. The minimum diameter of non-degenerate hexagons is 8 (see Fig. 4).

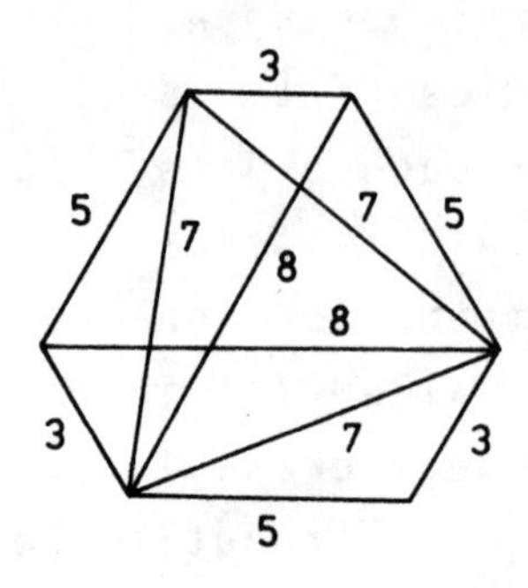

Figure 4

Minimal example of a non-degenerate integral hexagon.

The vertices of this hexagon are on a circle. As already mentioned, no example of a non-degenerate integral hexagon without four points on a circle was found till now.

We summarize our results for 4- and 5-gons:

THEOREM 1. *The minimum diameter of a non-trivial integral quadrangle is* 4, *the minimum diameter of a non-degenerate integral quadrangle is* 4, *the minimum diameter of a non-degenerate integral quadrangle where not all points lie on the same circle is* 8.

THEOREM 2. *The minimum diameter of a non-trivial integral pentagon is* 7, *the minimum diameter of a non-degenerate integral pentagon is* 8, *the minimum diameter of a non-degenerate integral pentagon without four points on a circle is* 73.

We now consider the existence of non-trivial n-gons.

THEOREM 3. *For the minimum diameter* $k(d,n)$ *of* n *points in* E^d $(n > d)$, *not all in the same hyperplane, and with pairwise integral distances it holds*

$$(3) \qquad k(d,n) \le \begin{cases} 2^{n-d+1} - 1, & \textit{if } n-d \equiv 0 (\bmod 2), \\ 3 \cdot 2^{n-d} - 3, & \textit{if } n-d \equiv 1 (\bmod 2), \end{cases}$$

and moreover for $d = 2$

$$(4) \qquad k(2,n) \le \begin{cases} 3 \cdot 2^{n-3} - 3, & \textit{if } n \equiv 0 (\bmod 2),\ n > 4, \\ 2^{n-2} - 2, & \textit{if } n \equiv 1 (\bmod 2),\ n > 5. \end{cases}$$

PROOF. For $t = 0,1,\ldots,m-1$ we consider $a = 2^{m+1}$, $b_t = 2^t(2^{2m-2t}-1)$, $c_t = 2^t(2^{2m-2t}+1)$ as sides of Pythagorean triangles (see [7, p. 30]). Thus

$$a^2 + b_t^2 = c_t^2 ,$$

$$b_0 > b_1 > \ldots > b_{m-1} , \text{ and } c_0 > c_1 > \ldots > c_{m-1} .$$

We choose $2m+1$ points $P_0, P_1, \ldots, P_{2m}$ on the x_1-axis of E^d such that P_{2m} is the origin, P_t has x_1-coordinates b_t, and P_{m+t} has x_1-coordinates $-b_t$ for $0 \le t < m$. In addition, we choose the vertices of a regular (d-1)-simplex in that hyperplane which is orthogonal to the x_1-axis such that its edge-length is a, and one vertex is P_{2m}. Then we have $n = 2m+d$ points with

$$k(d,n) = k(d,2m+d) \le \max(2b_0, c_0, a) = 2^{2m+1}-2 = 2^{n-d+1}-2 .$$

If we delete P_0 or P_m, we have $n = 2m+d-1$ points with

$$k(d,n) = k(d,2m+d-1) \le \max(b_0+b_1, c_0, a) = 3 \cdot 2^{2m-1}-3 =$$

$$= 3 \cdot 2^{n-d} - 3$$

for $m \geq 2$. Since for $m = 1$ the regular simplex with edge-length 1 yields $k(d,d+1) = 1$ the proof of (3) is complete.

For $d = 2$ we can choose two $(d-1)$-simplexes as above. Then the points $(0,a)$, $(0,-a)$, $P_0, P_1, \ldots, P_{2m}$ are $n = 2m+3$ points with

$$k(2,n) = k(2,2m+3) \leq \max(2b_0, c_0, 2a) = 2^{2m+1}-2 =$$
$$= 2^{n-2} - 2 \quad \text{for} \quad m \geq 2.$$

Without P_0 or P_m we have $n = 2m+2$ points with

$$k(2,n) = k(2,2m+2) \leq \max(b_0+b_1, c_0, 2a) = 3\cdot 2^{2m-1}-3 =$$
$$= 3\cdot 2^{n-3} - 3 \quad \text{for} \quad m \geq 2,$$

and (4) is proved.

We can improve the bound in (3) for many special pairs (d,n). We only give three examples here.

(I) Consider the vertices of a right prism in E^d of altitude 4 with a regular $(d-1)$-simplex of edge-length 3 as its base. If we delete t vertices from the base $(0 \leq t < d)$, then not all vertices lie in a hyperplane, no three vertices are collinear, and the diameter is 5, so that

$$k(d,n) \leq 5 \quad \text{for} \quad d < n \leq 2d.$$

(II) The vertices of two right prisms in E^d of altitude 11 and 80 with a common regular $(d-1)$-simplex of edge-length 60 as base have the distances 11, 60, 61, 80, 91, 100, and 109. Thus

$$k(d,n) \leq 109 \quad \text{for} \quad 2d < n \leq 3d,$$

if we delete t vertices from the base $(0 \leq t < d)$.

(III) Better than (I) is the following exact value $k(3,5) = 3$.

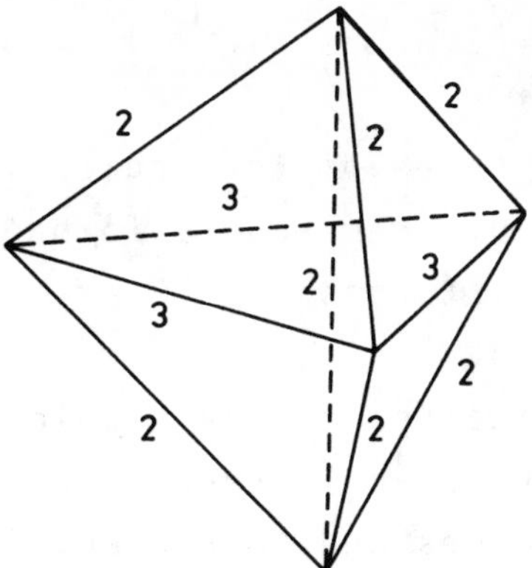

Figure 5

Example which proves $k(3,5) \leq 3$.

Fig. 5. proves $k(3,5) \leq 3$. We assume $k(3,5) \leq 2$. At least one pair A,B of points has distance 1 since otherwise five points in E^3 would have pairwise the same distance 2. For any further point both distances to A and to B are equal, either 1 or 2. Thus the three remaining points lie in a plane orthogonal to AB on two concentric circles C_1 of radius $h_1 = \frac{1}{2}\sqrt{3}$, and C_2 of radius $h_2 = \frac{1}{2}\sqrt{15}$. Two points on C_1 must have distance 1, and they determine an angle α with

$$70^0 < \alpha = 2 \operatorname{arc} \sin \frac{1}{3}\sqrt{3} < 71^0.$$

Two points on different circles must have distance 2, and they determine an angle β with

$$81^0 < \beta = \operatorname{arc} \cos \frac{1}{15}\sqrt{5} < 82^0.$$

Two points on C_2 with distance 1 and 2 determine γ and δ with

$$29^{\circ} < \gamma = 2 \arcsin \frac{1}{15} \sqrt{15} < 30^{\circ},$$

$$62^{\circ} < \delta = 2 \arcsin \frac{2}{15} \sqrt{15} < 63^{\circ}.$$

Neither the sum of three of the angles $\alpha, \beta, \gamma, \delta$ gives 2π, nor the sum of two of them equals to one of them. Thus three points with distances 1 or 2 are impossible, and $k(3,5) \geq 3$ is proved.

Another modification of the problems so far considered is the question for non-collinear point sets with pair-wise integral distances and, moreover, integral content of the convex hull of the set. In the plane this is a generalization of the question for integral Heron triangles.

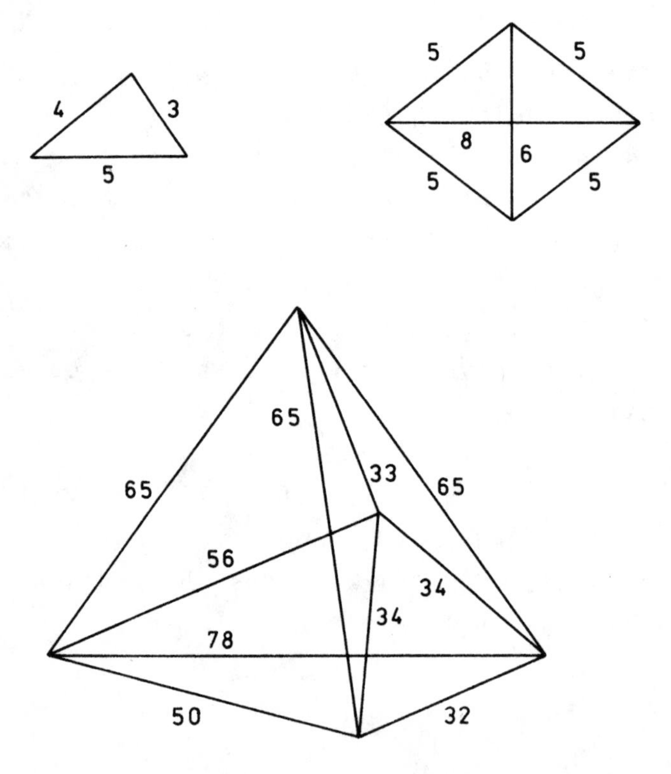

Figure 6

Examples which illustrate $k_3 \leq 5$, $k_4 \leq 8$, $k_5 \leq 78$.

For example, let k_n be the minimum diameter of an n-element point set in the plane, no three points collinear, no four points on a circle, all distances integral, and the area of the n-gon also integral. Then we have got as first values $k_3 = 5$, $k_4 = 8$ and $k_5 = 78$ (see Fig. 6).

ADDED IN PROOF: In the meantime we have determined 174 to be the minimum diameter of a non-degenerate integral hexagon without four points on a circle. Recently John Leech informed us, that our minimum example (see Figure 7) was already known to him to be an integral point set, and this was cited as Figure 3 in J. Lagrange, Points du plan dont les distances mutuelles sont rationnelles, Séminiaire de théorie des nombres, 1982-1983, Université de Bordeaux I, Talence, 1983.

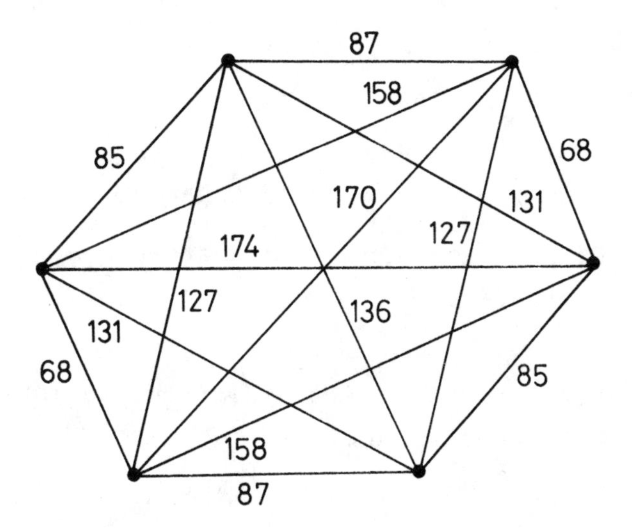

Figure 7

Six points, no three in line, no four on a circle, and all distances integral.

REFERENCES

[1] N.H. ANNING and P. ERDŐS, *Integral distances*. Bull. Amer. Math. Soc. 51(1945), 598-600.

[2] A. S. BESICOVITCH, *Rational polygons*, Mathematika 6(1959), 98.

[3] P. ERDÖS, *Integral distances*. Bull. Amer. Math. Soc. 51(1945), 996.

[4] H. HARBORTH, *On the problem of P. Erdős concerning points with integral distances*, Annals New York Acad. Sciences 175(1970), 206-207.

[5] H. HARBORTH, *Antwort auf eine Frage von P. Erdős nach fünf Punkten mit ganzzahligen Abständen*. Elem. Math. 26(1971), 112-113.

[6] A. MÜLLER, *Auf einem Kreis liegende Punktmengen ganzzahliger Entfernungen*, Elem. Math. 8(1953), 37-38.

[7] W. SIERPINSKI, *Pythagorean triangles*, New York 1962.

[8] J. C. SMITH, *A dense subset of the unit circle*, Problem E 2697, Amer. Math. Monthly 86(1979), 225.

[9] F. STEIGER, *Zu einer Frage über Mengen von Punkten mit ganzzahliger Entfernung*, Elem. Math. 8(1953), 66-67.

[10] S.M. ULAM, *A collection of mathematical problems*. New York, London 1960.

H. HARBORTH
Bienroder Weg 47, D-3300 Braunschweig, West Germany

A. KEMNITZ
Wümmeweg 10, D-3300 Braunschweig, West Germany

COLLOQUIA MATHEMATICA SOCIETATIS JÁNOS BOLYAI
48. INTUITIVE GEOMETRY, SIÓFOK, 1985.

A GENERALIZATION OF THE CONES OF ROTATION

W. JANK

In the threedimensional euclidean space R^3 any cone of rotation Δ with the vertex S has the following property: Let $\mathfrak{v}$ be an isometric mapping of Δ into a plane α. Then there exists a curve $k \not\ni S$ on Δ (namely a circle), which is mapped into a similar curve $\bar{k}$. For the purpose of finding cones in R^3 with an analogous property we formulate more percisely: For any cone of rotation $\Delta \subset R^3$ with the vertex S the following three theorems hold:

(1) For any isometric mapping $\mathfrak{v}:\Delta \to \alpha$ of Δ into a plane α there exists a similarity of a plane $\varepsilon \not\ni S$ onto α, so that $k = \varepsilon \cap \Delta$ is mapped onto $k^V = \mathfrak{v} \circ k \subset \bar{k} = \mathfrak{u} \circ k$.

(2) $S^V = \overline{S^n} = \mathfrak{u} \circ S^n$, where S^n means the orthogonal projection of S into ε.

(3) For any point $A \in k$ and its image point $\mathfrak{v} \circ A = A^V$ in the mapping $\mathfrak{v}$ there exists a similarity $\mathfrak{u}$ so that $\mathfrak{u} \circ A = \bar{A} = A^V$ holds.

(1), (2), (3) together characterize cones of rotation.

PROOF. From $S \notin \varepsilon$ follows $A \neq S$. Since $\mathfrak{v}$ is an injective mapping, $\bar{A} = A^V \neq S^V = \overline{S^n}$ holds and because $\mathfrak{u}$ is injective we find $A \neq S^n$: Therefore the three connecting lines AS,

$\overline{\bar{A}S^n}$ and AS^n exist and since $\mathfrak{v}$ and $\mathfrak{u}$ are conformal we have $\varphi = \sphericalangle AS, k = \sphericalangle \bar{A}\overline{S^n}$, $\bar{\kappa} = \sphericalangle AS^n, k$. Because S^n is the orthogonal projection of S into ε we find $\varphi = \pi/2$. Therefore any cone with the properties (1), (2), (3) contains a plane curve k which also lies on a concentric sphere and must be a cone of rotation.

If we substitute "any" by "at least one" in (3), there exists an infinite set of further solutions, where k is a cycloid (epicycloid or hypocycloid) or a hypercycloid and which independently from the author also had been discovered by W. Fuhs. The "cone property" of all these curves had also been known by C. Juel [2, p. 68], but he didn't have a completeness proof. In the case $S^v \neq \mathfrak{u} \circ S^n$ no solution is known.

Suppose, that k is represented in the form

$$(4) \qquad k \ldots r = f(s) > 0$$

where r means the distance between the fixed point S^n and the running point $P \in k$ and s the signed arc length of k with the initial point A. (4) may be called a "pseudonatural equation" of k [1], S^n and A the "associated fundamental points".

From the right angled triangle SS^nP we obtain by the theorem of Pythagoras

$$(5) \qquad \begin{aligned} &k^v \ldots R^2 = f^2(s) + c^2, \\ &\text{where } R = \overline{SP} > 0,\ r = f(s) = \overline{S^nP} > 0,\ c = \overline{SS^n} > 0, \\ &\qquad\qquad s \in I, \end{aligned}$$

that is a pseudonatural equation of k^v with the associated fundamental points $S^v = \overline{S^n}$ respectively $A^v = \bar{A}$.

The similarity $\mathfrak{u}$ leads to

$$(6)\qquad \begin{aligned} s &= a\bar{s} \\ r &= a\bar{r} \end{aligned} \qquad 0 < a < 1$$

and by substituting (6) into (4) we obtain a pseudo-natural equation of $\bar{k}$

$$(7)\qquad \bar{k} \ldots a\bar{r} = f(a\bar{s})$$

with the associated fundamental points $S^{\vee} = \overline{S^{n}}$ respectively $A^{\vee} = \bar{A}$.

Because of $k^{\vee} \subset \bar{k}$ (5) and (7) describe the same mapping $s \mapsto R$ respectively $\bar{s} \mapsto \bar{r}$, if s and $\bar{s}$ are members of the range I of the mapping (5). Therefore we can identify the variables s, $\bar{s}$ respectively R, $\bar{r}$ in (5) and (7). Elimination of r leads to

$$(8)\qquad a^2(f^2(s) + c^2) = f^2(as)$$

or

$$(9)\qquad F(ax) = a^2(F(x) + c^2), \text{ where } f^2 =: F,\ s =: x,\quad 0 < a < 1,\ c > 0,$$

that is a functional equation for the requested function f in (4).

From (9) follows, that any point $(x_0 | y_0 = F(x_0))$, $x_0, y_0 \in \mathbb{R}$ and its image $(ax_0 | a^2(y_0 + c^2))$ by the affine transformation

$$(10)\qquad \mathfrak{u}^{*} \ldots \begin{aligned} x^{*} &= ax \\ y^{*} &= a^2(y + c^2) \end{aligned}$$

are incident with the same solution curve. Hence any solution curve is a W-curve of the continuous one-parameter group which contains $\mathfrak{u}^*$.

Now we show by induction on n and investigation of (9), that all points of the set

$$(11)\qquad \begin{aligned} x_n &= a^n x_0 \\ y_n &= a^{2n} y_0 + a^2 c^2 (a^{2n} - 1)/(a^2 - 1) \end{aligned} \qquad n = 1,2,3,\ldots$$

lie on the same solution curve. Substitution of the natural number n by an arbitrary real number t leads to a parameter representation of all solution curves. We suppose $x_0 \neq 0$ ($x_0 \equiv 0$ yields no solution) and obtain the following pseudonatural equation of the solution curves

$$(12)\qquad \begin{aligned} &k \ldots r^2 = us^2 + bc^2(vs^2 - 1), \\ &\text{where } u = y_0/x_0^2 \in \mathbb{R},\ v = 1/x_0^2 \in \mathbb{R}^+, \\ &\qquad b = a^2/(a^2 - 1) \in \mathbb{R}^- . \end{aligned}$$

Hence k is symmetric with respect to the axis S^nA and

$$(13)\qquad \overline{S^nA}^2 = r^2(0) = -bc^2 .$$

From (12) and a formula of E. Cesàro [1]

$$(14)\qquad \rho^2 = r^2(1 - \dot{r}^2)/(1 - \dot{r}^2 - r\ddot{r})^2$$

we obtain the natural equation $\rho = g(s)$ of all solution curves

$$(15)\qquad \begin{aligned} &k \ldots s^2 d(d-1) + \rho^2 (d-1)^2 = -bc^2 > 0, \\ &\text{where } d = u + bvc^2,\ b \in \mathbb{R}^-,\ c \in \mathbb{R}^+ . \end{aligned}$$

$d = 1$ yields the contradiction $0 > 0$. In the case $d = 0$ we obtain circles k with the center S^n (13) and the radius $\rho = \ -bc^2$, which yield the expected cones of rotation Δ.

In the further cases the graph of (15), called MANNHEIM-curve of k, is a central conic and we obtain the following solution curves k [3]

$$
\begin{array}{rcl}
d < 0 & \Longleftrightarrow & \text{epicycloid } k \\
0 < d < 1 & \Longleftrightarrow & \text{hypercycloid } k \\
d > 1 & \Longleftrightarrow & \text{hypocycloid } k
\end{array}
\tag{16}
$$

Because of $d \neq d-1$ we obtain no common cycloid and because of $-bc^2 > 0$ no paracycloid [3].

(13) shows, that S^n is the focus of the solution curve k in any case. Because of (15) the height $c = \overline{SS^n}$ of any cone Δ is independent of the base k. Therefore one can demonstrate (in continuous manner) portions of infinitely many cones (with similar bases k) by using only one model. W. Wunderlich mentioned, that any solution curve k is a loxodrome with respect to two cones with imaginary vertices in SS^n (compare [4]). Further properties and a generalization of the cones above will be published in a following paper.

REFERENCES

[1] CESÀRO, E., *Vorlesungen über natürliche Geometrie.* Teubner, Leipzig (1926), 2. Aufl.

[2] FABRICIUS-BJERRE, Fr., *Über zykloidale Kurven in der Ebene und im Raum.* Dan. Mat. Fys. Medd. 26, No. 9 (1951).

[3] WIELEITNER, H., *Spezielle ebene Kurven*. Göschen, Leipzig (1908).

[4] WUNDERLICH, W., *Über die polykonischen Loxodromen*, Ann. di Math. 29, 177-186 (1949).

W. JANK
Institut für Geometrie
TU Wien
Wiedner Haupt str. 8-10.
A-1040. WIEN
ÖSTERREICH

COLLOQUIA MATHEMATICA SOCIETATIS JÁNOS BOLYAI
48. INTUITIVE GEOMETRY, SIÓFOK, 1985.

ON THE ISOPERIMETRIC MAPPING IN MINKOWSKI SPACES

K. JOHNSON, A.C. THOMPSON

Let X be a Minkowski space, i.e. a finite dimensional real normed linear space with unit ball B. It is, of course, possible to introduce a Euclidean metric on X defined by an ellipsoid E. The Euclidean structure, in turn, induces n-dimensional Lebesgue measure (which we shall refer to as Euclidean volume) on the measurable subsets of X and (n-1)-dimensional Lebesgue measure (Euclidean area) on the measurable subsets in each hyperplane in X. Following [8], we shall use $\varepsilon(\cdot)$ to denote both these measures and the Euclidean norm on X.

A *Minkowski* volume, i.e. a measure which is related to the metric induced by B, should be translation invariant (since all translations are isometries) and hence will be a multiple of Lebesgue measure. Thus, although we may wish to normalize the measure in some specific way related to B, n-dimensional content yields nothing essentially new. However, since (in general) there are few rotational isometries on X, the scalar multiple which converts Euclidean area to Minkowski area in a particular hyperplane H can depend on the linear functional which determines the family of hyperplanes parallel to H.

Let f be a Euclidean unit vector in the dual space X' and let

$$f^{\perp} = \{x \in X \mid f(x) = 0\} .$$

Let $\sigma(f)$ denote the ratio:

$$\sigma(f) = \text{Minkowski area in } f^{\perp}/\text{Euclidean area in } f^{\perp}.$$

Then, if S is a measurable subset of $H = \{x : f(x) = \alpha\}$ we have

$$\mu_B(S) = \int_S \sigma(f)d\varepsilon$$

where μ_B denotes the Minkowski measure derived from B (if no confusion is likely we write, simply, μ; as with ε we use the same notation for volume, area and length). This may be extended to smooth surfaces, Σ, ((n-1)-dimensional manifolds) by

$$\mu(\Sigma) = \int_{\Sigma} \sigma(f_x)d\Sigma_x$$

where f_x is the Euclidean unit outward normal at x and $d\Sigma_x$ is the Euclidean "area element" at x.

For this to be consistent with the structure of (X,B), the function σ cannot be entirely arbitrary. This has been discussed by H. Busemann and others [3], [4] and [8]. In the work of Minkowski and Busemann σ is defined by

$$\sigma^B(f) = \varepsilon_{n-1}/\varepsilon(f \cap B) .$$

In [8] it was shown that if, dually, $\sigma(f)$ is defined to be the (suitably normalized) "brightness" of the dual ball B^o in the direction f (see Firey [6]), i.e.

$$(*) \qquad \sigma_B(f) = \varepsilon_{n-1}^{-1}\ \varepsilon(P_f(B^o)) = \frac{1}{2}\ \varepsilon_{n-1}^{-1} \int_{\partial B^o} |f(x_g)|\, d\Sigma_g$$

then σ_B does indeed have all the necessary properties together with the additional pleasing property that $\mu_B(\partial B) = \mu_{B^o}(\partial B^o)$. In equation (*), $\varepsilon_{n-1} = \varepsilon(E_{n-1})$ is the (n-1)-dimensional Euclidean content of the (n-1)-dimensional Euclidean ball; P_f is the orthogonal projection along f, x_g is the unit outward normal to B^o and $d\Sigma_g$ is the area element at the variable point g of B^o.

In [3] it was shown that the function σ is the essential ingredient in the construction of the solution I(B) to the isoperimetric problem: that (unique up to homothety) convex set whose volume is maximal for a fixed Minkowski surface area. The relation between σ and I(B) is, in fact, very direct. Busemann showed that if σ is convex then I(B) is that convex body whose support function is σ. (Note that this means a normalization of I(B) too, which will be adopted in the sequel). The function σ_B thus leads to a more natural solution to the isoperimetric problem than does σ^B; it is immediate that, with σ_B, I(B) is the projection body of B^o (see Petty [9]).

From now on in this paper σ will indicate σ_B defined by (*) and we shall study further properties of the mapping I on the class of centrally symmetric convex sets. In particular we shall obtain solution to some of the questions raised in section 7 of [9].

The paper is organized as follows. In the first section we relate I to the work of E. Bolker [2] and W. Firey [6] on zonoids and projection bodies. In section 2 we consider some of the mapping properties of I, showing that it is injective and describing its image. In section 3 we restrict our attention to the special case in which B is a polytope, and establish results which we use in section 4 to show that I has no fixed points in this case. This is in contrast to the projection body map which Weil [12] showed to have numerous fixed points. Section 5 is devoted to studying how cartesian products and Blaschke sums behave with respect to I.

1. I(B) AS THE RANGE OF A VECTOR MEASURE

In this section we review part of the survey paper by Bolker [2] and place that material in the context of the mapping I.

PROPOSITION 1.1. (Bolker) *There is a vector measure* $\underline{\lambda}$ *on the Borel subsets of* ∂E *such that*

$$\sigma(f) = \sup\{f(y) : y = \underline{\lambda}(S),\ S \subset \partial E\}.$$

PROOF. In the case when B^0 is smooth and strictly convex the Gauss map $(g \to x_g)$ is a one-one mapping of ∂B^0 onto ∂E so that we may transfer the integral defining σ to ∂E, thus:

$$\sigma(f) = \frac{1}{2}\, \varepsilon_{n-1}^{-1} \int_{\partial E} |f(x)|\, d\lambda(x)$$

where λ is the measure induced on ∂E by the inverse Gauss map. Now $f(x) \geq 0$ on precisely one hemisphere H_f^+ of ∂E and, since B^0 is centrally symmetric, we have

$$\sigma(f) = \varepsilon_{n-1}^{-1} \int_{H_f^+} f(x)\, d\lambda(x).$$

Now define λ_f on the Borel subsets S of ∂E by

$$\lambda_f(S) = \varepsilon_{n-1}^{-1} \int_S f(x) d\lambda(x) .$$

Since $\lambda_f(S)$ is linear in f there is a vector measure $\underline{\lambda}$ defined by

$$\underline{\lambda}(S) = \varepsilon_{n-1}^{-1} \int_S x d\lambda(x) .$$

It is now evident that $f(\underline{\lambda}(S)) \leq f(\underline{\lambda}(H_f^+)) = \sigma(f)$. Hence $\sigma(f) = \max\{f(y) : y = \underline{\lambda}(S), S \subset \partial E\}$, the maximum being attained when $S = H_f^+$.

In the case when B^0 is not smooth and strictly convex it is still possible to define $\underline{\lambda}$. For each $x \in \partial E$ let F_x be the 'face' of B^0 corresponding to x; i.e. $F_x = \{f \in B^0 : f(x) = \mu(x)\}$. Then for each Borel set $S \in \partial E$ let $\lambda(S) = \varepsilon(\cup F_x : x \in S)$. Then λ is a measure on ∂E and

$$\sigma(f) = \frac{1}{2} \varepsilon_{n-1}^{-1} \int_{\partial E} |f(x)| d\lambda(x) .$$

The rest of the proof is as before.

We remark that the measure $\underline{\lambda}$ has an atom $\{x_0\}$ precisely when x_0 is a vertex of the ball B so that B^0 has a face F_{x_0} with positive area.

COROLLARY 1.2. *The function* σ *is the support function of the convex hull of the range of* $\underline{\lambda}$.

COROLLARY 1.3. I(B) *is the convex hull of the range of* $\underline{\lambda}$.

PROOFS. The first is a rewording of the Proposition. The second follows from the construction of I(B) in [3] and [8] as that convex body whose support function is σ.

REMARK. It is well-known that if $\underline{\lambda}$ is atomless then, by Lyapunov's theorem, the range of $\underline{\lambda}$ is convex. This, together with Proposition 1.1 gives a short (but less elementary) proof of the fact, proved rather laboriously in [8], that σ is convex.

2. ELEMENTARY MAPPING PROPERTIES OF I.

Let $\mathscr{C}$ ($\mathscr{C}_n$ if we need to specify the dimension) denote the set of all closed bounded convex bodies in X which are centrally symmetric about 0. Then I is a mapping on $\mathscr{C}$ which we shall call the *Isoperimetric mapping*. A *zonoid* may be defined either as the convex hull of the range of a vector measure (Bolker [2]) or as a limit, in the sense of the Hausdorff metric, of a sequence of *zonotopes* (i.e. polytopes all of whose faces are centrally symmetric). For the terminology and basic properties relating to zonotopes and, especially, zonohedra, we refer to Coxeter [5]; the survey article of Schneider and Weil [10] is also most useful.

PROPOSITION 2.1. *The mapping* I *is one-one and its range is the set of all zonoids in* $\mathscr{C}$.

PROOF. The first assertion is Theorem 2 of [6] while the second follows from Theorem 7.1 of [2].

REMARK. Central symmetry is essential in this proposition. To see this, let B be the cartesian product of a Reuleaux triangle (a plane region of constant width) of width

$$w = [2\pi/(\pi - \sqrt{3})]^{\frac{1}{2}}$$

with an interval of length w/2. A simple calculation,

using (*), shows that I(B) is the same as I applied to a right circular cylinder of unit height and radius (see [8], example 6.9).

DEFINITION 2.2. If $B_1, B_2 \in C$ let

$$d(B_1,B_2) = \inf\{\log\xi\eta : B_1 \subset \xi B_2,\ B_2 \subset \eta B_1\}.$$

It is easily seen (cf. [11]) that d is a semi-metric on $\mathscr{C}$ and that $d(B_1,B_2) = 0$ if $B_1 = \alpha B_2$. Thus, if we let $B_1 \sim B_2$ mean $B_1 = \alpha B_2$ for some α, then d is a metric on the equivalence classes of $\mathscr{C}$. If we restrict attention to a bounded subset of $\mathscr{C}$ having one representative from each equivalence class (e.g. $\mathscr{C}_0 = \{B \in \mathscr{C} : \varepsilon(B) = \alpha\}$ then d is equivalent to the Hausdorff metric.

PROPOSITION 2.3. *The mapping* I *on* $\mathscr{C}_n$ *has the following properties:*

(1) $I(\xi B) = \xi^{1-n}\, I(B)$;

(2) *if* $B_1 \subset B_2$ *then* $I(B_1) \supset I(B_2)$;

(3) $d(I(B_1), I(B_2)) \leq (n-1)d(B_1,B_2)$.

PROOF.

(1) Since $(\xi B)^o = \xi^{-1}B^o$ and area is homogeneous of degree (n-1) the result follows from the definition of I as the projection body of B^o.

(2) If $B_1 \subset B_2$ then $B_2^o \subset B_1^o$ and so $\sigma_2(f) \leq \sigma_1(f)$ from which we get $I(B_2) \subset I(B_1)$.

(3) If $d(B_1,B_2) = \log\delta$ $(\delta > 1)$ then we may assume (by considering suitable scalar multiples) that $B_1 \subset B_2 \subset$ $\subset \delta B_1$. Then

$$\delta^{1-n}\, I(B_1) = I(\delta B_1) \subset I(B_2) \subset I(B_1)$$

so that $d(I(B_1), I(B_2)) \leq \log(\delta^{n-1}) = (n-1)d(B_1,B_2)$.

REMARK. In R^3, the mapping I is not contractive with respect to d. If we take for B_1 the cylinder of height 2 over the unit disk and for B_2 the unit sphere, then $d(B_1,B_2) = \log_3 2$ whereas $d(I(B_1), I(B_2)) = \log_2 \frac{\pi}{2}$. However, in R^3, 2 does not seem to be the best Lipschitz constant. Furthermore, we have no counter example to the statement that I is contractive with respect to the Banach-Mazur distance ([1]) on $\mathcal{C}$. This metric, which identifies all ellipsoids, seems to be the 'right' one in this context.

COROLLARY 2.4. *The mapping* I *is continuous with respect to the Hausdorff metric on* $\mathcal{C}$.

It was pointed out in [8] that if B is smooth and strictly convex, then I(B) has these properties also. I does not preserve either of these properties separately however.

EXAMPLE 2.5. Let B be the convex hull in $\mathbb{R}^3$ of two unit spheres centered at (1,0,0) and (-1,0,0). B is clearly smooth but not strictly convex. A simple calculation shows that I(B) is not smooth at $\varepsilon_2^{-1}(\pm\pi,0,0)$.

Taking B' to be the polar dual of B yields an example of a strictly convex solid for which I(B) is not strictly convex. Directly from (*) we may compute that

$$\sigma_{B'}(f) = \varepsilon_2^{-1}(\pi + 4|\sin(\theta)|)$$

if f makes an angle θ with the x-axis. This function is not smooth at $\theta = 0$, which means that I(B'), whose support function this is, is not strictly convex.

It is also interesting to note that it is possible for I(B) to be smooth or strictly convex, even though B is neither.

3. I(B) AS A PROJECTION OF A CUBE

Throughout this section we assume that the unit ball B is a polytope. In this case the measure $\underline{\lambda}$ consists entirely of atoms. Let B^o have faces $F_1, F_2, \ldots, F_{2m}$ with unit outward normals $u_1, u_2, \ldots, u_{2m}$ respectively. These vectors are, of course, in the direction of the vertices of B. Let $\lambda_i = \varepsilon(F_i)$. Then $\underline{\lambda}(S) = \varepsilon_{n-1}^{-1} \Sigma \lambda_i u_i$ where the summation is taken over those i such that that $u_i \in S$. Moreover, if f is a linear functional such that $f^{\perp}$ contains no vertex of B, then $\underline{\lambda}(H_f^+) = \varepsilon_{n-1}^{-1} \Sigma \lambda_i u_i$ where the summation is over exactly half of the vertices of B (those with $f(u_i) > 0$). Now, as we saw above, f attains its supremum on I(B) at $\underline{\lambda}(H_f^+)$ and so do all those functionals sufficiently close to f. Thus $\underline{\lambda}(H_f^+)$ is a vertex of I(B).

We have, therefore, established the following proposition.

PROPOSITION 3.1. *If* B *is a polytope with normalized vertices* $\{u_i\}$ *then* I(B) *is a polytope whose vertices are given by* $v_k = \varepsilon_{n-1}^{-1} \Sigma \lambda_i u_i$ *where the summation is over those vertices contained in a half space and where* λ_i *is the Euclidean area of the corresponding face of* B^o.

If we make a predetermined choice $\{u_1, u_2, \ldots, u_m\}$ of one out of every pair of opposite vertices of B, then

$$v_k = \varepsilon_n^{-1} \sum_{i=1}^{m} (-1)^{k(i)} \lambda_i u_i$$

where the choice of signs $(-1)^{k(i)}$ is such that the set of vertices so chosen all lie in a half-space. Note that if we make an "illegitimate" choice of signs we still obtain a vector in the range of $\underline{\lambda}$ and hence a (non-extreme) point of I(B).

EXAMPLES. If B is the octahedron in R^3 with vertices $u_1 = (1,0,0)$, $u_2 = (0,1,0)$ and $u_3 = (0,0,1)$ then all 8 choices of sign are possible and I(B) is a cube.

If B is the cube in R^3 with vertices $u_1 = (1,1,1)$, $u_2 = (1,1,-1)$, $u_3 = (1,-1,1)$ and $u_4 = (-1,1,1)$ then only 14 choices of sign are possible (we exclude (-+++) and (+---) since the set $\{-u_1, u_2, u_3, u_4\}$ is not on one side of any hyperplane). It follows by calculation of the 14 sums that I(B) is a rhombic dodecahedron.

THEOREM 3.2. *If* B *is a polytope in* X *then* B *is an orthogonal projection of the unit octahedron in* R_m *for some* m *and* I(B) *is the image of the unit cube in* $\mathbb{R}^m$ *under the same projection.*

PROOF. Choose one from each pair of opposite vertices of B to form the set $\{v_1, v_2, \ldots, v_m\}$. These vertices can be used to define a bilinear form on X' which is in fact, an inner product. Let

$$\langle f,g\rangle_B = \sum_{i=1}^{m} f(v_i)g(v_i)$$

Then, with respect to this inner product, the star $\{v_1, v_2, \ldots, v_m\}$ is *eutactic* (see Coxeter [5], p. 251 and also [7]) and hence (Theorem 13.71 [5]) is the image under an orthogonal projection P of an orthonormal basis $\{b_1, b_2, \ldots, b_m\}$ in an m-dimensional inner product space. The usual identification with R^m concludes the first part of theorem.

Now, with the Euclidean structure ε derived from $\langle,\rangle_B$, let $\beta_i = \varepsilon_{n-1}^{-1} \lambda_i \varepsilon(v_i)^{-1}$ where λ_i is the area of the face of B^o corresponding to v_i. Then a vertex w_k of I(B) is of the form

$$w_k = \sum_{i=1}^{m} (\pm)\beta_i v_i = \sum_{i=1}^{m} (\pm)\beta_i P(b_i) .$$

Thus I(B) is the image under P of the rectangular parallelotope with vertices $\{\pm\beta_1 b_1, \pm\beta_2 b_2, \ldots, \pm\beta_m b_m\}$. If, now, we change the scale on each axis in R^m, I(B) is the image under P of the unit cube in R^m.

COROLLARY 3.3. *If* B *is a polytope in* R^n *then* I(B) *is a zonotope with edges parallel to the vertices of* B.

Again it follows from the work of Bolker [2] that every zonotope in R^n is the image under I of some polytope.

4. ITERATIONS AND FIXED POINTS OF THE MAPPING I

Since I is a mapping of $\mathcal{C}$ into $\mathcal{C}$ we may consider I^2 and so on. For ease of discussion we shall, in this section, suppose that X is 3-dimensional. Unfortunately, we have only been able to obtain results in the case that B is a polyhedron.

PROPOSITION 4.1. *If* B *is a polyhedron with* 2n *vertices, then* I(B) *has at least* 4(n-1) *vertices.*

PROOF. Each edge of I(B) is parallel to a vertex of B. Thus if a plane H (through the origin) contains $2n_1$ ($n_1 \geq 2$) vertices of B, then I(B) has two faces parallel to H (one on each side of H) each of which is a $2n_1$-gon. Let v and -v be two fixed vertices of B and consider a plane H through the origin containing these two vertices. If H is rotated about the line [-v,v] it will pass, in turn, through all the other vertices of B. Suppose there are r positions in which H contains other vertices and suppose that, in the ith position, the total number of vertices it contains (including v and -v) is $2n_i$. Then, on the one hand, we get

$$2n = 2 + \sum_{i=1}^{r} 2(n_i - 1); \text{ i.e. } (n-1) = \sum_{i=1}^{r} n_i - r .$$

On the other hand, to the ith position of H, there corresponds two faces of I(B) with $2n_i$ vertices each. The set of 2r such faces form a "zone" of I(B) (see Coxeter [5]) with each face joined to the next by an edge parallel to v. The total number of vertices in this zone is

$$\sum_{i=1}^{r} 4(n_i - 1) = 4(\sum_{i=1}^{r} n_i - r) = 4(n - 1) .$$

Hence the number of vertices of I(B) is at least $4(n - 1)$.

REMARK. We observe that in the case that B is a double pyramid over a $2(n - 1)$-gon then I(B) is a prism over a $2(n - 1)$-gon with exactly $4(n - 1)$ vertices. (In this case the zone of rectangular faces contains all the vertices of I(B).)

COROLLARY 4.2. *If* B *is a polyhedron then* $I(B) \neq B$.

COROLLARY 4.3. *If* B *is a polyhedron with* 2n *vertices then* $I^k(B)$ *has at least* $2^{k+1}(n-2) + 4$ *vertices.*

5. FUNCTIONAL PROPERTIES OF I

In this section we return to the general situation of an n-dimensional real normed space X. The definition of $\sigma((*))$ is (except for a scalar factor) the "brightness" function introduced by Firey [6]. In that paper, Firey also discussed the notion of the *Blaschke sum* $A \# B$ of two convex sets A and B.

PROPOSITION 5.1. *If* $A, B \in C$ *then* $I((A^o \# B^o)^o) =$ $= I(A) + I(B)$.

PROOF. Let $C = (A^o \# B^o)^o$, i.e. $C^o = A^o \# B^o$. The definition of $\#$ (see [6]) is precisely that the vector measure induced on E by C^o is the sum of the vector measures induced by A^o and B^o. Thus $I(C) = (A) + I(B)$.

DEFINITION 5.2. If $A \in \mathscr{C}_n$ $(A \subset X)$ and $B \in \mathscr{C}_m$ $(B \subset Y)$ define $A * B$ to be $A * B = \mathrm{co}\{A \times \{0\} \cup \{0\} \times B\} \subset X \times Y$.

PROPOSITION 5.3. *With the notation of* 5.2, *there are scalars* α *and* β *such that* $I(A * B) = \alpha(I(A)) \times \beta(I(B))$.

PROOF. We prove the result for polytopes; the general result follows by an approximation and continuity argument. Suppose A has vertices $\{\pm a_1, \pm a_2, \ldots, \pm a_k\}$ with the area of the face of A^o corresponding to a_i being λ_i. Then, as in the proof of Theorem 3.3, the star $\{\varepsilon_{n-1}^{-1} \lambda_i a_i\}$ is the image under an orthogonal projection P_1 of the unit vectors in R^k. Similarly, with an analogous notation, the star $\{\varepsilon_{m-1}^{-1} \mu_i b_i\}$ is the image under P_2 of the usual unit vectors in R^ℓ. Now $(A * B)^o = A^o \times B^o$. Moreover, the area of the face of $(A * B)^o$ corresponding to the vertex $a_i \times 0$ is $\lambda_i \varepsilon(B^o)$ and likewise for $0 \times b_i$. Now the star

$$\{\varepsilon_{n+m-1}^{-1} \lambda_i \varepsilon(B^o)(a_i \times 0),$$

$$\varepsilon_{n+m-1}^{-1} \mu_j \varepsilon(A^o)(0 \times b_j) \; i=1,2,\ldots,k, \; j=1,2,\ldots,\ell\}$$

is the image of the usual basis vectors in $R^{k+\ell}$ under the projection $P = \alpha P_1 \times \beta P_2$ where $\alpha = \varepsilon_{n-1}\varepsilon(B^o)\varepsilon_{n+m-1}^{-1}$ and $\beta = \varepsilon_{m-1}\varepsilon(A^o)\varepsilon_{n+m-1}^{-1}$ and where $(P_1 \times P_2)(a,b) = (P_1 a, P_2 b)$. Hence $I(A * B)$ is the image under P of the unit cube $C_{k+\ell}$ in $R^{k+\ell}$. But $C_{k+\ell}$ is the product of the cube C_k in R^k and the cube C_k in R^k and the cube C_ℓ in R^ℓ. Thus

$$I(A * B) = P(C_{k+\ell}) = P(C_k \times C_\ell) =$$

$$= \alpha P_1(C_k) \times \beta P_2(C_\ell) = \alpha I(A) \times \beta I(B) \; .$$

REFERENCES

[1] S. BANACH, *Opérations Linéaires* (Warsaw), 1932.

[2] E.D. BOLKER, *A class of convex bodies*. Trans. Amer. Math. Soc. 145 (1969) 323-345.

[3] H. BUSEMANN, *The isoperimetric problem for Minkowski area*, Amer. J. Math. 71 (1949) 743-762.

[4] H. BUSEMANN, *The foundations of Minkowski geometry*, Comm. Math. Helvetici 24 (1950) 156-187.

[5] H.S.M. COXETER, *Regular Polytopes* (London) 1948.

[6] W.J. FIREY, *The Brightness of convex bodies*, Technical Report 19, Dept. of Mathematics, Oregon State University, Corvallis, Oregon 1965.

[7] H. HADWIGER, *Über ausgezeichnete Vektorsterne und regulare Polytope*, Comm. Math. Helvetici 23 (1940) 90-107.

[8] R.D. HOLMES - A.C. THOMPSON, *n-dimensional area and content in Minkowski spaces*, Pac.J.Math. 85 (1979) 77-110.

[9] C.M. PETTY, *Projection bodies*, Proc. Colloq. Convexity Copenhagen 1967, 234-241.

[10] R. SCHNEIDER - W. WEIL, *Zonoids and related topics*, in Convexity and its applications, 296-317, Birkhäuser, Basel-Boston, 1983.

[11] A.C. THOMPSON, *On certain contraction mappings in a partially ordered vector space*, Proc. Amer. Math. Soc. 14 (1963) 438-443.

[12] W. WEIL, *Über die Projectionkörper konvexer Polytope*, Arch. Math. 22 (1971), 664-672.

K. JOHNSON - A.C. THOMPSON
Department of Mathematics, Statistics
and Computing Science
Dalhousie University
Halifax, Nova Scotia
Canada B3H 3T5

COLLOQUIA MATHEMATICA SOCIETATIS JÁNOS BOLYAI
48. INTUITIVE GEOMETRY, SIÓFOK, 1985.

ON PACKINGS OF CIRCULAR DISCS IN THE EUCLIDEAN PLANE WITH PRESCRIBED NEIGHBOURHOOD

E. JUCOVIČ - S. ŠEVEC

We consider packings of circular discs belonging to two disjoint finite non empty sets (white and black) in the euclidean plane. If in the packing P there are V white discs of radius 1, B black discs of radius $\rho > 0$ and such that every white disc has exactly v black neighbours, and every black disc has exactly b white neighbours, and no two discs of same colour are neighbours, then we call the sixtuple $((V,v,1)(B,b,\rho))$ realizable; the packing P itself is a realization of the sixtuple. (As usual, two discs are called neighbours if they have only boundary points in common.) Such a packing is in some sense a generalization of the k-neighbour packing of discs introduced in L. Fejes Tóth [1]. The obvious problem to be investigated is: Determine all realizable sixtuples of numbers. - A complete solution of this problem is not known so far. Partial results form the contents of this paper.

THEOREM 1. *A couple of numbers* (v,b), $v \geq b \geq 1$ *is contained in a realizable sixtuple* $((V,v,1)(B,b,\rho))$ *if and only if it is one of the following couples of integers*

(*a*) (v,q), v *arbitrary*

(*b*) $(v,2)$, $2 \leq v \leq 8$.

PROOF OF THEOREM 1.

1. We prove that $v \geq 2$ does belong to a realizable sixtuple $((V,v,1)(B,2,\rho))$ for no $v \geq 9$.

Assume the contrary, let P be a realization of $(V,v,1)(B,2,\rho))$, $v \geq 9$. We associate to P a graph $T(P) = T$ in the following way: Vertices of T are centers of the white circles of P. Two certices are joined by an edge if they are centers of two white circles touching the same black circle. So an edge of T is determined by one of two black circles and therefore the valencies of T lie in the closed interval $\langle \lceil \frac{v}{2} \rceil, v \rangle$. (For a real number a, $\lceil a \rceil$ denotes the smallest integer not less than a.)

As in Jucovič-Ševec [2], to T its block-cut vertex graph is associated and analogous considerations as in [2] lead to the existence of a circuit c in T bounding in the plane a region R which covers all vertices and edges of a block of T, i.e. also all edges incident to vertices of c with the possible exception of some edges incident with the cut-vertex C on c.

Now consider the inner angle a of the polygon c at a vertex A different from C. I affirm that $a \geq \pi$ under the assumption made (from which it follows that $\Sigma > (n-2)\pi$ for the sum Σ of the inner angles of c with n vertices). Indirectly, suppose $a < \pi$. In the angular region of a there are at least 7 black circles with radius ρ. Therefore,

$$\rho < \frac{\sin \pi/14}{1 - \sin \pi/14} \quad .$$

Let B and D be any two vertices of T adjacent to A and following each other when considering in cyclic order all vertices of T which are adjacent to the vertex A. The lengths of the segment AB and AD are greater than 2 and smaller that $2(1+\rho) < 2\frac{1}{1 - \sin \pi/14}$. As $\deg A \geq 5$ and $\sphericalangle BAD \leq \frac{\alpha}{4} < \frac{\pi}{4}$, there exist such B, D that $|BD| < 2$. P would not be a packing - and this is a contradiction.

2. We prove that $((V,v,1)(B,b,\rho))$ is not realizable if $v \geq 3$, $b = 3$. (Notice that not both numbers v and b can be greater than 3.) The proof is again performed so that if supposing the contrary, we get a g - gon in the euclidean plane with the sum of inner angles exceeding $(g-2)\pi$.

Associate to the packing P realizing $((V,v,1)(B,b,\rho))$, $v,b \geq 3$ a graph $G(P) = G$ as follows (cf. Linhart-Österreicher [3]): vertices of G are the centers of the circles, two vertices of G are joined by an edge if they are centers of neighbour circles, therefore their lengths are $1+\rho$. So G is a bipartite planar graph, all of its circuits have even lengths. As in the preceding part, there exists in G a circuit c of even length g such that the planar region R bounded by c covers all vertices and edges of a certain block K of G. Therefore, the number of edges decomposing the inner angle at any vertex of c is ≥ 1, the only exceptional situation can occur at a cut-vertex C. (This situation must be considered because it is not hard to construct for $b \in \{4,5\}$ a planar bipartite graph with valency 3 of vertices in one vertex part and b in the second one which does contain cut-vertices.) Let us consider the sum of the inner angles of c (see Fig. 1)

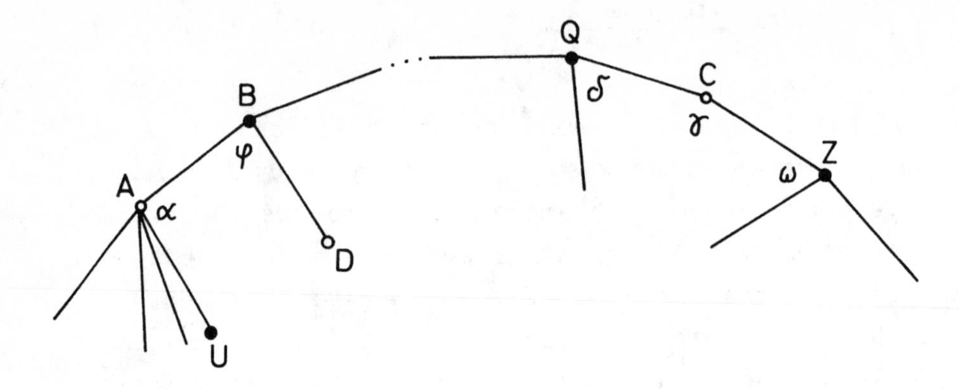

Figure 1

where, of course, the valency of A can be < 5, and C is supposed to be a cut-vertex. If A and B are neighbour vertices on c none of which is a cut-vertex of G, the sum $\alpha + \varphi \geq \pi$ because the vertices U and D do exist (both A and B have valencies ≥ 3!) and as they have different colours, they do not coincide and their mutual distance is $\geq 1 + \rho$. We have such edges on c at least $g - 2$. For the cut-vertex C consider the pairs Q, C and C, Z (Q, Z are vertices on c adjacent to C). If the valency of C in K is ≥ 3, the preceding argument works unchanged for the pairs Q, C and C, Z, too. Fig. 1 illustrates the opposite case; however, in this case holds $\gamma + \delta \geq \pi$, $\gamma + \delta + \omega > \pi$. In any case the sum of the inner angles of c would be more than $(g - 2)\pi + \pi = (g - 1)\pi$; this is a contradiction.

3. We have to show that the number couples mentioned in Theorem 1 do appear in realizable sixtuples of numbers. The situation is trivial in case (v,1). For (2,2) you can take the vertices of a regular 2k-gon with sides of length $1 + \rho$ as centers of the white as well as of the black circles. Fig. 2a or 2b make clear all couples of integers V,B contained in realizable sixtuples $((V,3,1)(B,2,\rho))$ or $((V,4,1)(B,2,\rho))$, namely $((2k,3,1)(3k,2,\rho))$ or $((k,4,1)(2k,2,\rho))$, respectively.

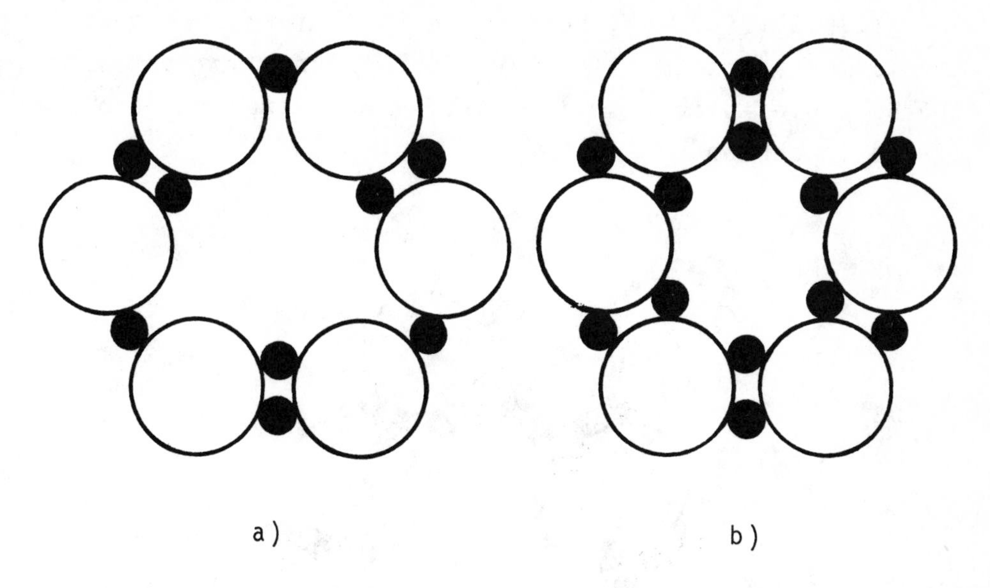

Figure 2

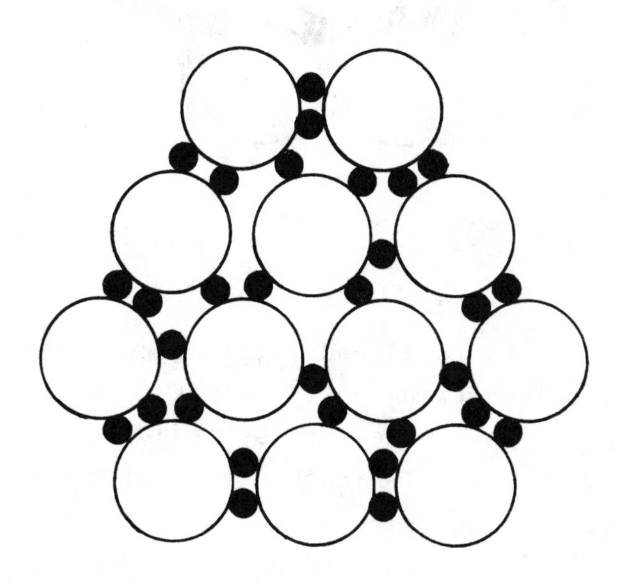

Figure 3

Fig. 3 or Fig. 4 is a realization of $((12,5,1)(30,2,\rho))$ or $((12,6,1)(36,2,\rho))$, respectively. Centers of the white circles are vertices of the regular triangular lattice, two suitable adjacent white circles have two or one black circle in common.

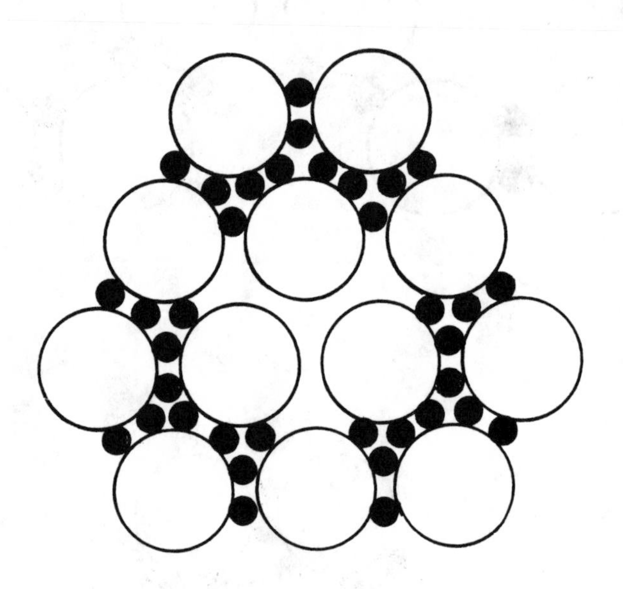

Figure 4

The couples (8,2) and (7,2): Take a part of the densest circle packing in the plane (see Fig. 5), decrease the radius of the white circles, add black circles so that every such circle touches two white ones and every white touches eight black circles; the straight lines m, n are parallel. The packing has many degrees of freedom (the radius of the black circles, the distances of the centers of the white ones etc.), therefore, the circles can be moved so that the lines m, n intersect in a point S such that SU and SU' are congruent segments and so are the segments SL and SL', and the angle USU' is an integer part of the full angle 2π. Therefore, the angular region determined by the lines m, n can be rotated so that a packing realizing some sixtuple $((V,8,1)(B,2,\rho))$ is reached. We get a realization of $((V,7,1)(B,2,\rho))$ if the angle USU' is an integer part of π and the rotation is performed with the angular region determined by the lines m and p as in Fig. 6. (Some of the black circles from the preceding case in Fig. 5 are removed.)

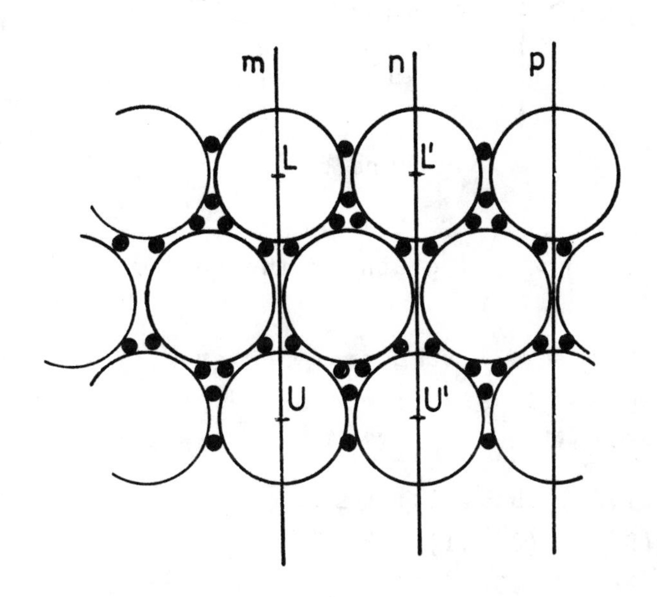

Figure 5

Our next theorem settles completely the problem posed for $\rho = 1$. To simplify the formulation, we will speak on minimally realizable sixtuples of numbers. This concept is introduced as follows: associate to a packing P of discs its graph G(P) as in part 2 of the proof of Theorem 1. A sixtuple $s = ((V,v,1)(B,b,\rho))$ is minimally realizable

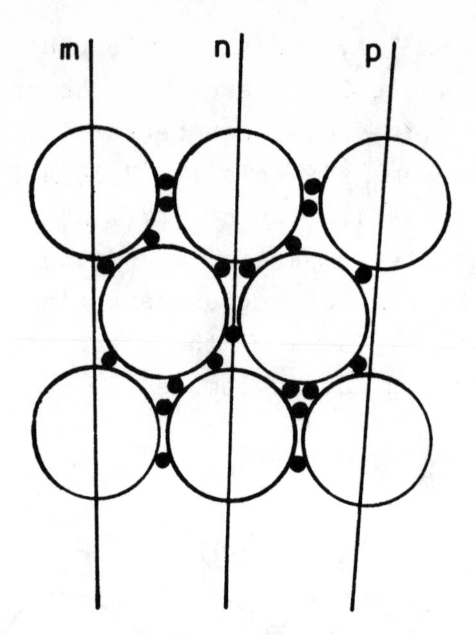

Figure 6

if G(P) is a connected graph of the realization P of s.

THEOREM 2. *The sixtuple of numbers* ((V,v,1)(B,b,1)) *is minimally realizable if and only if it is one of the following sixtuples of positive integers:*

(*a*) ((1,B,1)(B,1,1)), $B \leq 5$;
(*b*) ((B,2,1)(B,2,1)), $B \geq 2$;
(*c*) ((2k,3,1)(3k,2,1)), $k \geq 3$ *an integer;*
(*d*) ((V,4,1)(2V,2,1)), $V \geq 7$.

A complete proof of Theorem 2 is contained in Jucovič-Ševec [2] and we don't repeat it here. We remark only that in the proof of the negative part of the case (d), analogously it is proceeded as in part 2 of the proof of Theorem 1; of course, in details there are differences. New ideas are used for proving that

((4,3,1)(6,2,1)) is not realizable.

REMARK. Investigations described in this paper should be extended and generalized at least in the following directions:

a) Not to exclude neighbourhood of circles of same colour.

b) To increase the number of sets (colours) to which the circles belong.

c) Consider packings of circles with prescribed neighbourhoods on other surfaces of constant curvature.

d) Increasing the dimension of the space certainly requires new procedures.

REFERENCES

[1] L. FEJES TÓTH, *Scheibenpackungen konstanter Nachbarnzahl.* Acta Math. Acad. Sci. Hung. 20 (1969), 375-381.

[2] E. JUCOVIČ - S. ŠEVEC, *On generalized neighbour packings of domains in the euclidean plane.* Studia Sci. Math. Hung. (to appear).

[3] J. LINHART - F. ÖSTERREICHER, *Packungen kongruenter Stäbchen mit konstanter Nachbarnzahl.* Elemente d. Math. 37 (1982), 5-16.

E. Jucovič
Univerzita P.J. Šafárika
Jesenná 5
041 54 KOŠICE
Czechoslovakia

S. Ševec
Univerzita P. J. Šafárika
Jesenná 5
041 54 KOŠICE
Czechoslovakia

COLLOQUIA MATHEMATICA SOCIETATIS JÁNOS BOLYAI
48. INTUITIVE GEOMETRY, SIÓFOK, 1985.

DISCONTINUOUS GROUPS IN NORMED SPACES

B. KLOTZEK

First, I discussed some results about discontinuous groups in a normed plane in a paper which I had prepared for the 2nd Österreichisch-Ungarische Geometrie-Konferenz, Siófok 1983 (see [8]). Now, there have been other mathematicans whose research has dealt with this subject (see for instance [12] and [5]).

Here, we give results for dimension $n = 2,3$, for $2 \leq n < \infty$, and for an arbitrary metric space. The main question is the following: What conditions about the existence of discontinuous groups of isometries are sufficient that a normed space is Euclidean? On the other hand we suppose rosette groups $\mathscr{C}_p^{(1)}$ and $\mathscr{C}_q^{(2)}$ with respect to two norms, and study some assumptions about the transformation group generated by $\mathscr{C}_p^{(1)}$ and $\mathscr{C}_q^{(2)}$ since one can find this situation in some problems of physics.

Last, different conditions are used for the notion of a discrete group, for instance that the orbit of an arbitrary point is locally finite or that every point is isolated in its orbit. In the last section, we give a sufficient condition for the equivalence of the different definitions for discontinuous isometry groups in metric spaces, and also a corollary with respect to normed spaces.

This paper will be a contribution to intuitive geometry since the theory of normed spaces is a theory "nearest to the Euclidean geometry" in the sense of HILBERT's description (see [1]) and because of Theorem 1 each discontinuous motion group B of any finite dimensional normed space M^n is a discontinuous group of the real Euclidean space E^n in a suitable representation. Hence, we can use results and methods of the old fashionable Euclidean geometry.

1. BASIC NOTIONS

A *normed vector space* X is a vector space V with a function $||\cdot|| : V \to \mathbf{R}$, such that

$$\forall_{v\in V} \; ||v|| = o \Rightarrow v = o$$

$$\forall_{v\in V,\ \lambda\in\mathbf{R}} \; ||\lambda v|| = |\lambda| \; ||v||$$

$$\forall_{u,v\in V} \; ||u + v|| \leq ||u|| + ||v|| \; .$$

Now, we can define a *normed space* or *Minkowskian space* $\mathbf{M}^n$ as a tripel $(\mathbf{P},\mathbf{X},+)$ where $\mathbf{P}$ is a non-empty set of points, $\mathbf{X}$ a n-dimensional normed vector space, and

$$+ : \mathbf{P} \times \mathbf{X} \to \mathbf{P}$$

a function with the following properties:

(i) For every pair (P,Q) of points exists exactly one vector v in V such that P + v = Q.

(ii) The conditions (P + u) + v = P + (u + v) holds for all $P \in \mathbf{P}$ and $u,v \in V$.

Taken an origin O, the points of **P** can be described by position vectors in V. Then we speak about vectors p as points P as well. The latter are denoted by capitals.

The *distance* $\rho(P,Q)$ of points P,Q is defined by $\rho(P,Q) := \|q-p\|$. Thus, we can regard the unit sphere k with the centre O which is the *gauge figure* (*Eichfigur* by H. MINKOWSKI [11] - see also Fig. 1). We can also consider r-*neighbourhoods* $U_r(P)$ about the point P.
Any function $f:\mathbf{P} \to \mathbf{P}$ which preserves all distances is an *isometry*. The bijective isometries are called *motions* of the normed space.

There are different conditions to describe the notion of a *discontinuous isometry group* B: If P^B denotes the set of all points which are images P^τ of P under $\tau \in B$, i.e. the *orbit of the point* P *under the group* B, one uses

$$(1) \qquad \forall_{P,X,r>o} \; \mathrm{card}(U_r(X) \cap P^B) < \infty$$

i.e. each orbit is *locally finite*, or

$$(2) \qquad \forall_{P,X} \; \exists_{r>o} \; \mathrm{card}(U_r(X) \cap P^B) \leq 1 ,$$

or that each point of an orbit is *isolated*, i.e.

$$(3) \qquad \forall_{P} \; \exists_{r>o} \; \mathrm{card}\,(U_r(P) \cap P^B) = 1,$$

since

$$X \in P^B \Rightarrow X^B = P^B .$$

For instance, one can see the condition (1) by HILBERT and COHN-VOSSEN [6] and the condition (3) by L. FEJES TÓTH [4].

It is easy to see that

$$(1) \Rightarrow (2) \quad \text{and} \quad (2) \Rightarrow (3) .$$

On the other hand, if (3) holds for any group of isometries of a metric space no orbit under this group has a point of accumulation. But, if we have a normed space of finite dimension we have

$$(3) \Rightarrow (1) .$$

One can find more about this implication in the last part of this paper.

2. SOME IMPORTANT PROPOSITIONS

With a theorem of MAZUR and ULAM [10], each motion is affine.

The translations and the reflections in points are motions. If the dimension is finite each isometry is a motion.

Furthermore, each motion can be considered as a product of a motion for which a point 0 is fixed and of any translation. It is therefore important to describe the subgroup G_0 of such motions which have the fixed point O. These motions map the gauge figure k into itself. Moreover, if the space has a finite dimension n these motions also map the minimal ellipsoid e of LÖWNER into itself, i.e.

for each norm of a finite - dimensional space there is a unique Euclidean metric (with respect to e as gauge figure), and the motion group of this normed space is a subgroup of the Euclidean group of motions.

We ask now whether a group whose orbits are locally finite with respect to k has only locally finite orbits with respect to e and vice versa? We have to consider two alternatives for a neighbourhood

$$U_r(X) = \{Y \in R^n \mid d_e(X,Y) < r\}$$

$$V_s(X) = \{Y \in R^n \mid d_k(X,Y) < s\}$$

where d_e and d_k are distance functions of $\mathbf{E}^n$ and $\mathbf{M}^n$ defined by $\mathbf{R}^n$ and the unit sphere e and k, respectively. Since each neighbourhood $U_r(X)$ or $V_s(X)$ lies on a parallelepipedon we get

$$\forall_r \ \exists_s \ U_r(X) \subset V_s(X)$$

and

$$\forall_s \ \exists_r \ V_s(X) \subset U_r(X) \ .$$

We can now state that each motion group of a finite-dimensional Minkowskian space $\mathbf{M}^n$ is discontinuous according to HILBERT and COHN-VOSSEN iff it is discontinuous as a subgroup of the Euclidean motion group of E^n. As a result, we have

THEOREM 1. *Each discontinuous motion group* B *of any finite-dimensional normed space* $\mathbf{M}^n$ *is a discontinuous group of the real Euclidean space* $\mathbf{E}^n$ *in a suitable representation.*

The question is now whether all of the discontinuous groups of the real Euclidean space E^n of finite dimension will also be discontinuous groups of a normed space? We will see that certain minimal abundance of the motion group of a normed space is sufficient for its Euclidness.

3. THE DIMENSION 2

It is easy to see that we can realize all frize groups, containing only proper motions, in each normed plane. However, if the plane $\mathbf{M}^2$ is non-Euclidean we can not realize all rosette groups in it. More precisely, we have

THEOREM 2. (i) *In each normed plane there exist all frize groups which contain only translations and reflections in points.*

(ii) *In each non-Euclidean normed plane there is only a finite number of abstract rosette groups.*

COROLLARY. *We can not realize all ornamental groups in any normed plane.* (For details, see [5].)

CRITERION 1. *A normed plane is Euclidean iff one can realize an infinite set of the abstract proper rosette groups.*

These statements were presented at the Warsow University and at 2nd Österreichisch-Ungarische Geometrie-Konferenz in Siófok in 1983 (see [8]). The main idea of the proof of (ii) is that every rotation about O maps $e \cap k$ onto itself, and therefore, one can find a lower bound for angles which are possible for generating rotations (see Fig. 1).

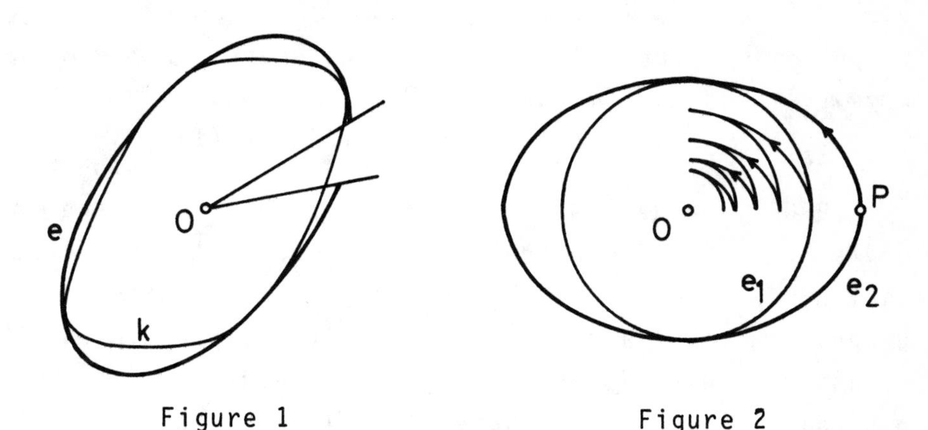

Figure 1 Figure 2

There is a lot of physical problems whose assumptions contain different metrics. Here, an answer is wanted for the following problem: Let $\|\cdot\|_1$ and $\|\cdot\|_2$ be two norms in the real co-ordinate plane R^2. What can one say about the transformation group $G = \langle \mathscr{C}_p^{(1)}, \mathscr{C}_q^{(2)} \rangle$ generated by a rosette group $\mathscr{C}_p^{(1)}$ in $(R^2, \|\cdot\|_1)$ and a rosette group $\mathscr{C}_q^{(2)}$ in $(R^2, \|\cdot\|_2)$ with the common fixed point O.

Without loss of generality, we may assume that the minimal ellipse e_1 with respect to $(\mathbf{R}^2, \|\cdot\|_1)$ is the ordinary circle.

First, we give an example for the case $p = q = 4$ where the minimal ellipse e_2 with respect to $(\mathbf{R}^2, \|\cdot\|_2)$ is not homothetic to e_1. It is easy to see that G fulfils neither (2) nor (1), independently of the metric with respect to $\|\cdot\|_1$ or $\|\cdot\|_2$ (see Fig. 2).

This result is only the simplest part of the following

PROPOSITION 1. *Let* $\|\cdot\|_1$ *and* $\|\cdot\|_2$ *be norms with the nonhomothetic minimal ellipses* e_1 *and* e_2. *If* $\mathscr{C}_p^{(1)}$ *is a rosette group in* $(\mathbf{R}^2, \|\cdot\|_1)$ *and* $\mathscr{C}_q^{(2)}$ *a rosette group in* $(\mathbf{R}^2, \|\cdot\|_2)$ *where as well as* p,q *the greatest common measure* (p,q) *are greater than* 2 *the group* $G = \langle \mathscr{C}_p^{(1)}, \mathscr{C}_q^{(2)} \rangle$ *fulfils neither* (2) *nor* (1).

PROOF. We denote the G.C.M. of p and q by d and the angle $\frac{2\pi}{d}$ by φ. There are natural numbers p' and q' such that $p = dp'$ and $q = dq'$. Let ρ_1 and ρ_2 be rotations generating $\mathscr{C}_p^{(1)}$ and $\mathscr{C}_q^{(2)}$, respectively. Therefore, $\rho_1^{p'}$ and $\rho_2^{q'}$ are rotations through φ. Without loss of generality, $\rho_1^{-p'}$ and $\rho_2^{q'}$ have the following matrices

$$A = \begin{pmatrix} \cos\varphi & \sin\varphi \\ -\sin\varphi & \cos\varphi \end{pmatrix}, \quad B = \begin{pmatrix} \cos\varphi & -\frac{1}{b}\sin\varphi \\ b\sin\varphi & \cos\varphi \end{pmatrix}, \quad 0<b<1,$$

respectively. Because of $b + \frac{1}{b} > 2$ and $\det A = \det B = 1$ their product

$$AB = \begin{pmatrix} \cos^2\varphi + b\sin^2\varphi & (1 - \frac{1}{b})\sin\varphi\cos\varphi \\ (b-1)\sin\varphi\cos\varphi & \cos^2\varphi + \frac{1}{b}\sin^2\varphi \end{pmatrix}$$

have two positive eigenvalues λ and μ ($> \lambda$) with $\lambda\mu = 1$, i.e. $0 < \lambda < 1$ and $\mu > 1$. Let v be an eigenvector for λ. We have

$$(AB)^i v = \lambda^i v ,$$

and $G = < \mathscr{C}_p^{(1)}, \mathscr{C}_q^{(2)} >$ fulfils neither (2) nor (1).

REMARK. The condition about the G.C.M., i.e. $d = (p,q) > 2$, is not necessary. We regard the example $p = 3$ and $q = 4$ and use the representation of the generating rotations ρ_i $(i=1,2)$ as products of two reflections in lines (see Fig. 3).

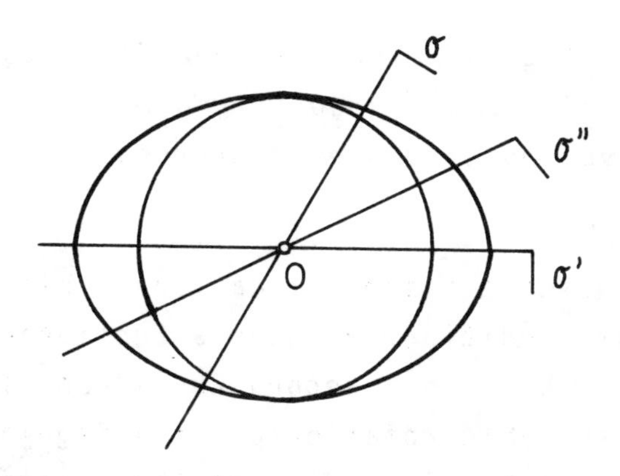

Figure 3

Without loss of generality, we find such reflections σ, σ' and σ'' that $\rho_1 = \sigma\sigma'$ and $\rho_2 = \sigma'\sigma''$. Hence, the kinds of the transform $\rho_1\rho_2 = \sigma\sigma''$ are

- elliptic rotation about 0 (elliptic case),
- shear or product of the reflection in point 0 and a shear with the axis through 0 (parabolic case),
- hyperbolic or pseudo-Euclidean rotation about 0 (hyperbolic case)

(see for instance [7], [9]). Using the analytic approach, one has the following necessary and sufficient conditions for the elliptic, parabolic, and hyperbolic case: $1 \geq b > \frac{\sqrt{3}}{3}$, $b = \frac{\sqrt{3}}{3}$, and $\frac{\sqrt{3}}{3} > b > o$, respectively. Finally, we state that $G' = \langle \rho_1 \cdot \rho_2 \rangle$ is always a discontinuous group in the parabolic case, and that $G = \langle \mathscr{C}_3^{(1)}, \mathscr{C}_4^{(2)} \rangle$ fulfils neither (1) nor (2) in the hyperbolic case.*

4. THE DIMENSION 3

First, SCHÖLZKE gave a generalization of Criterion 1 in Section 3 (see [12]). That is

CRITERION 2. *A normed 3-space is Euclidean iff one can realize all proper rosette groups of a plane by means of space rotations and the rotation group of a regular polytop.*

Here, we give a simplification of his proof. Because of the first condition, we have a Euclidean plane ε. Furthermore, there is a second Euclidean plane ε' as the image by a suitable rotation ρ' of a regular polytop. Let O be a common point of ε and the rotation axis, let g be the line $g := \varepsilon \cap \varepsilon'$. Hence $O \in g$. Since $\varepsilon' = \varepsilon^{\rho'}$ there is a rotation ρ'' of the normed space such that its axis is orthogonal to ε' and contains O, and that $\rho := \rho'\rho''$ has the axis g.

---

* The complete answer for our problem was given by H. Nimz in 1986: Since the commutator $\rho_1^{-1} \rho_2^{-1} \rho_1 \rho_2$ is always hyperbolic the group $G = \langle \mathscr{C}_p^{(1)}, \mathscr{C}_q^{(2)} \rangle$ fulfils neither (2) for (1) if both p and q are greater than 2.

Without loss of generality, we assume that the minimal ellipsoid e about 0 is the ordinary sphere. Since $\varepsilon' = \varepsilon^{\rho}$ is Euclidean $k \cap \varepsilon$ and $k \cap \varepsilon'$ where k denote the gauge figure about 0 lie in the same sphere s homothetic to e (see Fig. 4). With all rotations about the perpendicular to ε through 0, $k \cap \varepsilon'$ generates a spherical

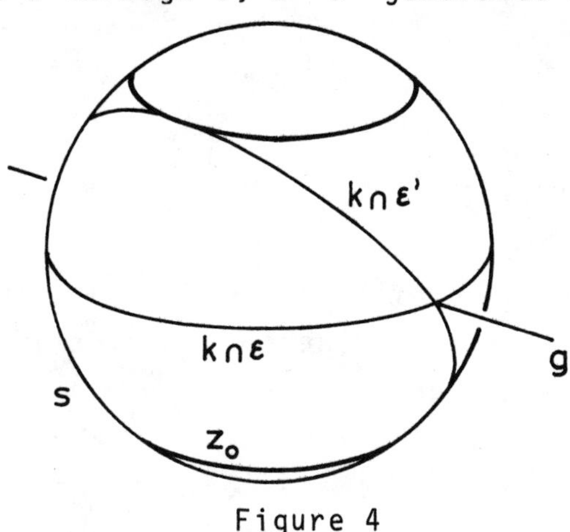

Figure 4

zone z_0 of s, $z_0{}^{\rho}$ generates a spherical zone z_1 of s, $z_1{}^{\rho^2}$ generates a spherical zone z_2 of s, and so on. Denoting the angle of z_0 by α_0, $\alpha_i := 2^i \alpha_0$ is the angle of z_i. Hence, $k = s = e$ is true. This proves our theorem.

F L A C H S M E Y E R found all motion groups of normed 2-spaces and 3-spaces and confirmed the above mentioned criterions all over again (see [5]).

5. FINITE DIMENSION n

The results of this Section were obtained together with SCHÖLZKE.

It is easy to see that *there exist all discontinuous*

motion groups of Euclidean geometry which have only translations or reflections in points as their elements in each normed space of the same finite dimension. But, what about rotational symmetries of a higher degree? A partial answer gives the following

CRITERION 3. *A normed space of a finite dimension* n *is Euclidean iff one can realize an infinite subset of abstract proper rosette groups of a plane* ε *by means of space rotations (about* $(n-2)$*-dimensional spaces) and the hyperplane reflections of a fixed* n*-cube onto itself.*

REMARK. Instead of the second condition, it is sufficient to formulate: There is an orthonormal Cartesian vector system $e_1,\ldots,e_n$ of the Euclidean space $\mathbf{E}^n$ with respect to the minimal ellipse e of LÖWNER, and there exist two types of hyperplane reflections where the elements of the first kind satisfy for a fixed index i

$$e_i \to -e_i, \text{ and } e_j \to +e_j, \; i \neq j,$$

and the elements of the second kind satisfy for two fixed indices i and j

$$e_i \to e_j, \; e_j \to e_i, \text{ and } e_k \to e_k, \; i,j \neq k.$$

First, to prove the Criterion 3, we state two propositions.

PROPOSITION 2. *All rotations about a* $(n-2)$*-dimensional space* $U \perp \varepsilon$ *in* $\mathbf{E}^n$ *are also motions of* $\mathbf{M}^n$.

Considering spheres s which are homothetic to e one proves this statement with the main idea of the proof of Criterion 1.

PROPOSITION 3. *Let* E^m, $m < n$ *be a* m*-dimensional plane of the Euclidean space* $\mathbf{E}^n$, *let* Σ *be a hyperplane reflection such that* $E^{m+1} := \overline{E^m \cup (E^m)^\Sigma}$ *is* $(m+1)$*-dimensional. Let* $O; e_1,\dots,e_n$ *denote an orthonormal Cartesian system such that* $O \in E^m \cap (E^m)^\Sigma$, e_i *belongs to* $E^m \cap (E^m)^\Sigma$ *if* $i < m$, e_m *to* E_m, *and* e_{m+1} *to* E^{m+1}. *Then, the orbit of* re_m *under the group generated by the rotations* $\rho_i(\varphi) = \rho(i,i+1;\varphi)$ *about* $\overline{\{e_1,\dots,e_n\}\setminus\{e_i,e_{i+1}\}}$ $(i = 1,\dots,m-1)$ *and by* Σ *is the whole sphere of* E^{m+1} *about* O *of radius* r.

PROOF. Changing eventually the signs of e_m and e_{m+1}, there is an angle β with $0 < \beta \le \frac{\pi}{2}$ such that

$$\begin{pmatrix} 1 & & & & & & & \\ & \ddots & & & & & & \\ & & 1 & & & & & \\ & & & \cos\beta & \sin\beta & & & \\ & & & \sin\beta & -\cos\beta & & & \\ & & & & & 1 & & \\ & & & & & & \ddots & \\ & & & & & & & 1 \end{pmatrix} \leftarrow \text{m-th row}$$

is a representation of Σ. The matrix of $\rho_i(\varphi) = \rho(i,i+1;\varphi)$ is

$$\begin{pmatrix} 1 & & & & & & & \\ & \ddots & & & & & & \\ & & 1 & & & & & \\ & & & \cos\varphi & -\sin\varphi & & & \\ & & & \sin\varphi & \cos\varphi & & & \\ & & & & & 1 & & \\ & & & & & & \ddots & \\ & & & & & & & 1 \end{pmatrix} \leftarrow \text{i-th row}$$

(All empty places of the matrices become zero.)

First, let x be an arbitrary radius vector of the spherical zone z_o of E^{m+1} symmetrically with respect to E^m where r is the radius and where $|x^{m+1}| \leq r \sin\beta$, i.e. the Cartesian co-ordinates are by polar co-ordinates:

$$\begin{aligned}
x^1 &= r \sin\varphi_m \ldots \sin\varphi_2 \sin\varphi_1 \\
x^2 &= r \sin\varphi_m \ldots \sin\varphi_2 \cos\varphi_1 \\
&\vdots \\
x^m &= r \sin\varphi_m \cos\varphi_{m-1} \\
x^{m+1} &= r \cos\varphi_m \qquad\qquad |\cos\varphi_m| \leq \sin\beta \\
0 & \\
\vdots & \\
0 &
\end{aligned}$$

Because of $|\cos\varphi_m| \leq \sin\beta$, there exist an angle α such that the image

$$\overline{e_m} := e_m^{\rho(m-1,m;\alpha)\Sigma}$$

of e_m has the polar co-ordinates 1; $0,\ldots,0$, φ, φ_m, $0,\ldots,0$.

Now, it is easy to see that

$$r\bar{e}_m^{\,\rho_{m-1}(\varphi_{m-1}-\varphi)\rho_{m-2}(\varphi_{m-2})\ldots\rho_1(\varphi_1)} = x .$$

The vectors

$$\cos\beta' \, e_m + \sin\beta' \, e_{m+1}, \quad |\beta'| \leq \beta$$

belong to the zone z_o. By the above machinery, the union

of their orbits is a zone z_1 symmetrically to E^m where $|x^{m+1}| \leq r \sin 2\beta$. Repeating this step, we obtain the whole sphere of E^{m+1} about 0 as soon as $(i+1)\beta \geq \frac{\pi}{2}$ and we are done.

PROOF OF CRITERION 3. Without loss of generality, we may assume that the minimal ellipsoid e is a sphere whose centre is the midpoint 0 of the fixed n-cube c and lies in ε.

Because of Proposition 2, $k \cap \varepsilon$ and $e \cap \varepsilon$ are homothetic with respect to 0. Let r denote the radius of $k \cap \varepsilon$ and s the sphere of $\mathbf{E}^n$ about 0 through $k \cap \varepsilon$.

First, we have $k \cap \varepsilon = s \cap \varepsilon$. There is a hyperplane reflection σ of $\mathbf{M}^n$ mapping c onto itself and $E^2 := \varepsilon$ such that $\dim \varepsilon \cap \varepsilon^\sigma = 1$. Because of Proposition 3, we have

$$k \cap \overline{\varepsilon \cup \varepsilon^\sigma} = s \cap \overline{\varepsilon \cup \varepsilon^\sigma} .$$

Second, supposing $k \cap E^m = s \cap E^m$, we find, in the same way, a hyperplane reflection Σ of M^n mapping c onto itself and E^m such that $\dim E^m \cap (E^m)^\Sigma = m-1$. Again because of Proposition 3, we have $k \cap \overline{E^m \cup (E^m)^\Sigma} = s \cap \overline{E^m \cup (E^m)^\Sigma}$. Finally, $s = k = e$ is true, i.e. $\mathbf{M}^n$ is Euclidean.

6. THE CONDITION (1) AND (3) IN AN ARBITRARY METRIC SPACE

Last, we point out a difficult problem in founding the main notion of this paper. A map group is called discontinuous by HILBERT and COHN-VOSSEN "wenn es zu jedem Punkt in jedem *endlichen* Gebiet nur endlich viele gebenüber der Gruppe äquivalente Punkte gibt" (see [6]), i.e. if there are only finite many points equivalent under the map group in any *finite* domain. Instead of "finite", we use the concept "bounded" (see the condition (1) in this paper or in [8]). But, the generalization of "finiteness" leads to "compactness" and to "total boundedness (precompactness)" in an arbitrary metric space. What concept is important so show the equivalence of the conditions (1) and (3)?

Let **X** be an arbitrary metric space and B a isometry group. We state that it is easy to see that

$$(1) \Rightarrow (2) \quad \text{and} \quad (2) \Rightarrow (3) .$$

On the other hand, if (3) holds for **X** and B then no orbit of any point P under the group B has any point of accumulation.
Furthermore, if each r-neighbourhood $U_r(X)$ is totally bounded, (1) is true. As a result, we have

THEOREM 3. *If the notions "boundedness" and "total boundedness" are equivalent in a metric space the conditions* (1) *and* (3) *also are equivalent.*

Applying this result to a normed space M^n is which the notions "boundedness" and "total boundedness" are equivalent we see that its Banach completion $\overline{M^n}$ is locally compact. By a theorem of F. RIESZ, the dimension n must be finite. But, it is well known that the conditions (1) and (3) are equivalent in each finite-dimensional normed space (see Section 2). Hence, Theorem 3 for normed spaces is no new assertion.

We do not know whether the sufficient condition of Theorem 3 will also be necessary.

REFERENCES

[1] ALEXANDROV, P.S., and others, *Die Hilbertschen Probleme*, Akademische Verlagsgesellschaft Geest & Portig K.-G., Leipzig, 1971; Problemy Gil'berta, Nauka, Moscow 1969 (in Russian).

[2] BOURBAKI, N., *Groupes et Algèbres de Lie* (*Éléments de Mathématique*), Hermann, Paris, 1968; Gruppy i algebry li, Mir, Moscow, 1972 (in Russian).

[3] COXETER, H.S.M., MOSER, W.O., *Generators and Relations for Discrete Groups*, Springer-Verlag, Berlin-Göttingen-Heidelberg, 1957.

[4] FEJES TÓTH, L., *Regular Figures*, Pergamon Press, Oxford, 1964; Reguläre Figuren, Akademiai Kiado, Budapest, B.G. Teubner Verlagsgesellschaft, Leipzig, 1965.

[5] FLACHSMEYER, J., *Zur Bewegungsgruppe von BANACH-Räumen*. In "Geometrie und Anwendungen 2, Potsdamer Forschungen, Wiss. Schriftenreihe der Päd. Hochsch. 'Karl Liebknecht' Potsdam, Reihe B, Heft 42", p. 103-110, Potsdam, 1984.

[6] HILBERT, D., and St. COHN-VOSSEN, *Anschauliche Geometrie*, Verlag von Julius Springer, Berlin, 1932.

[7] KLOTZEK, B., *Kegelschnitte und Büschel in äquiaffinen Ebenen*, Wiss. Z. Päd. Hochsch. "Karl Liebknecht" Potsdam 21 (1977), 133-144.

[8a] KLOTZEK, B., *Some Discontinuous Groups in non-Euclidean Geometries*, Blätter zu den Potsdamer Forschungen, Wiss. Schriftenreihe der Päd. Hochsch. "Karl Liebknecht" Potsdam, Preprint 1/83.

[8b] KLOTZEK, B., *Discontinuous Groups in some Metric and Nonmetric Spaces*, Beitr. Algebra Geom. 21 (1986), 57-66.

[9] KLOTZEK, B., and QUAISSER, E. *Büschel, Unterstrukturen und Orthogonalitätsrelationen in äquiaffinen Ebenen*, Math. Nachr. 58 (1973) 337-371.

[10] MAZUR, S. and ULAM, S., *Sur les Transformations Isométriques d'Espaces Vectoriels Normés*, Comptes Rendus Acad. Sci. Paris 194 (1932) 946-948.

[11] MINKOWSKI, H., *Geometrie der Zahlen*, B.G. Teubner, Leipzig, 1896 (1. Aufl.).

[12] SCHÖLZKE, J., *Diskrete Gruppen in normierten Räumen*. In "Geometrie und Anwendungen 1, Potsdamer Forschungen, Wiss. Schriftenreihe der Päd. Hochsch. 'Karl Liebknecht' Potsdam, Reihe B, Heft 41" p. 53-56, Potsdam 1984.

B. KLOTZEK
Pädagogische Hochschule
"Karl Liebknecht"
DDR-1500. Potsdam,
Am Neuen Palais

COLLOQUIA MATHEMATICA SOCIETATIS JÁNOS BOLYAI
48. INTUITIVE GEOMETRY, SIÓFOK, 1985.

ON PACKING THE PLANE WITH CONGRUENT COPIES OF A CONVEX BODY

W. KUPERBERG

1. INTRODUCTION AND PRELIMINARIES

A *convex body* is a compact convex set with non-empty interior. Given a convex body K in the plane R^2, a collection $\{K_i\}$ $(i = 1,2,3,\dots)$ of convex bodies in R^2 is called a *packing* of the plane with (copies of) K provided that each K_i is congruent to K and the interiors of the K_i's are mutually disjoint. A packing which entirely covers the plane is called a *tesselation*, or *tiling*, of the plane. A packing $\{K_i\}$ is said to be *uniform* if the exists a tiling $\{T_i\}$ such that $K_i \subset T_i$ for every i.

If S is a subset of the plane R^2, let -S denote the set symmetric to S about the origin, and let v + S denote the translate of S by the vector $v \in R^2$, i.e. $-S = \{x \in R^2 : -x \in S\}$ and $v + S = \{x \in R^2 : x - v \in S\}$. Also, if S is a measurable set, let $|S|$ denote its area.

Each packing can be assigned a real number between 0 and 1, called the density of the packing, which, intuitively speaking, represents the ratio between the sum of the areas of the bodies being packed and the area of the region in which they are packed. In general, a rigorous definition of the density of packing would involve the notion of limits, and it would require detailed investigation concerning existence and uniqueness. However, in case of uniform packings, it seems natural to define the density of the packing as the ratio between the area of the body being packed and the area of the body used for the tiling given by the definition of uniform packings. The uniqueness of the density defined in this manner follows immediately from the following

(1.1) PROPOSITION. If $\{K_i\}$ is a packing and if $\{T_i\}$ and $\{S_i\}$ are tiling with $K_i \subset T_i \cap S_i$ then $|T_i| = |S_i|$.

We omit the proof of this proposition, since it can be considered a simple exercise in analysis.

Consider the set S of real numbers r that have the property that every convex plane body admits a uniform packing of the plane with density $d \geq r$. Obviously, set S is bounded above by 1. As a matter of fact, the number $c = \sqrt{3}\pi/6 \approx 0.9069$ is an upper bound for that set, for it is known [2] that c is the greatest density of packing of congruent circular disks in the plane. Let d_0 be the least upper bound of S. In [1], G.D. Chakerian and L.H. Lange proved that every plane convex body K is contained in a quadrilateral of area at most $\sqrt{2}$ times that of K. Since every quadrilateral admits a tiling of the plane, they concluded that every convex body admits a (uniform) packing of the plane with density at least $\sqrt{2}/2 \approx 0.7071$. In other words, $d_0 \geq \sqrt{2}/2$. This inequality was later improved to $d_0 > 3/4 = 0.75$ (see [3]) by showing that every plane convex body K is contained in a certain type of hexagon, called a p-hexagon, of area smaller that 4/3 times that of K, which admits a tiling of the plane. In this note we expand the technique of [3] and obtain a further improvement, namely $d_0 \geq 25/32 = 0.78125$. The exact value of d_0 remains unknown.

2. TILING PROPERTIES OF p-HEXAGONS

DEFINITION. A p-*hexagon* is a hexagon with a pair of parallel opposite sides of equal length. "Opposite" here means separated by exactly two other sides in each, clockwise and counterclockwise, direction. Also, degeneracies are allowed under this definition; sides can be of length zero, even the parallel ones, three consecutive vertices can be collinear, etc. The degenerate p-hexagons are: pentagons with a pair of parallel sides, all quadrilaterals, and all triangles.

(2.1) THEOREM. *For every* p-*hexagon* H *there exists a tiling* $\{H_i\}$ *of the plane such that each* H_{2i} *is a translate of* H *and each* H_{2i-1} *is a translate of* $-H$.

For a proof of this theorem and a diagram showing the tiling pattern, see [3].

DEFINITION. A tiling $\{T_i\}$ of the plane is said to be *affine*, if for every affine transformation A of the plane onto itself, $\{A(T_i)\}$ is a tiling of the plane.

The following is a consequence of Theorem (2.1):

(2.2) COROLLARY. Every p-hexagon admits an affine tiling of the plane.

REMARK. One can prove that the above corollary cannot be strengthened, i.e. only p-hexagons admit affine tilings of the plane (assuming convexity of the tiles, of course). This statemenet can be proved by means of the following observations: (1) If $\{T_i\}$ is a tiling of the plane with copies of a convex body T, then T is a polygon with at most six sides; (2) If T and S are polygons with the property that, for each affine transformation A of the plane, the polygons A(T) and A(S) are congruent, then S is a translate of T or a translate of -T; and (3) If $\{T_i\}$ is a tiling such that, for each i and j, T_i is a translate of T_j or a translate of $-T_j$, then the model tile T is a p-hexagon.

3. THE MAIN RESULT

(3.1) THEOREM. *Every plane convex body* K *is contained in a* p-*hexagon of area at most* 32/25 *times that of* K.

We precede the proof of this theorem with the following lemmas.

(3.2) LEMMA. *Let* $\mathcal{K}$ *be the space of plane convex bodies furnished with the Hausdorff metric. Then the function* f *from* $\mathcal{K}$ *into the positive reals assigning to each convex body the minimum area of a* p-*hexagon containing the body is continuous.*

PROOF. Suppose $K_1, K_2, \ldots$ is a sequence of convex bodies converging to the convex body K_0 and let H_n be a minimum-area p-hexagon containing K_n $(n = 1,2,\ldots)$ so that $f(K_n) = |H_n|$. It is easy to notice that there exists a compact set in R^2, for instance a large enough circle, which contains all of the p-hexagons H_n. This implies, in particular, that the sequence $\{f(K_n)\}$ is bounded. We shall prove that it converges to $f(K_0)$ by showing that $f(K_0)$ is its only limit point. Let then $\{f(K_{n_i})\}$ be a subsequence of the sequence $\{f(K_n)\}$ with

$$\lim_{i\to\infty} f(K_{n_i}) = b \ .$$

Since all of the H_n's are contained in a compact set, the sequence $\{H_{n_i}\}$ has a subsequence $\{H_{k_i}\}$ which converges to a convex body H_0. Obviously, H_0 is a p-hexagon containing K_0, which implies that $f(K_0) \leq |H_0|$. We assert that, in fact, $f(K_0) = |H_0|$. Suppose not, i.e. suppose $f(K_0) < |H_0|$, which means there exists a p-hexagon H containing K_0 and with $|H| < |H_0|$. Then there exists a p-hexagon H^+ containing H in its interior and still with $|H^+| < |H_0|$. Such a p-hexagon H^+ can be obtained by a slight enlargement of H. But then almost all of the K_n's are contained in H^+, and that implies that there exists an integer s such that $|H_s| \leq |H^+|$ and $||H_s| - |H_0|| <$ $< |H_0| - |H^+|$ at the same time, which is impossible. This proves our assertion. Now, since $|H_0| = \lim |H_{k_i}| = b$, we get $\lim f(K_{n_i}) = f(K_0)$ and the proof of this lemma is complete.

(3.3) LEMMA. *Suppose* K_1 *and* K_2 *are affinely equivalent convex bodies and let* a *be a real number. If there exists a* p*-hexagon* H_1 *containing* K_1 *and such that* $|H_1| = a|K_1|$, *then there exists a* p*-hexagon* H_2 *containing* K_2 *and such that* $|H_2| = a|K_2|$.

PROOF. This lemma follows immediately from the fact that an affine image of a p-hexagon is a p-hexagon and affine transformations of the plane preserve ratios between areas.

PROOF OF THEOREM (3.1). By virtue of lemma (3.2), we can assume that the boundary of K is differentiable and contains no lime segments, since every convex body can be approximated by one with these special properties. This implies that, for every unit vector v in the plane, each of the two lines tangent to K and parallel to v touches K at one point only, and that each of the two points of tangency depends continuously on v. Let then, for an arbitrary unit vector v in the plane, ℓ_1 and ℓ_2 be the two lines parallel to v and tangent to K. Denote the points of tangency by P_1 and P_2, respectivelly. Denote by m the line defined by P_1 and P_2. Let m_1 and m_2 be the two lines tangent to K and parallel to m, and let Q_1 and Q_2 be the respective points of tangency. Let m_3 be the line parallel to m and in the middle distance between lines m and m_1, and let m_4 be the line parallel to m and in the middle distance between lines m and m_2. Let A_1 and A_2 be the points of intersection between m_3 and the boundary of K and let B_1 and B_2 be the points of intersection between m_4 and the boundary of K. Let the lines r_1,r_2, $s_1 s_2$ be tangent to K at the points A_1, A_2, B_1, B_2, respectively (see Fig. 1).

The lines m_1, m_2, r_1, r_2, s_1 and s_2 bound a hexagon H containing K. We assert that $|H| \leq 4|K|/3$. To prove this assertion, notice that H is contained in the union of two trapezoids: T_1 defined by the lines m, m_1, r_1 and r_2 and T_2 defined by the lines m, m_2, s_1 and s_2. Now, let $\bar{A}_1$, $\bar{A}_2$, $\bar{B}_1$, $\bar{B}_2$ be the projections on m of the points A_1, A_2, B_1, B_2, respectively, in the direction of v.

It is easy to calculate that the area of T_1 equals 4/3 times that of the pentagon $p_1 = \bar{A}_1 A_1 Q_1 A_2 \bar{A}_2$ and, analogously, the area of T_2 equals 4/3 times that of the pentagon $p_2 = \bar{B}_1 B_1 Q_2 B_2 \bar{B}_2$. (see Fig. 2.)

Since these two pentagons have disjoint interiors and are contained in K, we conclude $|H| \leq |T_1| + |T_2| =$ $= 4|p_1|/3 + 4|p_2|/3 \leq 4|K|/3$, as asserted.

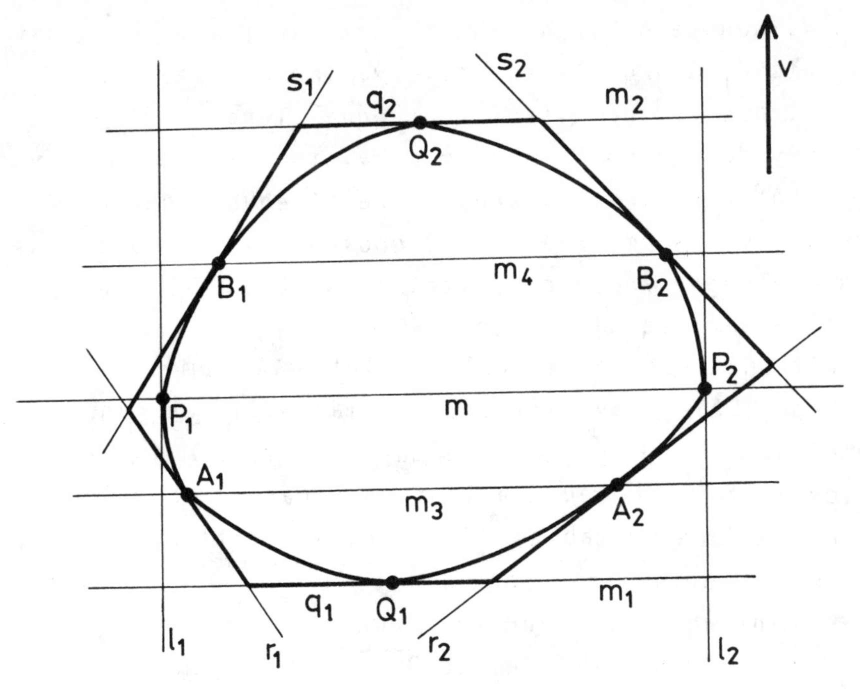

Figure 1

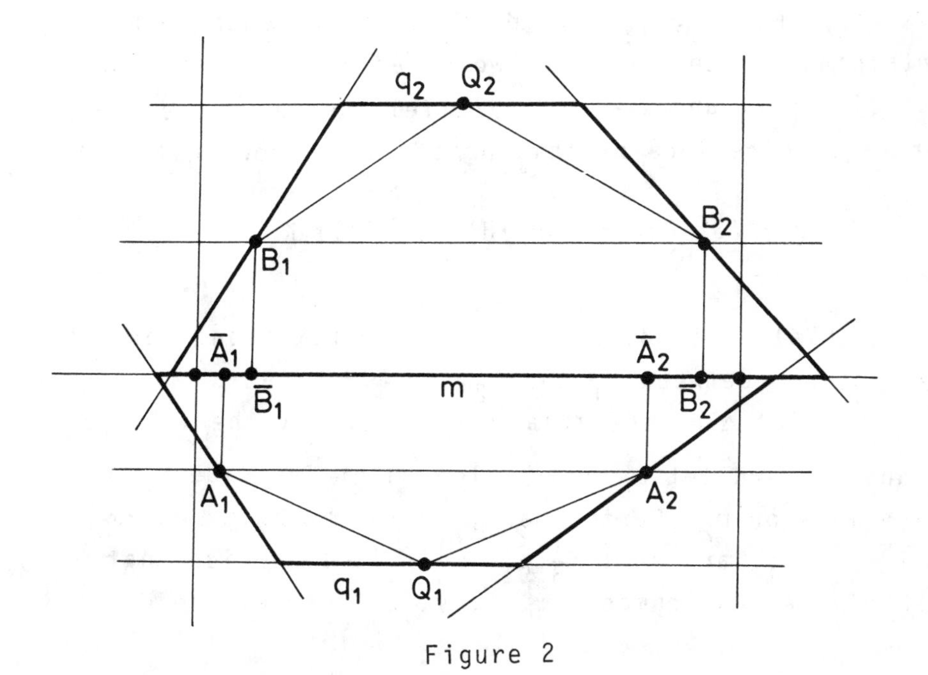

Figure 2

Now, denote by q_1 and q_2 the sides of H which contain the points Q_1 and Q_2, respectively. These sides are parallel, but not necessarily of equal length. However, since they depend continuously on v, the direction of v can be chosen so that q_1 and q_2 are of equal length. From now on let us assume that v was chosen just so, and thus H is a p-hexagon whose opposite sides q_1 and q_2 are parallel and of equal length.

Without loss of generality, let us assume that $|p_1| \le |p_2|$. Also, by virtue of lemma (3.3), we can assume that $|P_1| + |p_2| = 6$ and that the line m is perpendicular to v, because all of this can be obtrained by means of one suitable affine transformation.

Let t_1 be the line parallel to $\overline{B_1Q_2}$ and tangent to K at a point which lies on the same side of m as Q_2 and let t_2 be the line parallel to $\overline{B_2Q_2}$ and tangent to K at a point which lies on the same side of m as Q_2. Denote the points of tangency by C_1 and C_2, respectively. The five lines m_1, ℓ_1, ℓ_2, t_1 and t_2 bound a pentagon P containing K (see Fig. 3). Two sides of P, namely those that lie on ℓ_1 and ℓ_2, are parallel, hence P is a p-hexagon. The idea of this proof is to show that either

$$|P| \le (32/25)|K| \text{ or } |H| \le (32/25)|K|.$$

We shall introduce the following notation: Let d be the distance between P_1 and P_2; $d = \bar{P}_1\bar{P}_2$, let $b_1 = \bar{A}_1\bar{A}_2$, $b_2 = \bar{B}_1\bar{B}_2$, let Δ be the triangle bounded by the lines m_2, t_1 and t_2, and let b_0 be the length of the side of Δ which lies on m_2. Furthermore, let $w = b_0/b_2$; $x = (d-b_2)/b_2$; $y = (d-b_1)/b_1$; and $z = 1 - \frac{1}{3}|p_1| = \frac{1}{3}|p_2| - 1$. Notice that each of the real numbers w, x, y, z lies between 0 and 1.

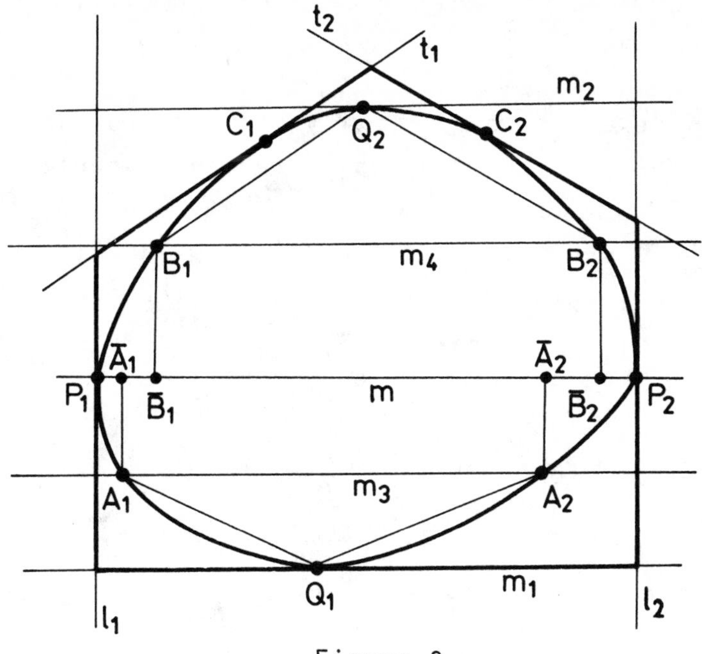

Figure 3

Notice now that the pentagon P is contained in the union of the following figures:

(i) The rectangle bounded by the lines ℓ_1, ℓ_2, m and m_1; its area equals $4(y+1)(1-z)$;

(ii) The pentagon p_2; its area equals $3(1+z)$;

(iii) The rectangle bounded by the lines ℓ_1, m, m_4, and $\overline{B_1\bar{B}_1}$, and the rectangle bounded by the lines ℓ_2, m, m_4 and $\overline{B_2\bar{B}_2}$; the total area of these two rectangles is $2x(1+z)$;

(iv) The parallelogram bounded by the lines m_2, m_4, t_1 and $\overline{B_1Q_2}$ and the parallelogram bounded by the lines m_2, m_4, t_2 and $\overline{B_2Q_2}$; the total area of these parallelograms is $2w(1+z)$;

(v) The triangle Δ mentioned above; its area equals $w^2(1+z)$.

The total of the areas of the figures named in (i) - (v) equals

$$[3 + w^2 + 2(x + w)](1 + z) + 4(y + 1)(1 - z) .$$

Denote this expression by $g(w,x,y,z)$, or g for short. Since the union of the figures listed above contains P, we get $|P| \leq g$.

On the other hand, K contains p_1, p_2 and the points P_1, P_2, C_1 and C_2. Therefore K contains the convex hull of $p_1 \cup p_2 \cup \{P_1, P_2, C_1, C_2\}$. A simple calculation shows that the area of that convex hull equals at least

$$|p_1| + |p_2| + w(1 + z) + y(1 - z) =$$

$$= (3 + x + w)(1 + z) + (3 + y)(1 - z) .$$

Denote this last expression by $h(w,x,y,z)$, or h for short. Thus we get $|K| \geq h$.

Recall that $|H| \leq (4/3)(|p_1| + |p_2|) = 8$.

Summarizing, we get the following inequalities:

$$\frac{|H|}{|K|} \leq 8/h(w,x,y,z) ;$$

and

$$\frac{|P|}{|K|} \leq g(w,x,y,z)/h(w,x,y,z) .$$

where $g(w,x,y,z) = [3 + w^2 + 2(x + w)](1 + z) + 4(y + 1)(1 - z)$, $h(w,x,y,z) = (3 + x + w)(1 + z) + (3 + y)(1 - z)$, and w,x,y,z are real numbers in the interval $[0,1]$.

We assert now that either $8/h \leq 32/25$ or $g/h \leq 32/25$. With this assertion proved the proof of the theorem will be complete.

Case I. $g \geq 8$.

In this case we get

$$[w^2 + 2(x + w)](1 + z) + (1 + 4y)(1 - z) \geq 2$$

which implies

$$[w^2 + 2(x + w)](1 + z) + 4y(1 - z) \geq 1 + z$$

and

$$(*) \qquad (\frac{w^2}{4} + \frac{x + w}{2})(1 + z) + y(1 - z) \geq \frac{1 + z}{4} \geq \frac{1}{4} .$$

Now, since $0 \leq w \leq 1$, and $0 \leq x \leq 1$, we get

$$(**) \qquad x + w \geq \frac{w^2}{4} + \frac{x + w}{2} .$$

Combining (*) with (**) we get

$$(x + w)(1 + z) + y(1 - z) \geq \frac{1}{4} .$$

Notice that $h = 6 + (x + w)(1 + z) + y(1 - z)$, therefore the above inequality implies $h \geq 25/4$, and, consequently, $8/h \leq 32/25$.

Case II. $g \leq 8$.

Let $f(w,x,y,z) = g(w,x,y,z)/h(w,x,y,z)$. In this case, the domain of the function f is the subset Ω of R^4 defined by the inequalities $0 \leq w \leq 1$, $0 \leq x \leq 1$, $0 \leq y \leq 1$, $0 \leq z \leq 1$ and $g(w,x,y,z) \leq 8$. All we need to show is that the maximum value of f over Ω is at most 32/25. Let S be the surface in R^4 whose equation is $g(w,x,y,z) = 8$.

ASSERTION. The maximum value of f over Ω is attained on the surface S. To prove this, observe the following facts:

(1) Each h and g is increasing with respect to each of the variables w, x and y;

(2) $g/h < 2$ over Ω;

(3) $\frac{\partial g/\partial w}{\partial h/\partial w} \geq 2$, $\frac{\partial g/\partial x}{\partial h/\partial x} \geq 2$ and $\frac{\partial g/\partial y}{\partial h/\partial y} \geq 2$ over Ω.

This implies that f is increasing with respect to each of the variables w, x and y.

Now, if $(w_0, x_0, y_0, z_0) \in \Omega \setminus S$, then either $w_0 < 1$, $x_0 < 1$, or $y_0 < 1$, because $g(1,1,1,z) = 16 > 8$. Therefore f is not maximum at (w_0, x_0, y_0, z_0). This proves the assertion.

On the surface S, $f = 8/h$, and the formula describing h can be simplified to

$$h(w,x,y,z) = \frac{1}{4}[24 + (1 + 2x + 3w - w^2)(1 + z)] \quad .$$

The maximum of f is reached at the point(s) where h raches its minimum. Obviously, this happens when $w = x = z = 0$, and the minimum value of h is 25/4. Therefore the maximum value of f is 32/25 . This completes the proof of the theorem.

(3.4) COROLLARY. Every plane convex body admits a uniform packing of the plane with density at least $25/32 = 0.78125$.

REFERENCES

[1] G.D. CHAKERIAN - L.H. LANGE, *Geometric extremum problems*, Math. Mag. 44 (1971), 57-69.

[2] L. FEJES TÓTH, *Lagerungen in der Ebene, auf der Kugel und im Raum*, Springer Verlag, Berlin 1953.

[3] W. KUPERBERG, *Packing convex bodies in the plane with density greater than 3/4*, Geometriae Dedicata 13 (1982), 149-155.

W. KUPERBERG
Department of Mathematics
Auburn University
Auburn, AL 36849
U.S.A.

COLLOQUIA MATHEMATICA SOCIETATIS JÁNOS BOLYAI
48. INTUITIVE GEOMETRY, SIÓFOK, 1985.

COVERING PLANE CONVEX BODIES WITH SMALLER HOMOTHETICAL COPIES

M. LASSAK

Let C be a plane convex body, this is, a compact convex set with non-empty interior. By $h_k(C)$ we denote the smallest positive ratio of k homothetical copies of C whose union covers C. A short argument [6] shows that the number $h_k(C)$ exists. Let $\hat{h}_k$ denote the smallest positive ratio of homothety under which every plane convex body can be covered by k homothetical copies. In other words

$$\hat{h}_k = \sup\{h_k(C);\ C \text{ is a plane convex body}\}.$$

Moreover, consider the number

$$\check{h}_k = \inf\{h_k(C);\ C \text{ is a plane convex body}\}.$$

In 1954 Levi [7] proved that every plane convex body can be covered by 7 homothetical copies with the ratio 1/2 (Fig. 1). Together with the example of disk this gives $\hat{h}_7 = 1/2$. He also showed [8] that every plane convex body different from a parallelogram can be covered by 3 smaller positive homothetical copies. Thus every plane convex body can be covered by 4 smaller positive copies. What is more [6], we have $\hat{h}_4 = \sqrt{2}/2$ (Fig. 2). Belousov [1] proved that no plane convex body can be covered with 3 homothetical copies of a positive ratio smaller than 2/3

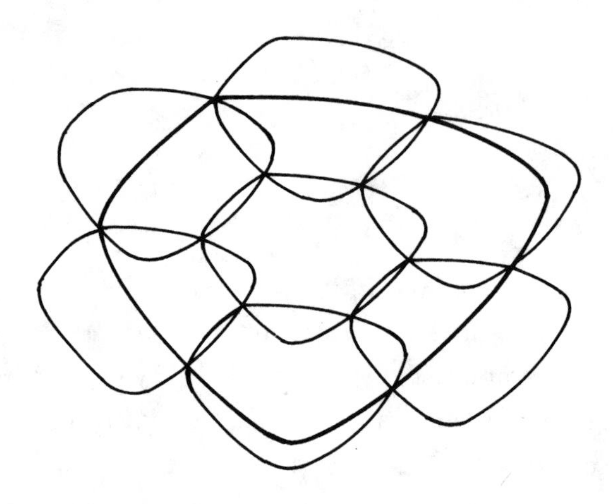

Figure 1

Figure 2

and that the extreme example is a triangle. So $\check{h}_3 = 2/3$. The above results lead to the question about the values of $\hat{h}_k$ and $\check{h}_k$ for other k.

Denote by D a disk, by T a triangle and by P a parallelogram. Table 1 presents the values or estimates of $\check{h}_k$, $h_k(D)$, $h_k(T)$, $h_k(P)$ and $\hat{h}_k$ for $3 \le k \le 9$ (for $k = 2$ all the numbers are equal to 1).

k	$\check{h}_k$	$h_k(D)$	$h_k(T)$	$h_k(P)$	$\hat{h}_k$
3	$\frac{2}{3}$	$\frac{\sqrt{3}}{2}$	$\frac{2}{3}$	1	1
4	$\frac{1}{2}$	$\frac{\sqrt{2}}{2}$	$\frac{4}{7}$	$\frac{1}{2}$	$\frac{\sqrt{2}}{2}$
5	$\frac{\sqrt{5}}{5} \le ? \le \frac{1}{2}$ conjecture: $\frac{1}{2}$	≈ 0.609	$\frac{8}{15}$	$\frac{1}{2}$	$0.609 < ? \le \frac{1}{2}\operatorname{ctg}\frac{\pi}{5}$
6	$\frac{\sqrt{6}}{6} \le ? \le \frac{1}{2}$ conjecture: $\frac{1}{2}$	≈ 0.556	$\frac{1}{2}$	$\frac{1}{2}$	$0.556 < ? \le \sin^2 \frac{3\pi}{10}$
7	$\frac{\sqrt{7}}{7} \le ? \le \frac{5}{11}$	$\frac{1}{2}$	$\frac{5}{11}$	$\frac{1}{2}$	$\frac{1}{2}$
8	$\frac{\sqrt{2}}{4} \le ? \le \frac{3}{7}$	? conjecture: $(1 + 2\cos\frac{2\pi}{7})^{-1}$	$\frac{3}{7}$	$\frac{1}{2}$	$\frac{1}{2}$
9	$\frac{1}{3}$	? conjecture: $\sqrt{2} - 1$	$\frac{2}{5}$	$\frac{1}{3}$	$\frac{2}{5} \le ? \le \frac{1}{2}$

Table 1

The values $h_3(D) = \sqrt{3}/2$, $h_4(D) = \sqrt{2}/2$ and $h_7(D) = 1/2$ are obvious. Neville [9] established that $h_5(D) \approx 0.609$ and his reasoning was completed by Bezdek [2]. Grünbaum (see [10], p. 106) conjectured that $h_6(D) \approx 0.557$. It was a surprise when Bezdek [2] showed that the best covering is quite different from that conjectured by Grünbaum and that $h_6(D) \approx 0.556$. For $k \geq 7$ the values of $h_k(D)$ are unknown. One may suppose that $h_8(D) = (1 + 2\cos\frac{2\pi}{7})^{-1} \approx 0.445$ and that $h_9(D) = \sqrt{2} - 1 \approx 0.414$, this is, that the best coverings occur when the center of a smaller disk D_1 is in the center c of D and the other smaller disks cover the ring $D \setminus D_1$ such that the arrengements are identical after the rotations around c on the angles $2\pi/7$, $\pi/4$, respectively.

The values of $h_k(T)$ were given by Belousov [1] for $3 \leq k \leq 6$ and by Fudali [4] for $7 \leq k \leq 15$.

It is easy to see that $h_k(P) = 1/n$ if $n \leq \sqrt{k} < n+1$, where n is a natural number. In fact, it is clear that $h_k(P) \leq 1/n$ for $k \geq n^2$. On the other hand, consider $(n+1)^2$ points $(i/n, j/n)$, $i,j \in \{0,\ldots,n\}$, in a coordinate system such that (0,0) and (1,1) are opposite vertices of P. Obviously, no copy of P with a positive ratio smaller than $1/n$ can cover two of those points simultaneously. Therefore $h_k(P) \geq 1/n$ for $k < (n+1)^2$.

Let us comment the values and estimates of $\hat{h}_k$ given in Table 1. From $h_3(P) = 1$ we obtain $\hat{h}_3 = 1$. Remember that $h_4 = \sqrt{2}/2$ and $\hat{h}_7 = 1/2$ are proved in [6] and [7]. The last equality, the obvious inequality $\hat{h}_8 \leq \hat{h}_7$ and the equality $h_8(P) = 1/2$ imply $\hat{h}_8 = 1/2$. The values of $\hat{h}_k$ for other k seem to be difficult to find. So in Table 1 we present only estimates of $\hat{h}_5$, $\hat{h}_6$ and $\hat{h}_9$. The left estimates result from the values of $h_5(D)$, $h_6(D)$ and $h_9(T)$. Let us motivate the right ones.

Böhme [3] and Grünbaum [5] proved that it is possible to incsribe an affine-regular pentagon G into each plane convex body C. By an affine-regular polygon we mean an affine image of a regular polygon. Denote by $d_1,\dots,d_5$ the vertices of G. Let c_i be the common point of the straight lines containing those disjoint sides of G whose end-points are different from d_i, $i=1,\dots,5$. Let H_i be the homothety with the center c_i and the ratio $\frac{1}{2}\cot\frac{\pi}{5}$, $i=1,\dots,5$. This is the smallest ratio under which $G\subset\cup\{H_i(G);\ i=1,\dots,5\}$. Consequently, $G\subset\cup\{H_i(C);\ i=1,\dots,5\}$. Consider an arbitrary point $x\in C$ outside G. Let for instance $x\in\Delta c_1d_3d_4$. Since $H_1(C)$ is convex and contains the points d_3, d_4 and $H_1(x)$, we have $\Delta d_3d_4H_1(x)\subset H_1(C)$. In particular, $x\in H_1(C)$. Thus we see that $C\subset\cup\{H_i(C);\ i=1,\dots,5\}$. Hence $\hat{h}_5\le\frac{1}{2}\cot\frac{\pi}{5}\approx 0.688$.

An analogous reasoning shows that C is covered by 6 homothetical copies with the ratio $\sin^2\frac{3\pi}{10}$ and with the homothety centers $c_1,\dots,c_6$, where c_6 is the center of G. Thus $\hat{h}_6\le\sin^2\frac{3\pi}{10}\approx 0.654$.

From $\hat{h}_9\le\hat{h}_8=1/2$ we get $\hat{h}_9\le 1/2$. A much better estimate is possible if we consider only centrally symmetric bodies. Namely, *we have* $h_9(C)\le\sqrt{2}-1$ *for every centrally symmetric plane convex body* C. The reader can show this by a similar method to the above considerations concerning the right estimates of $\hat{h}_5$ and $\hat{h}_6$, but this time using a theorem of Grünbaum [5] that an affine-regular octagon can be inscribed in each centrally symmetric plane convex body.

Finally, we comment on the numbers $\check{h}_k$. Remember that the value $h_3=2/3$ has been established by Beluosov [1]. Considering the areas of C and of its copies we observe that $\check{h}_k\ge\sqrt{1/k}$. Taking into account the values of $h_{n^2}(P)$ we see that $\check{h}_{n^2}=1/n$ for any natural n. Particularly, $\check{h}_4=1/2$ and $\check{h}_9=1/3$ are written in Table 1. The right estimates of $\check{h}_5, \check{h}_6, \check{h}_6, \check{h}_8$ result from $h_5(P) = h_6(P) = 1/2$, $h_7(T) = 5/11$ and $h_8(T) = 3/7$. One may conjecture that $\check{h}_5 = \check{h}_6 = 1/2$, this is, that $h_5(C) \geq 1/2$ and $h_6(C) \geq 1/2$ for each plane convex body C. If C is centrally symmetric, the following stronger property (connected with the result of Levi that $h_7(C) \leq 1/2$ for each plane convex body C) holds true.

For every centrally symmetric plane convex body C *we have* $h_7(C) = 1/2$.

Let us prove this property. Denote by c the center of C. Let $0 < \lambda < 1/2$. Suppose that a copy of C with ratio λ has more than one common point with the boundary bd C of C. It is easy to show that the intersection is connected, this is, that the intersection is an arc $\widehat{pr}$. Let s,t be the points of bd C such that the vectors $\vec{pr}$ and $\vec{sc} = \vec{ct}$ are of the same sense. Take points $u,w \in$ bd C such that $u \in sp$ and $\vec{uw} = \vec{sc}$. Moreover, take $y \in \widehat{wt}$ such that the segments cy and sp are parallel. Let z be the symmetric point. Since scwu is a parallelogram and $u \in \widehat{sp}$, we have $|sp| \geq |cy| = |yz|/2$. Since C is centrally symmetric, there is no segment inscribed in C and parallel to yz longer than the segment yz. So no homothetical copy of C with the ratio λ covers the segment sp and thus (by the convexity) the arc $\widehat{sp}$. Analogously, no homothetical copy of C with the ratio λ covers the arc $\widehat{rt}$. Hence no arc being a symmetric half of bd C can be covered by 3 copies of C with the ratio λ. Thus for the covering of bd C by copies of C with the ratio λ we need at least 7 of them. Observe that they do not cover the center c. This results from the fact that there is no segment of a given direction inscribed in C longer than the segment containing c. We see that no 7 copies of C with the ratio λ cover the whole C. This and $\check{h}_7 = 1/2$ imply $h_7(C) = 1/2$.

REFERENCES

[1] Yu.F. BELOUSOV, *Theorems on covering of plane figures.* Ukrain. Geom. Sb. 20(1977), 10-17.

[2] K. BEZDEK, *Über einige Kreisüberdeckungen,* Beitr. Algebra Geom. 14(1983), 7-13.

[3] W. BÖHME, *Ein Satz über ebene konvexe Figuren.* Math.-Phys. Semesterber. 6(1958), 153-156.

[4] S. FUDALI, *Homotetyczne pokrycie trójkata.* Matematyka 35(1982), 94-109.

[5] B. GRÜNBAUM, *Affine-regular polygons inscribed in plane convex sets,* Riveon Lematematika 13(1959), 20-24.

[6] M. LASSAK, *Covering a plane convex body by four homothetical copies with the smallest positive ratio.* Geom. Dedicata 21(1986), 155-167.

[7] F.W. LEVI, *Ein geometrisches Überdeckungsproblem.* Arch. Math. 5(1954), 476-478.

[8] F.W. LEVI, *Über eines Eibereiches durch Parallelverschiebung seines offenen Kerns,* Arch. Math. 6(1955), 369-370.

[9] E.H. NEVILLE, *On the solution of numerical functional equations,* Proc. London Math. Soc. (2), 14(1915), 318-326.

[10] D.O. SHKLIARSKY, N.N. TCHENCOV and I.M. YAGLOM, *Geometric estimates and problems in combinatorial geometry,* Nauka, Moscow 1974.

M. LASSAK
Institut Matematyki i Fizyki ATR,
ul. Kaliskiego 7, 85-790 Bydgoszcz,
Poland

COLLOQUIA MATHEMATICA SOCIETATIS JÁNOS BOLYAI
48. INTUITIVE GEOMETRY, SIÓFOK, 1985.

AN UPPER BOUND FOR THE INTRINSIC VOLUMES OF EQUILATERAL ZONOTOPES

J. LINHART

An equilateral zonotope [3] is a zonotope generated by line segments of equal length, which we may assume to be the unit. The intrinsic volumes $V_r(C)$ of a convex body C in d-dimensional Euclidean space E^d are defined by the formula

$$V_r(C) = \frac{1}{w_{d-r}} \binom{d}{r} W_{d-r}(C)$$

where $W_j(C)$ denote the well-known quermaß integrals ([1], [5]), and w_j the volume of the j-dimensional unit ball. For instance, $V_d(C)$ is the ordinary volume, $2V_{d-1}(C)$ is the surface area, $V_0(C)$ is constant = 1.

There are several inequalities for the intrinsic volumes of (convex) polytopes, especially for dimensions≤ 3, involving the inradius or circumradius and the number of faces and vertices of the polytopes [4]. In [8] two inequalities for equilateral zonotopes in E^3 are given, concerning the surface area and the inradius respectively. The first of these turned out to be a special case of a more general inequality, which shall be discussed in the present paper. The essential idea of this generalization (see formula (2)) originates from P. McMullen (oral communication).

THEOREM 1. *Let* Z *be a d-dimensional zonotope generated by* n *unit line segments. Then*

$$V_r^2(Z) = \binom{n}{r}\binom{d}{r}\left(\frac{n}{d}\right)^r. \tag{1}$$

REMARKS CONCERNING THE CASE OF EQUALITY

For $3 \le r \le d$, equality holds at least in the following cases:

(i) $n = d$ and Z is a cube,

(ii) $n = d+1$ and Z is generated by the segments connecting the center of a regular d-simplex with its vertices.

For $r = 0$ and $r = 1$, equality holds in any case.

For $r = 2$, equality may be characterized using the following notions:

1. Eutactic stars:

   Let $u_1,\dots,u_n$ be n vectors in E^d and denote the coordinates of u_i by $u_{i1},\dots,u_{id}$. Then $u_1,\dots,u_n$ are said to form a *eutactic star* iff the d columns of the matrix $U := (u_{ij})$ are pairwise orthogonal and of equal Euclidean norm.

   Up to a constant factor this is equivalent to saying that $u_1,\dots,u_n$ may be obtained by an orthogonal projection $E^n \to E^d$ of an orthonormal basis of E^n ([3], [11]).

2. Equiangular lines:

   A set of straight lines or line segments is called *equiangular*, if the angle between each pair of lines is the same [7].

3. Regular stars:

In this paper, a set of equiangular unit line segments with corresponding vectors forming a eutactic star will be called a *regular star*. The cases (i) and (ii) above give simple examples of regular stars, which will be called trivial. Now we have:

THEOREM 2. *For* $r = 2$, *equality holds in* (*i*) *iff the line segments form a regular star.*

REMARKS

There is a one-to-one correspondence between regular stars and "regular twographs" ([10], [11]). This made it possible to enumerate all regular stars up to dimension 14. There are exactly 10 non-trivial such stars. The following table (derived from [21]) shows the dimensions d and the numbers n of line segments of these stars:

d	3	5	6	7	7	9	13	13	13	13
n	6	10	16	14	28	18	26	26	26	26

For example, the 28 lines in E^7 may be obtained by connecting the center of a regular 7-simplex with the midpoints of its edges [9]. In E^{13} there are four non-congruent regular stars with the same number of lines segments. There are no additional regular stars in E^{14}. Furthermore, many regular stars are known in higher dimensions [2], e.g. in E^{15} there are at least 91 (non-congruent) regular stars with 36 line segments and 6 with 30 line segments each. Nevertheless, the upper bound in (1) seems to be rather coarse for larger n. It is not known whether the bound may be attained for $r > 2$ and $n > d+1$.

PROOF OF THEOREM 1.

The cases $r = 0$ and $r = 1$ are trivial, since $V_0(Z) = 1$ and $V_1(Z) = n$. To prove the theorem in the general case, we begin with the Steiner parallel formula

$$V(Z + \lambda B) = \sum_{r=0}^{d} w_{d-r} V_r(Z)\lambda^{d-r}$$

from which it is possible to deduce the following expression for $V_r(Z)$ ([6], S. 186):

$$V_r(Z) = \sum_{i=1}^{f_r} \alpha_i^{(r)} V_r(A_i^{(r)}) \quad .$$

Here $A_1^{(r)},\ldots,A_{f_r}^{(r)}$ are the r-faces of Z, and $\alpha_i^{(r)}$ denotes the (d-r-1)-dimensional external angle at the corresponding r-face. (For $r = d$ we have $f_r = 1$ and $\alpha_i^{(r)} = 1$).

Now, since Z is a zonotope, each r-face $A_i^{(r)}$ belongs to a class of parallel r-faces (a "zone"). Let L^{d-r} be a linear subspace of E^d totally orthogonal to such a zone, and φ the orthogonal projection $E^d \to L^{d-r}$. φ maps the r-faces of the zone onto the 0-faces (i.e. vertices) of the (d-r)-polytope $\varphi(Z)$. The external angles at these r-faces are equal to the external angles at the vertices of $\varphi(Z)$, so their sum must be 1. Thus, if we denote by m the number of zones of r-faces of Z and choose a representative $\tilde{A}_k^{(r)}$ out of each such zone, we may write

$$(2) \qquad V_r(Z) = \sum_{k=1}^{m} V_r(\tilde{A}_k^{(r)}) \quad .$$

Let $u_1,\dots,u_n$ be unit vectors lying in the line segments generating Z. Then (2) means

$$V_r(Z) = \sum_{i \le i_1 < \dots < i_r \le n} \| u_{i_1} \wedge \dots \wedge u_{i_r} \| , \tag{3}$$

where $\wedge$ denotes the usual exterior product: the coordinates of $u_{i_1} \wedge \dots \wedge u_{i_r}$ are the determinants of the (r x r)-submatrices of the rows with numbers $i_1,\dots,i_r$ of the matrix U described above (see the definition of "eutactic stars").

Let S be the value of the expression (3) with the norm replaced by its square. Then S is simply the sum of the determinants of *all* (r x r)-submatrices of U. If we denote by u_j^* the j-th column vector of U, we obtain exactly the same sum S, if we use the u_j^* instead of the u_j. Thus, using the convexity of the function $x \to x^2$,

$$V_r^2 \le \binom{n}{r} \Sigma \| u^*_{i_1} \wedge \dots \wedge u^*_{i_r} \|^2 \le \binom{n}{r} \Sigma a_{i_j} \cdot \dots \cdot a_{i_r} \tag{4}$$

where $a_j := \| u_j^* \|$ and the sum extends over all index-tuples $(i_1,\dots,i_r)$ with $i \le i_1 < \dots < i_r \le d$.

The sum in (4) is essentially the r-th elementary symmetric function of d variables.

$$\Sigma\, a_i = \Sigma\, u_{ik}^2 = \sum_{i=1}^{n} \| u_i \|^2 = n \quad ,$$

so the sum in (4) attains its maximum for $a_i = \dots = a_d = \frac{n}{d}$, which yields the desired inequality (1).

PROOF OF THEOREM 2.

For $r > 1$, the case of equality in (1) is characterized by equality in both inequalities of (4) and $a_1 = \dots = a_d$.

Now

$$\|u^*_{i_1} \wedge \ldots \wedge u^*_{i_r}\|^2 = a_{i_1} \cdot \ldots \cdot a_{i_r}$$

iff the vectors u^*_i are pairwise orthogonal. Since $a_1 = \ldots = a_d$, this is fulfilled iff the u_i form a eutactic star.

To obtain equality in the first part of (4), the volumes of all r-faces of Z have to be equal. For $r = 2$ of course, this means equiangularity of the line segments generating Z.

REFERENCES

[1] BETKE, U. - McMULLEN, P., *Estimating the sizes of convex bodies from projections*. J. London Math. Soc. (2), 27, 525-538 (1983).

[2] BUSSEMAKER, F.C., MATHON, R.A. and SEIDEL, J.J., *Tables of two-graphs*. Technological Univ. Eindhoven: Report 79-WSK-05 (1979).

[3] COXETER, H.S.M., *Reguler Polytopes*. 3rd ed., New York: Dower 1973.

[4] FEJES TÓTH, L., *Reguläre Figuren*. Budapest, Akadémiai Kiadó, 1965.

[5] HADWIGER, H., *Vorlesungen über Inhalt, Oberfläche und Isoperimetrie*. Berlin, Göttingen, Heidelberg. Springer, 1957.

[6] LEICHTWEIß, K., *Konvexe Mengen*. Berlin, Heidelberg, New York. Springer, 1980.

[7] LEMMENS, P.W.H. - SEIDEL, J.J. *Equiangular lines*. J. Alg. 24, 494-512. (1973).

[8] LINHART, J., *Extremaleigenschaften der regulären 3-Zonotope*. Studia Sci. Math. Hung. (to appear).

[9] LINT, J.H. VAN - SEIDEL, J.J., *Equilateral point sets in elliptic geometry*. Indag. Math. 28, 335-348. (1966).

[10] SEIDEL, J.J., *Eutactic stars*. Combinatorics (Coll. Keszthely, 1976). North Holland Publ. Comp.

[11] SEIDEL, J.J., *A survey of two-graphs*. Proc. Intern. Colloqu. Teorie Combinatorie (Roma, 1973), Tomo I., Acad. Naz. Lincei 481-511 (1976).

J. LINHART
Mathematisches Institut der Universität
Hellbrunnerstr. 34
A-5020. Salzburg

COLLOQUIA MATHEMATICA SOCIETATIS JÁNOS BOLYAI
48. INTUITIVE GEOMETRY, SIÓFOK, 1985.

NON-EUCLIDEAN CRYSTALLOGRAPHY

ALAN L. MACKAY

ABSTRACT

With the discovery of several experimental instances of the periodic minimal surfaces described by H.A. Schwarz, E.R. Neovius and later by A.H. Schoen, this field can now be seen to be crystallography (the arrangement of many units of a few types) on two-dimensional manifolds of non-positive curvature. The experimental observation of regions of icosahedral symmetry in quenched alloys also sharpens the interest in crystallography in three-dimensional manifolds of non-Euclidean metric. The structures observed represent compromises between the conflicting demands of short-range order and long-range order.

INTRODUCTION

A space is non-Euclidean when the distance between two points is no longer given by Pythagoras' theorem but involves a metric tensor. Thus, crystallography on the surface of a sphere, which is a space of constant Gaussian curvature $K = 1/R^2$, is familiar and is exemplified by the structures of the spherical viruses. Identical particles are related to each other by the symmetry operations of axes (and in the case of non-cheiral units also by mirror planes), but not by translations. In non-Euclidean space the concept of parallels is abandoned. The metric is here given by the rules of spherical trigonometry. We may note that the perimeter of a small circle of radius r increases more slowly than $2\pi r$, namely as $2\pi r - r^3/(6K) + O(r^5)$. Thus, six identical circles can no longer touch a seventh and, if it is required to map a hexagonal net on to a sphere, then 12 dislocations, where a point has five, instead of six, neighbours must be introduced. The symmetry groups on the surface of a sphere are well-known and include the icosahedral groups 53m and 532, respectively with and without mirror planes and having the orders of 120 and 60. Modifying exact equivalence to quasi-equivalence, larger lattices can be produced by tesselation. Many of these tesselations are known experimentally (and are also exemplified by the geodesic structures of Buckminster Fuller, although he, apparently, never found the skew tesselations). The space of a closed convex surface is clearly finite.

The curvatures arises from the laws of bonding between particles and, in this case, four particles joining together define a curvature. The curvature can be calculated for the N-dimensional case from the distances between the N + 2 points, using an expression which relates the radius of curvature R of the N-dimensional manifold to the N-dimensional content and to the "Ptolemaicity" P, which is a determinant formed from the distances between the points. This determinant formed from the distances between the points. This determinant is a generalisation of Ptolemy's theorem (Sommerville, 1929). For the case of four points:

$-\,2\,R^2 .\ 288\ V^2 = P$ where V is the volume of the simplex and where $-288\ V^2 =$

$$\begin{vmatrix} 0 & d_{12}^{\ 2} & d_{12}^{\ 2} & d_{14}^{\ 2} & 1 \\ d_{21}^{\ 2} & 0 & d_{13}^{\ 2} & d_{14}^{\ 2} & 1 \\ d_{31}^{\ 2} & d_{32}^{\ 2} & 0 & d_{34}^{\ 2} & 1 \\ d_{41}^{\ 2} & d_{42}^{\ 2} & d_{43}^{\ 2} & 0 & 1 \\ 1 & 1 & 1 & 1 & 0 \end{vmatrix}$$

and $P =$

$$\begin{vmatrix} 0 & d_{12}^{\ 2} & d_{13}^{\ 2} & d_{14}^{\ 2} \\ d_{21}^{\ 2} & 0 & d_{23}^{\ 2} & d_{24}^{\ 2} \\ d_{31}^{\ 2} & d_{32}^{\ 2} & 0 & d_{34}^{\ 2} \\ d_{41}^{\ 2} & d_{42}^{\ 2} & d_{43}^{\ 2} & 0 \end{vmatrix}$$

To see the effect of local demands for a coordination number less than 6, we may consider the geometrical problem of packing regular pentagons, and we find that there are two approaches. The first is, as indicated above, to curve the space so that there are no interstices

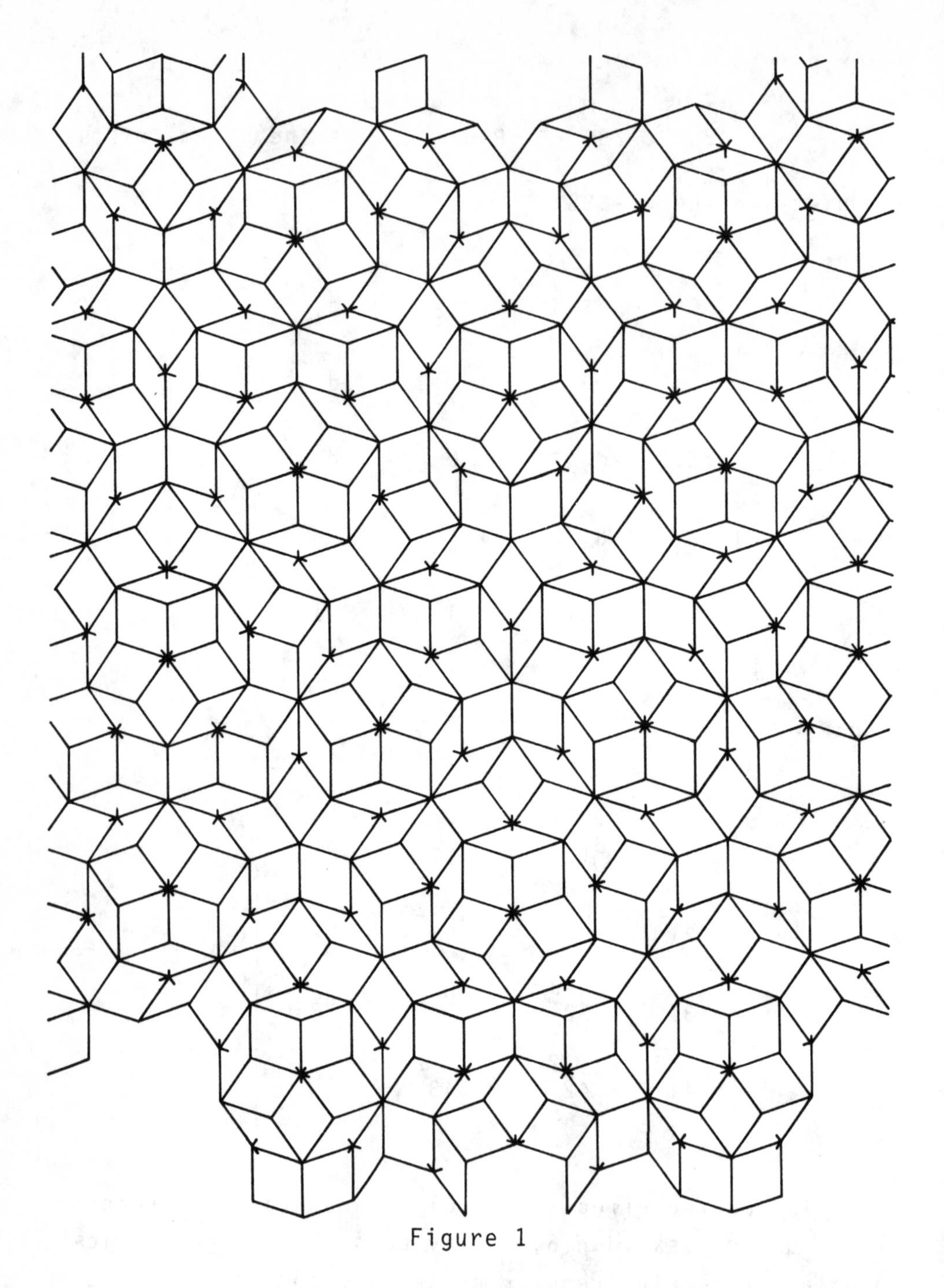

Figure 1

The 2-D Penrose pattern showing the two types of unit cell

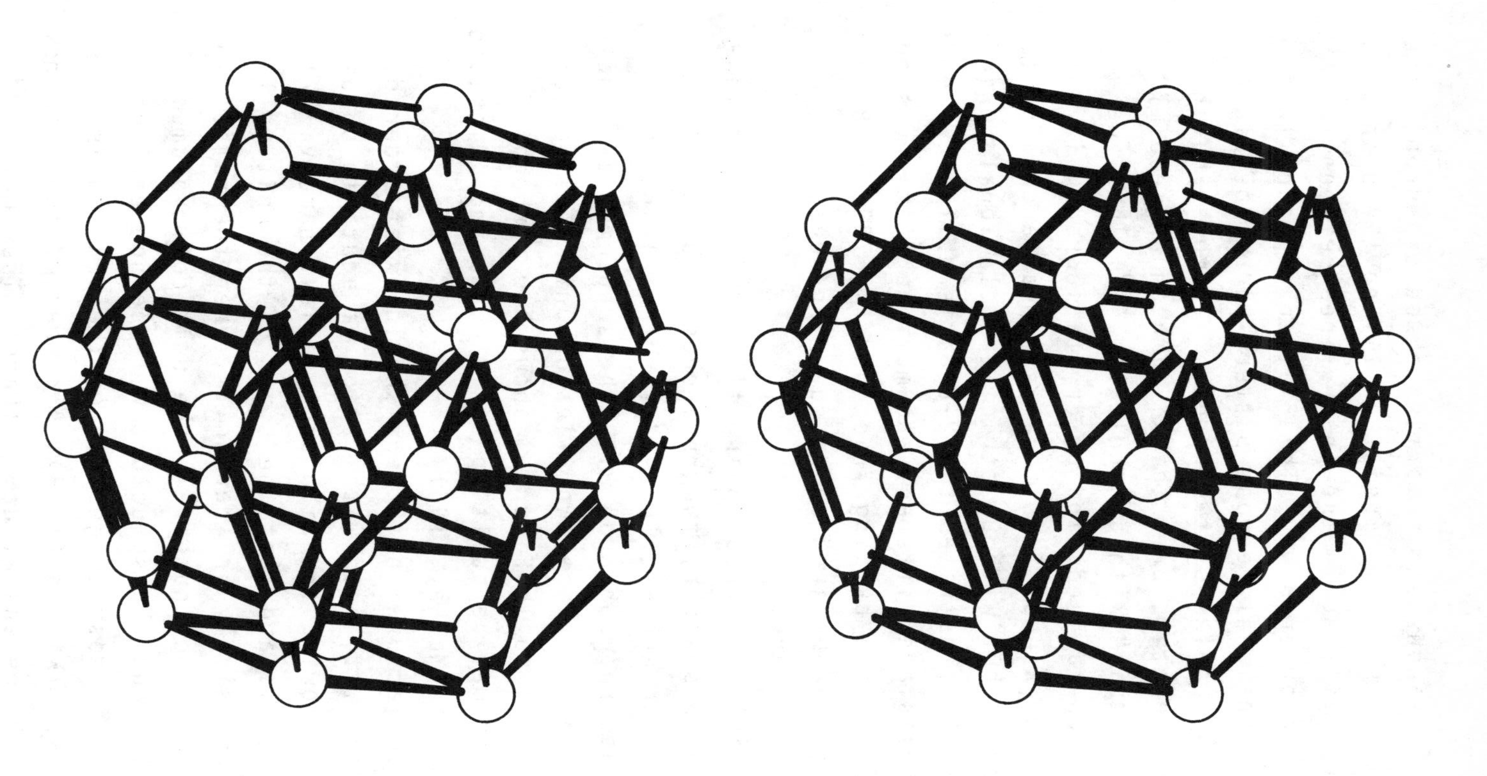

Figure 2

The 3-D Penrose pattern showing how a rhombic triacontahedron may be built up of 10 acute and 10 obtuse unit rhombohedra. (Stereo diagram for crossed--eyes viewing)

between the pentagons (thus forming a dodecahedron). The second approach is, if we insist on tiling the plane, to give rules for filling in the interstices between the pentagons in an ordered way. This latter approach leads to the two-dimensional Penrose pattern where the plane can be tiled by copies of two kinds of rhombus, having the same edge lengths, one "thick" with an angle of 72° and the other "thin" with an angle of 144°. Thus, we obtain a new generalised kind of crystallography in the plane, where two kinds of unit cells repeat according to a definite rule, but without strict translational periodicity. Figure 1 illustrates the 2-D pattern. The two ends of the rhombuses are different and must be distinguished. If we pack circles a packing fraction of 0.738633 (81% of close packing) is obtained and a covering density of 1.32131 is also found.

There exists a corresponding tiling of three-dimensional Euclidean space by two rhombohedral unit cells, one acute with an angle of 63.43^{o} (arctan 2) and the other an obtuse rhombohedron with the supplementary angle. Over a large volume these two tiles or unit cells occur in the ratio of τ to 1. Figure 2 shows how a rhombic triacontahedron may be built up of 10 acute and 10 obtuse unit rhombohedra (Mackay, 1981).

We may conjecture that there should exist ordered, crystallographic, tilings with ratios of acute to obtuse rhombohedra following the Fibonacci series, namely: 1/1, 3/2, 5/3, 8/5, ... and that the sizes of the unit cell will get progressively greater unitl, when the ratio approaches the Golden number τ, the unit cell volume will become infinite and the structure will be non-periodic - this process is most suggestive of the renormalisation technique of K.G. Wilson (1979). A number of these tilings (1/1, 3/2, 5/3) are easily demonstrated. They may be regarded also as packings of space, in the 2-D case by regular decagons and in the 3-D case by rhombic triacontahedra, with overlapping regions.

The 3-D Penrose pattern will be discussed later.

In the plane the mean coordinate number of a triangulated network is six. If the coordination number is less than six then the space must be curved positively and, if greater than six, the space must be curved negatively. If the physical units demand a coordination number of greater than six a two-dimensional space of negative curvature results.

TWO-DIMENSIONAL MANIFOLDS OF NON-POSITIVE CURVATURE

Manifolds of constant negative Gaussian curvature are known. An example is the pseudo-sphere (generated by revolving a tractrix about its asymptote). However, these surfaces are finite and are bounded by singularities.

If a soap-film is hung over four of the six edges of a regular tetrahedron (forming a skew quadrilateral) a saddleshaped surface of minimal curvature is obtained. Figures 3, 4, 5 and reproduce Schwarz' drawings. This surface has zero mean curvature J and, since the curvatures at every point are equal and opposite, a Gaussian curvature of K_1K_2 which is negative or zero (zero at the umbilic points) results. H.A. Schwarz (1890) showed that this element of surface could be repeated by rotations of 180° about its edges to give an infinite, periodic minimal surface (figure 4). This is called the F surface (F for face-centred-cubic). Recently, this surface has been found experimentally in glyceryl mono-oleate (GMO) (Longley and McIntosh, 1983, Mackay, 1985). The structure is such that molecules of GMO may be regarded as being packed in a

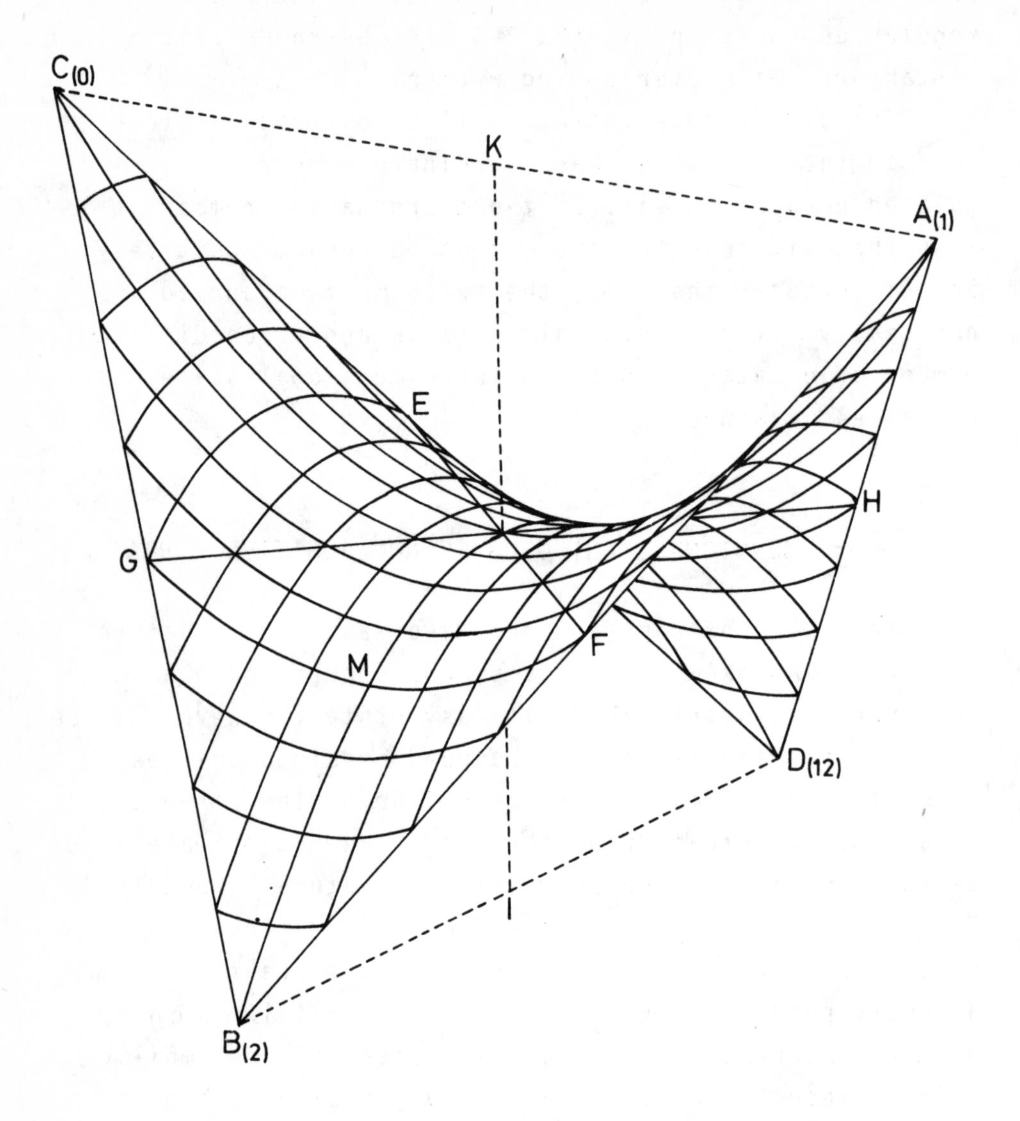

Figure 3

A soap film hung over a regular tetrahedral frame (from Schwarz, 1890) - the basic saddle making up the F-surface

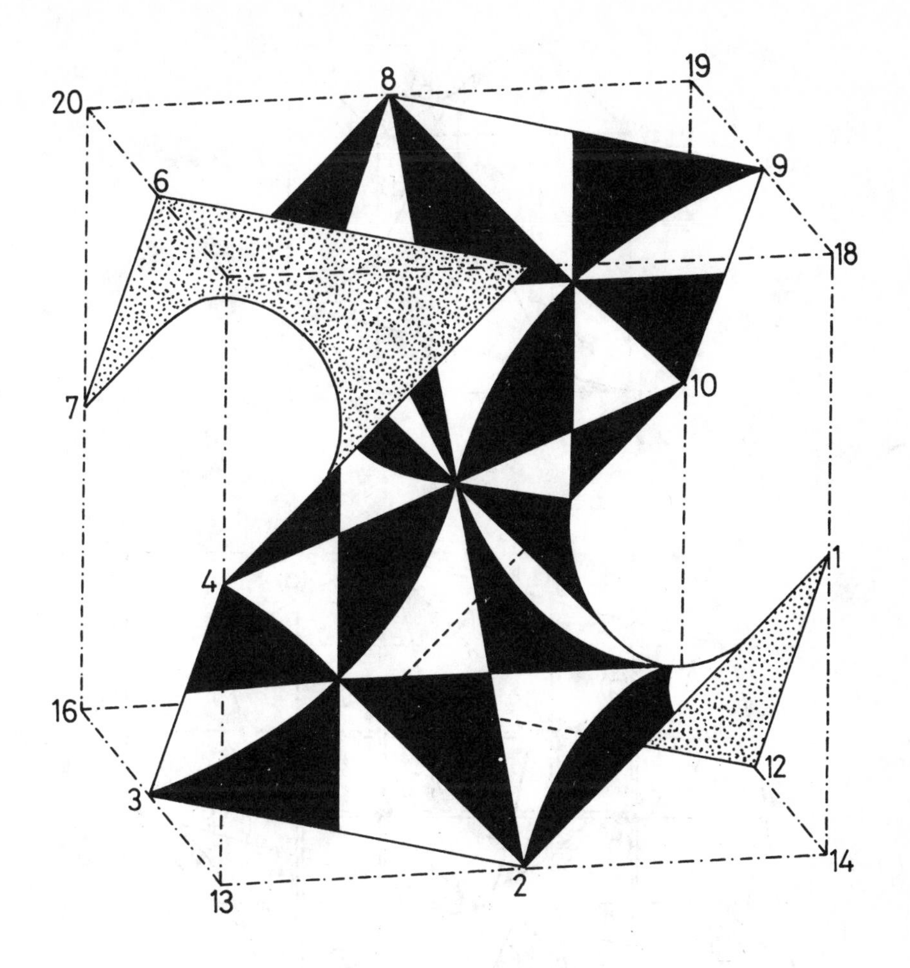

Figure 4

Schwarz' drawing (modified) of a unit cell of the F-surface which contains six of the tetrahedral saddles of fig. 3.

quasi-equivalent manner in a hexagonal mesh on this surface. Since the surface is of negative curvature, and the curved triangle which is the unit of pattern has an angular deficit of 15^0, dislocations of the six-connected surface lattice have to occur.

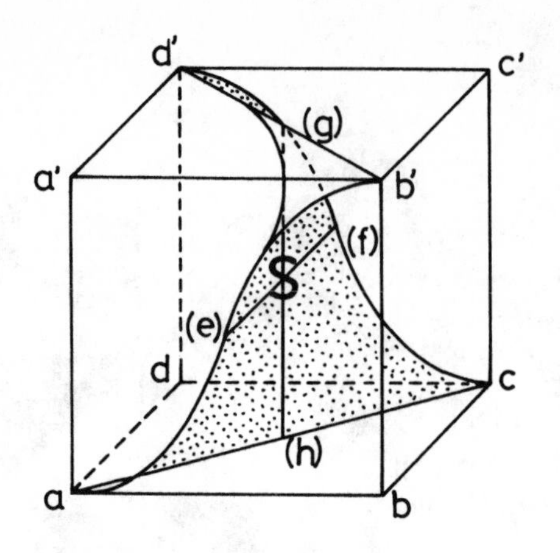

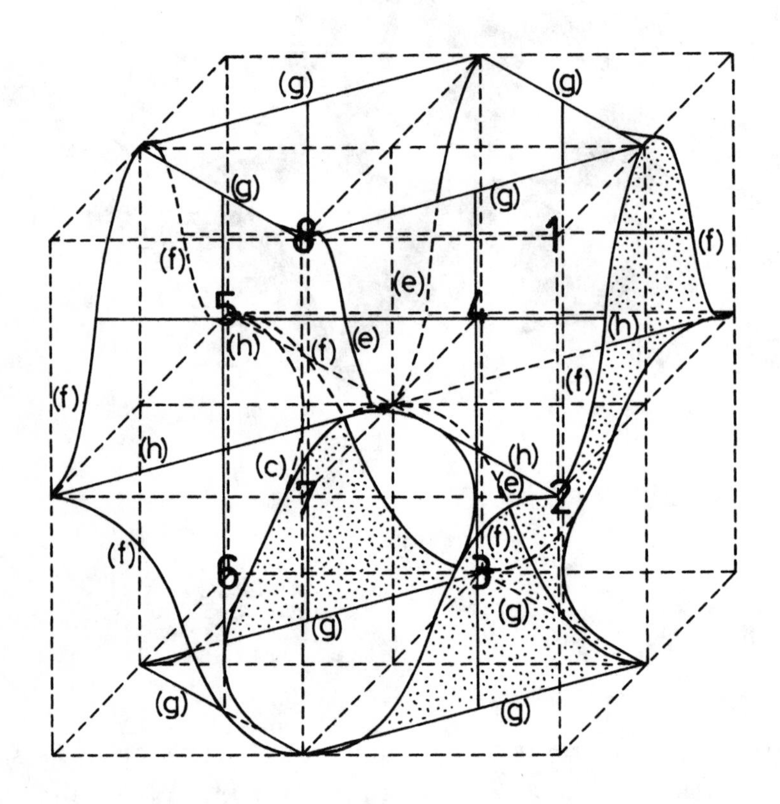

Figure 5

Schwarz' drawing of a unit cell of the T (tetragonal) surface

The F-surface may be seen to be a network of tubes, four tubes meeting at a tetrahedral joint, which parallels the tetrahedral network of the crystal structure of diamond, that is, it is an inflation of a ball-and-spoke model of the diamond structure. The surface has the property that it divides space into two congruent parts, the inside and the outside. The Laplace-Young differential equation which describes the surface is equivalent to prescribing a minimum area, which is the physical condition arising from the packing of the molecules.

According to a theorem of differential geometry, a surface of constant mean curvature has associated with it a surface of constant Gaussian curvature, although, if the mean curvature is zero, this parallel surface will be infinitely distant. However, the arrangement of the unit triangles in the F-surface can be mapped on to the hyperbolic plane (figure 6).

The figure (7) shows similarly a conformal representation of the space group 732 in the hyperbolic plane. The fundamental triangles have a smaller angular deficit (4.286^0) than for any other such space group and thus represent a minimum departure from planarity. There is an infinite number of such tilings of the hyperbolic plane and some others among them may be encountered in the mappings of elements on to the periodic minimal surfaces.

Schwarz and his colleague E.R. Neovius and later A.H. Schoen (1969) found altogether 18 such surfaces (figure 5 shows the tetragonal T surface). They are produced by arranging soap-film surfaces in the fundamental regions of certain of the 230 crystallographic space groups. Most occur in the kaleidoscope groups where the fundamental region is repeated purely by mirror planes to fill all space, but Schoen showed that there were some surfaces, in particular the gyroid surface G, which

Figure 6

The 2-dimensional hyperbolic space group 642 which shows the same arrangement of the triangular units of pattern as the F-surface

occured in other space groups. The gyroid surface divides all space into two congruent but enantiomorphous regions. (It is still an open question as to whether such surfaces can be found for the three-dimensional Penrose pattern).

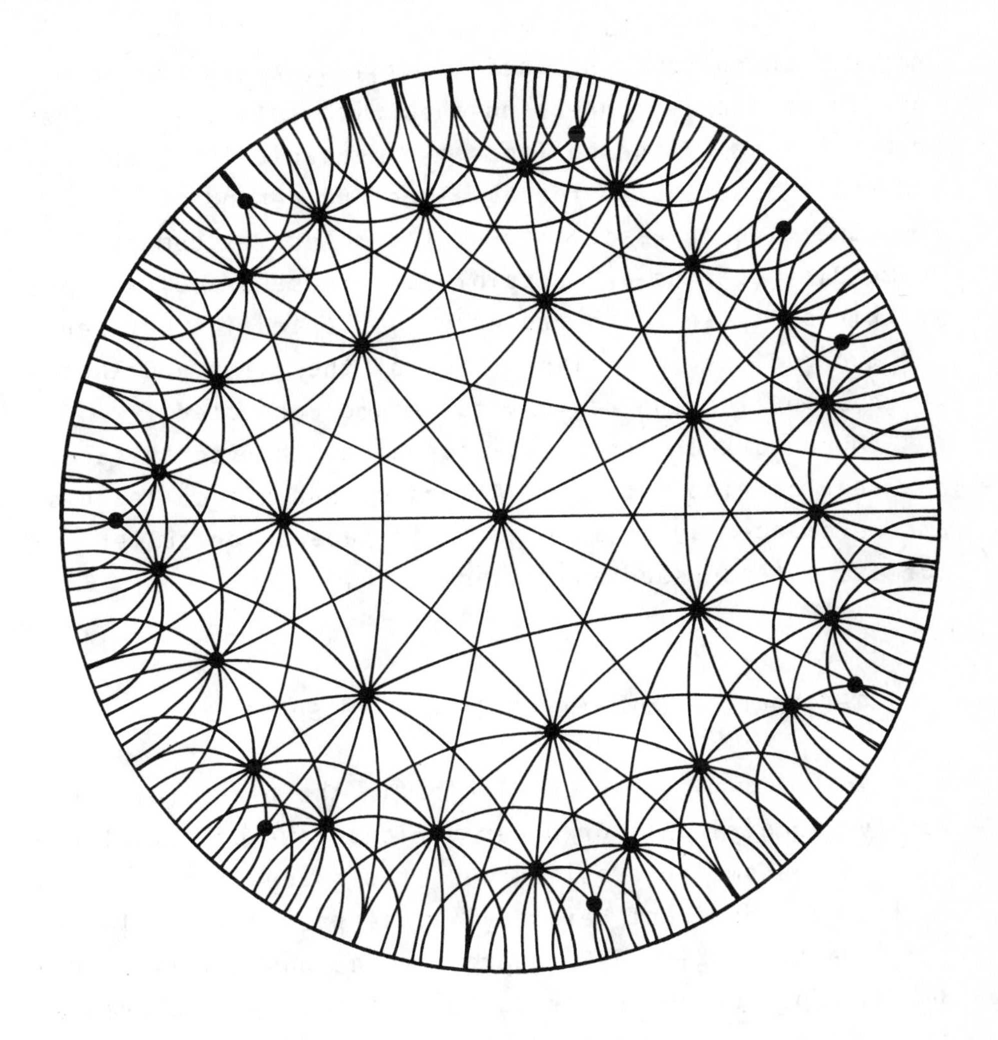

Figure 7

The 2-dimensional hyperbolic space group 732 (or 73m). This has the minimum curvature with a deficit of 4.286° per triangle

The other surfaces may be visualised in terms of regions of space connected by networks of tubes. Work is in progress on calculating their profiles, areas and curvatures and in connecting them with actual structures. Probably none of the surfaces are mechanically stable under surface-tension forces alone and depend on stiffening from the incompressibility of the molecules composing them. Dislocations are possible and these are of at least two kinds, those which preserve the separation between the two regions (one either side of the film) and those which allow the two regions to become connected.

Much work on these two-dimensional curved surfaces has been carried out by S. Andersson and his colleagues and reference may be made to their papers for fuller information (Andersson and others, 1984).

THE PACKING OF SPHERES IN CURVED SPACE

The packings of equal circles in the plane (either purely geometrically or under various laws of mutual repulsion) become interesting when the space is curved to exclude the dominant hexagonal close packing. Similarly the idea of packing 3-D spheres on the positively curved 3-D manifold which is the surface of 4-D hypersphere become attractive for the modelling of random packings of real spheres because it excludes the lattice repetition but automatically builds in the equivalent of periodic boundary conditions. Coxeter (1961) showed that the 120-cell represented the packing of 120 3-D spheres in a curved space so that the tetrahedral packing of nearest neighbours was everywhere preserved. The space is, however, rather highly curved. Coxeter suggested that, in a sense, the local demands, in the packing of equal spheres, for the formation of regular tetrahedra of neighbours, represented an attempt to take up a 4-D metric. It might be said that a real 3-D packing of spheres could be considered as a highly symmetrical arrangement of spheres in 4-D, "uncurved" to restore it to a 3-D metric. The "uncurved" structure is rendered stable in 3-D by filling in the interstices appropriately.

Coxeter's paper at the present colloquium puts forward a packing of 840 spheres on the surface of a hypersphere, which appears still more promising as a model since it is of lower curvature and builds in local five-fold symmetry. Kléman and Sadoc (1979) and Mackay (1980) introduced these concepts to model irregular packings.

The idea that 20 regular tetrahedra almost pack together at a point to make an icoshedron has given rise to the concept of a "frustration" (Nelson, 1983) which is the occurence of a configuration where the local metric demands cannot all be met simultaneously.

ICOSAHEDRAL PACKING

Orthodox crystallography has recently been startled by the observation of extended regions (in a splat-cooled alloy of composition about Al_3Mn) with full icosahedral diffraction symmetry (Shechtman et al. 1984). It is proved, on the first page of every textbook of crystallography, that five-fold symmetry cannot be exhibited by a regular lattice. Such crystals have been observed before (Adam and Hogan, 1972) but their significance was not realised.

We will maintain that this extraordinary type of structure appears because, as suggested first by Coxeter (1961), the local configuration results from preferred distances between atoms which are those which would be appropriate for a four-dimensional metric, but where the real structure, of course, has to be three-dimensional. The 3-D structure ought to be curved positively in the fourth dimension. It cannot be so and the resultant interstices have to be filled in in an ordered way. An ordering which describes the details of the filling in is the three-dimensional Penrose pattern.

In essence, the icosahedral phase is not a crystal but is as near to a crystal as the laws permit. As Steinhardt, Nelson and Ronchetti (1983) have shown, it appears as the most highly ordered example of bond-orientational ordering. The structure is driven by the local demands for icosahedral coordination but appears as if the units were being made to lie on sets of equally-spaced parallel planes running in six directions (perpendicular to the edges of a rhombic triacontahedron). These planes form the hexagrid used in the theory of Kramer and Neri (1984) based on an analysis of the two-dimensional case by de Bruijn (1981). The mutual phase relationships between these planes determine details of the ordering. The constituent units appear to be icosahedral clusters of Al atoms around Mn, the latter having 90% of the diameter of the former and thus packing closely. These constituent groups tend to promote a global environment of icosahedral symmetry but this can only be statistical, as it is incompatible with regular translational order.

DIFFRACTION FROM A PENTAGRID

The icosahedrally ordered structure, first reported by Shechtman and others (1984), has been examined by several others using electron microscipy and diffraction. Its diffraction pattern presents a number of features making it radically different from that of all crystals. For a thin crystal the electron diffraction pattern is the Fourier transform of the projected electron density, which is periodic, any point being given coordinates in terms of two non-parallel base vectors. Corresponding points are reached by integral translations parallel to these two base vectors. All such vectors are thus linearly dependent on the two base vectors.

It is a consequence of the very strong diffraction of electrons by matter that double diffraction occurs. Considering the thin crystal as the superposition of two layers each of half thickness, the first layer scatters the single incident beam into many diffracted beams (in the directions described by the reciprocal lattice, which is the Fourier transform of the direct lattice). Each of these may be of intensity comparable to that of the transmitted direct beam and consequently all these beams are again scattered by the second half of the crystal. Because ot the linear depencence just mentioned, no new directions of scattering arise. The intensities of the diffracted beams may be affected by the double diffraction (all tending to become more nearly equal to each other in intensity) but no new spots arise which are not on the same lattice.

The situation is different for a pentagonal arrangement. Suppose we take the simplest case where our structure is simply the superposition of five planar since waves of scattering matter, lying at 72° to each other. It is a basic theorem that a sinusoidal grating gives only one order of diffraction (its Fourier transform is one spot). Figure 8 shows such a transform. Imagine then that this pattern of beams is again incident on a similar slice of material and again on a third slice. Figure 9 shows the resulting multiplication of diffraction spots by convolution.

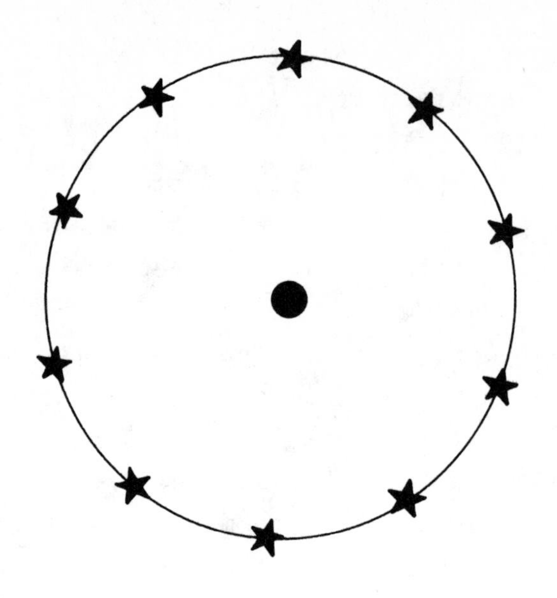

Figure 8

The Fourier transform of a "pentagrid" of five sinusoidal grids set at 72° to each other

Considering the Abbe theory of the microsope it is evident that the images formed by the superposition of each of the three orders of convolution of the scattered beams will be different. A micrograph rich in pseudo-detail would thus result by the mere superposition of five sinusoidal gratings (and would vary with their phase relationships). Rather careful analysis of the picture will be necessary, probably using the weak-beam technique of imaging to remove such effects. To account for the diffraction pattern it is sufficient to postulate a liquidcrystal type of structure where lines of units lie in some disorder on the pentagrid. It may be that microscopists are misled by the diffraction pattern into believing that there is more order in the structure than

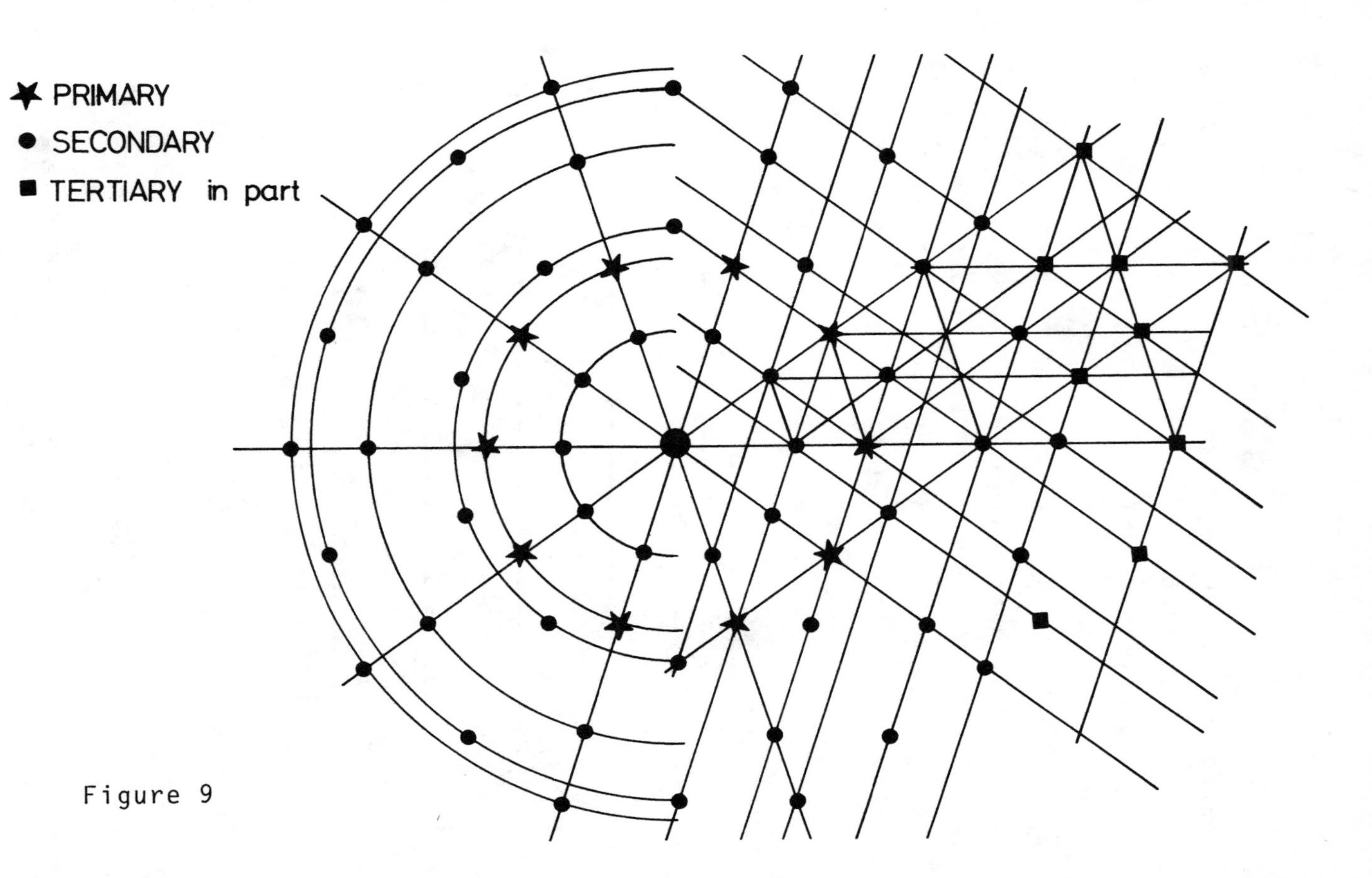

Figure 9

The result of double and triple diffraction of an electron beam passing through the pentagrid of figure 8. The pattern is the self-convolution of the primary ring of 10 points corresponding to the pentagrid

is necessary because of the curious self-convolution pattern of a ring of 10 points at the vertices of a decagon.

The repetion of the five-fold, three-fold and two-fold axes according to the symmetry of the icosahedron is more difficult to account for. In three dimensions six planes, making a hexagrid, are necessary. Within them there may be considerable disorder.

Many different types of disorder are known. A possibly similar situation is observed in the structure of the mineral Vaterite (Kamhi, 1963) where carbonate groups occur in positions which are inconsistent with the hexagonal symmetry of the material (figure 10). The plane

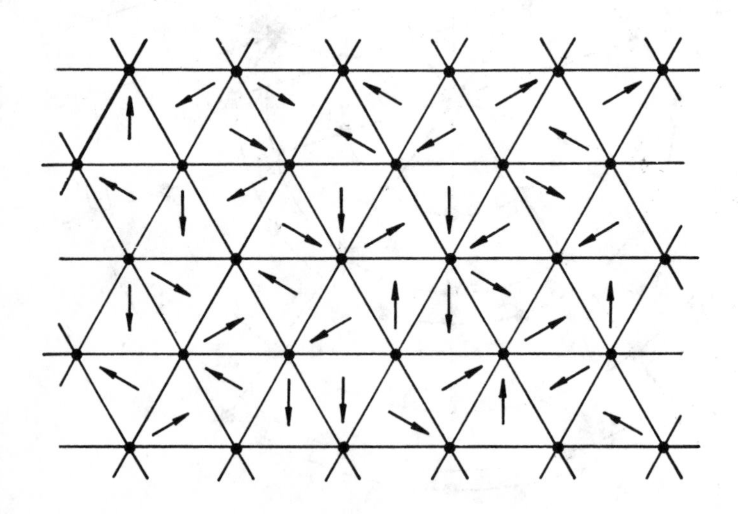

Figure 10

The disordering of the orientations of the carbonate groups in the mineral Vaterite ($CaCO_3$)

of the carbonate group is distributed statistically over three equivalent orientations. This might be taken as the simplest example of bond orientational order. If there were some systematic but noncrystallographic ordering of the directions of the group, then the situation would be similar to that observed in the icosahedral phases, although the relevant ratio would not be irrational.

A POSSIBLE SOLID WOVEN MODEL

Taking the two-dimensional Penrose pattern (figure 1), it is possible to draw five sets of lines across it in a kind of weaving. Each line travels from one tile to another across an edge, so that, perpendicular to each of the five axial directions there are sets of irregularly kinked threads which, on average, lie equally spaced and do not overlap. Two threads must cross in each tile and a rule, as to which is to be on top and which below, must be evolved. The result is a sheet of fabric with five directions of threads.

The same process can be imagined for the three-dimensional structure. Here there are fifteen sets of threads (which could be coloured differently). Five lie in each plane perpendicular to the six axes of the system. These planes form the hexagrid mentioned above. Three threads must enter and leave each obtuse and each acute rhombohedral cell and again a crossing rule must be articulated. There must be two possible hands of crossing and eight permutations taking account of orientation. Thus, we may imagine a woven solid, made of continuous but somewhat wavy threads, which has icosahedral symmetry. There will be strongly expressed planes, like the woven sheets of the 2-D pattern, which will be responsible for most of the scattering by the electron beam and which may have roughly sinusoidal variations from the mean electron density. Thus, we can postulate a "nematic" liquid crystal structure of considerable complexity but answering to the requirements. ["Nematic" means thread-like, and is a term applied to the simplest type of liquid crystal structure where thread-like molecules are roughly parallel to each other but otherwise completely disordered]. We may call this model the "Icosahedral nematic model" although the rhombic triacontahedron (RTH) is seen to be the more fundamental polyhedron.

THE ICOSAHEDRAL NEMATIC LIQUID CRYSTAL PHASE

Looking at one RTH in the proposed thread structure, it can be seen that one thread emerges from each of its 30 faces and connects two opposite faces. Thus each RTH has 30-fold coordination. The internal structure of 10 acute and 10 obtuse rhombohedra determines how the threads are transmitted inside the polyhedron. The whole 3-D structure can be regarded as a packing of RTH. Some of them touch and share a common face. Others overlap and share a single obtuse rhombohedron and others share a common rhombic dodecahedron (made of two acute and obtuse rhombohedra). In each case the RTH with the shared portion removed has the same number of faces (30) although now it has a concave region. Thus it continues to have 30 negih-bours connected by threads and the whole structure can be conceived of as made of 30-connected RTH somewhat over-lapped.

The elucidation of the structure of the icosahedral phases is now being very actively pursued in a number of centres and novel mathematical aspects of crystallography are being examined.

CONCLUSION

Questions of the geometry of curved spaces in two- and three-dimensions have thus taken on a new urgency because of their association with such physical problems as those mentioned here. Proclus (410-485) commented "It is well-known that the man who first made public the theory of irrationals perished in a shipwreck" in order that this embarassing circumstance should remain secret. Thus it is likely that the discovery of five-fold axes in extended crystals will meet with continued resistance from crystallographers entrenched in the beautiful edifice of classical crystallography.

REFERENCES

[1] ADAM, C. and HOGAN, L.M., (1972). *Photograph of 10-branched "snowflakes" of Al_3Fe reproduced in W.J. Moore, "Physical Chemistry",* (5th edn.) Longmans, London, 1972.

[2] ANDERSSON, S. and HYDE, S.T.,)1984). *"The intrinsic curvature of solids",* Z. f. Krist., 168, 1-7.

[3] COXETER, H.S.M., (1964). *"Regular compound tesselations of the hyperbolic plane",* Proc. Roy. Soc., A278, 147-167.

[4] COXETER, H.S.M., (1961). *"Introduction to Geometry",* Wiley, New York. pp. 411-412.

[5] COXETER, H.S.M., (1968). *"Twelve Geometric Essays",* S. Illinois University Press, Carbondale, Ill.

[6] COXETER, H.S.M., (1985). *"The packing of 840 balls of radius 9° 0' 19" on the 3-sphere".* (This volume).

[7] DE BRUIJN, N.G., (1981). *"Algebraic theory of Penrose's nonperiodic tilings of the plane I and II"*, Proc. Konig. Ned. Akad. Weten., A84, 39-65.

[8] HYDE, S.T., ANDERSSON, S., ERICSSON, B. and LARSSON, K., (1984). *"A cubic structure consisting of a lipid bilayer forming an infinite periodic minimum surface of the gyroid type in the glycerol mono-oleate-water system"*, Z. f. Krist., 168, 213-219.

[9] HYDE, S.T. and ANDERSSON, S., (1984). *"A systematic net description of saddle polyhedra and periodic minimal surfaces"*. Z. f. Krist., 168, 221-254.

[10] KAMHI, S.R., (1963). *"On the structure of Vaterite, $CaCO_3$."* Acta Cryst., 16, 770-772.

[11] KLEMAN, M. and SADOC, J-F., (1979). J. Phys. Paris, Lett. 40, L569.

[12] KRAMER, P. and NERI, R. (1984). *"On Periodic and Non-periodic Space Fillings of E^m Obtained by Projection"*, Acta Cryst., A40, 580-587.

[13] MACKAY, A.L., (1980). *"The packing of three-dimensional spheres on the surface of a four-dimensional hyper-sphere"*, J. Phys. A: Math. Gen., 13, 3373-3379.

[14] MACKAY, A.L., (1981). *"De Nive Quinquangula - On the Pentagonal Snowflake"*, Kristallografiya, 26, No. 5. 909-918. also Soviet Physics (Crystallography).

[15] MACKAY, A.L., (1985). *"Periodic minimal surfaces"*, Physica, 131 B, 300-305.

[16] NELSON, D.R. (1983). *"Order, frustration and defects in liquids and glasses"*, Phys. Rew., B28, 5515-5535. Sadoc, J-F. and Mosseri, R., (1984). *"Modelling of the structure of glasses"*, J. Non-cryst. Solids, 61 and 62, 487-498.

[17] SCHOEN, A.H., (1970). *"Infinite periodic minimal surfaces without self-intersections"*. NASA Techn. Note D-5541.

[18] SCHWARZ, H.A., (1890). *"Gesammelte Mathematische Abhandlungen"*, Springer, Berlin, (2 vols.).

[19] SOMERVILLE, D.M.Y., (1929). *"An Introduction to the Geometry of n Dimensions"*, Methuen, London.

[20] STEINHARDT, P.J., NELSON, D.R. and RONCHETTI, M., (1983). *"Bond orientational order in liquids and glasses"*, Phys. Rev., B28, 784-805.

[21] WILSON, K.G., (1979). *"Problems in Physics with Many Scales of Length"*, Sci. Amer., 241, Aug. 1979, 140-157.

ALAN L. MACKAY
Department of Crystallography,
Birkbeck College (University of London),
Malet Street, LONDON, WC1 7HX, U.K.

COLLOQUIA MATHEMATICA SOCIETATIS JÁNOS BOLYAI
48. INTUITIVE GEOMETRY, SIÓFOK, 1985.

FIVE-NEIGHBOUR PACKING OF CONVEX PLATES

E. MAKAI*, Jr.

Let D be a convex plate in the Euclidean plane. A packing $\{D_i\}$ of translates of D is called a *k-neighbour packing* if each D_i has at least k neighbours in $\{D_i\}$ (where D_j is a neighbour of D_i if D_j touches D_i). L. Fejes Tóth ([3], Theorem 2) proved that for a plate D other than a parallelogram the (lower) density d of a five-neighbour packing of translates of D is greater than a universal constant and conjectured $d \geq 3/7$. He exhibited a five-neighbour packing of translates of a triangle of density 3/7. Also he showed that the (lower) density of a six-neighbour packing of translates of any convex plate is at least 1/2 ([3], Theorem 3). Chvátal [2] proved that the density of a six-neighbour packing of translates of a parallelogram is at least 11/15. We will prove a theorem which includes the above mentioned conjecture and theorem of L. Fejes Tóth (restricted to D not a parallelogram).

THEOREM 1. *Let* D *be a convex plate which is not a parallelogram. Let* $\{D_i\}$ *be a five-neighbour packing of translates of* D, *and let* λ *denote the (lower) average*

*Supported by Hung. Nat. Found. for Sci. Research, grant no. 1238.

number of neighbours of the D_i's (*i.e. calculated with lim inf*). *Then the* (*lower*) *density* d *of the packing* $\{D_i\}$ *satisfies the inequality*

$$d \geq 1/(4 - \frac{\lambda}{3}).$$

This inequality is sharp for each possible λ. *Equality can hold* (*even for the infimum of* d *for all systems* $\{D_i\}$ *with* D *given*) *if and only if* D *is a triangle. In particular* $d \geq 3/7$ *for any five-neighbour packing, and* $d \geq 1/2$ *for any six-neighbour packing of translates of* D, *and equality can hold if and only if* D *is a triangle.*

The proof relies on several lemmas.

LEMMA 1. *Let* $\{D_i\}$ *be a five-neighbour packing of translates of a convex plate* D, *which is no parallelogram. Denote* $\{O_i\}$ *a set of homologous points of the plates* D_i. *Draw all the segments* O_iO_j *with* D_i, D_j *neighbouring. Then these segments do not have points in common except at common end-points. Further they decompose the plane into convex polygons which have uniformly bounded diameter and whose areas are greater than a positive constant.*

The intersection property and the decomposition of the plane into convex polygons were proved by L. Fejes Tóth [3], who also showed that the number of sides of these polygons is uniformly bounded. Hence the diameters of these polygons are also uniformly bounded. Lastly denote $E = \frac{1}{2}(D + (-D))$ (Minkowski sum). Then the corresponding translates $\{E_i\}$ of E form a packing, and E_i, E_j are neighbours if and only if D_i, D_j were (cf. [5], p. 217). Suppose O_i is the center of E_i. Then, as noted by L. Fejes Tóth, the area of a k-gon of the decomposition is easily seen to be at least $(\frac{k}{2} - 1)$ times the area of E.

LEMMA 2. *Let* D *be a convex plate which is not a parallelogram and let* $\ell \geq 1$ *be an integer. Let* $D_1, \ldots, D_{3\ell+2}$ *be translates of* D, *all touching* D, D_{i+1} *not overlapping* D_i. *Denote* $0, 0_1, \ldots, 0_{3\ell+2}$ *homologous points of the plates* $D, D_1, \ldots, D_{3\ell+2}$.
Then

$$\sum_{i=1}^{3\ell+1} \measuredangle 0_i 0 0_{i+1} > \ell\pi,$$

where all the angles $\measuredangle 0_i 0 0_{i+1}$ *are measured counter-clockwise (and are* > 0).

This is a generalisation of Grünbaum's theorem [5] that in a packing $\{D_i\}$ of translates of a convex plate D ($\neq$parallelogram) any D_i has at most six neighbours. Its proof see in [6].

LEMMA 3. *Under the conditions of Lemma 1 any polygon of the decomposition described there has at most six sides.*

PROOF. Suppose we have a polygon $0_1 \ldots 0_k$ in the decomposition, 0_{i+1} following 0_i in the positive sense, where $k \geq 7$. We show the sum of the exterior angles at $0_1, \ldots, 0_7$ is $>2\pi$, which is a contradiction. Among the neighbours of the plate D_i corresponding to 0_i there are five ones following each other in the positive sense of rotation, $D_{i,1} = D_{i-1}, D_{i,2}, D_{i,3}, D_{i,4}, D_{i,5} = D_{i+1}$, say (where $D_0 = D_k$). Indexing the homologous points in the same way as the domains we have to show that

$$\sum_{i=1}^{7} \left(\sum_{j=1}^{4} \measuredangle 0_{i,j} 0_i 0_{i,j+1} - \pi \right) > 2\pi,$$

i.e.

$$\sum_{i=1}^{7} \sum_{j=1}^{4} \measuredangle O_{i,j} O_i O_{i,j+1} > 9\pi .$$

Let us consider the vectors $(-1)^{i+1} \overrightarrow{O_i O_{i,j}} = \overrightarrow{O' O'_{(i-1)4+j}}$, $1 \le (i-1)4 + j \le 28$, $\overrightarrow{O_7 O_{7,5}} = \overrightarrow{O' O'_{29}}$. Then

$$\measuredangle O_{i,j} O_i O_{i,j+1} = \measuredangle O'_{(i-1)4+j} O' O'_{(i-1)4+j+1} .$$

The vectors $\overrightarrow{O' O'_1}$, ..., $\overrightarrow{O' O'_{29}}$ satisfy the hypotheses of Lemma 2, with $\ell = 9$, hence

$$\sum_{i=1}^{7} \sum_{j=1}^{4} \measuredangle O_{i,j} O_i O_{i,j+1} = \sum_{i=1}^{28} \measuredangle O'_i O' O'_{i+1} > 9\pi ,$$

as stated.

LEMMA 4. *Any convex plate* D *has an affine image* D' *whose area* A(D') *and diameter* δ(D') *satisfy* $A(D')/\delta^2(D') \ge \sqrt{3}/4$, *with equality if and only if* D *is a triangle.*

Cf. [1], p. 716 (II_3), pp. 745-746.

LEMMA 5. *Let* $D_1, \ldots, D_k$ *be translates of a convex plate* D, *touching each other in this cyclic order. Let their homologous points,* $O_1, \ldots, O_k$ *be the vertices of a convex* k-*gon* $O_1 \ldots O_k$. *Then the area of this* k-*gon is* $\le \frac{k}{\sqrt{3}} \cot \frac{\pi}{k} \cdot A(D)$. *Equality can hold only for triangles, for* k = 3 *and* k = 6.

PROOF. Since the statement is affine invariant, we may suppose by Lemma 4 that $\delta^2(D) \le 4A(D)/\sqrt{3}$. Note that $O_iO_{i+1} \le \delta(D)$. Hence

$$A(O_1 \ldots O_k) \le \frac{k}{4} \cot \frac{\pi}{k} \cdot \delta^2(D) \le \frac{k}{\sqrt{3}} \cot \frac{\pi}{k} \cdot A(D).$$

If D is no triangle, here strict inequality holds. If D is a triangle, by $\delta^2(D) \le 4A(D)/\sqrt{3}$ it is regular, and equality can hold only if $O_1 \ldots O_k$ is a regular k-gon, such that each side O_iO_{i+1} is a translate of a diameter of D. This is possible only for $k = 3$ and $k = 6$.

REMARK. An asymptotically sharp estimate is

$$A(O_1 \ldots O_k) \le k^2 A(D)/6.$$

(Roughly speaking "only so large area can be enclosed by k translates of D which form a convex ring". Such a way of posing the question originates from [4].) Its proof is more involved and will be given elsewhere.

PROOF OF THE THEOREM. Let us consider the decomposition of the plane from Lemma 1. Let G be a suitable large domain and let v, e and f denote the number of vertices, edges and faces of the decomposition lying in G. Taking in account the last statements of Lemma 1, the hypothesis of the theorem and Euler's theorem we see $e \sim \lambda' v/2$, $v+f \sim e$, where $\lambda' \ge \lambda$. Hence $v/f \sim 1/(\frac{\lambda'}{2} - 1)$ and the average number of sides of the polygons of the decomposition in G is $\sim 2e/f \sim \lambda'/(\frac{\lambda'}{2} - 1)$.

Since $\frac{k}{\sqrt{3}} \cot \frac{\pi}{k} \le \frac{5}{3} k - 4$ for $3 \le k \le 6$ (with equality for $k = 3,6$), the average area of the polygons is

$\lesssim (\frac{5}{3} \cdot \frac{\lambda'}{\frac{\lambda'}{2} - 1} - 4)A(D)$. Hence for the density d' in G we have $d' \gtrsim \frac{v\ A(D)}{f(\frac{5}{3} \cdot \frac{\lambda'}{\frac{\lambda'}{2} - 1} - 4)A(D)} \sim \frac{1}{4 - \frac{\lambda'}{3}} \geq \frac{1}{4 - \frac{\lambda}{3}}$.

Equality can hold only for triangles.

However for D = triangle equality can hold for any λ ($5 \leq \lambda \leq 6$, by [5], above cited). We may suppose D is regular. For $\lambda = 5$ we may take L. Fejes Tóth's example, with $d = \frac{3}{7}$, where the decomposition is the Archimedean tiling (6,3,3,3,3), with edges equal and parallel to those of D. This can be obtained from a lattice of regular triangles, with basic vectors two sides of the triangle, by deleting every seventh triangle in each row, the deleted triangles in the row neighbouring from below being translated 5/2 side lengths to the right (cf. Fig. 1, where the broken lines represent the deleted

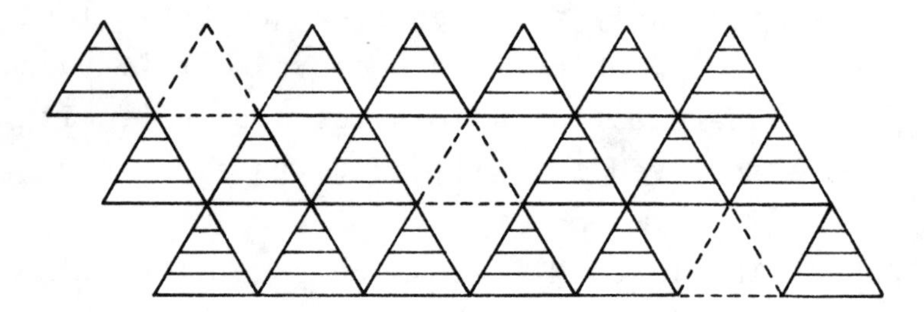

Figure 1

triangles). For $5 \leq \lambda \leq 6$ we perform the deletion of the triangles only on some part of the plane, the density of this part on the whole plane chosen suitably. Evidently this can be done so that λ will be an average and the density of $\{D_i\}$ will exist. Q.e.d.

REMARK. Let us relax the condition of $\{D_i\}$ being a five-neighbour packing to that $\{D_i\}$ is a packing and the conclusions of Lemmas 1 and 3 hold (which entails each O_i is a vertex of a polygon of the decomposition - i.e., if we suppose, as we may, D centrosymmetric, no D_i can be enclosed by at most six other D_j's none touching it, which follows by the methods of [5] and [6] using strictly convex approximation for D - and the average number of sides $\sim\lambda'/(\frac{\lambda'}{2}-1) \lesssim 6$ thus $\lambda \geq 3$). Then we still have the density estimate of the theorem, which will be sharp for each λ, $3 \leq \lambda \leq 6$. For this we consider the same lattice of regular triangles as above, but now we delete on some part of the plane triangles in such a way that the decomposition should consist of regular hexagons, with sides equal and parallel to those of D.

For the centrosymmetric case there holds the following

THEOREM 2. *Let* D *be a centrosymmetric convex plate which is not a parallelogram. Let* $\{D_i\}$ *be a five-neighbour packing of translates of* D, *and let* λ *denote the (lower) average number of neighbours of the* D_i*'s. Then the (lower) density* d *of the packing* $\{D_i\}$ *is at least* $\frac{3}{2}/(4-\frac{\lambda}{3})$. *This inequality is sharp for each possible* λ. *Equality can hold (even for the infimum of* d *for all systems* $\{D_i\}$ *with* D *given) if and only if* D *is an affine regular hexagon. In particular* $d \geq 9/14$ *for any five-neighbour packing and* $d \geq 3/4$ *for any six-neighbour packing of translates of* D *and equality can hold if and only if* D *is an affine regular hexagon.*

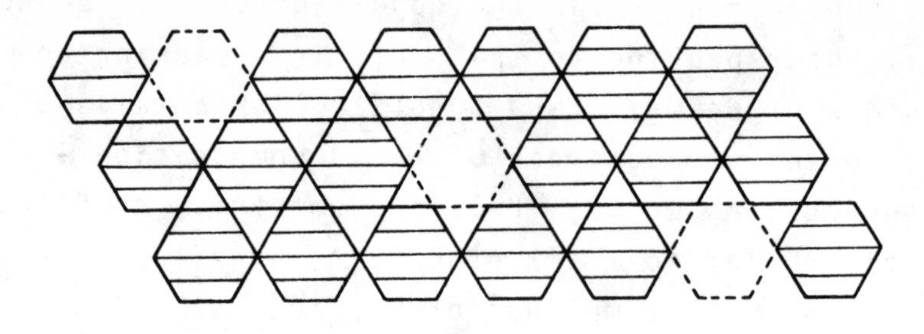

Figure 2

The case of equality $d = 9/14$ is shown in Fig.2. We note that Theorem 2 implies Theorem 1 by the difference body inequality $A(\frac{1}{2}(D+(-D))) \leq 3A(D)/2$, where equality holds only for $D =$ triangle, cf. [7] (note that D is a parallelogram if $\frac{1}{2}(D+(-D))$ is a parallelogram, [5], p. 217, (VI)). For Theorem 2 instead of analogues of Lemma 5 we need estimates of $A(O_1 \ldots O_k)$ in terms of $A(D)$ and the density of the thinnest six-neighbour lattice packing of translates of D. Details will be given elsewhere.

REFERENCES

[1] F. BEHREND, *Über einige Affininvarianten konvexer Bereiche*, Math. Ann. 113(1937), 713-747.

[2] V. CHVÁTAL, *On a conjecture of Fejes Tóth*, Period. Math. Hungar. 6(1975), 357-362.

[3] L. FEJES TÓTH, *Five-neighbour packing of convex discs*, Period. Math. Hungar. 4(1973), 221-229.

[4] L. FEJES TÓTH and A. HEPPES, *Regions enclosed by convex domains*, Stud. Sci. Math. Hungar. 1(1966), 413-417.

[5] B. GRÜNBAUM, *On a conjecture of Hadwiger,* Pacific J. of Math. 11(1961), 215-219.

[6] E. MAKAI, Jr., *On Hadwiger numbers and enclosing numbers,* (to appear)

[7] C.A. ROGERS and G.C. SHEPHARD, *The difference body of a convex body,* Arch. Math. 8(1957), 220-233.

E. MAKAI, Jr.
Mathematical Institute
of Hungarian Acad. Sci.
H-1364 Budapest
Pf. 127
Hungary

COLLOQUIA MATHEMATICA SOCIETATIS JÁNOS BOLYAI
48. INTUITIVE GEOMETRY, SIÓFOK, 1985.

SOME RESULTS AND PROBLEMS AROUND ZONOTOPES

H. MARTINI

Dedicated to Prof. L. Fejes Tóth

1. INTRODUCTION

In the last twenty years zonotopes and zonoids have intensively been studied because these special convex bodies have interesting properties and characterizations on the one hand and they are applied in solving various problems on the other hand.

In the present paper we deal with the second aspect and restrict our investigations to the so-called finite sums of line segments (zonotopes).

For these special polytopes several geometrical characterizations are known. For example they can be characterized as totally symmetric polytopes (that are polytopes all faces of which have a centre of symmetry), as shadows of measure polytopes (in the sense of a generalization of the well-known theorem of Pohlke), and as projection bodies of convex polytopes. Note, especially in the latter case, that zonotopes can be obtained by Minkowski addition of line segments.

The relations to arrangements of hyperplanes and to the investigations around convex bodies, which tile space by translation, are also to be mentioned.

A nearly complete survey on zonotopes give the papers of Coxeter [9], Bolker [3], McMullen [28, 29, 30], Shephard [36], Schneider-Weil [35] and Greene-Zaslavsky [11].

In the following some results, applications and open problems around zonotopes shall be presented.

2. DEFINITIONS AND PROPERTIES

By $\mathbb{E}^d$ we denote the d-dimensional Euclidean space ($d \geq 2$) with scalar product $<.,.>$, origin $\underline{o}$ and unit sphere $S^{d-1} := \{\underline{u} \in E^d : <\underline{u},\underline{u}> = 1\}$. $\mathbb{K}^d$ is written for the set of d-dimensional convex bodies (that are compact convex sets with non-empty interior), $\underline{\mathbb{K}}^d$ for the subset of smooth bodies and $\mathbb{P}^d$ for the set of convex d-polytopes (convex bodies, the extreme points of which form a finite set).

$\mathfrak{A}_n := \{\underline{a}_1,\dots,\underline{a}_n\}$ is an arbitrary finite vector star with $\dim \operatorname{lin} \{\underline{a}_1,\dots,\underline{a}_n\} = d-r$ ($r \in \{0,\dots,d-1\}$).

A point set

$$(1) \qquad Z := \left\{ \sum_{i=1}^{n} \lambda_i \underline{a}_i : 0 \leq \lambda_i \leq 1 \right\} = \sum_{i=1}^{n} \left[\underline{o}, \underline{a}_i\right]$$

($\left[\underline{o}, \underline{a}_i\right] = \operatorname{conv} \{\underline{o}, \underline{a}_i\}$) is called *zonotope* (or Minkowski sum of line segments) with the *generating star* $\mathfrak{A}_n$ and the *generating line segments* $\left[\underline{o}, \underline{a}_i\right]$, respectively ($i=1,\dots,n$). In

$$(2) \qquad \underline{a}_M := \frac{1}{2} \sum_{i=1}^{n} \underline{a}_i$$

the centre of Z is given, because for $\underline{x} \in Z$ with (1) and

$$\sum_{i=1}^{n} \underline{a}_i - \sum_{i=1}^{n} \lambda_i \underline{a}_i = \sum_{i=1}^{n} (1 - \lambda_i)\underline{a}_i =$$

$$= \sum_{i=1}^{n} \lambda_i^* \underline{a}_i \qquad (0 \leq \lambda_i;\ \lambda_i^* \leq 1)$$

it follows that $(2\underline{a}_M - \underline{x}) \in Z$.

A set will be called a *cell* of Z if it can be written as

$$(3) \qquad R = [\underline{o}, \underline{a}_1]_\varepsilon + \ldots + [\underline{o}, \underline{a}_k]_\varepsilon + (\delta_{k+1}\underline{a}_{k+1})_\varepsilon + \ldots + (\delta_n \underline{a}_n)_\varepsilon$$

with $0 \leq k \leq n$, ε as a permutation of $\{1,\ldots,n\}$ and $\delta_i \in \{0;\ 1\}$ $(i=k+1,\ldots,n)$. (Special cells are the faces of the zonotope. Z is also a face of itself.)

With the condition

$$\langle \underline{a}_i, \underline{u} \rangle > 0 \Longleftrightarrow i \in N_q \ , \quad \underline{a}_i \in \mathfrak{A}_n,$$

for every direction $\underline{u} \in S^{d-1}$ a subset N_q of $\{1,\ldots,n\}=:N_n$ is determined. Then with

$$(4) \qquad \underline{a}_S(\underline{u}) := \sum_{i \in N_q} \underline{a}_i$$

a vertex and by

(5) $$h(Z,\underline{u}) := \sum_{i \in N_q} \langle \underline{a}_i, \underline{u} \rangle = \langle \underline{a}_S(\underline{u}), \underline{u} \rangle, \quad \underline{u} \in S^{d-1},$$

the *support function* of Z is given.

A translate Z' of Z with centre $\underline{o}$, determined by

(6) $$Z' := \frac{1}{2} \sum_{i=1}^{n} [- \underline{a}_i, \underline{a}_i] ,$$

has the support function

(7) $$h(Z', \underline{u}) := \frac{1}{2} \sum_{i=1}^{n} |\langle \underline{a}_i, \underline{u} \rangle|, \quad \underline{u} \in S^{d-1} .$$

Because of the well-known transformation of the support function by translation of the corresponding convex body we get back the support function of Z from (7) with (2) in

(5*) $$h(Z,\underline{u}) := \frac{1}{2} \sum_{i=1}^{n} |\langle \underline{a}_i, \underline{u} \rangle| + \langle \underline{a}_M, \underline{u} \rangle, \quad \underline{u} \in S^{d-1} .$$

A *zone* of Z is the set of all edges in the intersection with Z of supporting hyperplanes with normal vectors orthogonal to a vector $\underline{a}_i \in \mathfrak{A}_n$. Consequently, the zonotope has only m zones ($m \leq n$), in which m is the number of lines in the system

$$G_m := \{\bar{g} : \bar{g} = \eta \underline{a}_i : -\infty < \eta < +\infty\}$$

for $\underline{a}_i \in \mathfrak{A}_n$.

Because of (3) the i-th zone (now $i=1,\ldots,m$) is represented by edges which are translates of the vector sum of all generating line segments parallel to the i-th line in G_m.

Vice versa: By an arbitrary orientation of one representing edge of the i-th zone we get the vector $\underline{b}_i$ ($i \in \{1,\ldots,m\}$), where $[o, \underline{b}_i]$ is a translate of this edge and, consequently, we obtain a vector star $\mathfrak{B}_m := \{\underline{b}_1,\ldots,\underline{b}_m\}$ for the whole zonotope.

For every pair of real numbers $\mu_i; \nu_i$ with $-1 \leq \mu_i; \nu_i \leq 1$ and $|\mu_i - \nu_i| = 1$ ($i=i,\ldots,m$) a star

$$\mathfrak{B}_p := \{\mu_1\underline{b}_1, \nu_1\underline{b}_1, \ldots, \mu_m\underline{b}_m, \nu_m\underline{b}_m\}$$

is called *zonal vector star* of the zonotope Z. (Integer p with $m \leq p \leq 2m$ denotes the number of all vectors different from the null vector.) Therefore, an arbitrary translate of Z containing $\underline{o}$ is given by

$$\bar{Z} := \sum_{i=1}^{n} [\mu_i\underline{b}_i, \nu_i\underline{b}_i]$$

with centre $\frac{1}{2} \sum_{i=1}^{m} (\mu_i + \nu_i)\underline{b}_i$. It should be noticed that the difference set of Z is represented by the convex hull of all these translates in

$$D(Z) := \sum_{i=1}^{m} [-\underline{b}_i, \underline{b}_i] .$$

Because of (2) in the set of all these translates the zonotope Z is determined by

$$(8) \qquad \sum_{i=1}^{m} (\mu_i + \nu_i)\underline{b}_i = 2\underline{a}_M \Longleftrightarrow \sum_{i=1}^{m} [\mu_i\underline{b}_i, \nu_i\underline{b}_i] = Z .$$

Since Z has segment summands parallel to each of its edges the spherical image of the zonotope is a dissection of S^{d-1} by (d-2)-dimensional great spheres (or a spherical mosaic of (d-2)-dimensional great spheres), where every great sphere corresponds to one zone and vice versa. Since there is a one-to-one correspondence between points of projective (d-1)-space and pairs of antipodal points of S^{d-1}, we see the combinatorial boundary structure of the zonotope in the central projection of the spherical mosaic, i.e. a projective (d-1)-arrangement $\bar{A}^{d-1}(m)$ of m hyperplanes.

In the following $\bar{A}^{d-1}(m)$ shall be called the *gnomonic projection* or, in the sense of Coxeter, the second projective diagram of Z. Further let $Z^{d-r}(m)$ denote the set of m-zonal (d-r)-zonotopes, $\underline{Z}^{d-r}(m)$ the subset of zonotopes with centre $\underline{o}$ and, clearly, $Z^{d-r}(d-r)$ the set of (d-r)-parallelotopes.

3. COVERING ZONOTOPES BY SMALLER HOMOTHETICAL COPIES

For a d-dimensional convex body K the integer a(K) shall denote the smallest number of translates of a homothetical copy νK ($\nu \in \mathbb{R}^+$, $\nu < 1$) such that K can be covered in $\mathbb{E}^d$ (compare Grünbaum [12]). It is obvious that for $K \in Z^d(d)$ this number is determined by $a(K) = 2^d$.

Hadwiger [15] conjectured that $a(K) \leq 2^d$ holds for all d-dimensional convex bodies and Boltyanskiĭ [5] gave his famous illumination problem as an equivalent problem. With respect to this illumination problem a(K) is the smallest number of directions which illuminate bd K. ($\underline{x} \in$ bd K is said to be illuminated by a direction $\underline{l}$ iff the ray $S := \{\underline{s} : \underline{s} = \underline{x} + \lambda \underline{l};\ \lambda > 0\}$ has nonempty intersection with int K.)

In the following list of known results to this topic four additional subsets of $\mathbb{K}^d$ are needed. We write

$\mathbb{K}_C^d$	for the set of convex bodies	of constant width,
$\mathbb{K}_M^d$		with a centre of symmetry,
$\mathbb{K}_1^d$		with no more than d singular boundary points,
$\mathbb{K}_2^d$		with the property that the projection cone of an arbitrary boundary point $\underline{x}$ contains a rotation cone with half opening angle α and $\cos \alpha < \frac{1}{d}$ as well as with apex $\underline{x}$.

On this base it is known:

d=2	$a(K)=3$	$K \in \mathbb{K}^2 \setminus Z^2(2)$,	Levi [20]
d=3	$a(K) \le 6$	$K \in \mathbb{K}_C^3$	Lassak [18], [19]
	$a(K) \le 8$	$K \in \mathbb{K}_M^3$	Lassak [18], [19]
	$a(K) \le 20$	$K \in \mathbb{K}^3$	Lassak [18], [19]
d≥2	$a(K) < 2^d$	$K \in Z^d(m : m > d)$,	Soltan [7]
	$a(K) \le 2^d(d \ln d + d \ln \ln d + 5d)$	$K \in \mathbb{K}_M^d$,	Rogers [6]
	$a(K) \le (d+1)^d$	$K \in \mathbb{K}_M^d$,	Levin-Petunin [6]
	$a(K) = d+1$	$K \in \mathbb{K}_1^d$,	Boltyanskiĭ [5]
	$a(K) = d+1$	$K \in \mathbb{K}_2^d$,	Weissbach [38].

In addition to this survey the author can make a supplement:

$$(9) \qquad a(K) \leq 3 \cdot 2^{d-2}, \ K \in Z^d(m:m > d).$$

The following theorem even refers to a more general class of polytopes than $Z^d(m:m > d)$. This class (its representatives are called "planets") has been introduced by Bolker [4]. A convex d-polytope is a planet iff it has segment summands parallel to each of its edges. (Obviously, every zonotope is a planet.)

At first some preliminary notes: It is an immediate consequence of the given definition that a d-planet has zones and a dissection of S^{d-1} by (d-2)-dimensional great spheres as its spherical image.
We take Q to be an m-zonal d-planet $(m > d)$.

For the illumination of bd Q (in the sense of Boltyanskiǐ [5]) we use only pairwise-opposite directions. Every pair of oppositely directed illuminations generates (as its "shadow-boundary") an additional (d-2)-dimensional great sphere on S^{d-1} with respect to the spherical image of Q. The gnomonic projection $\bar{A}^{d-1}(m:m > d)$ of the planet is a set of hyperplanes (with no point in common) in the real projective (d-1)-space together with the associated dissection of this space into cells of various dimensions. One may consider these cells to be relatively open (i.e., they have no points in common) or to be relatively bounded.

Clearly, the central projection of a shadow-boundary is a hyperplane of projective (d-1)-space, too. (We transfer the notation "shadow-boundary" also to these hyperplanes.)

So every relatively open (d-r-1)-cell in $\bar{A}^{d-1}(m)$ represents a pair of relatively open r-faces of Q ($r \in \{0,\dots,d-1\}$).
Such a pair of r-faces is illuminated by two opposite directions if the corresponding relatively *bounded* (d-r-1)-cell of $\bar{A}^{d-1}(m)$ has no points in common with the adequate shadow-boundary.
A (d-1)-arrangement $\bar{C}$ ($\bar{A}^{d-1}(m)$) of shadow-boundaries shall be called a *covering* of $\bar{A}^{d-1}(m)$ if every relatively bounded (d-r-1)-cell of the gnomonic projection has no points in common with at least one hyperplane of $\bar{C}$ ($\bar{A}^{d-1}(m)$).
Our interest refers to the lower bound of card $\bar{C}$ with respect to $\bar{A}^{d-1}(m)$.

An important conclusion shall be emphasized: Let $\bar{C}$ be a covering of $\bar{A}^{d-1}(m)$. If we adjoin to $\bar{A}^{d-1}(m)$ an additional set of hyperplanes, then the united arrangement is also covered by $\bar{C}$. (This follows from the simple fact that a cell which has no points in common with one shadow-boundary transfers this property to arbitrary dissections of itself.) Consequently, further on the consideration of (d-1)-arrangements with $d+1$ hyperplanes as gnomonic projections will suffice.

Secondly we point out (as an alleviation for the reader): Since every relatively bounded (d-r-1)-cell of $\bar{A}^{d-1}(d+1)$ belongs to relatively bounded (d-1)-cells of the same arrangement, we only need to consider cells of dimension $d-1$. We have

LEMMA 1. *Every projective 2-arrangement of four lines which have no point in common possesses a covering by three shadow-boundaries.*

PROOF. By studying all combinatorial types of these arrangements (Grünbaum [13 (ch. 18)]) we obtain for every type an arrangement $\bar{C}$ of three shadow-boundaries with the demanded property (dashed in Figure 1).

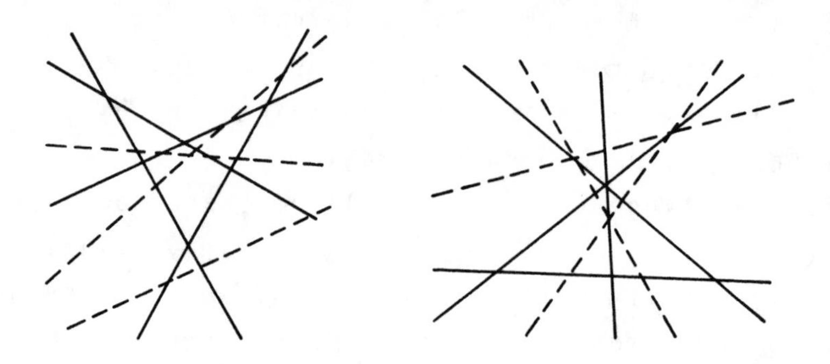

Figure 1

Among d-planets only parallelotopes have the property that every (d-1)-polytope obtained by orthogonal projection parallel to a segment summand is a (d-1)-parallelotope. Therefore, in terms of gnomonic projections, an arrangement $\bar{A}^{d-1}(d+1)$ contains at least one hyperplane H which includes a partial (d-2)-arrangement $\bar{A}^{d-2}(d)$. We call an *affinization* of $\bar{A}^{d-1}(d+1)$ a Euclidean arrangement $A^{d-1}(d)$ obtained by regarding H as the hyperplane at infinity. (To cover a Euclidean arrangement shall be the same as in the projective case.)

Finally, the union of all relatively bounded (d-1)-cells in $A^{d-1}(d)$ which have no points in common with H is said to be the *kernel* of this arrangement. Now we are ready for

LEMMA 2. *Let* $\bar{A}^{d-2}(d)$ *be covered by a* (d-2)-*arrangement of* k *shadow-boundaries, and let* $\bar{A}^{d-1}(d+1)$ *contain* $\bar{A}^{d-2}(d)$ *as a partial arrangement in* H, *where* H *is one of the hyperplanes in* $\bar{A}^{d-1}(d+1)$. *Then* $\bar{A}^{d-1}(d+1)$ *can be covered by a* (d-1)-*arrangement of* 2k *shadow-boundaries.*

PROOF. We want to show that the affinization $A^{d-1}(d)$ can be covered by a (d-1)-arrangement of 2k pairwise-parallel shadow-boundaries, the intersections of which in H are k covering shadow-boundaries regarding $\bar{A}^{d-2}(d)$ in the hyperplane at infinity. (From practical reasons, (d-2)-dimensional shadow-boundaries parallel to any hyperplane of $A^{d-1}(d)$ are excluded.)

The interior of the strip between a pair $\{S_1, S_2\}$ of parallel (d-2)-dimensional shadow-boundaries in $A^{d-1}(d)$ shall contain the kernel of this arrangement. Therefore, we can restrict our attention to the set of (d-1)-cells which do not belong to the kernel.

This set is also partitioned into two subsets by $\{S_1, S_2\}$: (d-1)-cells which have points in common with only one shadow-boundary of $\{S_1, S_2\}$ and with both shadow-boundaries, respectively. With the conditions of our illumination problem only the cells of the second type remain. From the characterizing property of these cells it follows immediately that, in each case, their and only their intersections with H contain points of $S_1 \cap S_2$.

The interpretation of $S_1 \cap S_2$ as (d-3)-dimensional shadow-boundary of $\bar{C}(\bar{A}^{d-2}(d))$ completes the proof.

Summarizing the statements, we formulate

THEOREM (3.1). *For a* d-*planet* Q *with* m *zones* (m > d) *we have*

$$a(Q) \leq 3 \cdot 2^{d-2}.$$

4. DETERMINATION OF d-ZONOTOPES BY THEIR SUPPORT NUMBERS

For a convex d-polytope P with n facets and $\underline{o} \in \operatorname{int} P$, the n facet hyperplanes shall be given by their distances from the origin. These distances together with the outer normal directions of the corresponding facets are called *support numbers* (or "Tangential parameter"; compare Minkowski [31]) of the polytope, and they replace its support function

$$h(P,\underline{u}) := \sup\{\langle \underline{x},\underline{u}\rangle : \underline{x} \in P\} \ , \ \underline{u} \in S^{d-1} \ .$$

Thus, for the determination of P, the n support numbers suffice, and for convex d-polytopes which are centred at the origin $\frac{n}{2}$ support numbers in directions of pairwise-distinct nonoriented facet normals are enough.

Since zonotopes are special centrally symmetric polytopes, the following problem is motivated: How many support numbers are sufficient for the determination of an arbitrary m-zonal d-zonotope Z with centre $\underline{o}$?

On account of (3) every zone of Z is represented by edges which are translates of a sum of parallel generating line segments or of one such line segment. A *meridian path* is a spatial closed polygon of edges in bd Z, containing exactly two such translates per zone and centred at the origin. It follows immediately that the determination of a half meridian path (consisting of pairwise-nonparallel edges) suffices for determining arbitrary zonal vector stars of Z. Then with (8) the line segments $[\underline{o}, \underline{b}_1]$,... ..., $[\underline{o}, \underline{b}_m]$, derived from an investigated half meridian path by translation, give us the zonotope by the formula

$$\frac{1}{2} \sum_{i=1}^{m} [-\underline{b}_i, \underline{b}_i] = Z .$$

Every vertex contained in the half meridian path is determined by d neighbouring facets (Fig. 2: shaded facets).

The correspondence between pairs of (d-r-1)-faces of Z and r-cells of the gnomonic projection $\bar{A}^{d-1}(m)$ ($r \in \{0,\dots,d-1\}$) permits an obvious illustration of the relationships in the projective arrangement: Every path in $\bar{A}^{d-1}(m)$ connecting neighbouring (d-1)-cells through the relative interior of their common facets and traversing every hyperplane of $\bar{A}^{d-1}(m)$, exactly once, represents a meridian path of Z (dashed in Fig. 2). Every (d-1)-cell traversed by such a path in our sense is "determined" by d of its vertices as projections of linearly independent facet normals of the zonotope (Fig. 2: small circles).

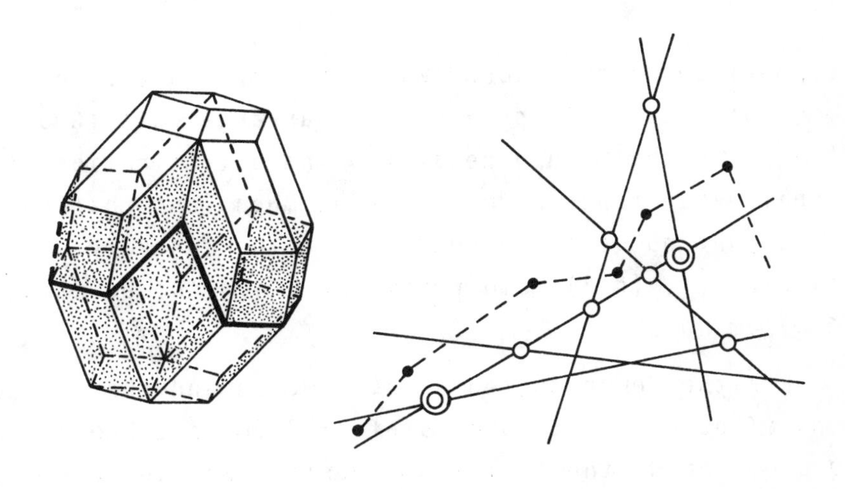

Figure 2

The two end points of a half meridian path are opposite points with respect to $\underline{o}$. This fact allows the elimination of one such point in our considerations. The two remaining end points of the spatial polygon have to be determined by d of their neighbouring facets in each case. Therefore, two of the 0-cells in fig. 2 are doubly designated.
We define $b(m,d)$ to be the smallest number of support numbers which determine Z.

Clearly, $b(m,d)$ depends on m, d and the incidence properties of $\bar{A}^{d-1}(m)$. Since, therefore, the determination of $b(m,d)$ is related to the hard problem of investigating combinatorial types of projective $(d-1)$-arrangements with m hyperplanes (Grünbaum [13 (ch. 18)]), a complete solution is probably very difficult.

Independently of these special incidence properties, one may find for $m \geq 5$ the nonsharp upper bound

$$\left[\frac{3m+3}{2}\right] > b(m,3) \ .$$

A comparison of this bound with the upper bound for the number of 0-cells in projective 2-arrangements (Grünbaum [14]) shows that, in general for the determination of m-zonal 3-zonotopes with centre $\underline{o}$, much less than half of all support numbers are needed. (On the contrary, regarding all $(d-2)$-fold prisms in $\underline{Z}^d(m)$ the equation $b(m,d) = m$ holds.)

For the demonstrated problem we demand that directions of outer facet normals of $Z \in \underline{Z}^d(m)$ are known. On the other hand, the following question was answered by Schneider [34]: Which subsets of S^{d-1} with corresponding support values of $Z \in \underline{Z}^d(m)$ suffice for the determination of the zonotope provided no information on Z is given? It is evident that such subsets cannot be finite sets. However, as an open question remains to be solved

PROBLEM (4.1). Find the minimal number of support numbers of an arbitrary m-zonal d-zonotope with centre $\underline{o}$ which are sufficient for the determination of this polytope.

5. ZONOTOPES AS PROJECTION BODIES OF CONVEX POLYTOPES

Because of the geometrical background of the following statements in an arbitrarily given star $\mathfrak{A}_n := \{\underline{a}_1,\ldots,\underline{a}_n\}$ vectors with equal directions shall be excluded.
If a_i is outwardly normal to the i-th facet $(i=1,\ldots,n)$ of an arbitrary convex d-polytope P and $\|\underline{a}_i\|$ represents the (d-1)-content of this facet, then we have

$$\sum_{i=1}^{n} \underline{a}_i = \underline{o} ,$$

and $P \in \mathbb{P}^d$ is (up to translation) determined by $\mathfrak{A}_n$ (Steinitz [37]).
The zonotope

$$\Pi^{d-1}(P) := \sum_{i=1}^{n} \left[\underline{o}, \underline{a}_i\right] = \frac{1}{2} \sum_{i=1}^{n} \left[- \underline{a}_i, \underline{a}_i\right]$$

is the *projection body* of P. Thus $h(\Pi^{d-1}(P), \underline{u})$ is the *outer (d-1)-quermass* or (d-1)-dimensional content of the orthogonal projection of P in direction $\underline{u} \in S^{d-1}$. (The outer (d-1)-quermass is also called brightness.) By

$$\sum_{i=1}^{n} \underline{a}_i = \underline{o}$$

and dim $\text{lin}\{\underline{a}_1,\ldots,\underline{a}_n\} = d$ the polytope $\Pi^{d-1}(P)$ is a d-zonotope and centred at the origin. Moreover, $\Pi^{d-1}(P)$ is m-zonal, where m is the number of the nonoriented facet normals of P. Thus $\Pi^{d-1}(P) \in \underline{Z}^d(m)$.

5.1. *Determination of convex* d-*polytopes by the* (d-1)-*contents of their orthogonal projections*

The question to what extent convex d-polytopes (up to translation) are determined by the (d-1)-contents of their normal projections leads with (8) to the characterization of special classes of polytopes (compare Martini [22]). Polytopes of such a class have the same projection body and shall be called polytopes of "constant relative brightness".

They belong to a polytope-species with the restriction that for every polytope parallel facets of i-th position ($i=1,\ldots,m$; $m \leq n$) have adequately the same sum of contents, and with the enlargement that in borderline cases one of two parallel facets can be lower-dimensionally degenerated.
(A polytope-species encloses all convex d-polytopes of the same, here in general centrally symmetric, system of outer facet normals; see Fedorov [10] or Steinitz [37].)

Further on, a class of polytopes with constant relative brightness shall be denoted by $\mathbb{P}^d(\Pi^{d-1})$.

In general, uncountably many translation classes belong to $\mathbb{P}^d(\Pi^{d-1})$. Only the d-parallelotopes are determined up to translation by their projection body.

Moreover, $\mathbb{P}^d(\Pi^{d-1})$ contains only finitely many or no translation classes of polytopes without pairs of parallel facets. Fig. 3 shows three representatives of constant relative brightness with the Archimedean rhombic dodecahedron as projection body.

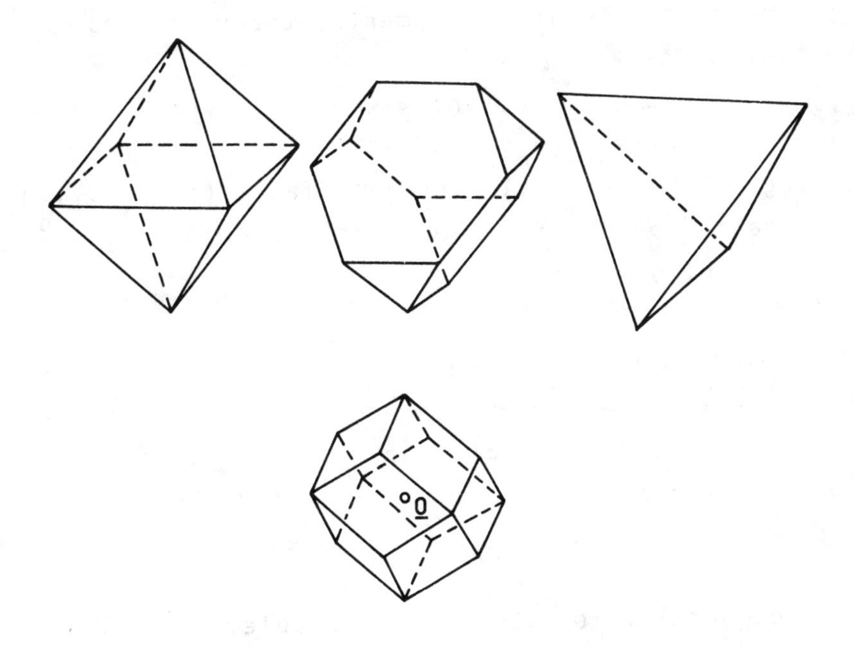

Figure 3

According to a theorem of Steinitz [37] and with $|\mu_i| = |\nu_i| = \frac{1}{2}$ $(i=1,\ldots,m)$ the equivalence (8) shows that a class $\mathbb{P}^d(\Pi^{d-1})$ contains exactly one translation class of polytopes with centre of symmetry, and these polytopes have (see, for example, Petty [33]) the maximal volume in $\mathbb{P}^d(\Pi^{d-1})$. Therefore, the following is motivated.

PROBLEM (5.1). Find a simple geometrical characterization of those polytopes in $\mathbb{P}^d(\Pi^{d-1})$ which have minimal volume.

5.2. *Optimization problems*

At first two necessary conditions for the determination of the extremal values of the function $h(\Pi^{d-1}(P), \underline{u})$, $\underline{u} \in S^{d-1}$, shall be given.
(Regarding the following statements, the reader is referred to Martini-Weissbach [26].)
$\underline{u}^*(\underline{u}_*)$ denotes the directions with the extremal outer (d-1)-quermasses $\bar{v}_{max}(\bar{v}_{min})$ of $P \in \mathbb{P}^d$.
According to theorem (5.1) the i-th facet ($i \in \{1,...,n\}$) of P is called *projecting facet* with respect to $\underline{u} \in S^{d-1}$ if $\langle \underline{a}_i, \underline{u} \rangle = 0$. We have

THEOREM (5.1). *If a convex* d-*polytope* P *has maximal (minimal) outer* (d-1)-*quermass regarding a direction* $\underline{u} \in S^{d-1}$, *then no facets (at least* d-1 *facets with linearly independent outer normal directions) of* P *are projecting facets relative to* $\underline{u}$.

Concequently, for the minimum problem $\binom{n}{d-1}$ directions need to be investigated.

For $\bar{v}_{max}$ and $\underline{u}^*$ even a necessary and sufficient condition is obtained in

THEOREM (5.2). *If for a proper subset* I^* *of* $\{1,...,n\} =: N_n$

$$\left\| \sum_{i \in I^*} \underline{a}_i \right\| \geq \left\| \sum_{i \in I} \underline{a}_i \right\|$$

holds with respect to every other subset I *of* N_n, *then all outer* (d-1)-*quermasses of a convex* d-*polytope* P, *represented by* $\mathfrak{A}_n := \{\underline{a}_1, ..., \underline{a}_n\}$ *up to translation, are not greater than*

$$\bar{v}_{max} := \left\| \sum_{i \in I^*} \underline{a}_i \right\|.$$

This bound is attained for the direction

$$\underline{u}^* := \frac{\sum_{i \in I^*} \underline{a}_i}{\bar{v}_{max}}.$$

Therefore, the maximum problem is restricted to $2^{n-1} - 1$ directions. Using theorem (5.2) for directly given polytopes with up to 50 facets one can determine $\bar{v}_{max}$ and $\underline{u}^*$ by means of "branch and bound" on the base of the Boolean optimization problem

$$(10) \qquad \left\| \sum_{i=1}^{n} \lambda_i \underline{a}_i \right\|^2 = \underline{\lambda}^T C \underline{\lambda} \rightarrow \underset{\lambda_i \in \{0;1\}}{\text{Max. !}}$$

(For instance in McBride– Yormark [27] techniques for solving such problems are given.)

The interpretation of the theorems (5.1) and (5.2) for the projection body $\Pi^{d-1}(P)$ leads to the half of the maximal and minimal width and to the corresponding support directions of this zonotope.

In Fig. 4 projections of the rhombocuboctahedron (with an equatorial plane of symmetry) for $\underline{u}^*$ and $\underline{u}_*$ are shown.

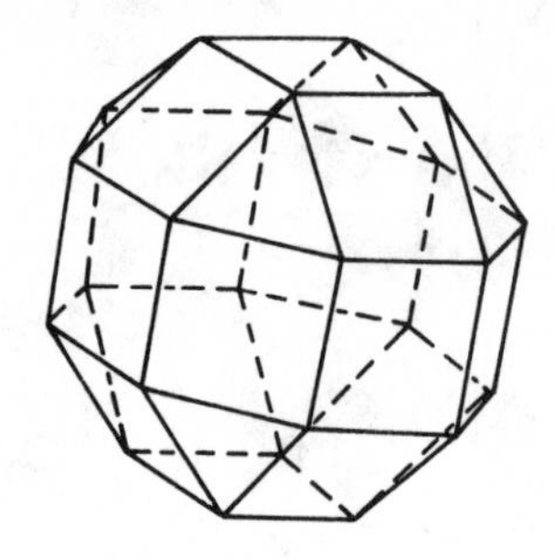

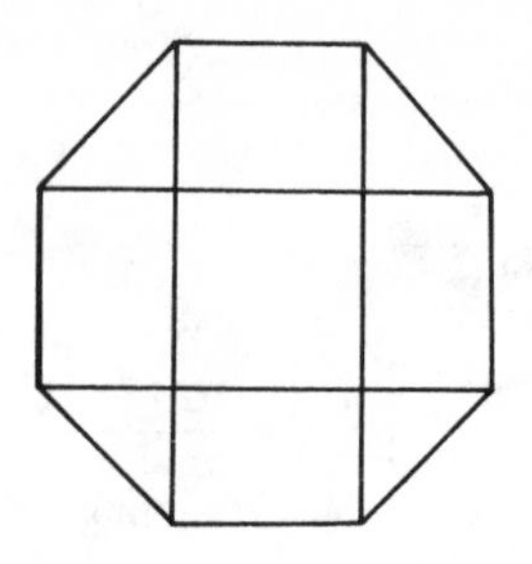

Figure 4

The comparison of the quotients $\frac{\bar{v}_{min}}{\bar{v}_{max}}$ for the Platonic solids is tabulated in:

n	6	4; 8	12	20
$\frac{\bar{v}_{min}}{\bar{v}_{max}}$	$\frac{\sqrt{3}}{3} \approx 0{,}577$	$\frac{\sqrt{2}}{2} \approx 0{,}707$	$\frac{5+\sqrt{5}}{10} \approx 0{,}851$	$\sqrt{\frac{5}{6}} \approx 0{,}913$

The regular polytopes in $\mathbb{E}^d$ $(d \geq 2)$ with 1 as length of one edge or v as content of one facet have the following extremal values:

	simplex	cross polytope	cube
$\bar{v}_{max}$	$\frac{1}{(d-1)!}\left(\frac{1}{\sqrt{2}}\right)^{d-1}\cdot\frac{1}{2}\begin{cases}(d+1)\ , & d \text{ odd}\\ \sqrt{d(d+2)}, & d \text{ even}\end{cases}$	$\frac{v}{\sqrt{d}}\cdot 2^{d-1}$ ⊕	$\sqrt{d}\cdot 1^{d-1}$
$\bar{v}_{min}$	$\frac{1}{(d-1)!}\left(\frac{1}{\sqrt{2}}\right)^{d-1}\cdot\sqrt{\frac{d+1}{2}}$		1^{d-1}

⊕ This conjecture announced in [26] was confirmed by Bittner [21].

The conjecture $\bar{v}_{min} = \frac{v}{\sqrt{d}}\cdot 2^{d-1}\cdot\frac{\sqrt{2}}{2}$ for the cross polytope (compare also [26]) could not be confirmed up to now. (Indeed, there exist estimations in another geometrical formulation of the problem (see Hall [16]).)

For $d = 3$ it is well known that one can prismatically perforate a regular hexahedron so that it is possible to slide throughout a congruent cube. In other words, we have two orthogonal 2-projections of the 3-cube one of which for a suitable embedding lies completely in the interior of the other (Fig. 5).

In figure 5 we show also a position of those two orthogonal 3-projections of one 4-cube which have maximal or minimal 3-quermasses. This representation gives rise to the following

CONJECTURE (5.1). For $d \geq 4$ there exist no two orthogonal (d-1)-projections of the d-cube one of which can be completely embedded in the interior of the other.

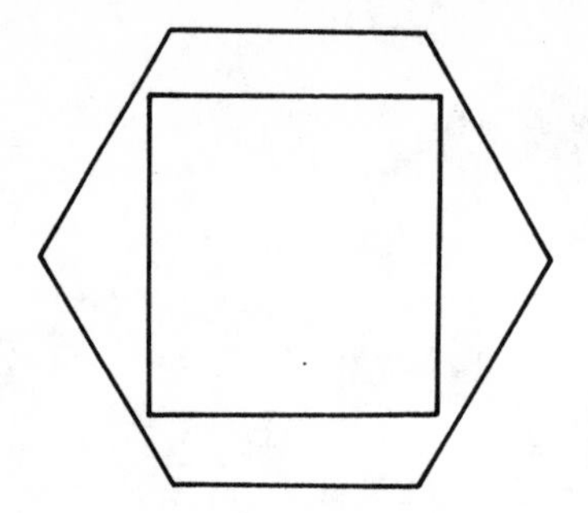

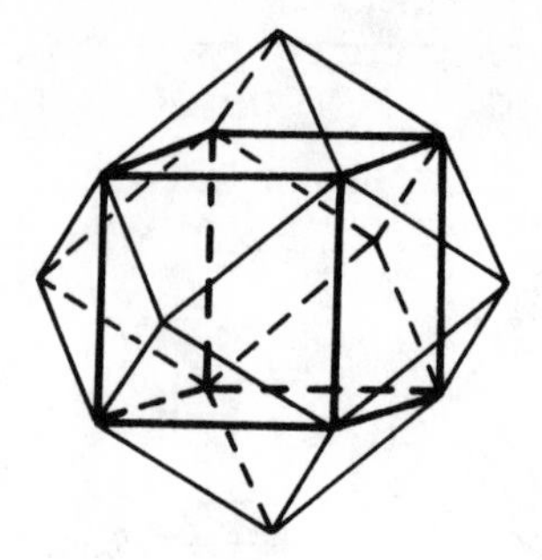

Figure 5

5.3. *Sharp shadow-boundaries of convex* d-*polytopes*

Using statements about the number of vertices of projection bodies one can derive upper and lower bounds for the number of sharp shadow-boundaries of convex d-polytopes with m nonoriented facet normals for parallel projection (Martini [23]).

A subcomplex of the boundary-complex of a convex d-polytope P which is obtained as intersection of P and of the projecting support cone regarding a point $\underline{y}$ outside of the polytope is called a shadow-boundary of P with respect to $\underline{y}$. Such a subcomplex is said to be a *sharp shadow-boundary* if it is homeomorphic to a (d-2)-sphere.

In [17] Kleinschmidt and Pachner investigated these subcomplexes for central projection and could give a sharp upper bound for the number $c_c(n,d)$ of sharp shadow-boundaries of convex d-polytopes with n facets. They characterized also those polytopes which attain this bound, whereas the lower bound to this problem is unknown. The corresponding number $c_p(m,d)$ of sharp shadow-boundaries for parallel projection depends only on

d, on the number m of nonoriented facet normals of P and on their spatial arrangement.
Contrary to central projection for parallel projection one succeeds in determining upper and lower bounds of $c_p(m,d)$.

From upper and lower bounds for the number of (d-1)-cells in projective (d-1)-arrangements of hyperplanes (Grünbaum [14], Zaslavsky [39]) can be derived

THEOREM (5.3). *If* $c_p(m,d)$ *denotes the number of sharp shadow-boundaries of a conves* d-*polytope* P *with* m *nonoriented facet normals* $(m \geq d)$ *for parallel projection, then we have*

$$\sum_{i=0}^{d-1} \binom{m-1}{i} \geq c_p(m,d) \geq 2^{d-2}(m-d+2) .$$

The upper bound is attained iff no d *of the* m *facet normals of* P *lie in one* (d-1)-*subspace of* $\mathbb{E}^d$. *The lower bound determines the* (d-2)-*fold prisms.*

It is remarkable that the combinatorial type of P and even the number n of facets (independently of m) do not influence $c_p(m,d)$. On the contrary, the global lower bound $\underline{\underline{c}}_p(d) = 2^{d-1}$ suffices to determine a class of affinely equivalent polytopes - the class of d-parallelotopes.

Comparing the numbers of vertices of suitable d-zonotopes and (d+1)-zonotopes which correspond with P in a special manner we get a characterization of the d-simplex in

THEOREM (5.4). *The following properties of a convex* d-*polytope* P *are equivalent:*

(1) $c_c(P) = c_p(P)$.

(2) P *is a* d-*simplex.*

The determination of $c_c(P)$ or $c_p(P)$ for a polytope P (for instance given by its support numbers) is possible by means of a procedure that depends on the sweep-plane method (Bieri– Nef [1]).
As an open question, already raised in [17], remains

PROBLEM (5.2). Find a lower bound for the number $c_c(n,d)$ of sharp shadow-boundaries of convex d-polytopes with n facets and characterize the corresponding polytopes.

6. ZONOTOPES IN THE THEORY OF GEOMETRICAL ILLUMINATION

Among the problems which the theory of geometrical illumination deals with, the analysis and representation of the illumination appearances of surfaces play an important role.

Hitherto, the parallel and central illumination of suitable surfaces in $\mathbb{E}^3$ from one direction or one point has been investigated (see Burmester [8], Bohne [2], Möller [32] and others).

For analogous examinations on the foundation of multiple illuminations of surfaces the utilization of zonotopes is useful, especially for the investigation of the illumination appearances on the boundaries of smooth convex bodies for multiple parallel illuminations (Martini [24]).

A configuration of n parallel illuminations is determined by a vector star $\mathscr{L}_n := \{\underline{l}_1, \ldots, \underline{l}_n\}$ in which an *illumination vector* $\underline{l}_i$ is oppositely directed regarding the direction of the i-th illumination and represents with $\|\underline{l}_i\| =: I_i$ the corresponding luminous intensity $(i=1,\ldots,n)$. (For $\mathscr{L}_n$ vectors of equal directions shall be excluded; moreover, full generality for $\mathscr{L}_n$ is one of our main aims.)

The illuminated convex body shall be given as a smooth one (in order to obtain results as general as possible). For such a body $K \in \mathbb{K}^d$ the point $\underline{x}$ represents an arbitrary boundary point, $\underline{u}$ the outer normal unit vector in $\underline{x}$, and $(\underline{x}, \underline{u})$ the *oriented element of area* with carrier point $\underline{x}$.

According to the well-known cosine law we define

$$\bar{b}_i := I_i \cdot \cos\varphi_i = \langle \underline{l}_i, \underline{u} \rangle, \; 0 \leq \bar{b}_i; \; \underline{u} \in S^{d-1},$$

for $\varphi_i = \sphericalangle(\underline{l}_i, \underline{u})$, $I_i \in \mathbb{R}^+$, and $i=1,\ldots,n$. The non-negative real number $\bar{b}_i$ is the i-th *intensity of illumination* of the oriented element of area $(\underline{x}, \underline{u})$, and, consequently, $(\underline{x}, \underline{u})$ is said to be illuminated by $\underline{l}_i$. (For $\cos\varphi_i < 0$ the element $(\underline{x}, \underline{u})$ is said to be non-illuminated by $\underline{l}_i$ and the intensity of illumination is assumed to be $\bar{b}_i = 0$.)

For multiple illuminations we have to sum up the single intensities of illumination in $(\underline{x}, \underline{u})$ resulting in a so-called composition intensity of illumination $\bar{b}(\underline{u})$. (This corresponds to physical experiences confirmed by experiments for $d = 3$.)

Replacing $\mathfrak{A}_n$ by $\mathscr{L}_n$ in (5) or (5*) we get the distribution of intensity of illumination on bd K with respect to $\underline{u} \in S^{d-1}$ as the support function

$$(11) \qquad \bar{b}(\underline{u}) := h(Z,\underline{u}) = \frac{1}{2} \sum_{i=1}^{n} |\langle \underline{l}_i, \underline{u}\rangle| + $$

$$+ \langle \underline{l}_M, \underline{u}\rangle, \quad \underline{u} \in S^{d-1},$$

of the (d-r)-zonotope ($r \in \{0,\ldots,d-1\}$)

$$Z = \sum_{i=1}^{n} [\underline{o}, \underline{l}_i] \quad .$$

Light poles are carrier points of oriented elements of area with maximal intensity of illumination in bd K. Using results of section 5, one can determine these points by means of the Boolean optimization problem

$$\left\| \sum_{i=1}^{n} \lambda_i \, \underline{l}_i \right\|^2 \rightarrow \underset{\lambda_i \in \{0;1\}}{\mathrm{Max}} \quad . \qquad \text{(See Fig. 6).}$$

Besides (8) shows that for an arbitrarily given distribution of intensity of illumination on bd K (produced by parallel illuminations) at the very least the m positions ($m \leq n$) of the single illuminations, the total luminous intensity per position, and the summation vector $2\underline{l}_M$ of all illumination vectors are determined. (It should be noticed that the vector $2\underline{l}_M$ plays an important role in the sense of the light field theory. For this topic the reader can consult Gershun [40].)

We can further deduce that there exists only one configuration of parallel illuminations to such a given distribution on bd K if $\underline{o} \in$ vert Z or if the origin is (for $s < d$ relatively) interior point of an s-dimensional

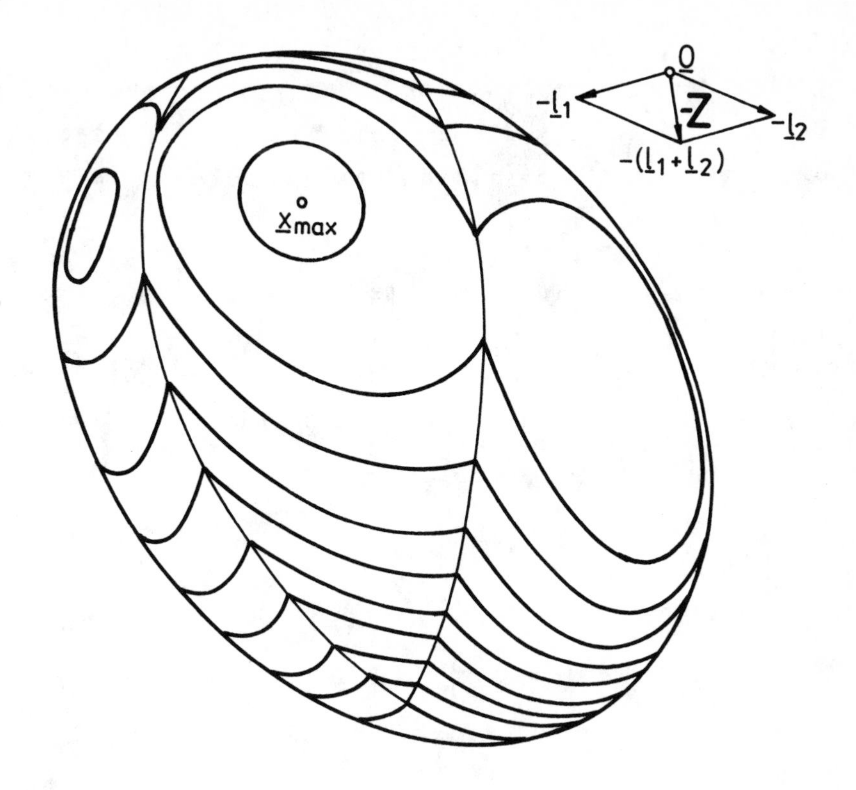

Figure 6

cubical face of the (d-r)-zonotope Z ($s \in \{1,...,d-r\}$). Such an example is shown in Figure 6.

Isophotical manifolds are point sets in bd K the elements of which are carrier points of oriented elements of area with a fixed intensity of illumination. Figure 6 (illumination of an ellipsoid in $\mathbb{E}^3$ from two directions) shows that the isophotes in the boundaries of smooth convex bodies are in general (d-2)-dimensional manifolds.

An interesting problem is the following one (see Martini [25]): Characterize configurations which isophotically illuminate the whole boundary of an arbitrary convex d-polytope P with h facets.

In terms of the classical theory of geometrical illumination, this problem can be solved only with multiple parallel illuminations. Clearly, points of one facet form an isophotical manifold. (The introduced notions involve only the illumination of relatively interior facet points of P.)

This leads one to formulate the following interpolation problem: Let $\mathfrak{N}_h := \{\underline{n}_1,\ldots,\underline{n}_h\}$ ($\|\underline{n}_j\| = 1$; $j=1,\ldots,h$) be the set of outer normal vectors of P. Find a vector star $\mathscr{L}_n$ (of the smallest possible cardinality) with the property

$$(12) \qquad \frac{1}{2}\sum_{i=1}^{n} |\langle \underline{l}_i, \underline{n}_j\rangle| + \langle \frac{1}{2}\sum_{i=1}^{n} \underline{l}_i, \underline{n}_j\rangle = \rho$$

for any fixed $\rho \in \mathbb{R}^+$. If we restrict our considerations to vector stars with $\underline{o}$ as centre of gravity then (12) will be of the simple form

$$(12^*) \qquad \frac{1}{2}\sum_{i=1}^{n} |\langle \underline{l}_i, \underline{n}_j\rangle| = \rho .$$

Consequently, one has to determine a zonotope the support values of which in all outer normal directions of P are equal to each other. Every generating star of such a sum of line segments then represents a configuration that illuminates bd P isophotically.

A result related to (12^*) is given in

THEOREM (6.1). *The whole boundary of an arbitrary convex polygon* $P \subset \mathbb{E}^2$ *with* m *nonoriented facet normals can be isophotically illuminated by* q *parallel illuminations, where* $q \in \{p, p+1, p+2\}$ *and*

$$p := \left[\frac{m+1}{2}\right] .$$

For $d \geq 3$ the interpolation problem seems to be unsolved. Partial results are known only for special classes of polytopes.

In Figure 7 one possibility for the homogeneous illumination of the regular icosahedron with six pairwise-opposite directions is presented: We take the Platonic solid to be centred at the origin.

Then the shown zonohedron $Z = \sum_{i=1}^{6} [\underline{o}, \underline{1}_i]$ has equal values $h(Z, \underline{n}_j)$ in all outer normal directions $\underline{n}_j$ ($j=1,\ldots,20$) of the icosahedron.

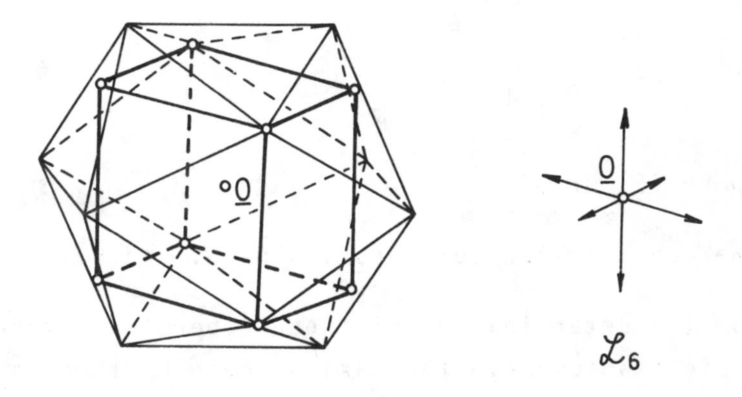

Figure 7

7. MEAN DISTANCES BETWEEN FINITE POINT SETS AND HYPERPLANES

Especially the metrical statements on sums of line segments give rise to an additional interpretation: Let $\mathfrak{A}_n$ represent an arbitrary finite point set in $\mathbb{E}^d$ where position vectors of equal directions are admitted.

$$H_{\underline{u}} := \{\underline{x} \in \mathbb{E}^d : \langle \underline{x}, \underline{u} \rangle = 0\} \ , \ \underline{u} \in S^{d-1},$$

shall denote a (d-1)-subspace of $\mathbb{E}^d$. Then by

$$\bar{q}(\underline{u}) := \frac{1}{n} \sum_{i=1}^{n} |\langle \underline{a}_i, \underline{u} \rangle|, \ \underline{u} \in S^{d-1} \ ,$$

the *mean distance* between those n points and $H_{\underline{u}}$ is defined.

Obviously, $\bar{q}(\underline{u})$ is the support function of the zonotope

$$(13) \qquad Z = \frac{1}{n} \sum_{i=1}^{n} \left[-\underline{a}_i, \underline{a}_i\right]$$

(compare (6) or (7)).

Now we can transfer some results of 5. and 6.:

For the determination of those hyperplanes through the origin which have the maximal mean distance $\bar{q}_{max}$ to $\mathfrak{A}_n$, once more the utilization of Boolean optimization is recommendable. By transcription of (10) for the zonotope given in (13), we may write the optimization problem in the form

$$\left\| \sum_{i=1}^{n} \lambda_i \, \underline{a}_i \right\|^2 = \underline{\lambda}^T \, C \, \underline{\lambda} \rightarrow \underset{\lambda_i \in \{-1;1\}}{\text{Max.!}}$$

Figure 8 shows representatives from the set of all hyperplanes through the centre of gravity regarding the 20 vertices of a Platonic dodecahedron, which have maximal or minimal mean distances to this point set.

Theorem (6.1) leads one to formulate the following statement:

Given a set of m lines through a fixed point in the Euclidean plane. Then there exist point sets of cardinality q ($q \in \{p, p+1, p+2\}$; $p := \left[\frac{m+1}{2}\right]$) with the *same*

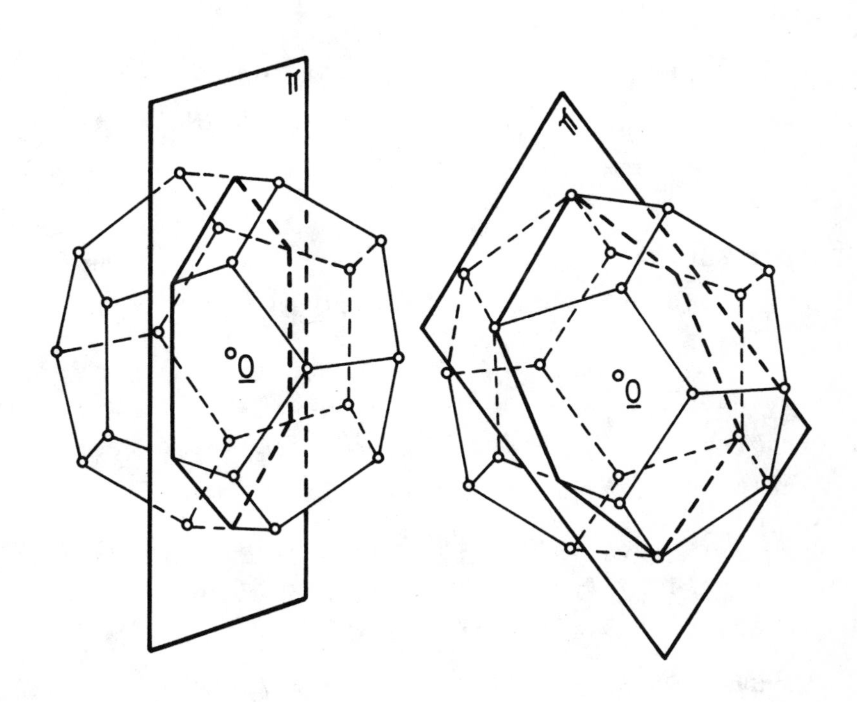

Figure 8

(positive) mean distance to *every* line of the given set. (Notice that the combinatorial statements under 5. have connections with problems around Radon's theorem and related topics (see Greene-Zaslavsky [11]).)

REFERENCES

[1] H. BIERI and W. NEF, *A recursive sweeep-plane algorithm, determining all cells of a finite division of* $\mathbb{R}^d$. Computing 28 (1982), 189-198.

[2] E. BOHNE, *Über Isophoten bei Drehquadriken*. Wiss. Zeitschr. TU Dresden 18 (1969), 477-486.

[3] E.D. BOLKER, *A class of convex bodies*. Trans. Amer. Math. Soc. 145 (1969), 323-345.

[4] E.D. BOLKER, *Centrally symmetric polytopes*. In: Proc. 12th Bienn. Internat. Seminar Canad. Math. Congr. (Vancouver, B.C., 1969), Canad. Math. Soc. Montreal (1970), 255-263.

[5] V.G. BOLTYANSKIĬ, *Zadača ob osveŝenii granicy vypuklogo tela*. Bull. Akad. Štiince RSS Moldoven 76 (1960), 77-84.

[6] V.G. BOLTYANSKIĬ and I.C. GOCHBERG, *Sätze und Probleme der kombinatorischen Geometrie*. Deutscher Verlag d. Wiss., Berlin, 1972.

[7] V.G. BOLTYANSKIĬ and P.S. SOLTAN, *Kombinatornaâ geometriâ različnyh klassov vypuklyh množestv*. Izdatelstvo Štiinca, Kižinev 1978.

[8] L. BURMESTER, *Theorie und Darstellung der Beleuchtung gesetzmäßig gestalteter Flächen*. Teubner, Leipzig, 1871.

[9] H.S.M. COXETER, *Regular Polytopes*. Macmillan, New York, 1963 (Reprint).

[10] E.S. FEDOROV, *Načala učeniâ o figurah*. Izdatelstvo Akademii Nauk SSSR, Moskva, 1953 (Reprint).

[11] C. GREENE and T. ZASLAVSKY, *On the interpretation of Whitney numbers through arrangements of hyperplanes, zonotopes, non-Radon partitions, and orientations of graphs*. Trans. Amer. Math. Soc. 280 (1983), 97-126.

[12] B. GRÜNBAUM, *Borsuk's problem and related questions*. Proc. Symposia Pure Math. 7 (1963), 271-284.

[13] B. GRÜNBAUM, *Convex Polytopes*. Wiley & Sons, New York, 1967.

[14] B. GRÜNBAUM, *Arrangements of hyperplanes*. In: Proc. 2nd Louisiana Conf. Combin., Graph Theory, and Computing. Louisiana State Univ., Baton Rouge (1971), 41-108.

[15] H. HADWIGER, *Ungelöste Probleme*. Nr. 20. Elemente Math. 12 (1957), 121.

[16] R.R. HALL, *On a conjecture of Littlewood*. Math. Proc. Cambridge Phil. Soc. 78 (1975), 443-445.

[17] P. KLEINSCHMIDT and U. PACHNER, *Shadow-boundaries and cuts of convex polytopes*. Mathematika 27 (1980), 58-63.

[18] M. LASSAK, *Covering three-dimensional convex bodies with smaller homothetical copies*. Lecture at Oberwolfach, Conf. Convexity, July, 1984.

[19] M. LASSAK, *Solution of Hadwiger's covering problem for centrally symmetric convex bodies in* $\mathbb{E}^3$. J. London Math. Soc. 30 (1984), 501-511.

[20] F.W. LEVI, *Überdeckung eines Eibereiches durch Parallelverschiebungen seines offenen Kerns.* Arch. Math. 6 (1955), 369-370.

[21] J. LINHART, *Private communication.*

[22] H. MARTINI, *Zur Bestimmung konvexer Polytope durch die Inhalte ihrer Projektionen.* Beitr. Algebra Geom. 18 (1984), 75-85.

[23] H. MARTINI, *Über scharfe Schattengrenzen und Schnitte konvexer Polytope.* Beitr. Algebra Geom. 19 (1985), 105-112.

[24] H. MARTINI, *Parallelbeleuchtung konvexer Körper mit glatten Rändern.* Beitr. Algebra Geom. 21 (1986), 109-124.

[25] H. MARTINI, *Zum homogenen Ausleuchten konvexer Polytope.* Beitr. Algebra Geom. 22 (1986), 71-79.

[26] H. MARTINI and B. WEISSBACH, *Zur besten Beleuchtung konvexer Polyeder.* Beitr. Algebra Geom. 17 (1984), 151-168.

[27] R.D. McBRIDE and J.S. YORMARK, *An implicit enumeration algorithm for quadratic integer programming.* Management Sci. 26 (1980), 282-296.

[28] P. McMULLEN, *On Zonotopes.* Trans. Amer. Math. Soc. 159 (1971), 91-110.

[29] P. McMULLEN, *Space tiling zonotopes.* Mathematika 22 (1975), 202-211.

[30] P. McMULLEN, *Transforms, diagrams and representations.* In: Contributions to Geometry. Proc. Geom. Sympos. Siegen 1978. Ed. by J. TÖLKE a. J.M. WILLS. Birkhäuser, Basel, 1979, 92-130.

[31] H. MINKOWSKI, *Allgemeine Lehrsätze über die konvexen Polyeder*. In: Ges. Abh., Bd. 2, Leipzig, 1921.

[32] R. MÖLLER, *Über Isophoten auf Drehquadriken bei zentraler geometrischer Beleuchtung*. Wiss. Zeitschr. PH Dresden 10 (1976) 3, 37-45.

[33] C.M. PETTY, *Projection bodies*. In: Proc. Coll. Convexity (Copenhagen, 1965), Københavns Univ. Mat. Inst. 1967, 234-241.

[34] R. SCHNEIDER, *On the projections of a convex polytope*. Pacific J. Math. 32 (1970), 799-803.

[35] R. SCHNEIDER and W. WEIL, *Zonoids and related topics*. In: Convexity and Its Applications. Ed. by P. GRUBER a. J.M. WILLS. Birkhäuser, Basel, 1983, 296-317.

[36] G.C. SHEPHARD, *Combinatorial properties of associated zonotopes*. Canad. J. Math. 26 (1974), 302-321.

[37] K. STEINITZ, *Polyeder und Raumeinteilungen*. Enzykl. math. Wiss., Bd. 3 (Geometrie), Teil 3 A B 12 (1922), 1-139.

[38] B. WEISSBACH, *Eine Bemerkung zur Überdeckung beschränkter Mengen durch Mengen kleineren Durchmessers*. Beitr. Algebra Geom. 11 (1981), 119-122.

[39] T. ZASLAVSKY, *The slimmest arrangements of hyperplanes. I : Geometric lattices and projective arrangements*. Geom. Dedicata 14 (1983), 243-259.

[40] A. GERSHUN, *The light field*. J. Math. Phys. 18 (1939), 51-151.

ADDED IN PROOF:

I wish to thank M. Lassak for the hint that theorem (3.1) is also obtainable (by means of other methods) as a corollary of a theorem of Boltyanskiĭ-Soltan.

With respect to the light field theory (see section 6) we remark that the so-called illumination distribution surface of one point $\underline{x} \in \mathbb{E}^d$ in the field of $\mathscr{L}_n$ is described by (11) (cf. [40]).

H. MARTINI
Pädagogische Hochschule Dresden
Sektion Mathematik
Wigardstraße 17
DDR-8060. Dresden

COLLOQUIA MATHEMATICA SOCIETATIS JÁNOS BOLYAI
48. INTUITIVE GEOMETRY, SIÓFOK, 1985.

ABOUT DISTRIBUTIONS OF POINTS IN A SQUARE AND A CIRCLE WITH SPECIAL DISTANCE

W. MÖGLING

In [1] the distance between two points $P_1 = (x_1, y_1)$ and $P_2 = (x_2, y_2)$ in a rectangular system of coordinates is defined by $a(P_1P_2) = |x_1 - x_2| + |y_1 - y_2|$, and distributions of points are investigated in a square Q using this distance. The parallels to the edges of Q are interpreted as "streets of a town".

In the first chapter of this paper a special problem of distribution will be treated, supposing that there are only a finite number of streets in Q. In the second chapter we present a problem analogous to [1] for a circle.

1. A SPECIAL PROBLEM OF DISTRIBUTION IN A SQUARE

In [1] L. Fejes Tóth supposes that in every point of Q there exists a pair of streets that meet otrhogonally, but now we assume that there are only a finite number of streets parallel to the edges of Q in distance 1 to each other, in further approximation to reality. We can imagine that Q is covered by a finite lattice consisting of unit squares. The points to be distributed may be regarded as shops or schools on the map of a town to be built. These points may lie only on the lines of the lattice, and the distance of two points has to be determined only on the lines of the lattice. Thus in Fig. 1:

$$d(P_1P_2) = d(P_1H_1) + d(H_1H_2) + d(H_2P_2) \quad .$$

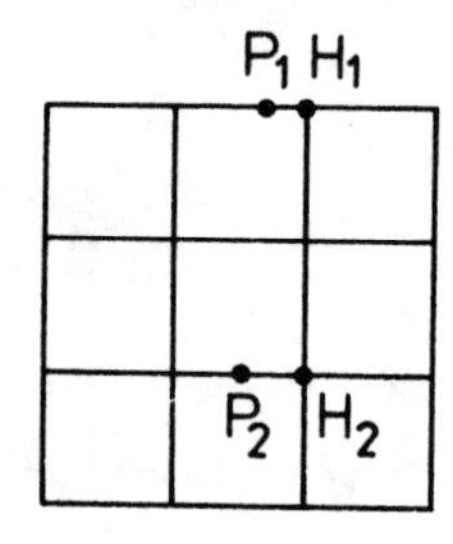

Figure 1

Supposing the length of the edge of Q is m (m=1,2,...) we can now ask for the solution of the following well-known problem: n points shall be distributed in Q so that the smallest distance between two of them will attain a maximum. From [1] we can conclude: For $m = 2k$ $(k = 1,2,...)$ we may place exactly $n = k^2 + (k+1)^2$ points in Q such that the smallest distance attains its maximum 2 (Fig. 2), and for $m = 2k - 1$ there are at least $2k^2$ points with smallest distance ≥ 2 (Fig. 3). On the boundary of every unit square in Q there are two points at most. Figures 3 and 4 show that the distribution of $2k^2$ points is not unique.

The special problem announced is the following: we ask for the greatest number n of points that can be placed in the streets of Q so that the smallest distance between two points is $2 + \varepsilon$ $(\varepsilon > 0)$. As a result we get the following

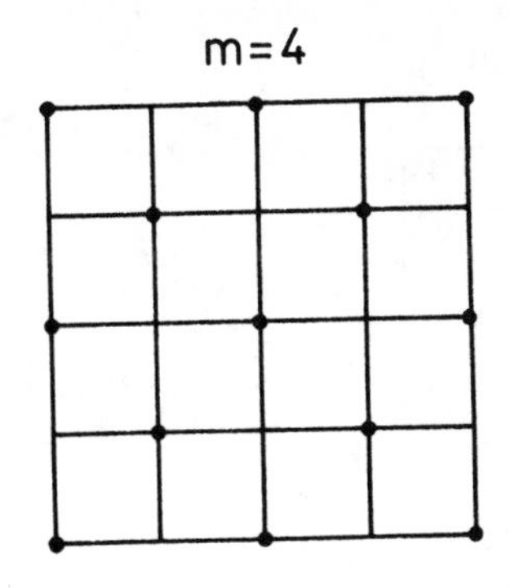

Figure 2

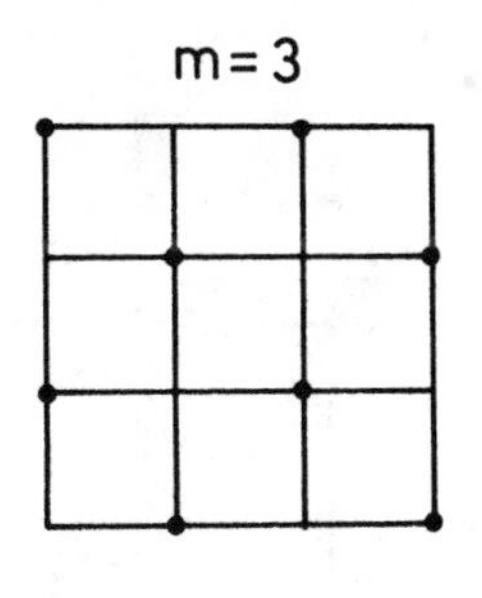

Figure 3

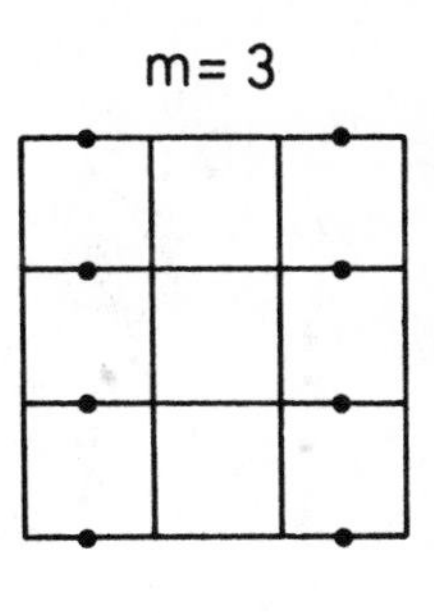

Figure 4

inequalities for n:

(1) $$(2k+1)k \le n \le 2k^2 + 2k - 1 < k^2 + (k+1)^2 \quad \text{for } m = 2k,$$

(2) $$(2k-1)k \le n \le 2k^2 - 1 < 2k^2 \quad \text{for } m = 2k-1 .$$

The proof will be given for $m = 2k$, the other inequality can be proved in the same way.

First we notice that in those streets which are on the boundary of a unit square only one point can be placed. We start the proof with the distribution of $n = k^2 + (k+1)^2$ points P_{ij} ($i = 1,2,\ldots,2k+1$; $j = 1,2,\ldots k+1$ for i odd, $j = 1,2,\ldots,k$ for i even) with the smallest distance 2. (See Fig. 5, which is drawn for $m = 4$.)

The points P_{ij} are moved in the direction of the ray $P_{11}P_{31}^{+}$ along the streets covering a way of $i\cdot\varepsilon$, and the points P_{2k+1j} are taken away. When ε is sufficiently small, $2k^2 + 2k + 1 - (k+1) = (2k+1)k$ points with the minimal distance $2+\varepsilon$ will remain, thus $(2k+1)k \le n$. In order to get an upper bound we try to place as many points as possible on the circumference $u = 4m = 8k$ of Q, as for

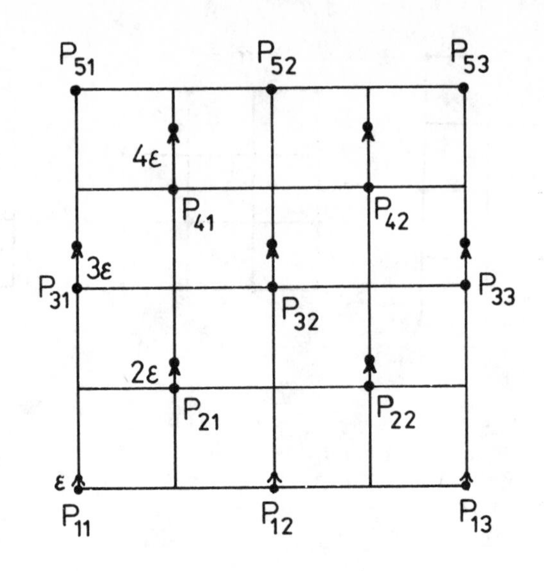

Figure 5

such a point only one unit square is required. These are $4k-1$ points at most. Thus $4k^2-(4k-1)=4k^2-4k+1$ unit squares without any points will remain. As a matter of fact, one point in the interior of Q will claim at least two unit squares, and therefore we can place only $2k^2-2k+4k-1=2k^2+2k-1$ points in Q. Thus (1) is proved. For $k=1$ we obtain $n=3$, for $k=2$ we have $10 \leq n \leq 11$, and by a special proof we can show, that only 10 points can be placed, thus we can conjecture $n=(2k+1)k$ still holds for further values of k. The conclusion is that $n=2k^2+2k$ points with the smallest distance $2+\varepsilon$ cannot be placed in Q. In this case the maximum for the smallest distance is 2, as well as for $n=k^2+(k+1)^2$.

Concerning inequality (2) we can prove that we have $n=6$ for $k=2$, and we can also conjecture that $n=(2k-1)k$ is true for other values of k.

2. DISTRIBUTIONS OF POINTS IN A CIRCLE

Instead of the square in [1] we will now consider a unit circle K with centre O and boundary k. Let O be the origin of a system of polar coordinates so that every point $P \neq O$ in the interior or on the boundary of K has the coordinates (r,φ) with $0 < r \leq 1$, $0 \leq \varphi < 2\pi$. Again we use the idea of streets. Here the radii of K are called main streets and the circles with the centre O lying in K are called circular roads. Through every point $P \in K$, $P \neq O$, there runs exactly one pair of streets, a main street and a circular road. The distance of two points $P_1 = (r_1,\varphi_1)$ and $P_2 = (r_2,\varphi_2)$ is defined as follows:

$$a(P_1P_2) := \min\{r_1 + r_2;\ |r_1 - r_2| + \varphi \cdot \min\{r_1,r_2\}\}$$

with

$$\varphi := \min\{|\varphi_1 - \varphi_2|;\ 2\pi - |\varphi_1 - \varphi_2|\} = |\sphericalangle P_1OP_2| \ .$$

Thus in order to determine the shortest distance between two points we use the shortest way between them along the main streets and the circular roads, respectively (Fig. 6).

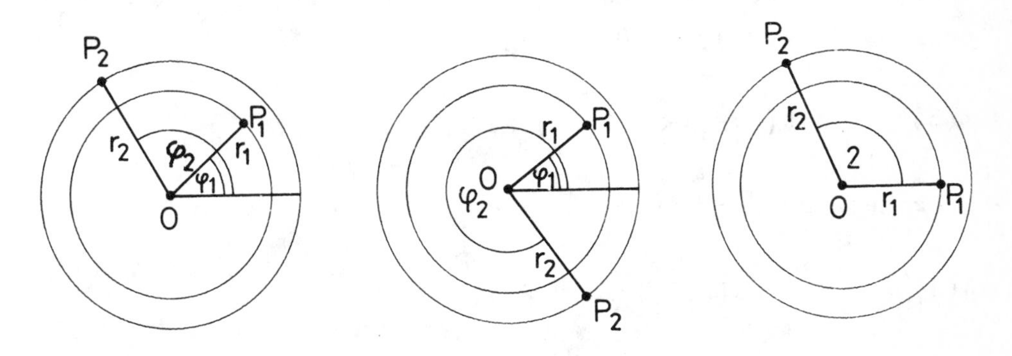

Figure 6

In generally the shortest way is unique except for $|\measuredangle P_1OP_2| = 2$ as we can easily see. For $|\measuredangle P_1OP_2| < 2$ the shortest way from P_1 to P_2 does not touch O, for $\pi \geq |\measuredangle P_1OP_2| > 2$ the shortest way from P_1 to P_2 runs via O which may be thought to be a central place in the town.

In analogy to [1] we have to distribute n points in K so that a_m, the minimal distance of two points, will become as large as possible. The result is the following

THEOREM. *If* n *points are distributed in a unit circle, in which the distance of two points is determined in the way described above, so that the minimal distance* a_m *between two points is as large as possible, then*

1. *for* $n = 2,3$ *we have* $a_m = 2 < \frac{2\pi}{n}$,
2. *for* $n = 4,5,6$ *we have* $1 < a_m = \frac{2\pi}{n} < 2$,
3. *for* $n = 7$ *we have* $\frac{2\pi}{n} < a_m = 1$,
4. *for* $n \geq 8$ *we have* $a_m < 1$.

For the proof of this theorem we need the following

LEMMA. *If there are two points* P_i *and* P_j *on the boundary* k *of* K *and if* $a(P_iP_j) < 1$ *respectively* $1 \leq a(P_iP_j) < 2$ *holds and if the point* P_j *is moved along the main street through* P_j *into the interior of* K *to* P_j', *so that* $a(P_jP_j') = \varepsilon$, $0 < \varepsilon < 1$, *then*

$$a(P_iP_j) < a(P_iP_j') < 1, \tag{3}$$

respectively

$$1 \leq a(P_iP_j') \leq a(P_iP_j), \tag{4}$$

and equality holds only for $a(P_iP_j) = 1$.

Proof of the inequality (3). Let be $a(P_iP_j) = 1 - \delta$ with $0 < \delta < 1$ (Fig. 7a). Then $a(P_iP_j') = \varepsilon + (1 - \varepsilon)(1 - \delta) = 1 - \delta(1 - \varepsilon) < 1$, and (3) is proved.

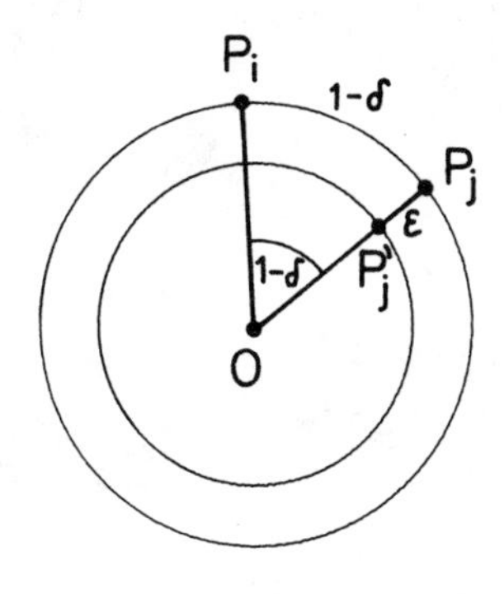

Figure 7a

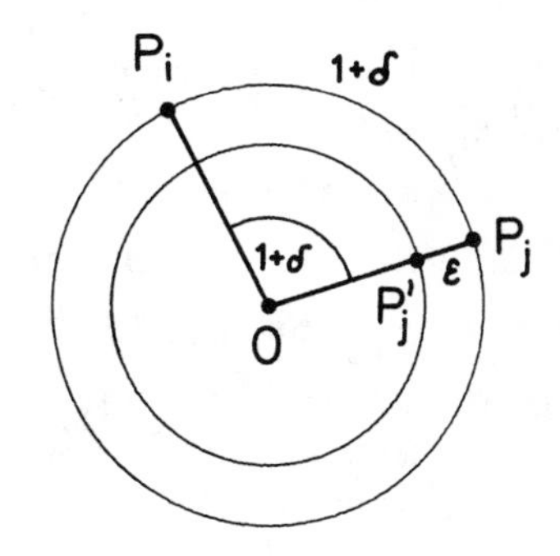

Figure 7b

Proof of the inequality (4). For $a(P_iP_j) = 1$ we have $a(P_iP_j') = \varepsilon + 1 - \varepsilon = 1$. For $1 < a(P_iP_j)$ (Fig. 7b), that means $a(P_iP_j) = 1 + \delta$, we have $a(P_iP_j') = \varepsilon + (1 - \varepsilon)(1 + \delta) = 1 + \delta - \varepsilon\delta < 1 + \delta = a(P_iP_j)$.

Thus the lemma is proved.

For proving the theorem we shall consider four different cases:

1. In the unit circle the distance of two points cannot be greater than 2. We can always find 2 or 3 points on k so that $a_m = 2$.

2. If the n points ($n = 4,5,6$) are regularly distributed on k, we have $1 < a_m = \frac{2\pi}{n} < 2$. If one of the points is moved along its main street into the interior of K, the distances to the next points will get smaller according to the lemma.

3. Six points P_h ($h = 1,2,...,6$) are distributed on k so that the minimal distance 1 is ensured, and P_7 is placed on O. Then we will prove: If points $P_1...P_6$ do not

lie on k with the minimal distance 1 or when $P_7 = O$ does not hold then there are always two points P_i and P_j from the set of the seven points whose distance is smaller than 1.

For the proof of this statement we have to consider three cases:

i) If one point P_h lies in the interior of K and if $P_7 = O$, then $a(P_hP_7) < 1$.

ii) Let be $P_7 \neq O$ and $a(OP_7) = \varepsilon$, $0 < \varepsilon < 1$, and let all points P_h be on k. If P_7 lies on a main street OP_h, then $a(P_7P_h) < 1$. If P_7 does not lie there then we consider the smallest sector P_iOP_j in which we find P_7. Let be $|\measuredangle P_jOP_i| = \alpha$, then we have either $|\measuredangle P_iOP_7| \leq \frac{\alpha}{2}$ or $|\measuredangle P_jOP_7| \leq \frac{\alpha}{2}$ with $1 \leq \alpha \leq 2\pi - 5$. Without loss of generality we assume $\measuredangle P_iOP_7 \leq \measuredangle P_jOP_7$ (Fig. 8).

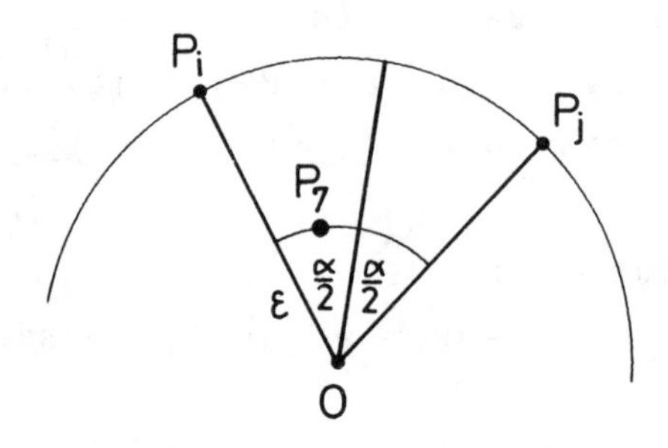

Figure 8

Then

$$a(P_iP_7) \leq 1 - \varepsilon + \frac{\alpha}{2} \cdot \varepsilon \leq 1 - \varepsilon + \frac{\varepsilon}{2}(2\pi - 5) =$$

$$= 1 - \frac{\varepsilon}{2}(7 - 2\pi) < 1 \ .$$

iii) If there is one point P_h in the interior of K and $P_7 \neq O$, then using the lemma we always can find two points of the set of the seven points whose distance is smaller than 1.

4. By the aid of the lemma we can show, in the same way as for $n = 7$, that $a_m < 1$ holds for $n \geq 8$. As an example we will regard the case $n = 8$. If we distribute $P_1 \ldots P_7$ regularly on k and if we put $P_8 = 0$, then we get $a_m(P_iP_j) = \frac{2\pi}{7} \approx 0.897\ldots$. A better distribution is represented in fig. 9. We get the following system of equations for $x = a(P_2P_3) = a(P_4P_6) = a(P_7P_8)$, $y = |\sphericalangle P_2P_8P_5|$ and $z = a(P_1H)$ (see Fig. 9):

$$x + 6y = 2\pi$$
$$x + z = 1$$
$$x = (1 - z)y + z .$$

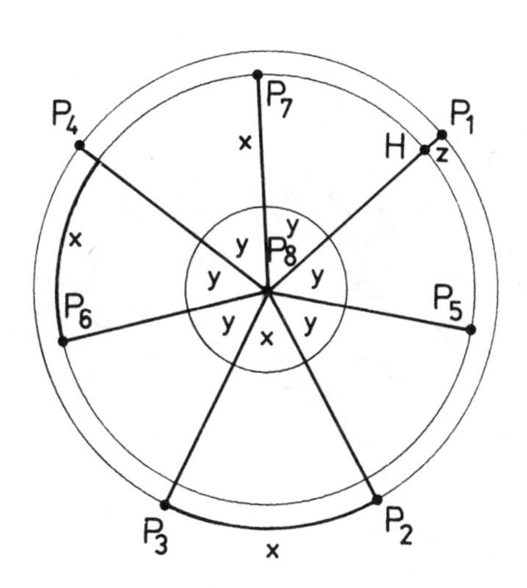

Figure 9

For this distribution we obtain

$$x = a_m(P_iP_j) = \pi - 6 + \sqrt{42 - 12\pi + \pi^2} \approx 0.905\ldots .$$

Because of the existence of inequality (3) this result is not surprising.

There is still the question, if the distribution in fig. 9 is the best possible one of 8 points.

In analogy to the square there will be further problems and tasks, if we admit only a finite number of main streets and circular roads.

REFERENCE

[1] L. FEJES TÓTH, *Punktverteilungen in einem Quadrat*, Studia Sci. Math. Hung. 6 (1971) 439-442.

W. Mögling
Pädagogische Hochschule
Erfurt/Mühlhausen
Nordhäuser Str. 63.
DDR - 5064 Erfurt

COLLOQUIA MATHEMATICA SOCIETATIS JÁNOS BOLYAI
48. INTUITIVE GEOMETRY, SIÓFOK, 1985.

COMPACT EUCLIDEAN SPACE FORMS PRESENTED BY SPECIAL TETRAHEDRA

E. MOLNÁR*

Dedicated to László Fejes Tóth on the occasion of his 70th birthday

In an earlier paper [6] I have constructed a surprising fundamental domain to present the compact euclidean space form $E^3/P3_1$. In a sense this is a "tetrahedron" (see also Fig. 7). It has two faces without common point, which are identified by translation p, and two faces with two edges in common, which are identified by 3_1 screw motion s. The transformations p and s generate the crystallographic group (space group) $G = P3_1$ which acts on the euclidean space E^3 freely (without fixed point). The identified tetrahedron, denoted by $\underset{\sim}{T}$, represents the orbit space $E^3/P3_1$ which is a complete connected 3-dimensional compact Riemannian manifold of zero sectional curvature, or briefly, a euclidean space form.

It is natural to ask which euclidean space forms can be presented by such tetrahedra. Among the 10 euclidean space forms [3, 6, 11] only those have to be considered, whose fundamental group G has two generators identifying

*Supported by Hung. Nat. Found. for Sci. Research, grant no. 1238.

the faces of the tetrahedron mentioned above. Thus only the space groups $P3_1$, $P4_1$, $P6_1$, Bb, $Pna2_1$ can play the role of G.

As a result we have got 8 such minimal fundamental tetrahedra, 1 tetrahedron for $P3_1$ and $P6_1$, and 2 tetrahedra for $P4_1$, Bb and $Pna2_1$, respectively (Fig. 1-8). Now the natural minimality condition means that the fundamental tetrahedron has the minimum number of edge segmenets. We shall see that the combinatorial structure of these tetrahedra is in close connection with the set of defining relations for G, which express that some products of the generators and their inverses are equal to the identity transformation of E^3 denoted by 1. We refer to [6] for general information about the theory of space forms, other details and a more complete list of references.

First, let us recall some basic definitions from [6] in terms of the group $G = \mathbf{P4}_1$. We shall elaborate only this example in detail (since the case of $G = \mathbf{P4}_1$ has also been studied in [6]). The other groups can be treated similarly.

1. THE TWO SPECIAL MINIMAL FUNDAMENTAL TETRAHEDRA FOR THE SPACE FORM $E^3/\mathbf{P4}_1$

We recall that the translational lattice $\mathbf{L}_G$ of the group $G = \mathbf{P4}_1$ is generated by the basis vectors $\overrightarrow{OE_1} = \mathbf{e}_1$, $\overrightarrow{OE_2} = \mathbf{e}_2$ and $\overrightarrow{OE_3} = \mathbf{e}_3$, whose Gram matrix is

$$((\mathbf{e}_i; \mathbf{e}_j)) := \begin{pmatrix} 1 & 0 & 0 \\ 0 & 1 & 0 \\ 0 & 0 & c \end{pmatrix}$$

where $c > 0$ is an arbitrary constant; $(;)$ denotes the euclidean scalar product in the vector space $\mathbf{E}^3$. The generating 4_1 screw motion $s(\mathcal{s}, \mathbf{s}) : \mathbf{x} := \overrightarrow{OX} \longrightarrow \mathbf{y} := \overrightarrow{OY}$ is given by the vector equation

$$(1) \qquad \mathbf{y} = \mathbf{x}\mathcal{s} + \mathbf{s} .$$

If $\mathbf{x} = x^i \mathbf{e}_i$, $\mathbf{y} = y^k \mathbf{e}_k$, $\mathbf{e}_i \mathcal{s} = s_i^k \mathbf{e}_k$, $\mathbf{s} = s^k \mathbf{e}_k$ are introduced, then (1) can be written in the form

$$(2) \qquad y^k \mathbf{e}_k = (x^i s_i^k + s^k) \mathbf{e}_k ,$$

or in our particular case by the matrix formula

$$(3) \qquad (y^1\ y^2\ y^3) = (x^1\ x^2\ x^3) \begin{pmatrix} 0 & 1 & 0 \\ -1 & 0 & 0 \\ 0 & 0 & 1 \end{pmatrix} + (0 \quad 0 \quad \tfrac{1}{4}) .$$

This means, the transformation $s(\mathcal{s}, \mathbf{s})$ has a linear part $\mathcal{s}$ with matrix (s_i^k) with respect to the base $\mathbf{e}_i$, and a translational part $\mathbf{s}$ with row matrix s^k with respect to $\mathbf{e}_i$ and to the origin O. We use the row-column multiplication; transformations and matrices operate on the right. Summation refers always to the index occuring in a term both as an upper and a lower index.

In this notation the generating translations for $\mathbf{L}_G$ are

$$(4) \qquad p_1(\mathit{1}, \mathbf{e}_1),\ p_2(\mathit{1}, \mathbf{e}_2) = s^{-1} p_1 s,\ p_3(\mathit{1}, \mathbf{e}_3) = s^4$$

(the 4-th power of s).

Hence the group $G = \mathbf{P4}_1$ is generated by the two generators s and $p = p_1$, and we get a presentation

(5) $$\mathbf{P4}_1 = (s,\ p - 1 = ps^{-1}psp^{-1}s^{-1}p^{-1}s = ps^{-2}ps^2).$$

From the second relation we deduce $s^{-1}ps = sp^{-1}s^{-1}$. Inserting this into the first relation, we get $1 = psp^{-1}s^{-1}p^{-1}s^{-1}p^{-1}s$. Hence by conjugation $spsp^{-1}s^{-1}p^{-1}s^{-1}p^{-1} = 1$ holds, and taking the inverse we get $pspsps^{-1}p^{-1}s^{-1} = 1$. Thus we have another presentation

(6) $$\mathbf{P4}_1 = (s,\ p - 1 = pspsps^{-1}p^{-1}s^{-1} = ps^{-2}ps^2.$$

In [6] we discussed the unique absolutely shortest presentation for $\mathbf{P4}_1$ and described its geometric realization up to some combinatorial equivalence by an interesting fundamental "simplex". This simplex can be chosen such that, with appropriate identifications, it is also the minimally presenting fundamental domain for the space forms $E^3/\mathbf{P2}_1\mathbf{2}_1\mathbf{2}_1$, $E^3/\mathbf{Bb}$, $E^3/\mathbf{Pna2}_1$ (see also [9]). In [6] we have algebraically proven that the presentation (5) and (6) are minimal (i.e., the length sum of the defining relations are the shortest) in the first type of presentation where a translation and a screw motion were selected as two generators for $G = \mathbf{P4}_1$. But an open question remained whether these presentations can be geometrically realized (by a fundamental domain) or not.

Now, in this paper we construct (up to some combinatorial equivalence) uniquely the corresponding two fundamental domains $\underset{\sim}{T}_1$ and $\underset{\sim}{T}_2$ (Fig. 1-2). Topologically, these are the tetrahedra mentioned in the title and in the introduction. In both cases there occur two cycles of segment equivalence classes, respectively. The first cycle consists of 8 segments according to the length and to the cyclic form of the first relation. The second one consists of 6 segments according to the second relation. These presentations will be minimal for $\mathbf{P4}_1$ with respect to the topological type of the fundamental tetrahedra described above.

All these considerations follow the way of [6], but after this intuitive explanation we recall more precisely the criteria required for the fundamental domains $\underset{\sim}{\mathbf{T}}$ of the groups G involved. Briefly, $\underset{\sim}{\mathbf{T}}$ is a 3-dimensional incidence-polyhedron (see e.g. [10]) endowed with some identifications.

1) The interior of the oriented compact fundamental domain $\underset{\sim}{\mathbf{T}}$ is homeomorphic to on open 3-dimensional simplex. The G-images of $\underset{\sim}{\mathbf{T}}$ tile the space E^3 without common inner points.

2) The oriented boundary of $\underset{\sim}{\mathbf{T}}$ consists of four oriented side faces. Each face is the common part of $\underset{\sim}{\mathbf{T}}$ and exactly one neighbouring G-image in the corresponding fundamental space tiling. The side faces may be curved, but each of them has a relative interior homeomorphic to an open 2-simplex.

3) If the domain $\underset{\sim}{\mathbf{T}}$ and one of its G-images $\underset{\sim}{\mathbf{T}}^g$ have, for some $g \in G$, the face f_g in common, then $\underset{\sim}{\mathbf{T}}$ and its image $\underset{\sim}{\mathbf{T}}^{g^{-1}}$ also have exactly one common face denoted by $f_{g^{-1}}$. We say that $f_{g^{-1}}$ is identified with f_g by the isometry g, or f_g is identified with $f_{g^{-1}}$ by g^{-1}. These identifying isometries generate the group G. The isometry g preserves the orientation if and only if the orientation of f_g on the surface of $\underset{\sim}{\mathbf{T}}$ is opposite to the orientation obtained from that of $f_{g^{-1}}$ by g. A flag (defined by a face and by an oriented edge on its boundary with assigned starting point) and its g-image, both lying on $\underset{\sim}{\mathbf{T}}$, determine the generating isometry g uniquely.

4) The boundary of each side face f of the domain $\underset{\sim}{\mathbf{T}}$ consists of finitely many oriented edge segments along each of which f joins exactly one adjacent side face. The segments may be curved, but each of them has a relative interior homeomorphic to an open interval. The side face identifications induce the G-equivalence of the edge segments. We require that a segment shall not have any G-equivalent points in its interior.

5) The boundary of each segment consists of two vertices. For any vertex A of $\underset{\sim}{\mathbf{T}}$ the segments and the faces, containing A, form cycles, i.e., the consecutive segments are contained in exactly one common face, the consecutive faces have exactly one common segment.

6) In addition, the tetrahedron $\underset{\sim}{\mathbf{T}}$ shall have the combinatorial and topological structure described in the Schlegel diagrams in Figures 1-8., i.e., the boundary of $\underset{\sim}{\mathbf{T}}$ is homeomorphic to the sphere on which four domains, the faces, are specified. Two of them have neither edges nor vertices in common. The other two faces have two components of common edges, both may be divided into segments by additional vertices. In the diagrams of Figures 1-8 the faces are only indicated by their suffix.

We remark that elements of finite order do not occur in G now, because of its free action on E^3.

Now we recall the procedure of Poincaré which shows that each segment equivalence class, induced by the generating face identifications of $\underset{\sim}{\mathbf{T}}$, involves a cycle transformation and a cycle relation of the generators. Instead of a general formulation we illustrate the algorithm for the particular fundamental domain $\underset{\sim}{\mathbf{T}}_1$ of $G = \mathbf{P4}_1$ (Figure 1).

Let us take an edge segment DE and choose a face from the two faces containing DE on the boundary, say $f_{p^{-1}}$. For this flag there exists ecactly one identifying generator of G. This is the translation p in our case, which maps $f_{p^{-1}}$ onto f_p and the edge segment DE onto the edge segment JK of f_p. There is exactly one other face, namely f_s, with JK on its boundary, and furthermore there is an identifying generator of G, namely the screw motion s^{-1}, which maps f_s onto $f_{s^{-1}}$ and JK onto CD on the boundary of $f_{s^{-1}}$.

The procedure ends, iff the starting flag, i.e., the segment DE and the face $f_{p^{-1}}$ reappear again.

We have the cycle scheme (imagine it written along a circle)

$$(DE:=f_s\cap f_{p^{-1}}) \xrightarrow{p} (JK:=f_p\cap f_s) \xrightarrow{s^{-1}} (CD:=f_{s^{-1}}\cap f_{p^{-1}}) \xrightarrow{p} (IJ:=f_p\cap f_{s^{-1}})$$

$$(7)\qquad \xrightarrow{s} (f_p\cap f_s=:GL) \xrightarrow{p^{-1}} (f_s\cap f_{p^{-1}}=:AF) \xrightarrow{s^{-1}} (f_p\cap f_{s^{-1}}=:HG) \xrightarrow{p^{-1}} (BA:=f_{p^{-1}}\cap f_{s^{-1}}) \xrightarrow{s} (DE:=f_s\cap f_{p^{-1}})$$

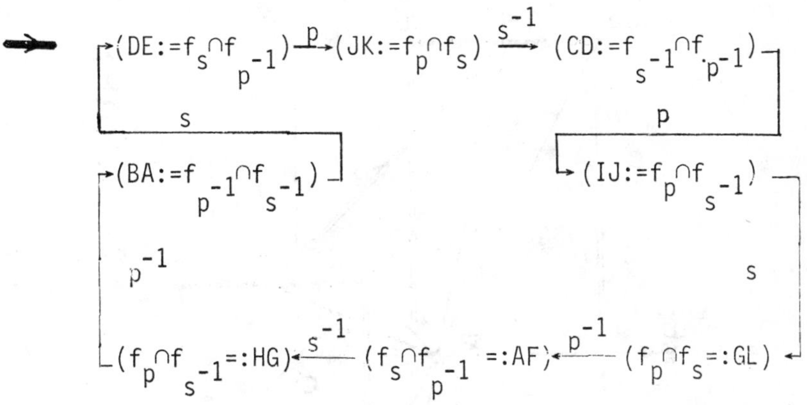

and the cycle transformation $ps^{-1}psp^{-1}s^{-1}p^{-1}s =: g_1$ according to the first segment class (now $g_1 = 1$ in $G = \mathbf{P4}_1$).

For the second segment class we similarly get the scheme

76. $P4_1$

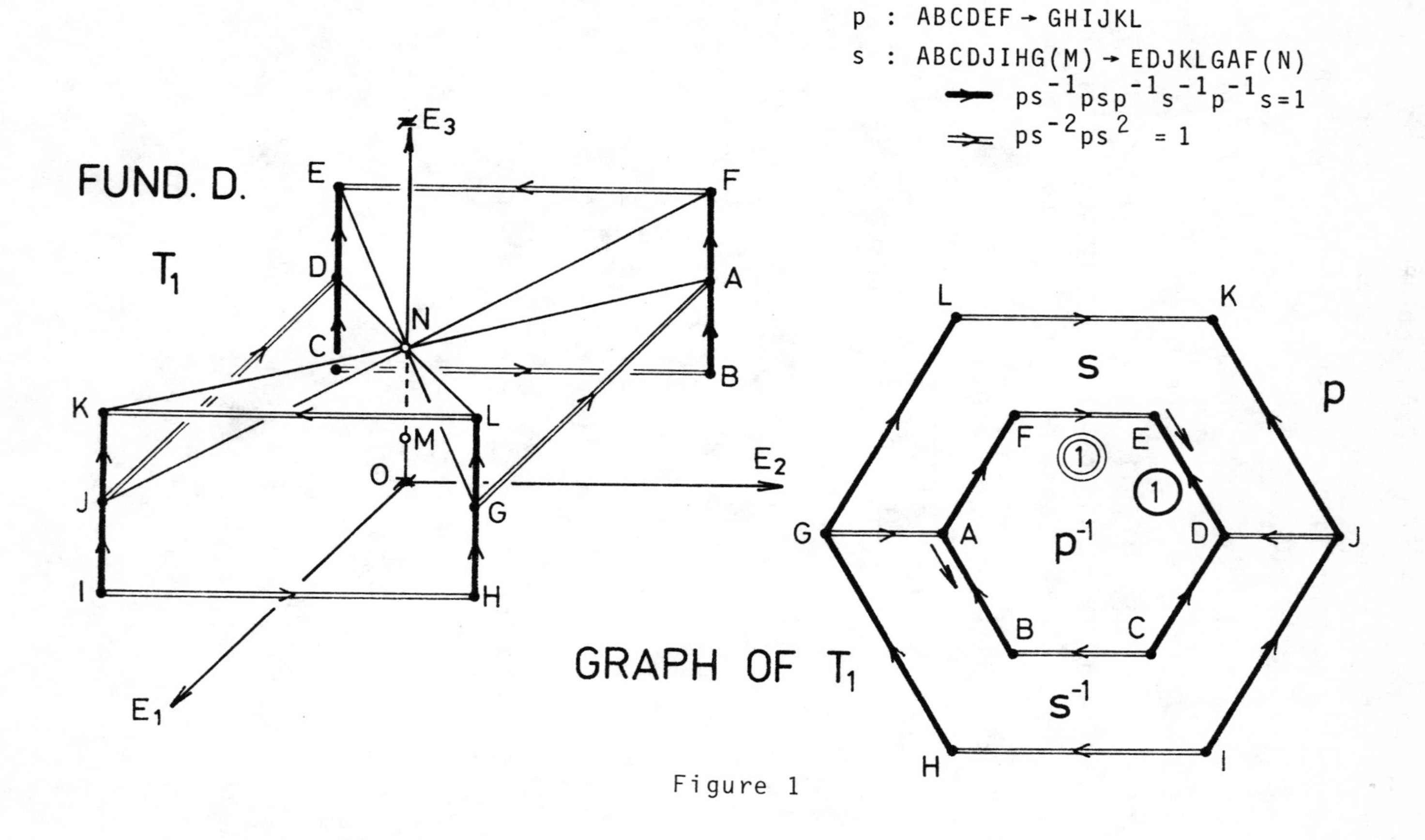

Figure 1

$$(8) \quad \Rightarrow \begin{array}{ccccc} \rightarrow(FE:=f_s \cap f_{p^{-1}}) & \xrightarrow{p} & (LK:=f_p \cap f_s) & \xrightarrow{s^{-1}} & (JD:=f_{s^{-1}} \cap f_s) \\ \uparrow s & & & & \downarrow s^{-1} \\ (f_{s^{-1}} \cap f_s=:GA) & \xleftarrow{s} & (f_{s^{-1}} \cap f_p=:IH) & \xleftarrow{p} & (f_{p^{-1}} \cap f_{s^{-1}}=:CB) \end{array}$$

and the cycle transformation $ps^{-2}ps^2 =: g_2$ ($g_2 = 1$ in $G = \mathbf{P4}_1$).

We recall the geometric meaning of the procedure only for the second class. The edge segment FE is consecutively surrounded in the fundamental space tiling by the domain $\underset{\sim}{\mathbf{T}}_1$ and its G-images

$$\underset{\sim}{\mathbf{T}}_1^{p^{-1}}, \ \underset{\sim}{\mathbf{T}}_1^{sp^{-1}}, \ \underset{\sim}{\mathbf{T}}_1^{s^2p^{-1}}, \ \underset{\sim}{\mathbf{T}}_1^{p^{-1}s^2p^{-1}}, \ \underset{\sim}{\mathbf{T}}_1^{s^{-1}p^{-1}s^2p^{-1}},$$

$$\underset{\sim}{\mathbf{T}}_1^{s^{-2}p^{-1}s^2p^{-1}} = \underset{\sim}{\mathbf{T}}_1^{g_2^{-1}}.$$

Now the crucial point, which is plausible intuitively, is the following.

First, if a finite presentation by generators and relations of a discontinuous transformation group G is given [1, 2, 4], then we can try to regard the relators as cycle transformations. That is, we can try to construct a fundamental domain $\underset{\sim}{\mathbf{T}}$, with faces identified by the generators such that Poincaré's process, applied to the edge segment classes (induced by the identification of faces), gives us the system of prescribed relators. This was our task and the method will be explained later on.

Second, if an appropriate domain is given together with a face pairing (identifications), so that for each class of equivalent edge segments the cycle transformation (or some of its powers) is equal to identity, then the domain is a fundamental domain of a transformation group G, whose presentation can be derived from the identifications and the cycle transformations. Roughly speaking this was Poincaré's program to construct discontinuous isometry groups, mainly in the Bolyai--Lobachevskian hyperbolic space (see also [5, 7, 8]).

We do not mention other generalizations, but we shall use both aspect of the problem.

Let us start with the presentation (5). We take the two relations and form the cycles just as in (7) and (8) by giving the intersection pattern of the faces, this time without indicating the vertices. We have to show that the edge segments join each other uniquely as it was given in (7) and (8) and in the graph of $\underset{\sim}{\mathbf{T}}_1$ (Figure 1). From the relations we see that f_p and $f_{p^{-1}}$ cannot have an edge segment in common, but f_s and $f_{s^{-1}}$ have two common segments, since in the second cycle (8) $f_{s^{-1}} \cap f_s$ occurs two times, because of the appearance of s^{-2} and s^2 in the second relation of (5). Each segment

$$f_{s^{-1}} \cap f_{p^{-1}},\ f_{s^{-1}} \cap f_p,\ f_s \cap f_{p^{-1}},\ f_s \cap f_p$$

appears three times in (7) and (8) but only once in the second segment class of (8). These facts imply two possibilities for the geometric structure of the fundamental domain $\underset{\sim}{\mathbf{T}}$. Figures 9.1.a and 9.2.a show the possible

Schlegel diagrams. The position of the remaining segments from the second class can be determined.

We shall see that the diagram of Figure 9.2.a, where f_p and $f_{p^{-1}}$ have a common vertex, is not compatible with the requirements 1) - 6) for $\underset{\sim}{\mathbf{T}}$ and for G. An easy check shows that any pairing $f_{p^{-1}} \to f_p$, $f_{s^{-1}} \to f_s$ on $\underset{\sim}{\mathbf{T}}$ fixes already the position of the segments as indicated in Figures 9.2.b and 9.2.c, respectively, but in both cases we would get only four segments in the second class. By the way, Figure 9.2.c realizes two inversive (conformal) space forms.

The possibility described in Figure 9.1.a fixes the positions of the second class segments as given by Figure 9.1.b. Otherwise a simple verification shows that any pairing $f_{p^{-1}} \to f_p$, $f_{s^{-1}} \to f_s$ would lead to a contradiction.

Up to this point we have considered only the way the two edge segment classes distribute on the faces of $\underset{\sim}{\mathbf{T}}$. Thus the geometric structure of Figure 9.1.b. holds also for $\underset{\sim}{\mathbf{T}}_1$ in cases of groups **Bb** (Figure 3) and $\mathbf{Pna2}_1$ (Figure 5). Now we take into account the orientation of the faces and see whether the identifying isometries are orientation preserving or not.

Since s and p preserve the orientation (by the way this would follow also from the second relation of (5) or from (8)), the diagram in Figure 9.1.b. uniquely becomes the graph of $\underset{\sim}{\mathbf{T}}_1$ in Figure 1 according to (7) and (8). We see that all vertices induced are equivalent under $G = \mathbf{P4}_1$. We have some freedom in choosing the vertex A in the coordinate system of Figure 1. Then we can derive the other vertices by (3) and (4). We particularly have

$$A(-\tfrac{1}{2};\tfrac{1}{2};\tfrac{1}{4}),\ G := A^p(\tfrac{1}{2};\tfrac{1}{2};\tfrac{1}{4}),\ E := A^s(-\tfrac{1}{2};-\tfrac{1}{2};\tfrac{1}{2}),$$

$$H := A^{s^{-1}}(\tfrac{1}{2};\tfrac{1}{2};\ 0),\ F := G^s(-\tfrac{1}{2};\tfrac{1}{2};\tfrac{1}{2}),\ I := G^{s^{-1}}(\tfrac{1}{2};-\tfrac{1}{2};0),$$

$$K := E^p(\tfrac{1}{2};-\tfrac{1}{2};\tfrac{1}{2}),\ B := H^{p^{-1}}(-\tfrac{1}{2};\tfrac{1}{2};0),\ L := F^p(\tfrac{1}{2};\tfrac{1}{2};\tfrac{1}{2}),$$

$$C := I^{p^{-1}}(-\tfrac{1}{2};-\tfrac{1}{2};0),\ D := K^{s^{-1}}(-\tfrac{1}{2};-\tfrac{1}{2};\tfrac{1}{4}) = B^s,$$

$$J := L^{s^{-1}}(\tfrac{1}{2};-\tfrac{1}{2};\tfrac{1}{4}) = C^s$$

for the vertices of $\underset{\sim}{\mathbf{T}}_1$. The faces $f_{p^{-1}}$ = ABCDEF and $(f_{p^{-1}})^p =: f_p$ = GHIJKL can be formed in an obviuous manner. But the face $f_{s^{-1}}$ is constructed with care; it is a star-shaped polygon with $M(0;\ 0;\ \frac{1}{8})$ as centre and the closed polygon ABCDJIHG as boundary. The face f_s = EDJKLGAF(N) is the s-image of $f_{s^{-1}}$. This way we ensure that the faces of $\underset{\sim}{\mathbf{T}}_1$ and their corresponding images do not intersect in their interiors. Finally we get a metric realization $\underset{\sim}{\mathbf{T}}_1$ in Figure 1. We see that $\underset{\sim}{\mathbf{T}}_1$ is determined up to the combinatorial equivalence mentioned in the construction above.

The presentation (6) gives us the realization $\underset{\sim}{\mathbf{T}}_2$ according to Poincaré's procedure. After having chosen the starting vertex, say $A(-\frac{1}{2};\ \frac{1}{2};\ 0)$, we derive the other vertices as images of A under G = $\mathbf{P4}_1$:

$$G := A^p(\tfrac{1}{2};\tfrac{1}{2};0),\quad B := A^s(-\tfrac{1}{2};-\tfrac{1}{2};\tfrac{1}{4}),\ F := G^s(-\tfrac{1}{2};\tfrac{1}{2};\tfrac{1}{4}),$$

$$L := F^p(\tfrac{1}{2};\tfrac{1}{2};\tfrac{1}{4}),\ H := B^p(\tfrac{1}{2};-\tfrac{1}{2};\tfrac{1}{4}),\ I := B^s(\tfrac{1}{2};-\tfrac{1}{2};\tfrac{1}{2}),$$

$$C := F^s(-\tfrac{1}{2};-\tfrac{1}{2};\tfrac{1}{2}),\ E := L^s(-\tfrac{1}{2};\tfrac{1}{2};\tfrac{1}{2}),\ K := H^s(\tfrac{1}{2};\tfrac{1}{2};\tfrac{1}{2}),$$

$$D := K^s(-\tfrac{1}{2};\tfrac{1}{2};\tfrac{3}{4}),\ J := I^s = D^p(\tfrac{1}{2};\tfrac{1}{2};\tfrac{3}{4})\ .$$

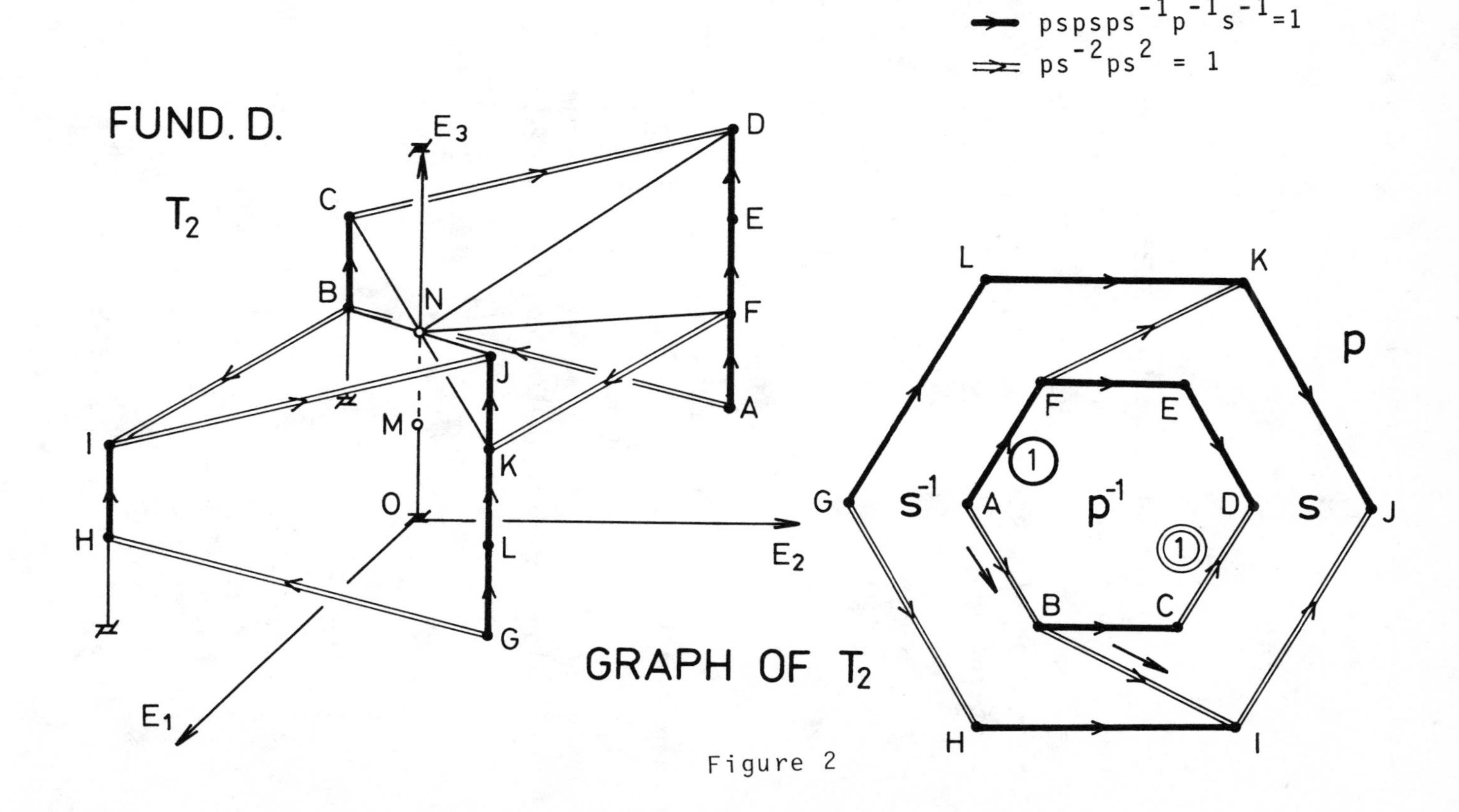
76. P4₁
p : ABCDEF → GHIJKL
s : ABIHGLKF(M) → BIJKFEDC(N)
pspsps⁻¹p⁻¹s⁻¹=1
ps⁻²ps² = 1
FUND. D.
T₂
E₃
D
C
E
B
N
F
J
A
M
I
K
O
H
L
E₂
G
E₁
GRAPH OF T₂
L
K
p
F
E
1
G
s⁻¹
A
p⁻¹
D
s
J
1
B
C
H
I

Figure 2

The faces of $\underset{\sim}{\mathbf{T}}_2$ are chosen with care again, as indicated in Figure 2, by the star-shaped polygon f_s = BIJKFEDC(N).

2. THE SPACE FORM $E^3/\mathbf{Bb}$

The B-centred monoclinic lattice $\mathbf{L}_G$ for the space group $G = \mathbf{Bb}$ is generated by the vectors $\mathbf{b}_1 = \overrightarrow{OB_1}$, $\mathbf{b}_2 = \overrightarrow{OB_2}$, $\mathbf{b}_3 = \overrightarrow{OB_3}$ in Figures 3-4. The generating glide reflection b(b, $\mathbf{b}$) is given by the linear transformation b defined by

$$(1)\qquad \begin{pmatrix} \mathbf{b}_1 & b \\ \mathbf{b}_2 & b \\ \mathbf{b}_3 & b \end{pmatrix} = \begin{pmatrix} 1 & 0 & 0 \\ 0 & 1 & 0 \\ 1 & 0 & -1 \end{pmatrix} \begin{pmatrix} \mathbf{b}_1 \\ \mathbf{b}_2 \\ \mathbf{b}_3 \end{pmatrix}$$

and by the vector $\mathbf{b} := \overrightarrow{OO^b}$. In Figure 3 $\mathbf{b} = \overrightarrow{BJ}$ has the coordinates $(-1; \frac{1}{2}; 2)$. In Figure 4 $\mathbf{b} = \overrightarrow{AF}$ has the coordinates $(0; \frac{1}{2}; 0)$. The generating translation p is determined by the vector $\mathbf{b}_3$ now. We have two presentations of G, which are realized by the tetrahedra $\underset{\sim}{\mathbf{T}}_1$ (Figure 3) and $\underset{\sim}{\mathbf{T}}_2$ (Figure 4), respectively.

The vertices of $\underset{\sim}{\mathbf{T}}_1$ are:

$B(0; 0; 0)$, $H := B^p(0; 0; 1)$, $J := B^b(-1; \frac{1}{2}; 2)$,

$G := H^b(0; \frac{1}{2}; 1)$, $D := J^{p^{-1}}(-1; \frac{1}{2}; 1)$, $C := D^{b^{-1}}((-1; 0; 1)$,

$I := C^p(-1; 0; 2)$, $A := G^{p^{-1}}(0; \frac{1}{2}; 0)$, $K := A^b(-1; 1; 2)$,

$L := G^b(0; 1; 1)$, $F := J^b(0; 1; 0)$, $E := D^b = K^{p^{-1}}(-1; 1; 1)$.

9. Bb

p : ABCDEF → GHIJKL

b : ABCDJIHG(M) → KJDEFAGL(N)

$pb^{-1}pbp^{-1}b^{-1}p^{-1}b=1$

$b^{2}pb^{-2}p^{-1} = 1$

FUND. D.

T_1

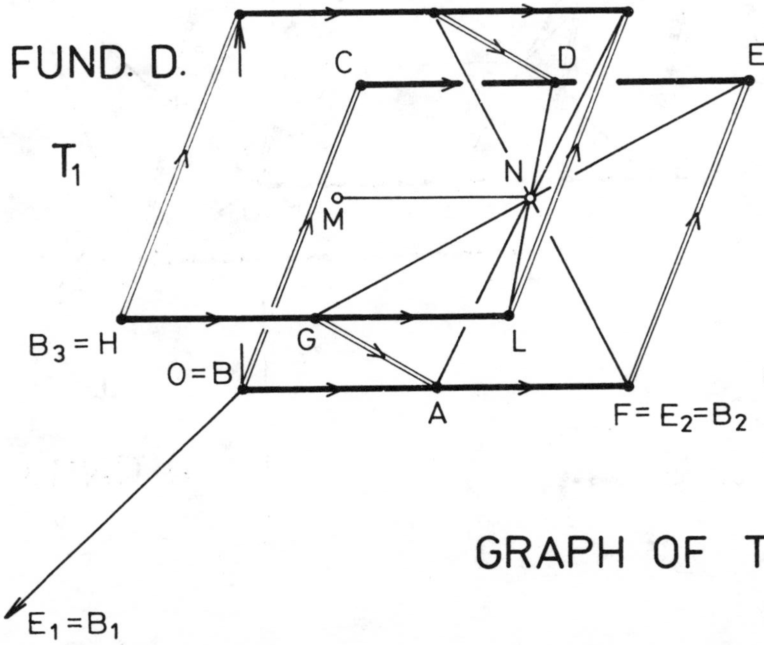

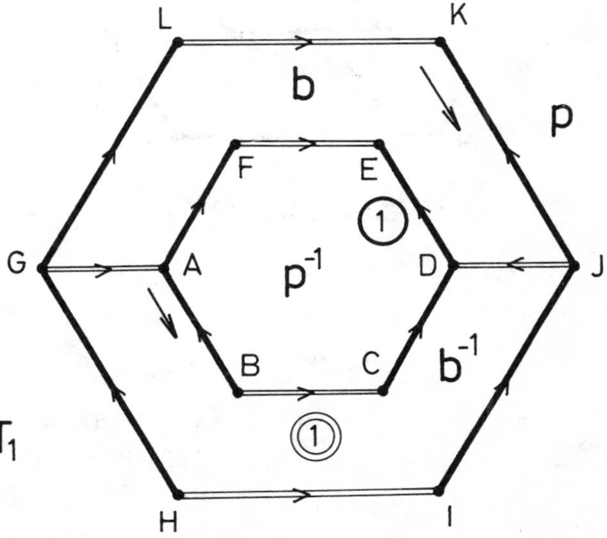

GRAPH OF T_1

Figure 3

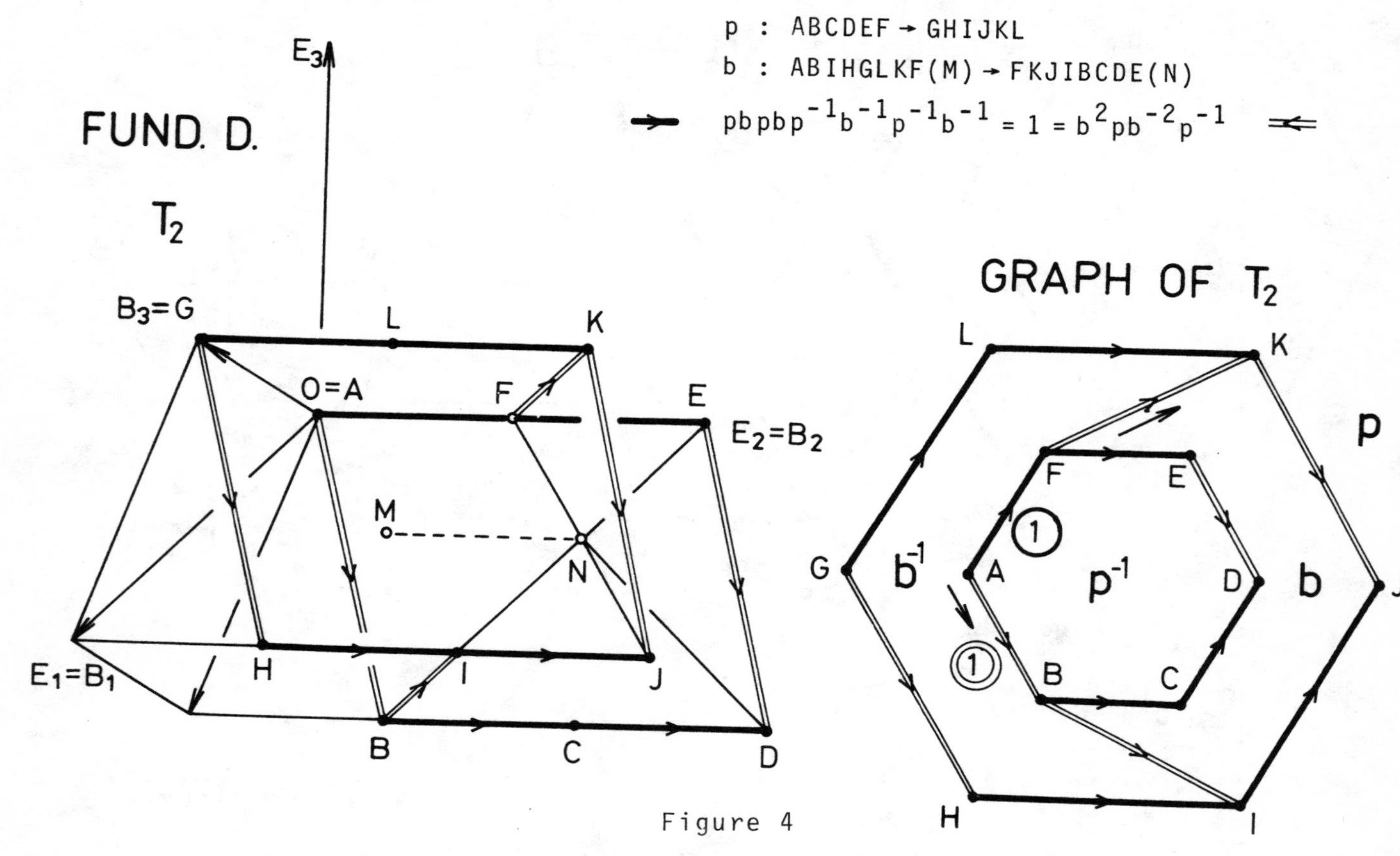
9. Bb
FUND. D.
T_2
p : ABCDEF → GHIJKL
b : ABIHGLKF(M) → FKJIBCDE(N)
$pbpbp^{-1}b^{-1}p^{-1}b^{-1} = 1 = b^2pb^{-2}p^{-1}$
E_3
B_3=G
L
K
O=A
F
E
E_2=B_2
M
N
E_1=B_1
H
I
J
B
C
D
GRAPH OF T_2
L
K
p
F
E
1
G
b^{-1}
A
p^{-1}
D
b
J
1
B
C
H
I

Figure 4

The points $M(-\frac{1}{2}; \frac{1}{4}; 1)$ and $N := M^b(-\frac{1}{2}; \frac{3}{4}; 1)$ are the centres of the star-shaped polygons $f_{b^{-1}} = ABCDJIHG(M)$ and $f_b = KJDEFAGL(N)$, respectively. $\underset{\sim}{\mathbf{T}}_2$ has the vertices

$A(0; 0; 0)$, $G := A^p(0; 0; 1)$, $F := A^b(0; \frac{1}{2}; 0)$,

$B := G^b(1; \frac{1}{2}; -1)$, $L := F^p(0; \frac{1}{2}; 1)$, $H := B^p(1; \frac{1}{2}; 0)$,

$E := F^b(0; 1; 0)$, $C := L^b(1; 1; -1)$, $K := E^p(0; 1; 1)$,

$I := C^p = H^b(1; 1; 0)$, $D := K^b(1; \frac{3}{2}; -1)$, $J := D^p = I^b(1; \frac{3}{2}; 0)$.

Now $M(\frac{1}{2}; \frac{1}{2}; 0)$ and $N := M^b(\frac{1}{2}; 1; 0)$ centre the star-shaped polygons $f_{b^{-1}}$ and f_b, respectively. Both $\underset{\sim}{\mathbf{T}}_1$ and $\underset{\sim}{\mathbf{T}}_2$ have the graph required. Each metric realization depends on a starting vertex as before.

3. THE SPACE FORM $E^3/\mathbf{Pna2_1}$

The orthorhombic lattice $\mathbf{L}_G$ for the space group $G = \mathbf{Pna2_1}$ is generated by pairwise orthogonal vectors $\overrightarrow{OE_1}$, $\overrightarrow{OE_2}$, $\overrightarrow{OE_3}$. The generating glide reflection a $(a, \mathbf{a})$ is defined by

$$(1) \qquad \begin{pmatrix} 1 & 0 & 0 \\ 0 & -1 & 0 \\ 0 & 0 & 1 \end{pmatrix}, \quad (\tfrac{1}{2} \quad \tfrac{1}{2} \quad 0) \quad ,$$

the other generating glide reflection $n(n, \mathbf{n})$ is given by

33. $Pna2_1$

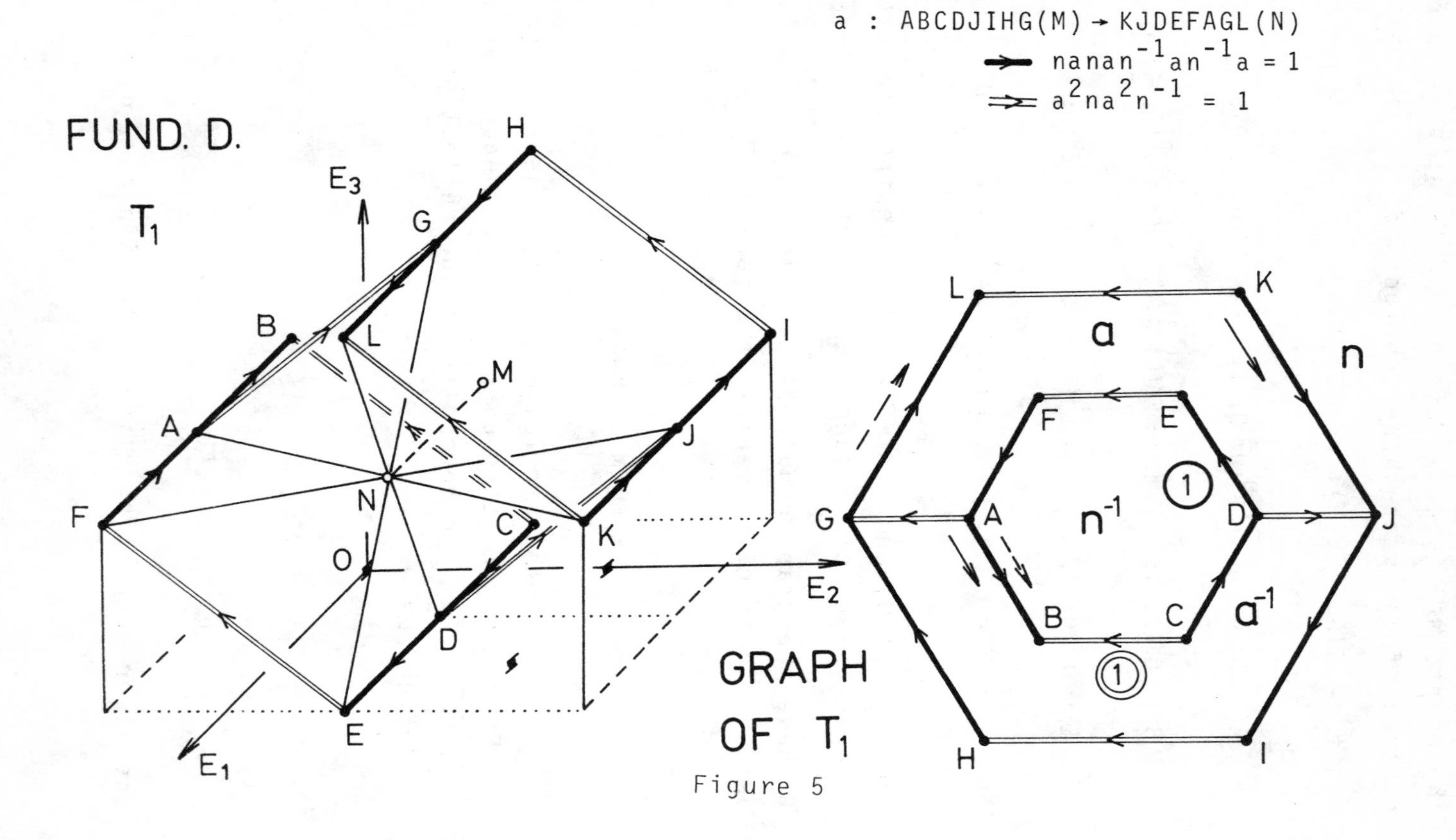

Figure 5

$$(2) \qquad \begin{pmatrix} -1 & 0 & 0 \\ 0 & 1 & 0 \\ 0 & 0 & 1 \end{pmatrix}, \quad \left(\frac{1}{2} \quad \frac{1}{2} \quad \frac{1}{2}\right) \quad .$$

These two generators give us the presentations in Figures 5-6, which are again realizable by tetrahedra. $\underset{\sim}{\mathbf{T}}_1$ and $\underset{\sim}{\mathbf{T}}_2$ have unique combinatorial structures determined by the relations as before. Both generators are orientation reversing isometries. In spite of this we have analogous structures for the fundamental domain. In Figure 5 the vertices of $\underset{\sim}{\mathbf{T}}_1$ are as follows:

$$C(-\tfrac{1}{4}; \tfrac{1}{4}; 0),\ K := C^n(\tfrac{3}{4}; \tfrac{3}{4}; \tfrac{1}{2}),\ D := C^a(\tfrac{1}{4}; \tfrac{1}{4}; 0),$$

$$A := K^{a^{-1}}(\tfrac{1}{4}; -\tfrac{1}{4}; \tfrac{1}{2}),\ J := D^n(\tfrac{1}{4}; \tfrac{3}{4}; \tfrac{1}{2}),\ E := D^a(\tfrac{3}{4}; \tfrac{1}{4}; 0),$$

$$I := A^{a^{-1}}(-\tfrac{1}{4}; \tfrac{3}{3}; \tfrac{1}{2}),\ G := A^n(\tfrac{1}{4}; \tfrac{1}{4}; 1),\ B := J^{a^{-1}}(-\tfrac{1}{4}; -\tfrac{1}{4}; \tfrac{1}{2}),$$

$$F := J^a(\tfrac{3}{4}; -\tfrac{1}{4}; \tfrac{1}{2}),\ L := G^a(\tfrac{3}{4}; \tfrac{1}{4}; 1),\ H := G^{a^{-1}} = F^n(-\tfrac{1}{4}; \tfrac{1}{4}; 1).$$

The star-shaped polygons $f_{a^{-1}}$ and f_a have the centres $M(0; \tfrac{1}{4}; \tfrac{1}{2})$ and $N := M^a(\tfrac{1}{2}; \tfrac{1}{4}; \tfrac{1}{2})$, respectively. In Figure 6 the polyhedron $\underset{\sim}{\mathbf{T}}_2$ has the vertices

$$A(0; \tfrac{1}{4}; 0),\ F := A^a(\tfrac{1}{2}; \tfrac{1}{4}; 0),\ K := A^n(\tfrac{1}{2}; \tfrac{3}{4}; \tfrac{1}{2}),$$

$$E := F^a(1; \tfrac{1}{4}; 0),\ L := F^n(0; \tfrac{3}{4}; \tfrac{1}{2}),\ B := K^{a^{-1}}(0; -\tfrac{1}{4}; \tfrac{1}{2}),$$

$$D := K^a(1; -\tfrac{1}{4}; \tfrac{1}{2}),\ G := E^n(-\tfrac{1}{2}; \tfrac{3}{4}; \tfrac{1}{2}),\ C := L^a(\tfrac{1}{2}; -\tfrac{1}{4}; \tfrac{1}{2}),$$

$$J := B^n(\tfrac{1}{2}; \tfrac{1}{4}; 1),\ H := D^n(-\tfrac{1}{2}; \tfrac{1}{4}; 1),\ I := C^n(0; \tfrac{1}{4}; 1).$$

The centres M and $N := M^a$ have the coordinates as before.

33. $Pna2_1$

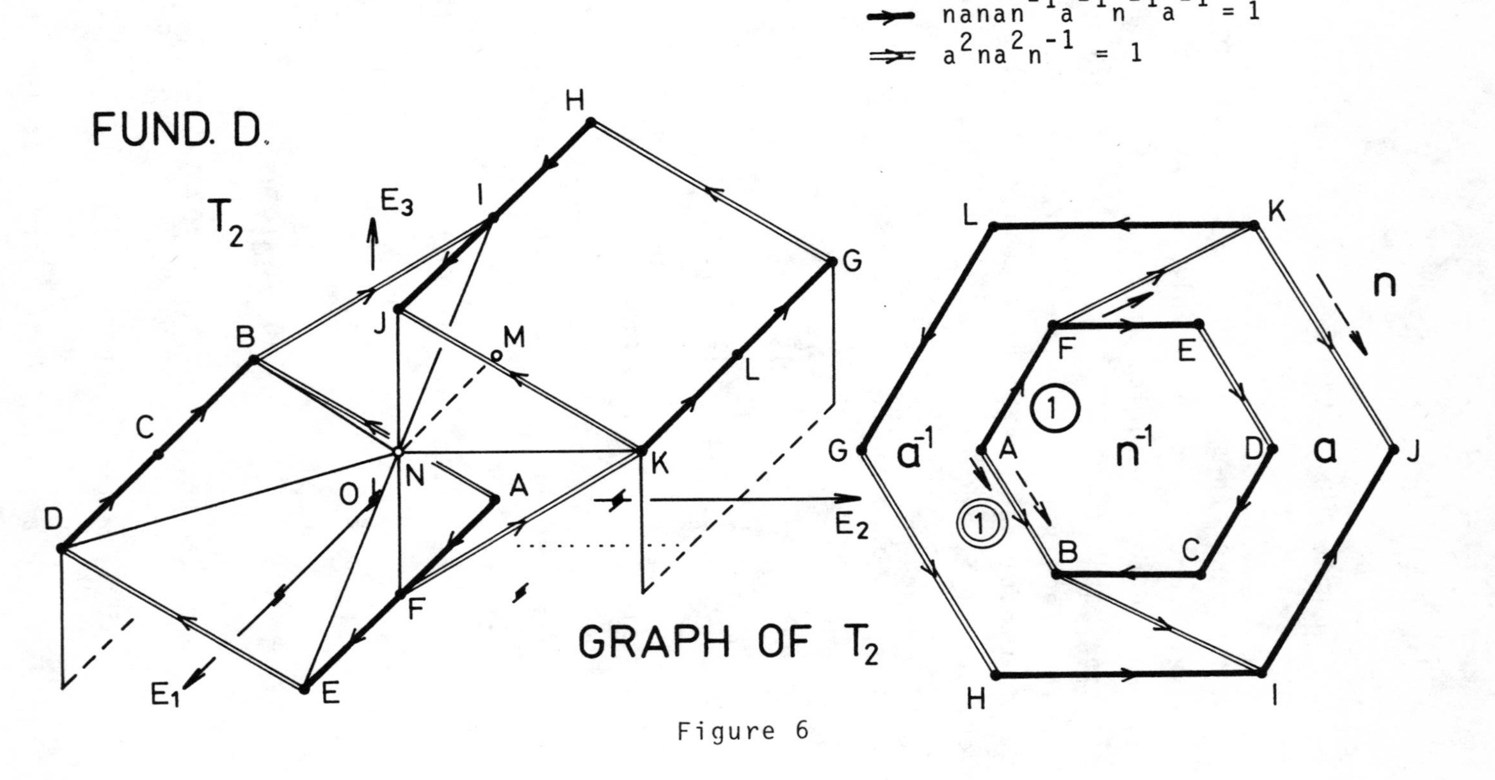

Figure 6

4. THE SPACE FORMS $E^3/\mathbf{P3}_1$ and $E^3/\mathbf{P6}_1$

The minimal presentations of the space group $\mathbf{P3}_1$ have been given in detail in [6]. We reproduce only the resulting fundamental domain with the combinatorial structure required in Figure 7.

The space group $G = \mathbf{P6}_1$ has the same hexagonal lattice $\mathbf{L}_G$ as $\mathbf{P3}_1$, which is generated by vectors $\mathbf{e}_1 = \overrightarrow{OE_1}$ $\mathbf{e}_2 = \overrightarrow{OE_2}$, $\mathbf{e}_3 = \overrightarrow{OE_3}$ with Gram matrix

$$(1) \qquad ((\mathbf{e}_i, \mathbf{e}_j)) = \begin{pmatrix} 1 & -\frac{1}{2} & 0 \\ -\frac{1}{2} & 1 & 0 \\ 0 & 0 & c \end{pmatrix},$$

where $c > 0$ is an arbitrary constant (Figure 8).

The generating 6_1 screw motion $s(s, \mathbf{s})$ is defined by

$$(2) \qquad \begin{pmatrix} 1 & 1 & 0 \\ -1 & 0 & 0 \\ 0 & 0 & 1 \end{pmatrix}, \quad \begin{pmatrix} 0 & 0 & \frac{1}{6} \end{pmatrix}.$$

Choosing the translation $p = p_1(I, \mathbf{e}_1)$ as second generator, we have the presentation of Figure 8 by a fundamental domain $\underset{\sim}{T}$ with the desired combinatorial structure.

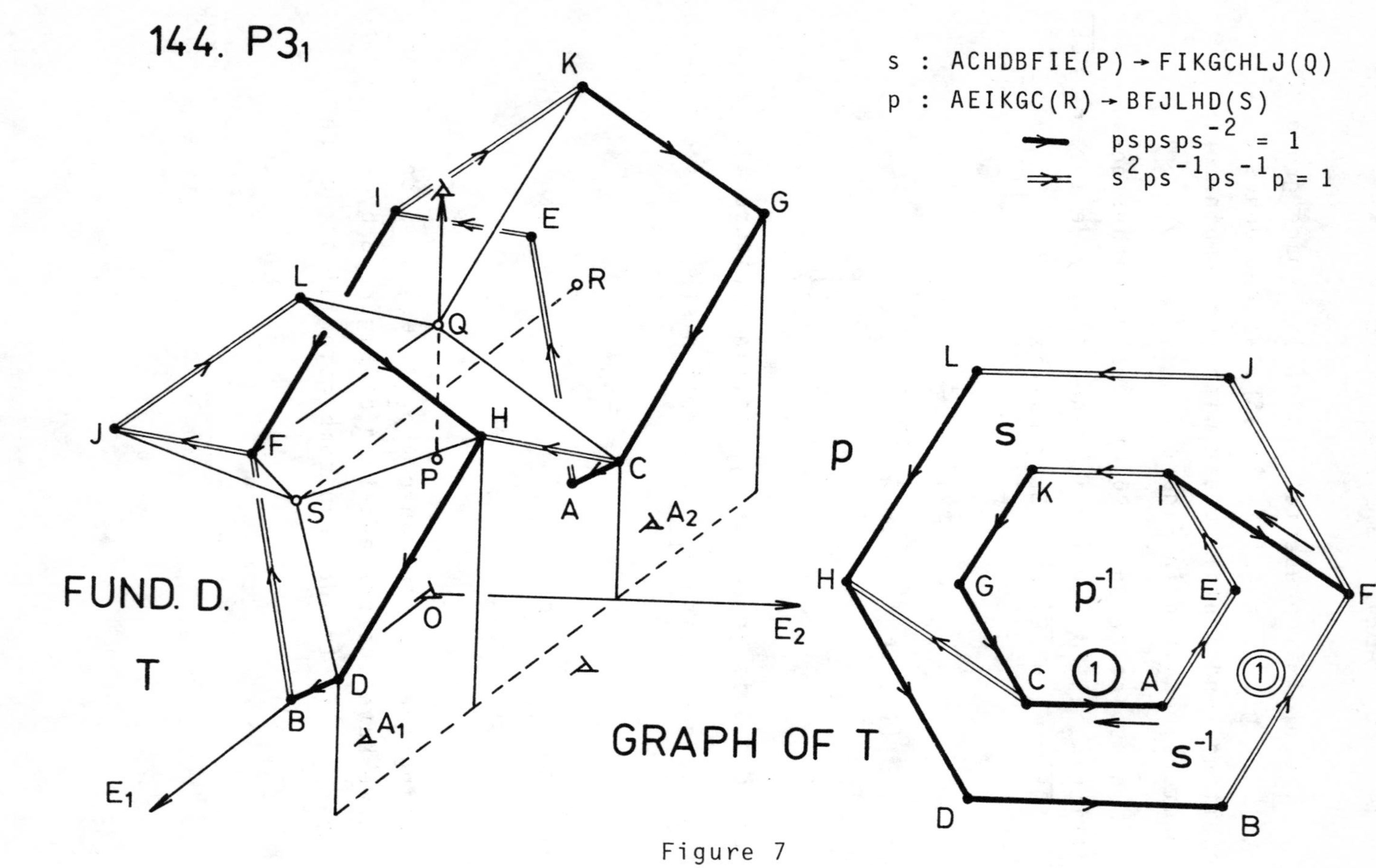
144. P3₁
s : ACHDBFIE(P) → FIKGCHLJ(Q)
p : AEIKGC(R) → BFJLHD(S)
pspsps⁻² = 1
s²ps⁻¹ps⁻¹p = 1
FUND. D.
T
GRAPH OF T
K
G
I
E
R
L
Q
H
J
F
P
C
A
S
O
D
B
E1
E2
A1
A2
p
s
p-1
s-1
1

Figure 7

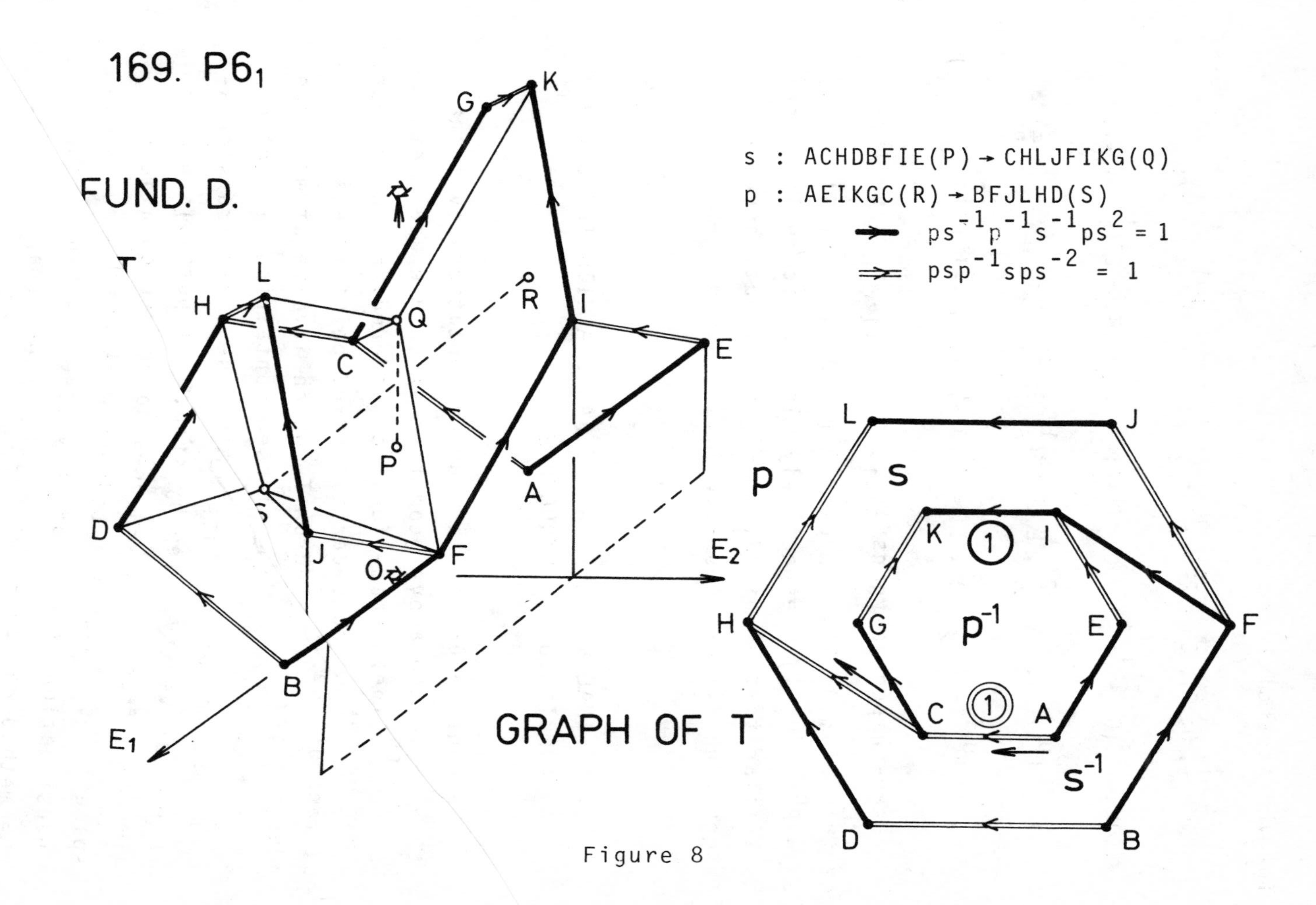
169. P6$_1$
FUND. D.
T
s : ACHDBFIE(P) → CHLJFIKG(Q)
p : AEIKGC(R) → BFJLHD(S)
$ps^{-1}p^{-1}s^{-1}ps^2 = 1$
$psp^{-1}sps^{-2} = 1$
A B C D E F G H I J K L
P Q R S O
E$_1$
E$_2$
GRAPH OF T
p
s
p^{-1}
s^{-1}
1
1

Figure 8

The vertices of $\underset{\sim}{\mathbf{T}}$ are

$$A(-\tfrac{1}{2};\ 0;\ 0),\ C := A^{s}(-\tfrac{1}{2};\ -\tfrac{1}{2};\ \tfrac{1}{6}),\ B := A^{p}(\tfrac{1}{2};\ 0;\ 0),$$

$$H := C^{s}(0;\ -\tfrac{1}{2};\ \tfrac{2}{6}),\ D := C^{p}(\tfrac{1}{2};\ -\tfrac{1}{2};\ \tfrac{1}{6}),\ F := B^{s}(\tfrac{1}{2};\ \tfrac{1}{2};\ \tfrac{1}{6}),$$

$$G := H^{p^{-1}}(-1;\ -\tfrac{1}{2};\ \tfrac{2}{6}),\ L := H^{s}(\tfrac{1}{2};\ 0;\ \tfrac{3}{6}),\ J := D^{s}(1;\ \tfrac{1}{2};\ \tfrac{2}{6}),$$

$$E := F^{p^{-1}}(-\tfrac{1}{2};\ \tfrac{1}{2};\ \tfrac{1}{6})\ ,\ I := F^{s}(0;\ \tfrac{1}{2};\ \tfrac{2}{6}),\ K := I^{s} = L^{p^{-1}}(-\tfrac{1}{2};0;\tfrac{3}{6}).$$

The star-shaped polygons $f_{s^{-1}}$, f_s, $f_{p^{-1}}$, f_p have the centres $P(0;\ 0;\ \frac{1}{6})$, $Q := P^{s}(0;\ 0;\ \frac{2}{6})$, $R(-\frac{1}{2};\ 0;\ \frac{1}{4})$, $S := R^{p}(\frac{1}{2};\ 0;\ \frac{1}{4})$, respectively. The metric realization is similar to that of $\mathbf{P3}_1$ in Figure 7, but there are interesting differences as well.

5. MINIMAL FUNDAMENTAL TETRAHEDRA WITH THE GIVEN COMBINATORIAL STRUCTURE

We want to prove that our presentations by tetrahedra with the above geometric structure are minimal for the e[illegible]lidean space forms considered. For a geometric proof we hav[illegible] to consider all fixed point free identifications of the [illegible]trahedron $\underset{\sim}{\mathbf{T}}$, which involve at most two edge segment equiv[illegible]nce classes containing altogether at most 14 segments. We [illegible]ermine the cycle transformations and so the defining [illegible]tions of possible groups G. One can verify, step by st[illegible] presentations different from the given ones do n[illegible] to euclidean space forms. This is a tedious work wh[illegible]ds the investigation of not too many cases. We on[illegible]strate the method and some interesting phenomena with [illegible]les in Figure 10. Here we indicate the induced orientation of the faces and edges. The generators are always denoted by x and y, the cycle transformations are taken by Poincaré's algorithm.

Figures 10. a-d show different (fixed point) free identifications of the same polyhedron $\underset{\sim}{\mathbf{T}}$. Now, we only remark that $\underset{\sim}{\mathbf{T}}$ can be explicitely constructed in the inversive space and Figure 10.d describes a spherical space form. The corresponding relations are as follows:

$$\text{a)} \Rrightarrow \; x^2 = 1 \; ; \quad \rightarrow \; yx^{-1}yx = 1 \, .$$

$$\text{b)} \Rrightarrow \; x^2 = 1 \; ; \quad \rightarrow \; yxy^{-1}x = 1 \, .$$

$$\text{c)} \Rrightarrow \; x^2 = 1 \; ; \quad \rightarrow \; yxy^{-1}x^{-1} = 1 \, .$$

The groups in b) and c) are obviuously isomorphic.

$$\text{d)} \Rrightarrow \; x^1y^{-1} = 1 \; ; \quad \rightarrow \; x^2y = 1 \, .$$

The identifications on $\underset{\sim}{\mathbf{T}}$ in Figure 10.a do not generate a freely acting group. Forming the Poincaré cycle, the starting vertex reappears in the process. For example, the vertex A is fixed under the transformation yx, although the cycle transformation is

$$\text{e)} \Rrightarrow \; yx^2yx^{-2} \, .$$

In Figure 10.f an interesting example is described:

$$\text{f)} \Rrightarrow \; x^2y^{-1}xy^{-1} = 1 \; ; \quad \rightarrow \; x^2yxy = 1 \, .$$

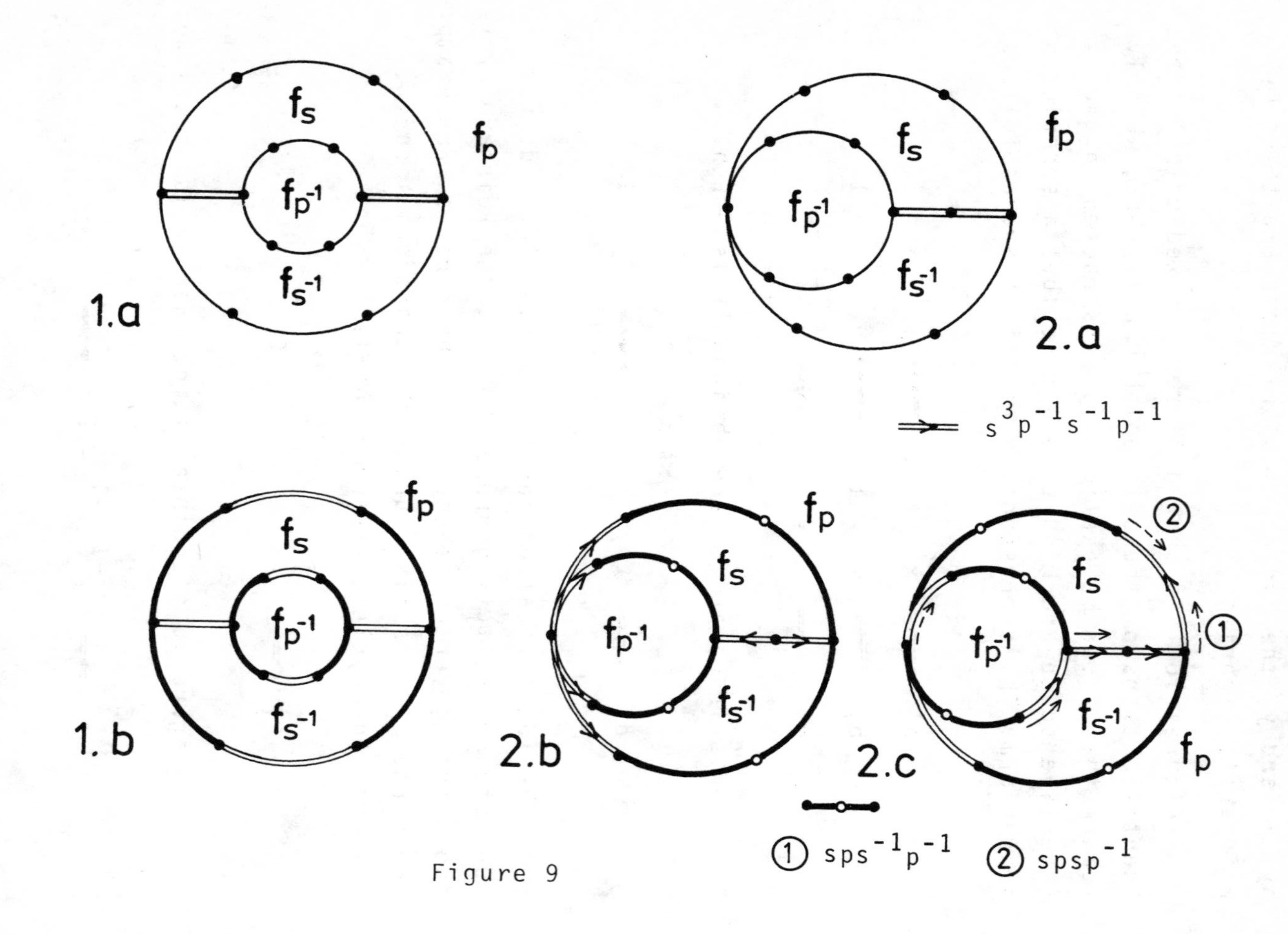

Figure 9

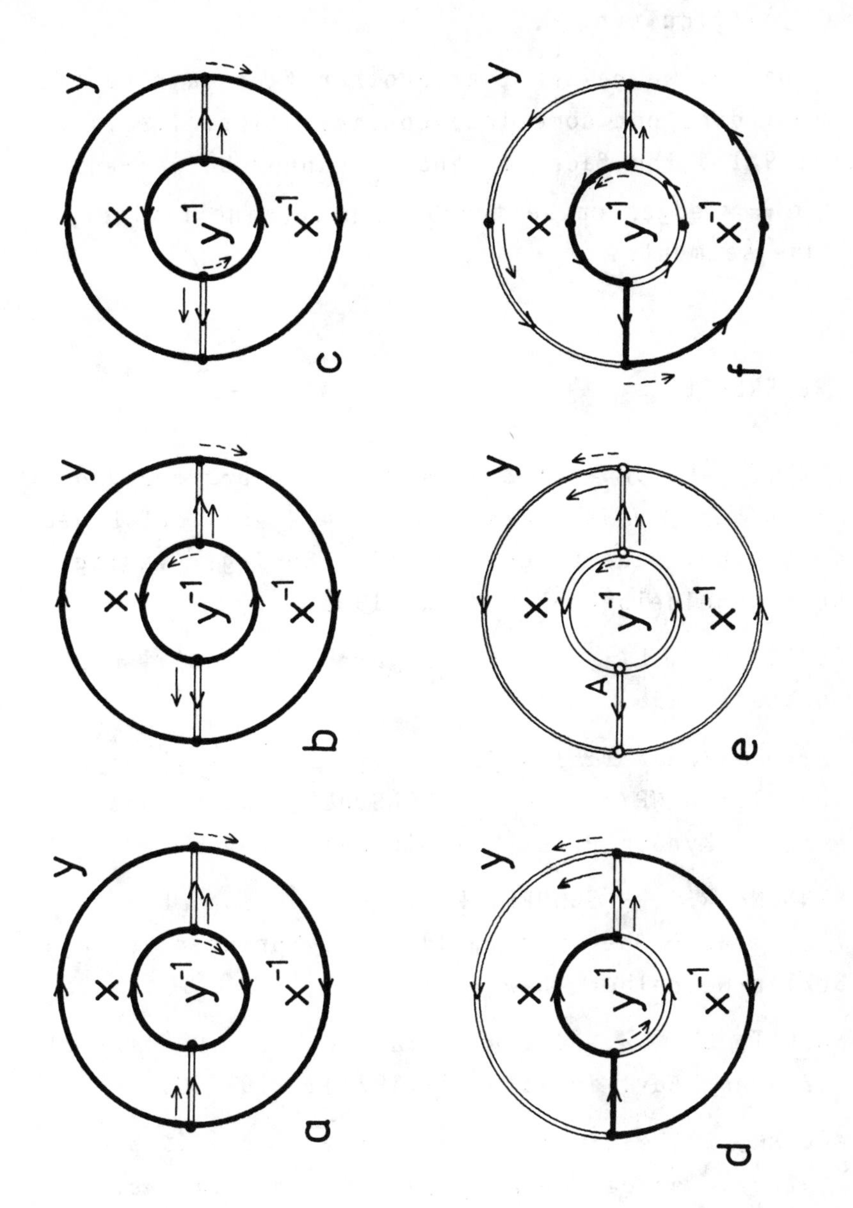

Figure 10

The group, generated by the identifications, acts freely on $\underset{\sim}{T}$. We have got an inversive space form again, but the explicit construction of the fundamental domain $\underset{\sim}{T}$ seems to be a difficult task.

Finally, we remark that another face pairing of the diagram $\underset{\sim}{T}$ does not come into consideration, i.e., in diagram 9.1.a the faces f_s and f_p cannot be paired.

The ⟹ edges cause trouble in forming finitely many edge segments.

REFERENCES

[1] COXETER, H.S.M. - MOSER, W.O.J., *Generators and relations for discrete groups.* 4th edition, Ergeb. der Math., Neue Folge, Bd. 14., Springer Verlag, Berlin-Heidelberg-New York, 1980.

[2] FEJES TÓTH, L., *Reguläre Figuren.* Akadémiai Kiadó, Budapest, 1965.

[3] *International Tables for X-ray Crystallography.* Vol. I.: HENRY, N.F.M. - LONSDALE, K., *Symmetry groups,* Kynoch Press, Birmingham, 1969.

[4] LYNDON, R.C. - SCHUPP, P.E., *Combinatorial group theory.* Erg. der Math., Bd. 89., Springer Verlag, Berlin-Heidelberg-New York, 1977.

[5] MASKIT, B., *On Poincaré's theorem for fundamental polygons.* Adv. in Math. 7 (1971), 219-230.

[6] MOLNÁR, E., *Minimal presentation of the 10 compact euclidean space forms by fundamental domains.* Studia Sci. Math. Hung. to appear in 1986.

[7] MOLNÁR, E., *Space forms and fundamental polyhedra. Proceedings of the Conference on Differential Geometry and Its Applications.* Nové Město na Moravě, Czechoslovakia, 1983., Part 1. Differential Geometry, 91-103.

[8] MOLNÁR, E., *Twice punctured compact euclidean and hyperbolic manifolds and their two-fold coverings.* Proceedings of Colloquium on Differential Geometry. Hajdúszoboszló, Hungary, 1984, to appear in 1987.

[9] MOLNÁR. E., *Über Polyederdarstellung der kristallographischen Gruppen.* "Geometrie in der Technik und Kunst" Praha-Dobřichovice, ČSSR, 1985, 126-135.

[10] SCHULTE, E., *On arranging regular incidence-complexes as faces of higher-dimensional ones.* Europ. J. Combinatorics 4 (1983), 375-384.

[11] WOLF, J.A., *Spaces of constant curvature.* University of California, Berkeley, California, 1972. Russian translation: Izd. "Nauka" Moskow, 1982.

EMIL MOLNÁR
Eötvös Loránd University, Institute of Mathematics,
Department of Geometry, Budapest, VIII. Rákóczi út 5.
H-1088.

COLLOQUIA MATHEMATICA SOCIETATIS JÁNOS BOLYAI
48. INTUITIVE GEOMETRY, SIÓFOK, 1985.

PENTAGONAL FACES OF SIMPLE ARRANGEMENTS

I. PALÁSTI

ABSTRACT

Let the number of k-gonal faces of a simple arrangement $\mathcal{A}_n$ of n lines be denoted by $p_k(\mathcal{A}_n)$. Here we give a construction with a lot of pentagonal faces which implies $\max p_5(\mathcal{A}_n) = n^2/4 + O(n)$.

1. INTRODUCTION

By the arrangement $\mathcal{A}_n$ of n lines we mean a finite family of lines $L_1,\ldots,L_n$ in a real projective plane. An Arrangement is called simple provided every vertex is a point of intersection of exactly two of the lines. So the plane is divided into $(n^2-n+2)/2$ faces. $p_k(\mathcal{A}_n)$ denotes the number of k-gons among these faces, $p_k^s(n) = \max\{p_k(\mathcal{A}_n) : \mathcal{A}_n \text{ is simple}\}$.Estimations for $p_k(\mathcal{A}_n)$ are known in cases $k = 3,4,6$ ([1], [2], [3], [4], [5] and [8]).

Research supported by Hungarian National Foundation for Scientific Research grant no. 1808.

The question occurs what bounds can be given for $P_5(\mathscr{A}_n)$. To answer this we use the same way of generating a simple arrangement with a large number of pentagons as it was done in [1] for triangles.

2. CONSTRUCTION

By P(0) let us denote any fixed point on the circumference of a unit circle with centre c. $P(\alpha)$ means the point obtained by rotating P(0) around c with angle α (α is a real number). Denote by $L(\alpha)$ the straight line through the points $P(\alpha)$ and $P(\pi - 2\alpha)$. In the case $\alpha \equiv \pi - 2\alpha \pmod{2\pi}$, $L(\alpha)$ is a tangent to the unit circle at the point $P(\alpha)$.

For any integer $n \geq 7$ let the arrangement of n lines be

$$\mathscr{A}_n = \begin{cases} L((4k-1)\frac{\pi}{N}); & 1 \leq k \leq \left[\frac{n+1}{6}\right], \\ L((4\left[\frac{n+1}{6}\right] + 2k - 1)\frac{\pi}{N}); & 1 \leq k \leq n - 2\left[\frac{n+1}{6}\right] \\ L((2n + 4k - 1)\frac{\pi}{N}); & 1 \leq k \leq \left[\frac{n+1}{6}\right], \end{cases}$$

where $N = n + \left[\frac{n}{3}\right]$. See Figures 1 and 2. Then we have the

THEOREM. $\mathscr{A}_n$ *is a simple arrangement of lines such that*

$$p_5(\mathscr{A}_n) \geq n(n-5)/4 + \varepsilon, \quad (-3/2 \leq \varepsilon \leq 2).$$

Moreover we have

$$p_5^s(n) = n^2/4 + O(n).$$

PROOF. To prove that $\mathcal{A}_n$ is simple we can apply the following lemma an elementary proof of which is given in [1]:

LEMMA. *The lines* $L(\alpha)$, $L(\beta)$ *and* $L(\gamma)$ *are concurrent if and only if* $\alpha + \beta + \alpha \equiv 0 \pmod{2\pi}$.

We outline here how this lemma can be used for our case. We observe that $\mathcal{A}_n$ consists of lines of the form $L_i = L((2i+1)\pi/N)$. If the lines $L_i = L((2i+1)\pi/N)$ and $L_j = L((2j+1)\pi/N)$ belong to $\mathcal{A}_n$ then the only line $L(\alpha)$ which goes through the point of intersection of L_i and L_j is the line $L((2N-2i-2j-2)\pi/N)$. However this line does not belong to $\mathcal{A}_n$ by definition. This shows that $\mathcal{A}_n$ is simple.

We recall that the lines $L((2i-1)\pi/N)$, $i = 1,\dots,N$, divide the plane into polygons the bulk of which cosists of triangles and hexagons (see Table 1 in [1]). Moreover, this subdivision of the plane "has almost" the structure of the Archimedean tiling (3,6,3,6). The number of pentagonal faces of $\mathcal{A}_n$ can be computed easily using this fact and observing that $\mathcal{A}_n$ arises from the above arrangement by omitting appropriate lines. The table gives the number of triangles, quadrangles, pentagons and hexagons in $\mathcal{A}_n$ for some small values of n.

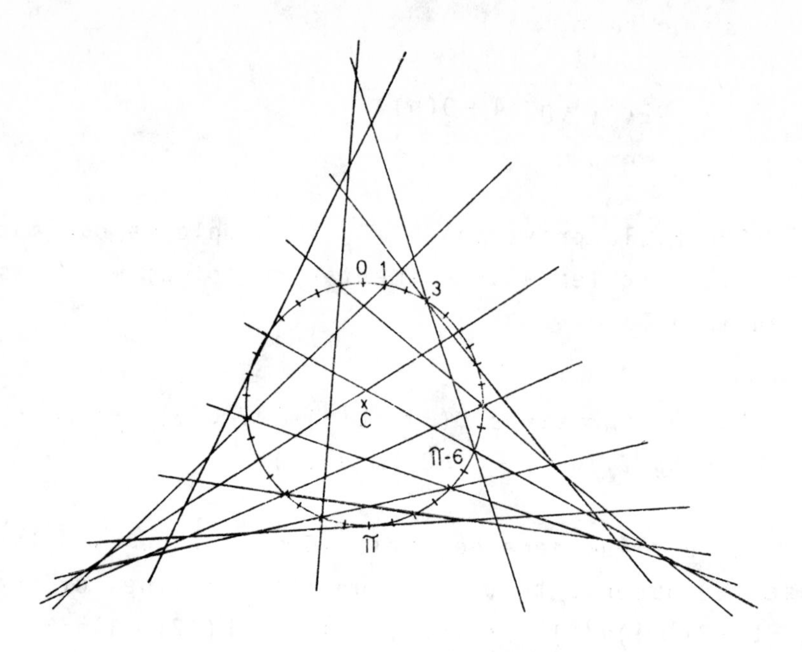

Figure 1

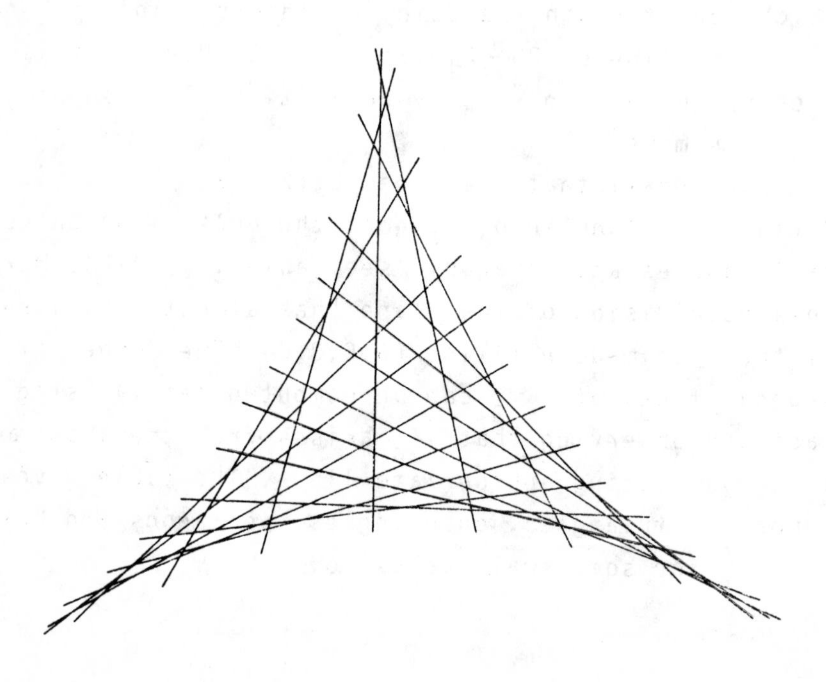

Figure 2

We still show that $p_5^s(n) < (n^n - n)/4$. Let $\mathcal{B}_n$ be an arbitrary simple arrangement of lines and $p_i = p_i(\mathcal{B}_n)$. Then $\Sigma p_i = (n^2 - n+2)/2$ and $\Sigma i p_i = 2(n^2 - n)$. Hence $\Sigma(i-3)p_i = (n^2 - n - 6)/2$, i.e. $p_k < (n^2 - n)/2(k-3)$ holds for $k \geq 4$.

n	$p_3(\mathcal{A}_n)$	$p_4(\mathcal{A}_n)$	$p_5(\mathcal{A}_n)$	$p_6(\mathcal{A}_n)$
5	$(n^2-n)/4$	n	$(n^2-5n+4)/4 = 1$	0
6	$(n^2+4)/4$	0	$(n^2-2n)/4 = 6$	0
7	$(n^2-5)/4$	$n-2$	$(n^2-5n+6)/4 = 5$	1
8	$(n^2-n)/4$	$n-2$	$(n^2-5n+8)/4 = 8$	1
9	$(n^2-2n+1)/4$	n	$(n^2-4n+3)/4 = 12$	0
10	$(n^2-2n)/4$	$n+2$	$(n^2-5n-2)/4 = 12$	2
11	$(n^2-2n+1)/4$	$n+1$	$(n^2-5n+2)/4 = 17$	2
12	$(n^2-2n)/4$	$n+1$	$(n^2-5n+8)/4 = 23$	1
13	$(n^2-2n+1)/4$	n	$(n^2-5n+4)/4 = 27$	2
14	$(n^2-2n+4)/4$	n	$(n^2-5n+2)/4 = 32$	3
.				
.				
.				
18	$(n^2-2n-4)/4$	$n+3$	$(n^2-5n-6)/4 = 57$	4

Table

REFERENCES

[1] Z. FÜREDI - I. PALÁSTI, *Arrangement of lines with large number of triangles*. Proc. Amer. Math. Soc. 92 (4), (1984) 561-566.

[2] B. GRÜNBAUM, *Arrangements and spreads*. Amer. Math. Soc. Providence, R.I. 1972.

[3] B. GRÜNBAUM, *Two-colouring the faces of arrangements*, Period. Math. Hungar. 11 (1980) 181-185.

[4] H. HARBORTH - I. MENGERSEN, *Geradeordnungen mit maximaler Anzahl vierseitiger Flächen*. Period. Math. Hungar. 12(2), (1981) 93-98.

[5] I. PALÁSTI, *The maximal number of quadrilaterals bounded by general straight lines in a plane*. Period. Math. Hungar. 6(1975), 323-341.

[6] G.B. PURDY, *Triangles in arrangements of lines*. Discrete Math. 25(1979), 157-163.

[7] G.B. PURDY, *Triangles in arrangements of lines II*. Proc. Amer. Math. Soc. 79(1980), 77-81.

[8] S. ROBERTS, *On the figures formed by the intercepts of a system of straight lines in a plane and on analogous relations in space of three dimensions*. Proc. London Math. Soc. 19(1887-1888), 405-422.

[9] I. PALÁSTI, *A construction for arrangements of lines with vertices of large multiplicity*. Studia Sci. Math. Hung. 21(1986).

I. PALÁSTI
Math. Institute of Hungarian Academy of Sci.,
Budapest, Reáltanoda u. 15-17.
1053. Hungary

COLLOQUIA MATHEMATICA SOCIETATIS JÁNOS BOLYAI
48. INTUITIVE GEOMETRY, SIÓFOK, 1985.

APPLICATION OF TWO-DIMENSIONAL AND THREE-DIMENSIONAL CRYSTALLOGRAPHIC P-SYMMETRY GROUPS TO THE STUDY OF MULTI-DIMENSIONAL SYMMETRY GROUPS

A.F. PALISTRANT

1. The concept of P-symmetry was introduced by Zamorzaev, A.M. [1], and essentially it is defined as follows. We ascribe to each of the points of a figure at least one of the 1, 2, ..., p indices and we shall the P-symmetry transformation of the figure, that isometric transformation which carries each point with the index i to the point with the index k_i, so that the index permutation $\begin{pmatrix} 1 & 2 \dots & p \\ k_1 & k_2 \dots & k_p \end{pmatrix}$ belongs to the given substitution group P of indices. If the substitutions included in the transformation group G exhaust the group P then the group G is defined a group of complete P-symmetry. This group can always be derived from a classical symmetry group S by means of searching in S and P for normal subgroups H and Q for which the isomorphism $S/H \simeq P/Q$ holds, by paired multiplication of the cosets adequate in this isomorphism and by unification of the products obtained. The groups of complete P-symmetry fall into senior (S=H and G=SxP), junior ($S/H \simeq P$ and $G \simeq S$) and middle groups for Q=P, Q=e and $e \subset Q \subset P$, respectively (cf. [2, 3]).

The P-symmetry is a broad extension of the classical theory of symmetry and it includes all antisymmetry and colour symmetry generalizations [2, 3] which directly combine the change of some features (indices or colours) ascribed to the points of a figure with the isometric transformations of its points, where the change of features does not depend on the particular choice of points. The P-symmetry especially includes the Shubnikov antisymmetry, the Zamorzaev multiple antisymmetry ((2,...,2)-symmetry), the Belov colour symmetry (p-symmetry corresponding to the cyclic group $P = \{(1,2,...,p)\}$ given by the generating cyclic permutation), the Pawley colour symmetry ((p/)-symmetry corresponding to the group $P = \{(1,...,p)(\bar{p},...,\bar{1}), (1,\bar{1})...(p,\bar{p})\}$ with 2p transformed features, i.e. p "positive" i and p "negative" $\bar{i}$, etc.) [2, 3].

These cases of P-symmetry have a simple pictorial symmetrical scheme. Thus the substitution group P for 2-symmetry, (2,2)-, p- and (p/)-symmetry is respectively represented by replacing the vertices of a segment, a rectangle, an oriented regular p-gon or a semi-regular 2p-gon when considering their symmetries (see also Fig. 1-3).

In [4] a geometric method for classification of P-symmetry was suggested for the so-called crystallographic P-symmetries when the group P is isomorphic to one of the 32 crystal classes. This method allows us precisely to distinguish the 32 P-symmetries having the 32 crystallographic point groups as their geometrical scheme.

The indices ascribed to the points of a figure have an extrageometric sense with respect to the space in which the figure is considered. In additional dimensions these index permutations can be geometrically interpreted.

This enables us to apply the two- and three-dimensional crystallographic P-symmetries (studied by the Kishinev geometricians) to the investigation of multi-dimensional discrete symmetry groups, viz. the n-dimensional Fedorov groups G_n and their "minor" subgroups G_{nm} (with an invariant m-dimensional plane) and subgroups $G_{nm...k}$ (with invariant m-dimensional ... and k-dimensional planes successively embedded into each other). This application could advance in principal the geometric solution of same problems of n-dimensional crystallography for n = 5 and 6. The present paper deals with determining the numbers of all various groups of some categories of "small" subgroups and 5- and 6-dimensional Fedorov groups.

2. In [2] some applications of simple and multiple antisymmetry are worked out when we investigate n-dimensional symmetry groups for n = 2,3,4. Thus, for example, the zero dimensional symmetry and antisymmetry groups of class G_0^1 (the generating group 1 and senior group $\underline{1}$) model one-dimensional point groups of class G_{10} (these denoted by 1 and m) if we interpret the permutations of + and - signs assigned to an assymetric point as the discrete dispositions of the points of a straight line over one or the other side of its invariant point. Further, the zero-dimensional symmetry and double antisymmetry groups of class G_0^2 (consisting of the generating group 1, senior group of the first kind $\underline{1}$; senior of the second kind $1'$; senior of the (1,2) kind $\underline{1}'$ and senior of the third kind $\underline{1} \times 1'$) completely describe all various symmetry groups of class G_{210} (1, 1m, m1, 2, mm). These groups leave invariant in the plane a point and two perpendicular lines which pass through the given invariant point. Here we interpret the first + and - signs assigned to an assymetric point as belonging to the upper or the lower semi-planes devided by the invariant line; and the

second + and - signs we interpret as belonging to one or the second direction of the invariant line. Similarly the zero dimensional symmetry and three-fold antisymmetry groups of class G_0^3 completely model all various symmetry groups of class G_{3210} (finite bands), if we interpret the first + and - signs assigned to an assymetric point as belonging to one or the second normal direction of its first invariant plane; and the second + and - signs, assigned to an assymetric point, as belonging to one or the second normal direction of its third invariant plane. Thus the zero dimensional symmetry and three-fold antisymmetry groups G_0^3 can model symmetry and two-fold antisymmetry groups of segments G_{10}^2 and symmetry and antisymmetry groups of finite edgings G_{210}^1 and symmetry groups of finite bands G_{3210}. Table I compares the above four point groups (in the order of their enumeration). The groups compared have been written in the international symbols. The bar under the symbol of a symmetry element in column 1 indicates the replacement by the antisymmetry element of type 1. The prime above to the right of a symmetry element (or antisymmetry element of type 1) indicates the replacement by an antisymmetry element of type 2 (or type (1,2)) in the first column and by an antisymmetry element of type 1 in the second column respectively. The asterisks above to the left of a symmetry element (or antisymmetry element of type 1, type 2 or type (1,2)) indicates the replacement by an antisymmetry element of type 3 (or type (1,3), (2,3) or (1,2,3) in the first column, by an antisymmetry element of type 2 (or type (1,2) in the second column and by an antisymmetry element of type 1 in the third column respectively (cf. [5]).

Table I

The comparison of groups in classes G_0^3, G_{10}^2, G_{210}^1

Zero-dimensional symmetry and three-fold antisymmetry groups G_0^3	Linear point groups		
	Symmetry and two-fold antisymmetry groups of segment G_{10}^2	Symmetry and antisymmetry groups of finite edgings G_{210}^1	Symmetry groups of finite bands G_{3210}
1	2	3	4
1, *1, 1'	1, *1, 1'	1, *1, 1m	1, 11m, 1m1
*1', 1' x *1, $\underline{1}$	*1', 1' x *1, m	1*m, 1m*1, m1	211, 2mm, m11
*$\underline{1}$, $\underline{1}$ x *1, $\underline{1}$'	*m, m*1, m'	*m1, m1*1, 2	121, m2m, 112
*$\underline{1}$', $\underline{1}$'x*1, $\underline{1}$x1'	*m', m'*1, m1'	*2, 2*1, mm	$\bar{1}$, 112/m, mm2
$\underline{1}$x*1', *$\underline{1}$x1'	m*1', *m1'	m*m, *mm	2/m11, 12/m1
*$\underline{1}$x*1', $\underline{1}$x1'x*1	*m*1', m1x1'x*1	*m*m, mm*1	222, mmm

Table I clearly illustrates that there is a simple isomorphic correspondence between all various zero-dimensional groups of three-fold antisymmetry G_0^3 and symmetry groups G_{3210} (finite bands). That is why the number of various zero-dimensional symmetry and three-fold antisymmetry groups G_0^3 fully coincides with the number of various symmetry groups of class G_{3210}, and the structure of a separate group of class G_0^3 determines the structure of the corresponding group of class G_{3210}.

The same relation, taking place between groups of classes G_0^3 and G_{3210}, stands e.g. between two-dimensional symmetry and three-fold antisymmetry groups G_2^3 and five-dimensional symmetry groups G_{5432}. Thus by a detailed count of three-dimensional symmetry and two-fold antisymmetry groups $G_{3...}^2$ we can find the numbers of all various groups of class $G_{543...}$. This method was used to find that there are 4920 G_{5432}, 1379 G_{54321}, 624 G_{5430}, 671 G_{54320} and 374 G_{543210} groups (according to the numbers of symmetry and two-fold antisymmetry groups G_{32}^2 (layer type), G_{321}^2 (band type), G_{30}^2 (point type), G_{320}^2 (two-sided rosettes) and G_{3210}^2 (finite bands) [5]).

It should be noted that there are not 17807 and 2784 groups of classes G_{543} and G_{5431}, respectively (corrresponding to the numbers of various G_3^2 and G_{31}^2 groups [5]), but considerably less (for G_{543} this number is 17410 and for G_{5431} it is 2597), because among symmetry and two-fold antisymmetry groups mentioned there are such ones, which differ only due to enantiomorphism and wo their corresponding groups in five-dimensional space do not differ. So to find the number of various symmetry groups of class G_{5431}, with the help of rod symmetry and two-fold antisymmetry groups, by the formula $P_2 = 5N_o + 6N_1 + N_2$ of paper [5] one must remember that the number of generating groups $N_o = 67$ (but not 75), the number of derivative groups of one type $N_1 = 226$ (not 244) and the number of derivative groups of three types $N_2 = 906$ (not 945). Consequently, the number of rod symmetry and two-fold antisymmetry groups, interpreting only different groups of class G_{5431}, will be 5 x 67 + 6 x 226 + 906 = = 2597. Similarly, when we count various groups of class G_{543} with the help of space symmetry and two-fold antisymmetry groups G_3^2 it is necessary to substitute $N_o = 219$ (not 230), $N_1 = 1156$ (not 1191) and $N_2 = 9379$ (not 9511) into the formula $5N_o + 6N_1 + N_2$. Thus only 17410 (= 5 x 219 + 6 x 1156 + 9379) groups from the class G_3^2 can be put into one-to-one correspondence with the groups in the class G_{543}.

In its turn a detailed count of the three-dimensional and three-fold antisymmetry $G^3_{3...}$ helps us to find the numbers of all various symmetry groups of the class $G_{6543...}$. In this way we found that there are 25677 G_{65431} groups; 64924 G_{65432} groups; 4885 G_{654320}, 14419 G_{654321} and 2825 $G_{6543210}$ groups (corresponding to the numbers of G^3_{31} (disregarding enantiomorphism), G^3_{32}; G^3_{320}; G^3_{321} and G^3_{3210} [5]).

3. It should be noted that 10 symmetry groups of one-sided G_{20} rosettes can be described via zero-dimensional G_0 symmetry, G^1_0 antisymmetry, G^P_0 and $G^{P'}_0$ p- and (p')-symmetry, respectively, for p = 2,3,4,6. In fact the extension of G_0 to antisymmetry, p- and (p')-symmetry for p=2,3,4,6 gives one generating group and nine senior groups $1 \times \underline{1}$, $1 \times 1^{(2)}$, $1 \times 1^{(3)}$, $1 \times 1^{(4)}$, $1 \times 1^{(6)}$, $1 \times 1^{(2.\,\underline{1}')}$, $1 \times 1^{(3.\,1')}$, $1 \times 1^{(4.\,1')}$, $1 \times 1^{(6.1')}$, that is 10 P-symmetry groups of indices (including P=e) in the class G^P_0. But the pictorial geometrical schemes of these 10 P-groups on the plane (i.e. an asymmetric point with index 1, two points with indices 1 and $\bar{1}$, symmetrc to an axis, two points with indices 1 and 2, symmetric to a centre stands for 1-, (1')- and 2-symmetry, respectively (Fig. 1), an

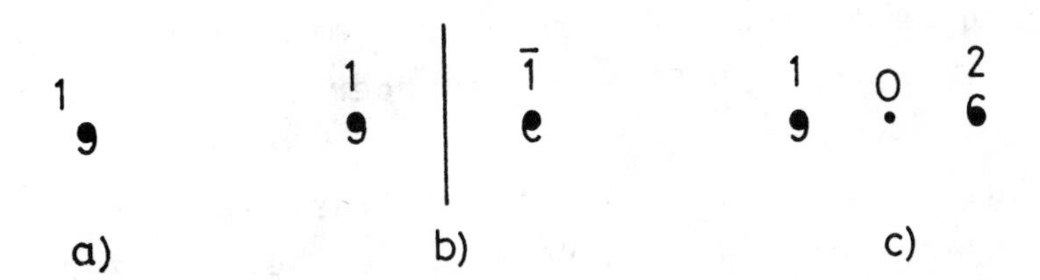

Figure 1. Graphic geometric schemes a) I-symmetry, b) (I/)-symmetry, c) 2-symmetry

oriented regular p-gon with vertices 1,2,...,p stands for p-symmetry when p > 2 (Fig. 2) and the (p')-symmetry is modelled by a semiregular 2p-gon with vertices 1,2,...,p

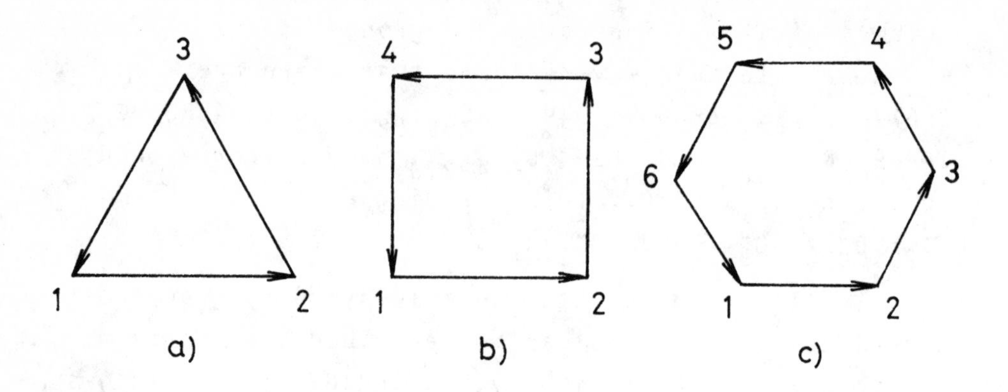

Figure 2. Graphic geometric schemes of P-symmetries: a) 3-symmetry ($P=\{(1,2,3)\}$); b) 4-symmetry ($P=\{(1,2,3,4)\}$); c) 6-symmetry ($P=\{(1,2,3,4,5,6)\}$).

and $\bar{1},\bar{2},\ldots,\bar{p}$, so that the substitution group $P=\{(1,2,\ldots,p)\ (\bar{p},\ldots,\bar{2},\bar{1}),\ (1,\bar{1})\ (2,\bar{2})\ldots(p,\bar{p})\}$ of their vertices describes its complete symmetry group when $p \geq 2$ (Fig. 3) apparently interpret them as one-sided rosette symmetry group G_{20}.

Table 2 compares the above zero-dimensional P-symmetry groups to one-sided rosette symmetry groups in Shubnikov's symbols [5] (in the order of their enumeration).

Thus in this case, as it is illustrated in table 2, there is an isomorphic correspondence between zero-dimensional groups G_0^P (symmetry, antisymmetry, p- and (p')-symmetry for $p = 2,3,4,6$) and rosette symmetry groups G_{20}. There is the same correspondence (as between zero-dimensional gropus G_0^P of the above particular P-symmetry cases and rosette symmetry groups G_{20}) between symmetry, antisymmetry, p- and (p')-symmetry groups for $p = 2,3,4,6$, for example, between the class $G_{3...}^{p}$ and five-dimensional symmetry groups of class $G_{53...}$. In other words, the numbers of all various dimensional crystallographic symmetry groups of class $G_{53...}$ can be found via three-dimensional P-symmetry groups $G_{3...}^{P}$, symmetry $G_{3...}$, antisymmetry $G_{3...}^{1}$, p- and (p/)-symmetry $G_{3...}^{p}$ and $G_{3...}^{P/}$ for $p = 2,3,4,6$ (where only complete P-symmetry groups are written) (disregarding enantiomorphism). As it was found by this method, there are 33075 G_{53} groups (corresponding to the numbers of G_3 groups (classical space groups), G_3^1 (antisymmetry) G_3^P (p-symmetry) and $G_3^{P'}$ ((p')-symmetry) disregarding enantiomorphism); 1208 G_{530} groups (corresponding to the numbers of G_{30}; G_{30}^1, G_{30}^P and $G_{31}^{P'}$); 5177 G_{531} groups (corresponding to the numbers of G_{31}, G_{31}^1, G_{31}^P and $G_{31}^{P'}$ groups disregarding enantiomorphism); 1274 G_{5320}; 9282 G_{532}; 2597 G_{5321} groups (compare with [3,6]).

Table 2

The comparison of zero-dimensional P-symmetry groups to rosette symmetry groups

Zero-dimendional P-symmetry groups G_0^P	Rosette symmetry groups G_{20}
1, $1x1^{(2)}$, $1x1^{(3)}$, $1x1^{(4)}$, $1x1^{(6)}$	1, 2, 3, 4, 6
$1x1'$), $1x1^{(2.1')}$, $1x1^{(3.1')}$	m, 2·m, 3·m
$1x1^{(4.1')}$, $1x\bar{1}^{(6.1')}$	4·m, 6·m

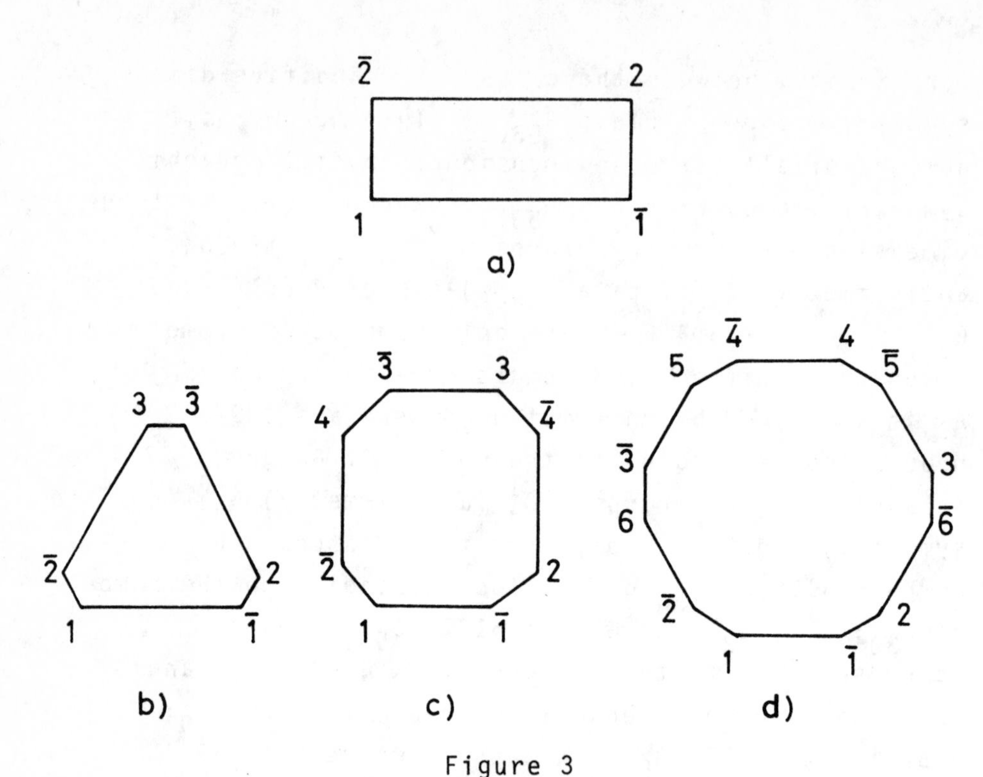

Figure 3

Graphic geometric schemes of (p′)-symmetries:

a) (2′)-symmetry ($P=\{(1,2)(\bar{2},\bar{1}),(1,\bar{1})(2,\bar{2})\}$);

b) (3′)-symmetry ($P=\{(1,2,3)(\bar{3},\bar{2},\bar{1}),(1,\bar{1})(2,\bar{2})(3,\bar{3})\}$);

c) (4′)-symmetry ($P=\{(1,2,3,4)(\bar{4},\bar{3},\bar{2},\bar{1}),\ (1,\bar{1})(2,\bar{2})\ \times$ $\times\ (3,\bar{3})(4,\bar{4})\}$);

d) (6′)-symmetry ($P=\}(1,2,3,4,5,6)(\bar{6},\bar{5},\bar{4},\bar{3},\bar{2},\bar{1}),$ $(1,\bar{1})(2,\bar{2})(3,\bar{3})(4,\bar{4})(5,\bar{5})(6,\bar{6})\}$).

4. Finally, the 32 crystallographic point groups G_{30} can be described via 32 zero-dimensional groups of crystallographic P-symmetry $G_0{}^P$ as introduced in [7]. We want to prove that all three-dimensional crystallographic point symmetry and antisymmetry groups G_{30}^1 can be described via one-dimensional point groups of 32 crystallographic

P-symmetries in geometrical classification introduced in [4]. This class G_{10}^{P} consists of generating groups I and m, senior groups $1 \times 1^{(3)}$, $m \times 1^{(3)}$ for 3-symmetry, derivative group $m^{(2)}$ for 2-symmetry, 2-middle group $m^{(4)}$ for 4-symmetry and etc. (Table 3).

Indeed, let there be a section group of crystallographic P-symmetry, interpreted on a one-dimensional figure $F^{(M)}$, where M is a great number of qualities assigned to the points of figure F. Let us imagine that the geometrical figure F is placed on the one-dimensional plane E_1 in the four dimensional Euclidean space E_4, and an isogon, inscribed into a sphere and giving a geometrical scheme of the considered P-symmetry itself, is placed on the three-dimensional plane E_3, perpendicular to E_1 and crossing it in a particular point of figure F; the sphere centre must be coincident with the point of F considered. It is evident, that under such special arrangement of F and the isogon in E_4, the group considered in the class G_{10}^{P} can be interpreted as a group of the class G_{410} (= G_{430}) and it is isomorphic to a group in the class G_{30}^{1}.

Table 3 compares the 122 segment groups G_{10}^{P} of complete 32 crystallographic P-symmetry to point symmetry and antisymmetry groups G_{30}^{1} in Shubnikov's symbols described in detail in [5] (in the order of their enumeration). The symbols $1^{(2)}, \ldots, 1^{(2.1^{2})}, \ldots, (1^{(2}, 1^{3)})$, $1^{(2)} \times \underline{1}, \ldots, (1^{(4}, 1^{3)}) \times \underline{1}$, $\underline{1}^{(2)}, \ldots, (\underline{1}^{(4}, 1^{3)})$ in the first column of the table indicate groups of index permutations giving 2-symmetry,..., (22)-, ..., (23)-,..., ..., (2$\underline{1}$)-, ..., (43$\underline{1}$)-, $\underline{2}$-,..., ($\underline{4}$3)-symmetry, respectively introduced in [4].

Table 3

The comparison of groups G_{10}^{P} to groups G_{30}^{1}

Segment groups of 32 crystallographic P-symmetries G_{10}^{P}	Three-dimensional symmetry and antisymmetry point groups G_{30}^{1}
$1,\ 1x\underline{1};\ m,\ mx\underline{1},\ \underline{m};$	$1,\ \underline{1};\ \tilde{2},\ \tilde{2}x\underline{1},\ \underline{\tilde{2}};$
$1x1^{(2)},\ 1x1^{(2)}x\underline{1},\ 1x\underline{1}^{(2)};$	$2,\ 2x\underline{1},\ \underline{2};$
$m^{(2)},\ m^{(2)}x\underline{1},\ \underline{m}^{(2)};\ mx1^{(2)},$	$m,\ mx\underline{1},\ \underline{m};\ 2{:}m,$
$mx1^{(2)}x\underline{1},\ \underline{m}\cdot\underline{1}^{(2)},\ \underline{m}\cdot 1^{(2)},$	$2{:}mx\underline{1},\ \underline{2}{:}m,\ 2{:}\underline{m},$
$mx\underline{1}^{(2)},\ m^{(2.1^{2})},\ m^{(2.1^{2})}x\underline{1},$	$\underline{2}{:}\underline{m};\ 2\cdot m,\ 2\cdot mx\underline{1},$
$m^{(2\cdot\underline{1}^{2})},\ \underline{m}^{2)}\cdot 1^{(2};\ 1x1^{(2.1^{2})},$	$\underline{2}\cdot m,\ 2\cdot\underline{m};\ 2{:}2,$
$1x1^{(2\cdot 1^{2})}x\underline{1},\ 1x1^{(2.\underline{1}^{2})};\ 1x1^{(4)},$	$2{:}2x\underline{1},\ 2{:}\underline{2};\ 4,$
$1x1^{(4}\cdot\underline{1},\ 1x\underline{1}^{(4)};\ m^{(4)},\ m^{(4)}x\underline{1},$	$4x\underline{1},\ \underline{4};\ \tilde{4},\ \tilde{4}x\underline{1},$
$\underline{m}^{(4};\ mx1^{(4)},\ mx1^{(4)}x\underline{1},$	$\underline{\tilde{4}};\ 4{:}m,\ 4{:}mx\underline{1},$
$mx\underline{1}^{(4)},\ \underline{m}x1^{(4)},\ \underline{m}\cdot\underline{1}^{(4)};$	$\underline{4}{:}m,\ 4{:}\underline{m},\ \underline{4}{:}\underline{m};$
$m^{2)}.1^{(4},m^{2)}.1^{(4}x\underline{1},m^{2)}.\underline{1}^{(4},\underline{m}^{2)}.1^{(4};$	$4\cdot m,\ 4\cdot mx\underline{1},\ \underline{4}\cdot m,\ 4\cdot\underline{m};$
$1x1^{(4.1^{2})},1x1^{(4.1^{2})}x\underline{1},1x\underline{1}^{(4.1^{2})},1x1^{(4.\underline{1}^{2})};$	$4{:}2,\ 4{:}2x\underline{1},\ \underline{4}{:}2,\ 4{:}\underline{2};$
$1x1^{(3)},1x1^{(3)}x\underline{1};mx1^{(3)},mx1^{(3)}x\underline{1},\underline{m}x1^{(3)};$	$3,\ 3x\underline{1};\ \tilde{6},\ \tilde{6}x\underline{1},\ \underline{\tilde{6}};$
$m^{2)}\cdot 1^{(3},\ m^{2)}\cdot 1^{(3}x\underline{1},\ \underline{m}^{2)}\cdot 1^{(3};$	$3\cdot m,\ 3\cdot mx\underline{1},\ 3\cdot\underline{m};$
$1x1^{(3\cdot 1^{2})},\ 1x1^{(3.1^{2})}x\underline{1},\ 1x1^{(3.\underline{1}^{2})};$	$3{:}2,\ 3{:}2x\underline{1},\ 3{:}\underline{2};$
$m^{(6)},\ m^{(6)}x\underline{1},\ \underline{m}^{(6)};$	$3{:}m,\ 3{:}mx\underline{1},\ 3{:}\underline{m};$
$1x1^{(6)},\ 1x1^{(6)}x\underline{1},\ \underline{1}^{(6)};\ mx1^{(6)},$	$6,\ 6x\underline{1},\ \underline{6};\ 6{:}m,$
$mx1^{(6)}x\underline{1},\ mx\underline{1}^{(6)},\ \underline{m}x1^{(6)},\ \underline{m}x\underline{1}^{(6)};$	$6{:}mx\underline{1},\ \underline{6}{:}m,\ 6{:}\underline{m},\ \underline{6}{:}\underline{m};$
$m^{2)}.1^{(6},\ m^{2)}.1^{(6}x\underline{1},\ m^{2)}.\underline{1}^{(6},\ \underline{m}^{2)}.1^{(6};$	$6\cdot m,\ 6\cdot mx\underline{1},\ \underline{6}\cdot m,\ 6\cdot\underline{m};$

Table 3 (cont.)

1	2
$1x1^{(6.1^2)}, 1x1^{(6.1^2)}x\underline{1}, \underline{1}^{(6.1^2)}, 1^{(6.\underline{1}^2)};$	$6{:}2, 6{:}2x\underline{1}, \underline{6}{:}2, 6{:}\underline{2};$
$1x(1^{(2}, 1^{3)}, 1x(1^{(2},1^{3)})x\underline{1};$	$3/2, 3/2x\underline{1};$
$m^{(4.1^3)}, m^{(4.1^3)}x\underline{1}, \underline{m}^{(4.1^3)};$	$3/\tilde{4}, 3/\tilde{4}x\underline{1}, 3/\underline{\tilde{4}};$
$1x(1^{(4},1^{3)}), 1x1^{(4},1^{3)})x\underline{1}, 1x(\underline{1}^{(4},1^{3)});$	$3/4, 3/4x\underline{1}, 3/\underline{4};$
$mx1^{(2.1^2)}, mx1^{(2.1^2)}x1, mx1^{(2.\underline{1}^2)},$	$2\cdot m{:}2, 2\cdot m{:}2x\underline{1}, \underline{2}\cdot m{:}2,$
$\underline{m}\cdot 1^{(2.1^2)}, \underline{m}\cdot 1^{(2.\underline{1}^2)}; m^{(4.1^2)},$	$2\cdot\underline{m}{:}\underline{2}, \underline{2}\cdot\underline{m}{:}2; \tilde{4}{:}2,$
$m^{(4.1^2)}x\underline{1}, \underline{m}^{(4.1^2)}, m^{(4.\underline{1}^2)}, \underline{m}^{(4.\underline{1}^2)};$	$\tilde{4}{:}2x\underline{1}, \underline{\tilde{4}}{:}2, \tilde{4}{:}\underline{2}, \underline{\tilde{4}}{:}\underline{2};$
$mx1^{(4.1^2)}, mx1^{(4.1^2)}x\underline{1}, mx\underline{1}^{(4.1^2)},$	$4\cdot m{:}2, 4\cdot m{:}2x\underline{1}, \underline{4}\cdot m{:}2,$
$\underline{m}\cdot 1^{(4.1^2)}, mx1^{(4.\underline{1}^2)}, \underline{m}\cdot\underline{1}^{(4.1^2)},$	$4\cdot\underline{m}{:}2, 4\cdot m{:}\underline{2}, \underline{4}\cdot\underline{m}{:}2,$
$\underline{m}\cdot 1^{(4.\underline{1}^2)}; mx1^{(3.1^2)}, mx1^{(3.1^2)}x\underline{1},$	$4\cdot\underline{m}{:}\underline{2}; 3\cdot m{:}2, 3\cdot m{:}2x\underline{1},$
$\underline{m}\cdot 1^{(3.1^2)}, mx1^{(3.\underline{1}^2)}, \underline{m}\cdot 1^{(3.\underline{1}^2)};$	$3\cdot\underline{m}{:}2, 3\cdot m{:}\underline{2}, 3\cdot\underline{m}{:}\underline{2};$
$m^{(6.1^2)}, m^{(6.1^2)}x\underline{1}, \underline{m}^{(6.1^2)},$	$3{:}m\cdot 2, 3{:}m\cdot 2x\underline{1}, 3{:}\underline{m}\cdot 2,$
$m^{(6.\underline{1}^2)}, \underline{m}^{(6.\underline{1}^2)}; mx1^{(6.1^2)},$	$3{:}m\cdot\underline{2}, 3{:}\underline{m}\cdot\underline{2}; 6{:}m\cdot 2,$
$mx1^{(6.1^2)}x\underline{1}, mx\underline{1}^{(6.1^2)}, \underline{m}\cdot 1^{(6.1^2)},$	$6{:}m\cdot 2x\underline{1}, \underline{6}{:}m\cdot 2, 6{:}\underline{m}\cdot 2,$
$mx1^{(6.\underline{1}^2)}, \underline{m}\cdot\underline{1}^{(6.1^2)}, \underline{m}\cdot 1^{(6.\underline{1}^2)};$	$6{:}m\cdot\underline{2}, \underline{6}{:}\underline{m}\cdot 2, 6{:}\underline{m}\cdot\underline{2};$
$mx(1^{(2},1^{3)}), mx(1^{(2},1^{3)}x\underline{1}, \underline{m}\cdot(1^{(2},1^{3)});$	$3/2\cdot m, 3/2\cdot mx\underline{1}, 3/2\cdot\underline{m};$
$mx(1^{(4},1^{3)}), mx(1^{(4},1^{3)})x\underline{1}, mx(1^{(4},1^{3)}),$	$3/4\cdot m, 3/4\cdot mx\underline{1}, 3/\underline{4}\cdot m,$
$\underline{m}\cdot(1^{(4},1^{3)}), \underline{m}\cdot(\underline{1}^{(4},1^{3)})$	$3/4\cdot\underline{m}, 3/\underline{4}\cdot\underline{m}$.

Thus in this case, as it is clearly illustrated in table 3, there is a simple isomorphic correspondence between classes G_{10}^{P} and G_{410}. For example there is an analogous connection between groups of classes $G_{2...}^{P}$ and $G_{52...}$ as that between groups G_{10}^{P} and G_{410}. Consequently, the number of various five-dimensional symmetry groups of class $G_{52...}$ is equal to that of various two-dimensional groups $G_{2...}^{P}$ of 32 crystallographic P-symmetries in the geometrical classification. With the help of the classic border groups G_{21} and their generalizations with the mentioned 32 P-sy-metries in the geometrical classification it was found that there are 2304 various symmetry groups in the class G_{521}; while using symmetry groups G_{320}, G_{321} and G_{30} and their extension to the same 32 P-symmetries, it was shown that there are 7835 groups of the class G_{6320}, 20839 groups of the class G_{6321} and 7177 groups of the class G_{630} and not 7229 as was indicated in [7]*.

5. The analysis of the methods above for the description of some classes of five- and six-dimensional symmetry groups by using two-dimensional crystallographic P-symmetry groups leads us to the following conclusions: 1) the class $G_{(r+\ell)(r+\ell-1)...(r+1)r...}$ contains as many various symmetry groups as there are various symmetry and ℓ-fold antisymmetry groups in the class $G^{\ell}_{r...}$; 2) there are as many symmetry groups in the class $G_{(r+2)r...}$ as there are various symmetry $G_{r...}$, antisymmetry $G^{1}_{r...}$, p-symmetry $G^{p}_{r...}$ and (p')-symmetry $G^{p'}_{r...}$ groups taken together for p=2,3,4,6; 3) the class $G_{(r+3)r...}$ has as many various symmetry groups as is generated by class $G^{P}_{r...}$ when it is generalized with 32 crystallographic P-symmetries in the geometrical classification.

---

* In paper [7] it was indicated by mistake that the number of 2-middle groups G_{30}^{P} for $(\dot{2}2\underline{1})$-symmetry is 116 instead of 64.

REFERENCES

[1] ZAMORZAEV, A.M. *On the groups of quasisymmetry (P-symmetry)*. Kristallografiya, 12, 819-835, 1967. (In Russian).

[2] ZAMORZAEV, A.M., GALYARSKII, E.I. and PALISTRANT, A.F., *Colour symmetry, its generalizations and applications*. Publishing House "Shtiintsa", Kishinev, 1978, 276 p. (In Russian).

[3] ZAMORZAEV, A.M., PALISTRANT, A.F., *Antisymmetry, its generalizations and geometrical applications*. Zeitschrift für Kristallographie, 151, 231-248, 1980.

[4] ZAMORZAEV, A.M., PALISTRANT, A.F., *Geometric classification of P-symmetry*, Doklady AN of the USSR, 256, No. 4, 856-859, 1981. (In Russian).

[5] ZAMORZAEV, A.M., *The theory of simple and multiple antisymmetry*. Publishing House "Shtiintsa", Kishinev, 1976, 283 p. (In Russian).

[6] PALISTRANT, A.F., *The space groups of* (p')*-symmetry (of Pawley) and their application to derivation of five-dimensional crystallographic groups*. Doklady AN of the USSR, 254. No. 5, 1126-1130, 1980. (In Russian).

[7] PALISTRANT, A.F., *Application of three-dimensional point groups of P-symmetry to derivation of six-dimensional symmetry groups*, Doklady AN of the USSR, 260, No. 4, 884-888, 1981. (In Russian).

A.F. PALISTRANT
Kishinev, State University
USSR

COLLOQUIA MATHEMATICA SOCIETATIS JÁNOS BOLYAI
48. INTUITIVE GEOMETRY, SIÓFOK, 1985.

EQUIDECOMPOSABLE POLYHEDRA

R. SCHNEIDER

1. INTRODUCTION

According to the Wallace-Bolyai-Gerwien theorem, two polygons in the plane are equivalent by geometric dissection (scissors congruent) if and only if they have the same area. In this theorem, as well as in the whole higher-dimensional equidissection theory which emerged from Hilbert's third problem (see the books by Hadwiger [9], Boltianskii [2], [3], Sah [20] and the survey articles by Hadwiger [11], [12], Hertel [14], McMullen-Schneider [19]), dissection is understood in the sense of elementary geometry: Two polyhedra P,Q are equidissectable if there are representations $P = \cup_{i=1}^{k} P_i$, $Q = \cup_{i=1}^{k} Q_i$ with polyhedra P_i, Q_i such that P_i and Q_i are congruent ($i = 1,\dots,k$) and the intersections $P_i \cap P_j$, $Q_i \cap Q_j$ ($i \neq j$) do not have interior points. Thus, the components of such a dissection are polyhedra which may have common boundary points. While this notion of decomposition appears natural for historical reasons and for its applications in elementary geometry, from a more general geometric viewpoint it is interesting to consider also other types of decomposition, in particular those into pairwise disjoint components. (For still other notions of decomposition, see Aumann [1], Dubins-Hirsch-Karush [6], Kántor [16]). The character of such a decomposition theory will heavily depend on the class of point sets from which the components may be chosen. If no restrictions are imposed and if the axiom of choice is used, it is well known that one can get highly nonintuitive, "paradoxical" decompositions (see Hadwiger [9], chap. 3, and the literature quoted there, Wagon [22] and the recent book [23] by the latter author). Hadwiger [8] has studied disjoint decompositions of plane polygons into sets consisting of open polygons together with arbitrary subsets of their boundaries. He talks of multicongruence relative to $\mathcal{M}$ if, as in this case, the pairwise congruent components of decompositions have to be elements of a given class $\mathcal{M}$ of subsets. In the geometry of polyhedra, it appears particularly natural to study multicongruence relative to the system of relatively open convex polytopes. It is the purpose of the following to describe the elements of such an equivalence theory for polyhedra under decomposition into disjoint unions of relatively open polytopes. A close connection with valuations is established, just as classical equidissectability is tied up with simple valuations.

2. VALUATIONS AND RO-DECOMPOSITIONS.

We begin with some notation. E^n is the n-dimensional Euclidean vector space. For a subset $A \subset E^n$, we denote by $\bar{A}$ its closure, by int A its interior, and by A_0 its relative interior, that is, the interior with respect to the affine hull, aff A, of A. For a hyperplane $H \subset E^n$, let H^+, H^- be the two closed halfspaces bounded by H. $\mathcal{P}^n$ is the set of compact convex polytopes in E^n (including the empty polytope $\emptyset$), and $\mathcal{P}^n_{ro}$ is the set of relatively open convex polytopes. By $U(\mathscr{P}^n)$, $U(\mathscr{P}^n_{ro})$ we denote the system of all finite unions of elements of $\mathscr{P}^n$ or $\mathscr{P}^n_{ro}$, respectively. The elements of $\mathscr{P}^n$, $\mathscr{P}^n_{ro}$, $U(\mathscr{P}^n)$, $U(\mathscr{P}^n_{ro})$, in this order, will be called polytopes, ro-polytopes, polyhedra, ro-polyhedra. Note that each polytope is a ro-polyhedron.

Let G be a group of rigid motions of E^n containing the translation group T. As usual, two polyhedra $P,Q \in U(\mathscr{P}^n)$ are called G-*equidissectable* if they admit representations

$$P = \bigcup_{i=1}^{k} P_i, \quad Q = \bigcup_{i=1}^{k} Q_i \quad \text{such that}$$

$$P_i, Q_i \in \mathscr{P}^n, \quad P_i \cong_G Q_i \qquad \text{for } i = 1,\dots,k,$$

$$\text{int}\,(P_i \cap P_j) = \emptyset = \text{int}\,(Q_i \cap Q_j) \quad \text{for } i \neq j,$$

where $\cong_G$ denotes congruence under G. We shall say that two ro-polyhedra $P,Q \in U(\mathscr{P}^n_{ro})$ are G-*ro-equidecomposable* if there are representations

$$P = \bigcup_{i=1}^{k} P_i, \quad Q = \bigcup_{i=1}^{k} Q_i \quad \text{such that}$$

$$P_i, Q_i \in \mathscr{P}^n_{ro}, \; P_i \cong_G Q_i \qquad \text{for } i = 1,\dots,k,$$

$$P_i \cap P_j = \emptyset = Q_i \cap Q_j \qquad \text{for } i \neq j.$$

Thus, if G is the entire motion group, G-ro-equidecomposable means the same as multicongruent relative to $\mathscr{P}^n_{ro}$ in the sense of Hadwiger [8]. The situation

$$P = \bigcup_{i=1}^{k} P_i, \quad P_i \cap P_j = \emptyset \quad \text{for } i \neq j$$

will briefly be expressed (here differing from the usage in [19]) by

$$P = \biguplus_{i=1}^{k} P_i.$$

It is well known (see, e.g., Hadwiger [9] and chap. II of McMullen-Schneider [19] and the literature quoted there) that the investigation of G-equidissectability is closely tied up with (in a certain sense, equivalent to) the study of simply additive G-invariant functionals of polyhedra. In a similar way, G-ro-equidecomposability is connected with general (not necessarily simple) G-invariant functionals. Let us first collect a few observations concerning these functionals.

In the following, by a *functional* we understand a mapping φ from an intersection-stable family $\mathcal{S}$ of subsets in E^n into an abelian group with the property that $\varphi(\emptyset) = 0$ (if $\emptyset \in \mathcal{S}$). Such a functional is called *additive* if

$$\varphi(A \cup B) = \varphi(A) + \varphi(B) - \varphi(A \cap B)$$

for all $A,B \in \mathcal{S}$ with $A \cup B \in \mathcal{S}$, and *simply additive* if, moreover, $\varphi(A) = 0$ if A lies in some hyperplane. A (simply) additive functional is also called a (*simple*) *valuation*. The functional φ on $\mathcal{S}$ will be called *weakly additive* or a *weak valuation* if either $\mathcal{S} = \mathcal{P}^n$ and

$$\varphi(P) = \varphi(P \cap H^+) + \varphi(P \cap H^-) - \varphi(P \cap H) \quad \text{for } P \in \mathcal{P}^n$$

or $\mathscr{S} = \mathscr{P}^n_{ro}$ and

$$\varphi(P) = \varphi(P \cap H_0^+) + \varphi(P \cap H_0^-) + \varphi(P \cap H) \quad \text{for } P \in \mathscr{P}^n_{ro},$$

where $H \subset E^n$ is an arbitrary hyperplane. The following extension result plays a central role.

(2.1) THEOREM. *Every weak valuation on either* $\mathscr{P}^n$ *or* $\mathscr{P}^n_{ro}$ *can be uniquely extended to a valuation on* $U(\mathscr{P}^n_{ro})$.

That a valuation on $\mathscr{P}^n$ has a unique extension to a valuation on $U(\mathscr{P}^n)$, was first proved by Volland [21]. As remarked in [19], p.192, the special case of (2.1) which refers to weak valuations on $\mathscr{P}^n$ with values in a real vector space, can be deduced from a result of Groemer [7]. By a modification of Volland's method, we shall obtain a simple direct and geometric proof of (2.1) in the next section.

If a valuation on $\mathscr{P}^n$ is given, one may ask how its additive extension to $U(\mathscr{P}^n_{ro})$, which exists by (2.1), can be computed explicitly. To see this, we extend a remark made in [19], p.193. In the following, for $P \in \mathscr{P}^n$, sums of the form

$$\sum_{F \subseteq P}, \quad \sum_{G \subseteq F \subseteq P}$$

extend over all faces F of P, respectively over all faces F containing the given face G. (The notion of a face is still only used for closed polytopes, and faces are closed.)

(2.2) LEMMA. *For arbitrary functionals* φ, Ψ *on* $\mathscr{P}^n$ *the relations*

(2.3) $\varphi(P) = \sum_{F \subseteq P} \Psi(F)$ *for* $P \in \mathscr{P}^n$

and

(2.4) $\Psi(P) = \sum_{F \subseteq P} (-1)^{\dim P - \dim F} \varphi(F)$ *for* $P \in \mathscr{P}^n$

are equivalent.

This is a known consequence of the Euler relation

$$\sum_{G \subseteq F \subseteq P} (-1)^{\dim F} = 0 \qquad \text{for } G \neq P$$

(for which one may see, e.g., Brøndsted [5], p.103): If (2.3) holds, then

$$\sum_{F \subseteq P} (-1)^{\dim P - \dim F} \varphi(F) = \sum_{F \subseteq P} (-1)^{\dim P - \dim F} \sum_{G \subseteq F} \Psi(G)$$

$$= \sum_{G \subseteq P} \Psi(G) \sum_{G \subseteq F \subseteq P} (-1)^{\dim P - \dim F} = \Psi(P),$$

which is (2.4), and the other direction is obtained in the same way.

If now φ is a valuation on $U(\mathscr{P}^n_{ro})$ and if $P \in \mathscr{P}^n_{ro}$, then the decomposition

$$\bar{P} = \biguplus_{F \subseteq \bar{P}} F_0$$

together with the additivity of φ implies

$$\varphi(\bar{P}) = \sum_{F \subseteq \bar{P}} \varphi(F_0) = \sum_{F \subseteq \bar{P}} \Psi(F)$$

with $\Psi(Q) := \varphi(Q_0)$ for $Q \in \mathscr{P}^n$. Using Lemma (2.2) with $\Psi(\bar{P}) = \varphi(P)$, we get

$$(2.5) \qquad \varphi(P) = \sum_{F \subseteq P} (-1)^{\dim P - \dim F} \varphi(F) \quad \text{for } P \in \mathscr{P}^n_{ro}.$$

For a general ro-polyhedron $P \in U(\mathscr{P}^n_{ro})$ we may now choose a representation

$$P = \bigcup_{i=1}^{k} P_i \qquad \text{with } P_1,\ldots,P_k \in \mathscr{P}^n_{ro}$$

(which exists by Lemma (3.1) below); by the additivity of φ on $U(\mathscr{P}^n_{ro})$ we then have

$$(2.6) \qquad \varphi(P) = \sum_{i=1}^{k} \varphi(P_i).$$

The fact that weak valuations on $\mathscr{P}^n$ have additive extensions to $U(\mathscr{P}^n_{ro})$, makes them important for ro-equidecomposability. Let φ be a G-invariant weak valuation on $\mathscr{P}^n$; then it is clear from (2.5) and (2.6) that its additive extension to $U(\mathscr{P}^n_{ro})$ is G-invariant, too. Hence, for the G-ro-equidecomposability of two ro-polyhedra P and Q it is clearly necessary that $\varphi(P) = \varphi(Q)$.

Before proceeding further, let us point out a consequence. Consider two closed convex polytopes $P,Q \in \mathscr{P}^n$ which are T-ro-equidecomposable. For a given unit vector u of E^n and for $K \in \mathscr{P}^n$ we denote by $\varphi_u(K)$ the (n-1)-dimensional volume of the intersection of K with its supporting hyperplane with exterior normal vector u. Then φ_u is a T-invariant weak valuation on $\mathscr{P}^n$. By (2.1) it has an additive extension to $U(\mathscr{P}^n_{ro})$, hence the assumption implies that $\varphi_u(P) = \varphi_u(Q)$. Since this holds for all u, P and Q are translates of each other, by Minkowski's theorem (Bonnesen-Fenchel [4], p.116). Thus it is only for general ro-polyhedra that a non-trivial T-ro-equidecomposability theory can be expected.

For a general motion group G (containing T) Hadwiger ([9], p.58) has proved that two polytopes $P,Q \in \mathcal{P}^n$ are G-equidissectable if and only if $\varphi(P) = \varphi(Q)$ for all real-valued G-invariant simple valuations φ. Hadwiger calls this result "formales Hauptkriterium"; its proof needs the axiom of choice, and the theorem does not provide an effective criterion for checking whether two given polytopes are G-equidissectable. An analogous result holds for ro-equidecomposability and general valuations:

(2.7) THEOREM. *Two ro-polyhedra* $P,Q \in U(\mathcal{P}^n_{ro})$ *are* G-*ro-equidecomposable if and only if* $\varphi(P) = \varphi(Q)$ *for all real-valued* G-*invariant valuations* φ *on* $U(\mathcal{P}^n_{ro})$.

An effective criterion for T-equidissectability is due to Hadwiger-Glur [13] for $n = 2$, to Hadwiger [10] for $n = 3$, and to Jessen-Thorup [15] for general n. From this result, we shall deduce an effective criterion for T-ro-equidecomposability. Since it requires some more definitions, we postpone its formulation to §4, where the proof will be given.

One can deduce Theorem (2.7) from Hadwiger's result in a similar way as Theorem (4.5) below is obtained from the Jessen-Thorup theorem; this will not be carried out here. Alternatively, it is probably possible to develop a theory of decompositions into pairwise disjoint ro-polytopes which parallels Hadwiger's [9] dissection theory.

In an obvious way, the method of deducing decomposition from dissection results which yields Theorem (4.5), leads also to some further results. Let G_n denote the full motion group of E^n.

(2.8) THEOREM. *Two closed convex polygons* $P, Q \in \mathscr{P}^2 \setminus \{\emptyset\}$ *are* G_2*-ro-equidecomposable if and only if they have the same area and the same perimeter.*

Performing the inductive procedure of §4 in this case, one has to observe that P and Q, being closed, convex and non-empty, have the same Euler characteristic.

For the G_3-equidissectability of two closed convex polytopes $P, Q \in \mathscr{P}^3 \setminus \{\emptyset\}$ of equal volume, the Dehn conditions are necessary and sufficient (see, e.g., Hadwiger [11] or §9 of McMullen-Schneider [19]). Together with the equality of the surface area and the Minkowski functional W_2 (or the mean width), they are equivalent to the G_3-ro-equidecomposability of P and Q.

3. PROOF OF THEOREM (2.1). Although the following assertion seems obvious, we indicate a proof for the sake of completeness.

(3.1) LEMMA. *Let* $P_1, \ldots, P_k \in U(\mathscr{P}^n_{ro})$. *Then there is a finite set* $\{Q_i : i \in I\}$ *of pairwise disjoint ro-polytopes such that*

$$P_j = \dot{\bigcup_{i \in I_j}} Q_i \qquad \text{for } j = 1, \ldots, k$$

with suitable subsets $I_1, \ldots, I_k \subseteq I$.

PROOF. Each P_j has a representation as a finite union of ro-polytopes. To each of these ro-polytopes we choose a representation as a finite intersection of hyperplanes and open halfspaces. With each of these hyperplanes we associate the two open halfspaces which it bounds, and with each of the open halfspaces we associate the bounding hyperplane and the other open halfspace bounded by it. Let $\mathscr{H}$ denote the resulting system of finitely many hyperplanes and open halfspaces. If we call two points of E^n equivalent if they belong to the same elements of $\mathscr{H}$, then the bounded equivalence classes are ro-polytopes. Let Q_i, $i \in I$, be those of these ro-polytopes which meet $\cup_{j=1}^{k} P_k$, and write $I_j = \{i \in I : Q_i \cap P_j \neq \emptyset\}$ for $j = 1,\dots,k$. By construction we have $Q_i \subseteq P_j$ for $i \in I_j$. Evidently this proves the lemma.

Now we prove "one half" of Theorem (2.1).

(3.2) THEOREM. *Every weak valuation on* $\mathscr{P}^n_{ro}$ *has a unique extension to a valuation on* $U(\mathscr{P}^n_{ro})$.

PROOF. By Lemma (3.1), a given $P \in U(\mathscr{P}^n_{ro})$ can be represented as a disjoint union $P = \uplus_{i\in I} Q_i$ of ro-polytopes Q_i. If φ is a valuation on $U(\mathscr{P}^n_{ro})$, then $\varphi(p) = \Sigma_{i \in I} \varphi(Q_i)$, which proves the uniqueness asserted in the theorem.

Now let φ be a weak valuation on $\mathscr{P}^n_{ro}$. Suppose, first, that P and P_i ($i \in I$, I finite) are ro-polytopes such that

$$(3.3) \qquad P = \biguplus_{i\in I} P_i .$$

We show that this implies

$$(3.4) \qquad \varphi(P) = \sum_{i\in I} \varphi(P_i).$$

For the proof, we use induction with respect to $\dim P$. For $\dim P = 0$, there is nothing to prove. Suppose that (3.3) holds with $\dim P = k > 1$ and that the assertion is true for ro-polytopes of smaller dimension. We prove (3.4) by induction with respect to the number m of k-dimensional ro-polytopes among the P_i, $i \in I$. For $m = 1$, there is nothing to prove (since in that case no lower-dimensional ro-polytopes can occur in representation (3.3)). Suppose that $m > 1$ and that the assertion is true for all representations of k-dimensional ro-polytopes with less than m k-dimensional components. Let H be a hyperplane which strictly separates two k-dimensional ro-polytopes in the representation (3.3). Since φ is weakly additive on $\mathcal{P}^n_{ro}$, we get

$$\varphi(P) = \varphi(P \cap H_0^+) + \varphi(P \cap H_0^-) + \varphi(P \cap H)$$

$$= \varphi(\bigcup_{i\in I} (P_i \cap H_0^+)) + \varphi(\bigcup_{i\in I} (P_i \cap H_0^-)) + \varphi(\bigcup_{i\in I} (P_i \cap H)).$$

Since at least one of the sets $P_i \cap H_0^+$ is empty, the (second) induction hypothesis yields

$$\varphi(\bigcup_{i\in I} (P_i \cap H_0^+)) = \sum_{i\in I} \varphi(P_i \cap H_0^+),$$

similarly with H_0^- instead of H_0^+. Since $\dim (P \cap H) < \dim P$, the first induction hypothesis gives

$$\varphi(\bigcup_{i\in I} (P_i \cap H)) = \sum_{i\in I} \varphi(P_i \cap H).$$

Hence we get

$$\varphi(P) = \sum_{i\in I} [\varphi(P_i \cap H_0^+) + \varphi(P_i \cap H_0^-) + \varphi(P_i \cap H)] = \sum_{i\in I} \varphi(P_i).$$

This completes the inductive proof of the fact that (3.3) implies (3.4).

Now let $P \in U(\mathcal{P}^n_{ro})$ be an arbitrary ro-polyhedron. By (3.1), there is a representation $P = \dot{\bigcup}_{i\in I} P_i$ with pairwise disjoint ro-polytopes $P_i \in \mathcal{P}^n_{ro}$ ($i \in I$, I finite).

Suppose that also $P = \uplus_{j\in J} Q_j$ with $Q_j \in \mathscr{P}^n_{ro}$ ($j\in J$, J finite). Then we have $P_i = \uplus_{j\in J}(P_i \cap Q_j)$ for $i \in I$, hence $\varphi(P_i) = \Sigma_{j\in J}\, \varphi(P_i \cap Q_j)$ by the result proved above, and similarly $\varphi(Q_j) = \Sigma_{i\in I}\, \varphi(P_i \cap Q_j)$ for $j \in J$. Hence we get

$$\sum_{j\in J} \varphi(Q_j) = \sum_{j\in J}\sum_{i\in I} \varphi(P_i \cap Q_j) = \sum_{i\in I} \varphi(P_i).$$

Thus it is unambiguous to define

$$\varphi(P) := \sum_{i\in I} \varphi(P_i),$$

which extends φ to $U(\mathscr{P}^n_{ro})$. To prove that this extension is additive, let $P_1, P_2 \in U(\mathscr{P}^n_{ro})$. By (3.1) there is a set $\{Q_i : i \in I\}$ of pairwise disjoint ro-polytopes and there are subsets $I_1, I_2, I_3 \subseteq I$ such that

$$P_\alpha = \biguplus_{i\in I_\alpha} Q_i, \quad \alpha = 1,2; \quad P_1 \cap P_2 = \biguplus_{i\in I_3} Q_i.$$

Since $I_1 \cap I_2 = I_3$, we get

$$\varphi(P_1 \cup P_2) + \varphi(P_1 \cap P_2) = \sum_{i\in I_1\cup I_2} \varphi(Q_i) + \sum_{i\in I_1\cap I_2} \varphi(Q_i)$$

$$= \sum_{i\in I_1} \varphi(Q_i) + \sum_{i\in I_2} \varphi(Q_i) = \varphi(P_1) + \varphi(P_2),$$

which completes the proof of Theorem (3.2).

Now for an arbitrary functional φ on $\mathscr{P}^n$ we define

$$\text{(3.5)} \qquad \varphi(P) := \sum_{F\subseteq \bar{P}} (-1)^{\dim P - \dim F}\, \varphi(F) \quad \text{for } P \in \mathscr{P}^n_{ro};$$

then Lemma (2.2) implies that

$$(3.6) \qquad \varphi(P) = \sum_{F \subseteq \bar{P}} \varphi(F_0) \qquad \text{for } P \in \mathscr{P}^n.$$

We can now prove the second half of Theorem (2.1).

(3.7) THEOREM. *Every weak valuation on* $\mathscr{P}^n$ *has a unique extension to a valuation on* $U(\mathscr{P}^n_{ro})$.

PROOF. Let φ be a weak valuation on $\mathscr{P}^n$. First we define φ on $\mathscr{P}^n_{ro}$ by (3.5) and assert that

$$(3.8) \qquad \varphi(P) = \varphi(P \cap H_0^+) + \varphi(P \cap H_0^-) + \varphi(P \cap H)$$

for $P \in \mathscr{P}^n_{ro}$ and every hyperplane H. For $\dim P = 0$ this is trivial. Suppose that $\dim P > 1$ and the assertion is true for ro-polytopes of smaller dimension. Using (3.6) and the weak additivity of φ on $\mathscr{P}^n$, we get

$$\varphi(P) + \sum_{F \subseteq \bar{P}} \varphi(F_0) = \varphi(\bar{P})$$

$$= \varphi(\bar{P} \cap H^+) + \varphi(\bar{P} \cap H^-) - \varphi(\bar{P} \cap H).$$

Application of (3.6) to $\bar{P} \cap H^+$ yields

$$\varphi(\bar{P} \cap H^+) = \varphi(P \cap H_0^+) + \varphi(P \cap H) + \sum_{F \subset \bar{P}} [\varphi(F_0 \cap H_0^+) + \varphi(F_0 \cap H)]$$

and a similar relation for $\varphi(\bar{P} \cap H^-)$, further

$$\varphi(\bar{P} \cap H) = \varphi(P \cap H) + \sum_{F \subset \bar{P}} \varphi(F_0 \cap H).$$

From the induction hypothesis we have

$$\varphi(F_0) = \varphi(F_0 \cap H_0^+) + \varphi(F_0 \cap H_0^-) + \varphi(F_0 \cap H)$$

for every proper face F of $\bar{P}$. Inserting this into the above equalities, we get (3.8).

Thus we have proved that φ, restricted to $\mathscr{P}^n_{ro}$, is weakly additive. By Theorem (3.2), it has an extension to a valuation on $U(\mathscr{P}^n_{ro})$. Formula (3.6) shows that on $\mathscr{P}^n$ this extension coincides with the original functional and thus extends it.

Suppose that Ψ is another additive extension of φ from $\mathscr{P}^n$ to $U(\mathscr{P}^n_{ro})$. Writing $\Psi_0(P) := \Psi(P_0)$, we have

$$\varphi(P) = \Psi(P) = \sum_{F \subseteq P} \Psi_0(F) \qquad \text{for } P \in \mathscr{P}^n$$

by the additivity of Ψ. Lemma (2.2) then shows that

$$\Psi(P_0) = \Psi_0(P) = \sum_{F \subseteq P} (-1)^{\dim P - \dim F} \varphi(F) \quad \text{for } P \in \mathscr{P}^n,$$

which coincides with $\varphi(P_0)$ as defined above. Thus the extension is unique on $\mathscr{P}^n_{ro}$ and hence on $U(\mathscr{P}^n_{ro})$.

4. TRANSLATIVE RO-EQUIDECOMPOSABILITY.

We begin with some notation and auxiliary results. Let $\mathscr{A}$ be a translation class of flats (affine subspaces) in E^n. By $\mathscr{P}(\mathscr{A}) \subset \mathscr{P}^n$ we denote the set of polytopes lying in the elements of $\mathscr{A}$. Let h be a translation invariant real valuation on $\mathscr{P}(\mathscr{A})$ which is $\mathscr{A}$-simple, that is, which vanishes on polytopes of dimension less than $\dim \mathscr{A}$ ($= \dim A$ for any $A \in \mathscr{A}$). For $P \in \mathscr{P}^n$ we define

$$(4.1) \qquad \varphi_h(P) := \sum_{\substack{F \subseteq P \\ \mathrm{aff}\, F \in \mathscr{A}}} \gamma(F,P)h(F),$$

where $\gamma(F,P)$ denotes the (normalized) external angle of P at its face F. By a result due to Hadwiger (see McMullen [18], Theorem 4; put $\Psi_{\mathscr{A}'} = 0$ for each translation class $\mathscr{A}'$ different from $\mathscr{A}$), φ_h is a translation invariant valuation on $\mathscr{P}^n$. By Theorem (2.1), it has an additive extension to $U(\mathscr{P}^n_{ro})$, also denoted by φ_h. The explicit form of this extension on ro-polytopes is obtained as follows.

Let $P \in \mathscr{P}^n_{ro}$. Using (2.5) we get from (4.1)

$$\varphi_h(P) = \sum_{F \subseteq \bar{P}} (-1)^{\dim P - \dim F} \varphi_h(F)$$

$$= \sum_{\substack{G \subseteq \bar{P} \\ \text{aff } G \in \mathscr{A}}} \sum_{G \subseteq F \subseteq \bar{P}} (-1)^{\dim P - \dim F} \gamma(G,F) h(G).$$

A result of McMullen ([17], Lemma 4, $\bar{\gamma} = \gamma \circ \mu$ in his terminology) says that

$$\sum_{G \subseteq F \subseteq \bar{P}} (-1)^{\dim P - \dim F} \gamma(G,F) = (-1)^{\dim P - \dim G} \gamma(G,\bar{P}),$$

hence

$$\varphi_h(P) = \sum_{\substack{G \subseteq \bar{P} \\ \text{aff } G \in \mathscr{A}}} (-1)^{\dim P - \dim G} \gamma(G,\bar{P}) h(G)$$

$$= (-1)^{\dim P - \dim \mathscr{A}} \varphi_h(\bar{P}).$$

This will now be applied to the so-called basic Hadwiger functionals (compare [19], p.198). By $\mathscr{U}^k$ we denote the Stiefel manifold of orthonormal k-frames in E^n. For a polytope $P \in \mathscr{P}^n$, let P_u the face of P corresponding to the exterior unit normal vector u. For $U = (u_1, \ldots, u_k) \in \mathscr{U}^k$ we inductively define P_U by

$$P_{(u_1,\dots,u_k)} = (P_{(u_1,\dots,u_{k-1})})_{u_k}.$$

Now for a translation class $\mathscr{A}$ of flats we say that $U \in \mathscr{U}^k$ is *$\mathscr{A}$-adapted* if all vectors of U are parallel to $\mathscr{A}$. For each $\mathscr{A}$ (of dimension $k \in \{0,\dots,n\}$) and each $\mathscr{A}$-adapted $U = (u_1,\dots,u_{k-r}) \in \mathscr{U}^{k-r}$ (where $r \in \{0,\dots,k\}$) we define a functional $\varphi_{\mathscr{A},U}$ on $\mathscr{P}^n$ by means of

$$(4.3) \qquad \varphi_{\mathscr{A},U}(P) = \sum_{\substack{F \subseteq P \\ \text{aff } F \in \mathscr{A}}} \gamma(F,P)h(F)$$

with

$$(4.4) \qquad h(F) = \sum_{\varepsilon_i = \pm 1} \varepsilon_1 \dots \varepsilon_{k-r} V_r(F_{(\varepsilon_1 u_1,\dots,\varepsilon_{k-r} u_{k-r})}),$$

where V_r is the r-dimensional volume (for $r = k$, we have to interpret this as $U = \emptyset$ and $h(F) = V_k(F)$). On the class of polytopes F with aff $F \subseteq A \in \mathscr{A}$, (4.4) defines a basic Hadwiger functional, hence h is a translation invariant $\mathscr{A}$-simple valuation. By the remarks made above, $\varphi_{\mathscr{A},U}$ extends to a translation invariant valuation on $U(\mathscr{P}^n_{ro})$, also denoted by $\varphi_{\mathscr{A},U}$. Now we can formulate our main result:

(4.5) THEOREM. *The ro-polyhedra* $P,Q \in U(\mathscr{P}^n_{ro})\setminus\{\emptyset\}$ *are* T-*ro-equidecomposable if and only if* $\varphi_{\mathscr{A},U}(P) = \varphi_{\mathscr{A},U}(Q)$ *for all* $\mathscr{A}$ *and* U.

To compute $\varphi_{\mathscr{A},U}(P)$ for a given ro-polyhedron P, one has to represent P as a disjoint union of ro-polytopes. For the latter, $\varphi_{\mathscr{A},U}$ is obtained from (4.2) and (4.3). Finally, for a polytope $P \in \mathscr{P}^n$ it is clear that there are only finitely many translation classes $\mathscr{A}$ and frames

U for which $\varphi_{\mathscr{A},U}(P) \neq 0$. Thus, (4.5) provides (in principle) an effective criterion for checking the T-ro-equidecomposability of two given polyhedra.

The necessity of the condition in (4.5) is clear, so let us prove the sufficiency. Let $P,Q \in U(\mathscr{P}^n_{ro}) \setminus \{\emptyset\}$ be given such that

$$(4.6) \qquad \varphi_{\mathscr{A},U}(P) = \varphi_{\mathscr{A},U}(Q)$$

for all translation classes $\mathscr{A}$ of flats and all $\mathscr{A}$-adapted frames U.

The case $\mathscr{A} = \{E^n\}$ of (4.6) shows that any basic Hadwiger functional h (additively extended to $U(\mathscr{P}^n_{ro})$) attains the same value at P and Q. The closure $\bar{P}$ of P can be expressed as the disjoint union of P and a finite number of ro-polytopes of dimensions less than n. Since basic Hadwiger functionals are simple valuations and thus vanish on lower dimensional ro-polytopes, we have $h(\bar{P}) = h(P)$ and hence $h(\bar{P}) = h(\bar{Q})$. By the theorem of Jessen-Thorup [15] (see [19], §6, for a survey), $\bar{P}$ and $\bar{Q}$ are T-equidissectable, hence there exist n-dimensional polytopes $P_i, Q_i \in \mathscr{P}^n$ $(i = 1,\ldots,m)$ such that

$$\bar{P} = \bigcup_{i=1}^{m} P_i, \quad \bar{Q} = \bigcup_{i=1}^{m} Q_i,$$

$$\operatorname{int}(P_i \cap P_j) = \emptyset = \operatorname{int}(Q_i \cap Q_j) \quad \text{for } i \neq j,$$

$$P_i = \tau_i Q_i \quad \text{with} \quad \tau_i \in T,\ i = 1,\ldots,m.$$

We have representations

$$(\operatorname{int} P_i) \cap P = \bigcup_{k=0}^{n} \bigcup_{r} P_{kr}, \quad P_{kr} \in \mathscr{P}^n_{ro}, \quad \dim P_{kr} = k,$$

$$(\text{int } Q_i) \cap Q = \dot{\bigcup}_{k=0}^{n} \dot{\bigcup}_{s} Q_{ks}, \quad Q_{ks} \in \mathscr{P}^n_{ro}, \quad \dim Q_{ks} = k.$$

Denote by $A_1,\ldots,A_p$ the non-empty ones among the intersections $P_{nr} \cap \tau_i Q_{ns}$. Then $A_1,\ldots,A_p$ are open convex polytopes, and $(\text{int } P_i) \cap P$ is the disjoint union of $A_1,\ldots,A_p$ and finitely many lower-dimensional ro-polytopes. Moreover, $(\text{int } Q_i) \cap Q$ is the disjoint union of $\tau_i^{-1} A_1,\ldots,\tau_i^{-1} A_p$ and finitely many lower-dimensional ro-polytopes. Since $i \in \{1,\ldots,m\}$ was arbitrary, it follows that there are representations

$$P = P^{(n)} \dot{\cup} \dot{\bigcup} P_i^{(n-1)}, \quad Q = Q^{(n)} \dot{\cup} \dot{\bigcup} Q_i^{(n-1)}$$

with T-ro-equidecomposable ro-polyhedra $P^{(n)}$ and $Q^{(n)}$ and with finitely many ro-polytopes $P_i^{(n-1)}$, $Q_i^{(n-1)}$ of dimensions at most n-1.

Now we make the inductive assumption that for some $k \in \{0,,,,.n\}$ there exist representations

$$(4.7) \qquad P = P^{(k)} \dot{\cup} \dot{\bigcup_i} P_i^{(k-1)}, \quad Q = Q^{(k)} \dot{\cup} \dot{\bigcup_i} Q_i^{(k-1)}$$

with T-ro-equidecomposable ro-polyhedra $P^{(k)}$ and $Q^{(k)}$ and with finitely many ro-polytopes $P_i^{(k-1)}$, $Q_i^{(k-1)}$ of dimensions at most k-1. For k=0, this means that P and Q are T-ro-equidecomposable, hence we may assume that $k \geq 1$.

Consider a translation class $\mathscr{A}$ of (k-1)-dimensional flats and some $\mathscr{A}$-adapted frame U. From (4.6) and (4.7) we have

$$\varphi_{\mathscr{A},U}(\dot{\bigcup_i} P_i^{(k-1)}) = \varphi_{\mathscr{A},U}(\dot{\bigcup_i} Q_i^{(k-1)}).$$

Let $A \in \mathscr{A}$ be a fixed flat, and to each $P_i^{(k-1)}$ with $\text{aff } P_i^{(k-1)} \in \mathscr{A}$ choose a translate in A such that all these translates are pairwise disjoint. Let $\tilde{P}$ be the union of these translates, and define $\tilde{Q}$ similarly, starting from the $Q_i^{(k-1)}$. From the definition of $\varphi_{\mathscr{A},U}$ it follows that $\varphi_{\mathscr{A},U}(P_i^{(k-1)}) = 0$ unless aff $P_i^{(k-1)} \in \mathscr{A}$, hence we have $\varphi_{\mathscr{A},U}(\tilde{P}) = \varphi_{\mathscr{A},U}(\tilde{Q})$. For the polyhedra in the affine space A, the functionals $\varphi_{\mathscr{A},U}$ are just the basic Hadwiger functionals. We can, therefore, argue as above in the case $A = E^n$ and thus obtain representations

$$\tilde{P} = \tilde{P}^{(k-1)} \uplus \biguplus_i \tilde{P}_i^{(k-2)}, \quad \tilde{Q} = \tilde{Q}^{(k-1)} \uplus \biguplus_i \tilde{Q}_i^{(k-2)}$$

with T-ro-equidecomposable ro-polyhedra $\tilde{P}^{(k-1)}$ and $\tilde{Q}^{(k-1)}$ and with finitely many ro-polytopes $\tilde{P}_i^{(k-2)}$, $\tilde{Q}_i^{(k-2)}$ of dimensions at most k-2. Performing this procedure for the finitely many translation classes to which the affine hulls of the (k-1)-dimensional ones among the ro-polytopes $P_i^{(k-1)}$, $Q_i^{(k-1)}$ belong, we arrive at representations

$$P = P^{(k-1)} \uplus \biguplus_i P_i^{(k-2)}, \quad Q = Q^{(k-1)} \uplus \biguplus_i Q_i^{(k-2)}$$

with T-ro-equidecomposable ro-polyhedra $P^{(k-1)}$, $Q^{(k-1)}$ and with finitely many ro-polytopes $P_i^{(k-2)}$, $Q_i^{(k-2)}$ of dimensions at most k-2. This completes the induction step, and the case $k = 0$ of (4.7) proves the theorem.

REFERENCES

[1] G. AUMANN, *Sind die elementargeometrischen Figuren Mengen?* Elem. Math. 7 (1952) 25-28.

[2] V.G. BOLTIANSKII, *Equivalent and equidecomposable figures*. D.C. Heath Comp., Boston 1963 (Russian original 1956).

[3] V.G. BOLTIANSKII, *Hilbert's third problem*. John Wiley & Sons, New York etc. 1978.

[4] T. BONNESEN and W. FENCHEL, *Theorie der konvexen Körper*. Springer-Verlag, Berlin 1934.

[5] A. BRØNDSTED, *An introduction to convex polytopes*. Springer-Verlag, New York etc. 1983.

[6] L. DUBINS, M.W. HIRSCH and J. KARUSH, *Scissor congruence*. Israel J. Math. 1 (1963) 239-247.

[7] H. GROEMER, *On the extension of additive functionals on classes of convex sets*. Pacific J. Math. 75 (1978) 397-410.

[8] H. HADWIGER, *Multikongruenz ebener Mengen und pythagoreischer Lehrsatz*. Bull. Ecole Polytechn. Jassy 2 (1947) 98-105.

[9] H. HADWIGER, *Vorlesungen über Inhalt, Oberfläche und Isoperimetrie*. Springer-Verlag, Berlin etc. 1957.

[10] H. HADWIGER, *Translative Zerlegungsgleichheit der Polyeder des gewöhnlichen Raumes*. J. reine angew. Math. 233 (1968) 200-212.

[11] H. HADWIGER, *Neuere Ergebnisse innerhalb der Zerlegungstheorie euklidischer Polyeder*. Jber. Deutsche Math.-Ver. 70 (1968) 167-176.

[12] H. HADWIGER, *Zerlegungsgleichheit euklidischer Polyeder bezüglich passender Abbildungsgruppen und invariante Funktionale*. Math.-Phys. Sem.-Ber. 22 (1975) 125-133.

[13] H. HADWIGER and P. GLUR, *Zerlegungsgleichheit ebener Polygone*. Elem. Math. 6 (1951) 97-106.

[14] E. HERTEL, *Neuere Ergebnisse und Richtungen der Zerlegungstheorie von Polyedern*. Mitt. Math. Ges. DDR 1977, Heft 4, 5-22.

[15] B. JESSEN and A. THORUP, *The algebra of polytopes in affine spaces*. Math. Scand. 43 (1978) 211-240.

[16] S. KÁNTOR, *Über die Zerlegungsgleichheit*. Publ. Math. Debrecen 23 (1976) 255-261.

[17] P. McMULLEN, *Non-linear angle-sum relations for polyhedral cones and polytopes*. Math. Proc. Camb. Phil. Soc. 78 (1975) 247-261.

[18] P. McMULLEN, *Valuations and Euler-type relations on certain classes of convex polytopes*. Proc. London Math. Soc. (3) 35 (1977) 113-135.

[19] P. McMULLEN and R. SCHNEIDER, *Valuations on convex bodies. In: Convexity and its Applications*. Eds. P.M. Gruber and J.M. Wills, Birkhäuser Verlag, Basel etc. 1983, 170-247.

[20] C.H. SAH, *Hilbert's third problem: Scissors congruence*. Pitman Advanced Publishing Program, San Francisco etc. 1979.

[21] W. VOLLAND, *Ein Fortsetzungssatz für additive Eipolyederfunktionale im euklidischem Raum*. Arch. Math. 8 (1957) 144-149.

[22] S. WAGON, *Circle-squaring in the twentieth century*. Math. Intelligencer 3 (1980) 176-181.

[23] S. WAGON, *The Banach-Tarski paradox*. Cambridge University Press, Cambridge etc. 1985.

R. SCHNEIDER
Math. Institut der Universität
Albertstr. 23 b
D-7800 Freiburg i. Br.

COLLOQUIA MATHEMATICA SOCIETATIS JÁNOS BOLYAI
48. INTUITIVE GEOMETRY. SIÓFOK, 1985.

ANALOGUES OF STEINITZ'S THEOREM ABOUT NON-INSCRIBABLE POLYTOPES

E. SCHULTE

ABSTRACT

As long ago as 1832 Steiner asked whether each isomorphism type of convex 3-polytopes is inscribable, that is, has a representative with all vertices in a Euclidean 2-sphere. Only in 1928 Steinitz proved by a particularly elegant geometric argument that the general answer is in the negative. The duals of his counterexamples are non-circumscribable, that is, they are not realizable with all 2-faces tangent to a sphere.

The problem of realizing each isomorphism type of 3-polytopes with all *edges* tangent to a 2-sphere has resisted all efforts so far. Surprisingly, the analogous problems for m-faces of polytopes in dimensions $d \geq 4$ can be solved without great effort, thereby leaving the case $d = 3$, $m = 1$ as the outstanding problem.

1. INTRODUCTION

A convex d-polytope P in the Euclidean d-space $\mathbb{E}^d$ is called *inscribable* (or *of inscribable type*) if there is a combinatorially equivalent convex d-polytope that has all its vertices in a Euclidean $(d-1)$-sphere. Analogously, P is called *circumscribable* (or *of circumscribable type*) if there is an isomorphic copy of P that has all its facets tangent to a Euclidean $(d-1)$-sphere.

In 1832 Steiner asked whether each convex 3-polytope is of inscribable type (cf. [7]). Only in 1928 Steinitz proved by a particularly elegant geometric argument that the general answer to Steiner's question is in the negative (cf. [8], or Grünbaum [3]). Making use of the dual relation between inscribability and circumscribability Steinitz obtained a simple criterion for a convex 3-polytope to be non-circumscribable, leading to infinitely many non-circumscribable polytopes as well as infinitely many non-inscribable polytopes. For instance, starting from a 3-polytope Q with at least as many vertices as facets, and cutting off all vertices of Q by planes which have no common points in Q, gives a non-circumscribable 3-polytope P; its dual is of non-inscribable type. Among the non-inscribable 3-polytopes are also simplicial ones. (Recall that a d-polytope is called simplicial if all its facets are $(d-1)$-simplices, that is, triangles if $d = 3$.) This is of interest, since a hasty consideration of Brückner may leave the impression that simplicial 3-polytopes are inscribable in general (cf. [1, p. 163, footnote 4]).

Steinitz's results give also a negative solution for the corresponding problem in higher dimensions. In fact, if for $d \geq 4$ a convex d-polytope P has one 3-dimensional face of non-inscribable type, then P is necessarily non-inscribable too, and consequently, by duality, its dual non-circumscribable.

Another proof for Steinitz's theorem was given by Grünbaum (cf. [2]). He established the existence of 3-polytopes P which have even no representative P' with circumcircles, thereby settling a problem of Motzkin. (Recall that a 3-polytope is said to possess circumcircles provided each facet has a circumcircle.)

Steinitz's methods were extended by Grünbaum and Jucovič who investigated how badly non-inscribable convex 3-polytopes may be (cf. [4]). They proved the existence of simplicial 3-polytopes, which behave very poorly with respect to the possibility of inscribing isomorphic copies into a sphere.

In this paper we consider analogues of Steinitz's results in higher dimensions. Generalizing the concept of inscribability and circumscribability we investigate the problem, whether each isomorphism type of convex polytopes can have a representative that has all faces of a given dimension m tangent to a Euclidean sphere.

Let d and m be natural numbers with $d \geq 2$ and $0 \leq m \leq d-1$. A convex d-polytope P is called (m,d)-*scribable* (or *of* (m,d)-*scribable type*) if there is an isomorphic copy P' of P such that all faces of P' of dimension m are tangent to some Euclidean $(d-1)$-sphere S. In this case the m-faces of P' must touch S in relative interior points. For $m = 0$ and $d-1$, (m,d)-scribability coincides with inscribability and circumscribability, respectively. Sometimes, if the dimension d is obvious from the context, we will also say *escribable* instead of $(1,d)$-*scribable* (*e* for edges).

Generalizing Steinitz's results we will establish the existence of infinitely many convex d-polytopes which are not (m,d)-scribable, or (m,d)-nonscribable as we will say, for each $d \geq 4$ and each m with $0 \leq m \leq d-1$. Our methods are based on the duality between (m,d)-scribability and $(d-1-m,d)$-scribability, on a reduction theorem connecting (m,d)-scribability to $(m-1,d-1)$- and $(m,d-1)$-scribability, and on Steinitz's results in three dimensions.

The only case that has resisted all efforts so far is the escribability in three dimensions. Somehow it is strange that all higher-dimensional analogues turn out to be solvable, while the 'elementary' three-dimensional case of escribability seems to be intractable. Also, it is annoying that it is by no means clear what to do, either to find counterexamples or to prove that every 3-polytope is escribable. In fact, all our present knowledge about inscribability, circumscribability, and related problems is compatible with both possibilities.

We remark that many open problems about escribability and related matters are discussed in Grünbaum-Shephard [5]. One of the most interesting is the question if, for each given 3-polytope, there is an isomorphic polytope P and a dual polytope Q such that corresponding edges of P and Q intersect perpendicularly. In many naive discussions of duality it is simply asserted that such pairs exist in all cases; however, this is far from being proved. Note that, clearly, each escribable 3-polytope gives rise to such a pair, even one where the polytopes are reciprocals of each other.

For notation and basic results about convex polytopes the reader is referred to Grünbaum [3].

2. NONSCRIBABLE POLYTOPES

We start our investigations with a duality theorem, which relates (m,d)-scribability to $(d-1-m,d)$-scribability. For the case $m=0$, this relation was already successfully used in Steinitz's proof. Before we state the result, we remind the reader to some facts about reciprocation of polytopes with respect to spheres (cf. Grünbaum [3, p. 46]).

Let P be a convex d-polytope containing the origin 0 as an interior point, and let $\mathbb{S}^{d-1}$ be the unit-sphere centered at 0. Then, the isomorphism type of the polytopes dual to P can be realized by the *polar set*

$$P^* = \bigcap_{x \in P} \{y \mid \langle x,y \rangle \leq 1\}$$

of P (with respect to $\mathbb{S}^{d-1}$), where $\langle , \rangle$ denotes the inner product. P^* is a convex d-polytope containing 0 as an interior point, and the $(d-1-m)$-faces $\hat{F}$ of P^* corresponding to an m-face F of P is given by

$$\hat{F} = P^* \cap \bigcap_{x \in F} \{y \mid \langle x,y \rangle = 1\},$$

for $m = 0,\ldots,d-1$. This correspondence between the faces of P and the faces of P^* is one-to-one and inclusion reversing. Furthermore, the polar set of P^* with respect to $\mathbb{S}^{d-1}$ is P, that is, $(P^*)^* = P$, and $\hat{\hat{F}} = F$ for all proper faces F of P.

In the proof of Theorem 1 we make also use of the following property of projective transformations of $\mathbb{E}^d$. For each interior point y of the Euclidean unit-ball in $\mathbb{E}^d$, there is a projective transformation T of $\mathbb{E}^d$ carrying

$\mathbb{S}^{d-1}$ onto itself and mapping y onto 0. For example, if $y = (a, 0, \ldots, 0)$ for $|a| < 1$, then we can choose the transformation

$$T_a(x_1, x_2, \ldots, x_d) := \frac{1}{1 - ax_1}(x_1 - a, bx_2, \ldots, bx_d)$$

where $b := \sqrt{1 - a^2}$. In general, an orthogonal transformation followed by a transformation T_a will have the required properties (cf. Grünbaum [3, p. 285]).

THEOREM 1. *Let* $d \geq 2$ *and* $0 \leq m \leq d-1$. *A convex d-polytope P is* (m,d)*-scribable, if and only if its dual d-polytope* P^* *is* $(d-1-m, d)$*-scribable.*

PROOF. By our above considerations, it is sufficient to prove only one direction of the assertion. Therefore, let us assume that the convex d-polytope P is (m,d)-scribable, and has all its faces of dimension m tangent to $\mathbb{S}^{d-1}$. Replacing P by a projective copy if need be we may suppose that 0 is an interior point of P. We will show that the polar set P^* of P with respect to $\mathbb{S}^{d-1}$ has all its $(d-1-m)$-faces tangent to $\mathbb{S}^{d-1}$ (even with the same points of contact), and is therefore a $(d-1-m)$-scribable d-polytope.

Let F be an m-face of P touching $\mathbb{S}^{d-1}$ in x_F, and let $\hat{F}$ be the respective $(d-1-m)$-face of P^*. The tangent hyperplane H to $\mathbb{S}^{d-1}$ at x_F is a supporting hyperplane for P intersecting P in F. Therefore, $\langle x, x_F \rangle \leq 1$ for all x in P implies $x_F \in P^*$, and $\langle x, x_F \rangle = 1$ for all x in F implies $x_F \in \hat{F}$. Furthermore, if y is in $\hat{F}$, then $x_F \in F$ implies $\langle x_F, y \rangle = 1$. Consequently, $\hat{F}$ lies in H and touches $\mathbb{S}^{d-1}$ in x_F. But that completes the proof.

In the construction of nonscribable polytopes we make essentially use of the following reduction theorem. Recall that a vertex-figure of a d-polytope P at a vertex z of P is the intersection of P with a hyperplane which strictly separates z form the other vertices of P. Note that, for any two isomorphic d-polytopes, the vertex-figures at corresponding vertices are isomorphic convex $(d-1)$-polytopes.

THEOREM 2. *Let* $d \geq 3$, $0 \leq m \leq d-1$, *and* P *an* (m,d)-*scribable convex* d-*polytope.*

(*a*) *If* $m \leq d-2$, *then each facet of* P *is* $(m,d-1)$-*scribable.*

(*b*) *If* $m \geq 1$, *then each vertex-figure of* P *is* $(m-1,d-1)$-*scribable.*

PROOF. Assume without loss of generality that P has all m-faces tangent to a $(d-1)$-sphere S, and let $m \leq d-2$. Then, for each facet G of P, the hyperplane H containing G cuts out the $(d-2)$-sphere $S \cap H$. Now, all m-faces of G touch S, and are contained in H, and so touch $S \cap H$. This proves part (*a*).

Part (*b*) follows immediately from Theorem 1. In fact, P^* is $(d-1-m,d)$-scribable, and so, by part (*a*), each facet of P^* $(d-1-m,d-1)$-scribable. As each vertex-figure of P is dual to a facet of P^*, another application of Theorem 1 proves part (*b*).

REMARK. It is noteworthy that we can equally well give a direct proof of part (*b*) without applying Theorem 1. In fact, if all m-faces of P touch a sphere S, then the contact points of S with the m-faces of P containing a given vertex z must lie in one hyperplane H. Then, projecting P centrally from z onto H gives an isomorphic copy of the vertex-figure of P at z, whose $(m-1)$-faces are tangent to the $(d-2)$-sphere $S \cap H$.

With the help of Theorems 1 and 2 it is now possible to state our main result. Due to the fact that all convex 2-polytopes, that is, convex polygons, are inscribable as well as circumscribable, our result will not cover escribability in three dimensions.

THEOREM 3. *Let* $d \geq 3$, $0 \leq m \leq d-1$, *and* $(m,d) \neq (1,3)$. *Then, there are infinitely many* (m,d)*-nonscribable convex* d*-polytopes.*

PROOF. From Steinitz's results we know that our assertion is true for dimension $d = 3$ as well as dimensions $d \geq 4$ and $m = 0$ or $d-1$. Therefore, we may assume $d \geq 4$ and $1 \leq m \leq d-2$.

Let $d = 4$. By part (b) of Theorem 2, the pyramid over a non-inscribable 3-polytope cannot be escribable, since the vertex-figure at its apex is isomorphic to its base. From Theorem 1 we can conclude that the dual of a non-escribable 4-polytope is necessarily $(2,4)$-nonscribable. Hence, our assertion follows from Steinitz's results in three dimensions. Note, in particular, that in four dimensions our theorem holds without restrictions on the dimension m of faces.

For $d \leq 5$, we can proceed inductively by Theorem 2

(b), making use of the fact that in $d-1$ dimensions the theorem holds for all dimensions m of faces. In fact, the pyramid over an $(m-1,d-1)$-nonscribable $(d-1)$-polytope must be (m,d)-nonscribable. That completes the proof.

Our methods can be modified in order to prove the existence of simplicial and simple nonscribable polytopes. (Recall that a convex d-polytope is called simple if all its vertices have valence d, that is, if the dual polytope is simplicial.) In fact, if we replace the pyramid in our proof by a bipyramid, then we obtain simplicial (m,d)-nonscribable d-polytopes from simplicial $(m-1,d-1)$-nonscribable $(d-1)$-polytopes.

From Grünbaum-Jucovič [4] we know that, for each dimension $d \geq 3$, there are infinitely many simplicial non-inscribable d-polytopes. Among them is the Kleetope over the d-simplex (cf. Grünbaum [3, p. 217]). Therefore, our methods provide simplicial (m,d)-nonscribable d-polytopes at least for $d \geq 4$ and $0 \leq m \leq d-3$, and by duality, simple (m,d)-nonscribable d-polytopes for $d \geq 4$ and $2 \leq m \leq d-1$.

In a similar fashion we can obtain simplicial non-circumscribable d-polytopes and simple non-inscribable d-polytopes for all $d \geq 3$ from simplicial non-circumscribable polytopes in three dimensions; the existence of simplicial non-circumscribable 3-polytopes was already proved in Grünbaum [2]. However, it is undecided yet if simplicial $(d-2,d)$-nonscribable and simple non-escribable d-polytopes exist for $d \geq 3$.

3. WEAK SCRIBABILITY

Generalizing the concept of a weakly escribable 3-polytope (cf. Grünbaum-Shephard [5]) we call a convex d-polytope P *weakly* (m,d)*-scribable* if for some isomorphic convex d-polytope the affine hulls of its m-faces are all tangent to a Euclidean $(d-1)$-sphere $(0 \le m \le d-1)$. In case such a polytope fails to exist P is said to be *strongly* (m,d)*-nonscribable*. Clearly, strong (m,d)-nonscribability implies (m,d)- nonscribability, but a priori the reverse statement need not be true unless $m = 0$ (see Problem 2 in Grünbaum-Shephard [5]). Now, extending our problem we may ask if there exist strongly (m,d)-nonscribable d-polytopes.

The attempt to generalize our methods immediately reveals the main difference between (m,d)-scribability and weak (m,d)-scribability. In fact, a polytope with all m-faces tangent to $\mathbb{S}^{d-1}$ has necessarily an interior point in the open unit-ball, while this need not be true if only the affine hulls of all m-faces are tangent to $\mathbb{S}^{d-1}$. Therefore, Theorem 1 might fail.

However, Theorem 2 can be salvaged for weak scribability. In fact, if P is a weakly (m,d)-scribable d-polytope, then each facet of P is weakly $(m,d-1)$-scribable if $m \le d-2$, and each vertex-figure of P is weakly $(m-1,d-1)$-scribable if $m \ge 1$. The first assertion follows in the same manner as in the proof of Theorem 2, while the second can be proved as in the remark to Theorem 2. In particular, this implies that the pyramid (or bipyramid) over a strongly $(m-1,d-1)$-nonscribable $(d-1)$-polytope is a strongly (m,d)-nonscribable d-polytope.

Therefore, taking into account that inscribability and weak inscribability are the same for each dimension (case $m = 0$), we find strongly (m,d)-*nonscribable* d-polytopes (even simplicial ones) at least for $d \geq 3$ and $0 \leq m \leq d-3$. For example, the bipyramid over a (simplicial) non-inscribable $(d-1)$-polytope is a (simplicial) strongly non-escribable d-polytope. The remaining values for (m,d) except for $(1,3)$ could also be covered by the analogue of Theorem 2 (*a*) provided the existence of strongly non-circumscribable polytopes were known for all dimensions greater than or equal to 3. However, this is open even for dimension 3. Note that, again, the case $(m,d) = (1,3)$ is exceptional.

4. RELATED PROBLEMS

(1) Another interesting direction of research could be the investigation of polytopes, which behave very poorly with respect to realizing any isomorphic copy with all m-faces tangent to a sphere $(m = 0,\dots,d-1)$; for the case of inscribability see Grünbaum-Jucovič [4]. For a convex d-polytope P let $v(P)$ denote the total number of vertices, and $s_m(P)$ the largest integer s with the following property: there is an isomorphic copy P' of P, and a sphere S, such that s m-faces of P' are tangent to S.

Following the terminology introduced in Grünbaum-Walther [6] the number

$$s_m^{(d)} := \liminf \frac{\log s_m(P)}{\log v(P)},$$

where P ranges over all d-polytopes, could be called the (m,d)-*scribability exponent* of the family of all d-polytopes. Analogously, we could define the (m,d)-scribability exponent $\bar{s}_m^{(d)}$ of the family of all simplicial d-polytopes (that is, P ranges only over simplicial polytopes). For $m = 0$, partial results about the values of $s_m^{(d)}$ and $\bar{s}_m^{(d)}$ are known.

For $d \geq 3$, Grünbaum and Jucovič prove

$$s_0^{(d)} \leq \bar{s}_0^{(d)} \leq \log(d-1)/\log d,$$

and conjecture equality for the right side for all $d \geq 3$, and also for the left side for $d = 3$ (cf. [4]). However, for other values of d and m the exponents are completely unexplored.

(2) Concluding we would like to focus attention to the following open question. It would be interesting to know to what extent our results hold if we replace the Euclidean sphere in our considerations by other convex bodies. It seems that completely different methods are necessary to prove the

CONJECTURE. Let K be a convex body in dimension $d \geq 3$, and let $0 \leq m \leq d-1$. Then, there are convex d-polytopes, for which no isomorphic copy has all its m-faces tangent to K.

Finally I would like to thank Professors B. Grünbaum and G. C. Shephard for stimulating discussion on the subject of this paper.

NOTE ADDED IN PROOF. As I have heard recently, some results of this paper were also obtained in independent work by P. McMullen.

REFERENCES

[1] M. BRÜCKNER, *Vielecke und Vielflache*, Teubner, Leipzig, 1900.

[2] B. GRÜNBAUM, *On Steinitz's theorem about non-inscribable polyhedra*, Proc. Ned. Akad. Wetenschap. Ser. A, 66 (1963) 452-455.

[3] B. GRÜNBAUM, *Convex polytopes*, Interscience, New York, 1967.

[4] B. GRÜNBAUM and E. JUCOVIČ, *On non-inscribable polytopes*, Czechoslovak Mathematical Journal, 24 (99) 1974, Praha.

[5] B. GRÜNBAUM and G. C. SHEPHARD, *Some problems on polyhedra*, Proc. of the Conf. on Polyhedra held in Northampton (Mass.) in April 1984.

[6] B. GRÜNBAUM and H. WALTHER, *Shortness exponents of families of graphs*, J. Comb. Theory, 14 (1973) 364-385.

[7] J. STEINER, *Systematische Entwicklung der Abhängigkeit geometrischer Gestalten voneinander*, Fincke, Berlin, 1832 (= Gesammelte Werke, Vol. 1, Reimer, Berlin, 1881, 229-458).

[8] E. STEINITZ, *Über isoperimetrische Probleme bei konvexen Polyedern*, J. reine angew. Math. 159 (1928) 133-143.

E. SCHULTE
Mathematisches Institut
Universität Dortmund
Postfach 50 05 00
D-46. Dortmund
West-Germany

COLLOQUIA MATHEMATICA SOCIETATIS JÁNOS BOLYAI
48. INTUITIVE GEOMETRY, SIÓFOK, 1985.

EMBEDDING 2-MANIFOLDS INTO THE BOUNDARY COMPLEXES OF SEWN 4-POLYTOPES

CH. SCHULZ

1. INTRODUCTION

A *polyhedral 2-manifold* (or briefly a *polyhedron*) is a geometric complex M (in the sense of Grünbaum [7]) in eudlidean d-space E^d, whose underlying point set is a closed, compact topological 2-manifold. So, the cells of M are plane convex polygons, and their edges and vertices.

One open problem concerning polyhedral 2-manifolds is the minimal number of vertices of a polyhedron M of given topological type in a space of given dimension d. Here we may of course assume that M is a triangulation, and since any 2-dimensional abstract simplical complex has a geometric realization in E^5 the problem is solved by the corresponding results on abstract triangulations ([10], [11]) in case of $d \geq 5$. For $d = 3,4$ however the minimal number is known only in few cases.

In this paper we focus on realizations in E^3. This implies of course that M is orientable. The minimal number v of vertices of an abstract triangulation of a closed orientable surface of genus g is ([10], [11])

$$(*) \qquad v = \lceil (7 + \sqrt{48g+1})/2 \rceil$$

with the exception of $g = 2$, $v = 10$. ($\lceil x \rceil$ denotes the integer ceiling of x.) Polyhedra in E^3 with the same number of vertices are only known up to $g = 3$. The case $g = 1$ is often referred to as Császár's torus [6] although it already goes back to Möbius [9], and so should rather be called Möbius' torus. The cases $g = 2,3$ are due to U. Brehm [5]. While (*) gives a lower bound on the minimal number of vertices an upper bound is given by a series of polyhedra in E^3 with $v = O(g/\log g)$ for all genera g constructed in [8].

Of particular interest are those values of g for which the quotient on the right hand side of (*) is an integer, since they are *neighborly*, i.e. each pair of vertices is joined by an edge. The first non-trivial neighborly case is Möbius' torus, while in the next case $g = 6$, $v = 12$ the existence of a neighborly polyhedron in E^3 is still open.

One way to construct a polyhedral manifold M in E^3 is to embed M as a subcomplex into the 2-skeleton of some 4-polytope P. A projection of the boundary complex of P into E^3 as a Schlegel-diagram (see [7]) yields the desired polyhedral realization of M in 3-space. In case of M being neighborly P also must be neighborly, since one may assume that M covers all the vertices of P. A. Altshuler has shown in [1] that Möbius' torus is contained in the cyclic 4-polytope $C(7,4)$ on 7 vertices, but for all $n > 7$ there is no polyhedron M with

$$\mathrm{skel}_1 C(n,4) \subset M \subset \mathrm{skel}_2 C(n,4).$$

However this does not mean that it is impossible to find new neighborly polyhedra in E^3 via neighborly 4-polytopes since there are many more neighborly 4-polytopes than just the cyclic ones. Especially the so-called *sewing-construction* of I. Shemer [13] yields new combinatorial types of neighborly 4-polytopes whose number is growing superexponentially with respect to the number of vertices.

It is the aim of this paper to show that most of the 4-polytopes P obtained by Shemer sewing do not admit a polyhedron M with $\mathrm{skel}_1 P \subset M \subset \mathrm{skel}_2 P$. In particular we show that no sewn 4-polytope on 12 vertices is appropriate for constructing a polyhedron M in E^3 with $g = 6$, $v = 12$.

In the next section we give a brief description of some basic concepts we need in the sequel, section 3 contains the results, and section 4 the proofs. Our notation follows Grünbaum [7], but we shall denote the cell spanned by x,y,z by xyz.

2. UNIVERSAL EDGES AND SHEMER SEWING

By a *universal edge* (*u.e.*) of a neighborly 4-polytope P we mean an edge ab of P such that abc is a 2-face of P for each $c \in \mathrm{vert}\, P - \{a,b\}$. link(b,P) is (isomorphic to) the boundary complex of some simplical 3-polytope, and a is a *universal vertex* of link(b,P), i.e. a vertex joined to all other vertices by an edge. A 3-polytope Q on at least 5 vertices has at most 2 universal vertices, and the cyclic 3-polytopes are the only ones having 2 universal vertices. This yields:

REMARK 1. A vertex x of a neighborly 4-polytope P is contained in at most two u.e. 's. If x is contained in exactly two u.e.'s then link(x,P) is cyclic.

It follows that the graph of all u.e.'s consists of paths and circuits, and it is shown in [13] that it consists of paths only unless P is cyclic.

In case of $d = 4$ (and this is the only case we shall need) *Shemer's sewing construction* may be described combinatorially in the following way. If ab is a u.e. then a is a universal vertex of the simplical 2-sphere link(b,P), and so

$$C = \text{link}(a,\text{link}(b,P))$$

is a circuit containing all the vertices of link(b,P) but a. Replacing an edge xy of C by the two edges ax and ay one obtains a Hamiltonian circuit H on link(b,P). H separates link(b,P) into two triangulated disks D_1, D_2. We now introduce a new vertex c, and replace each face of P spanned by b and some cell F of D_1 by the cell spanned by c and F. Furthermore we add the edge bc and all cells spanned by b,c or bc and some vertex or edge of H to the complex. This gives a neighborly triangulated 3-sphere S' on n+1 vertices. For the proof that c may be chosen such that S' is the boundary complex of a neighborly 4-polytope P' the reader is referred to [13].

We observe that ab and bc are u.e.'s of P'. So, each sewn 4-polytope contains two consecutive u.e.'s. For an arbitrary neighborly 4-polytope P' with u.e.'s ab and bc one may consider $P = \text{conv}(\text{vert } P' - \{c\})$. It is shown in [13] that the combinatorial structure of P is uniquely determined by the combinatorial structure of P', and that P' may be obtained from P by sewing. This gives:

REMARK 2. A neighborly 4-polytope P' on n+1 vertices is sewn, i.e. may be obtained from a neighborly 4-polytope P on n vertices by Shemer sewing, if and only if it contains two consecutive u.e.'s ab and bc.

If ab is a u.e. of P, and $P' = \mathrm{conv}(\mathrm{vert}\ P - \{b\})$ then again by [13] the combinatorial structure of P determines the combinatorial structure of P' and we have

$$st(a,P') = (st(a,P) - st(ab,P)) \cup \{\mathrm{conv}((\mathrm{vert}\,F - \{b\}) \cup \{a\}) :$$

$$: F \in st(b,P)\}.$$

In other words st(a,P') arises from $st(a,P) \cup st(b,P)$ by collapsing each tetrahedron spanned by ab and an edge xy of link(ab,P) to a triangle axy. Since st(a,P') is a 3-ball this implies that $st(a,P) \cup st(b,P)$ also is a 3-ball, and by induction we get:

REMARK 3. If $x_i x_{i+1}$, $i = 1,\dots,k-1$ are u.e.'s of a neighborly 4-polytope P then

$$\bigcup_{i=1}^{k} st(x_i,P)$$

is a triangulated 3-ball.

3. RESULTS

THEOREM 1. *A neighborly 4-polytope* P *on* $n > 7$ *vertices admitting a polyhedral 2-manifold* M *with* $\mathrm{skel}_1 P \subset M \subset \mathrm{skel}_2 P$ *satisfies the following conditions:*

a) P *does not have three consecutive u.e.'s* wx, xy *and* yz.

b) If P *has two consecutive u.e.'s* xy *and* yz *then the simplical 2-spheres* link(x,P) *and* link(z,P) *each have exactly two 3-valent vertices.*

THEOREM 2. *There is no sewn 4-polytope* P *on* 12 *vertices that admits a polyhedron* M *with* $\mathrm{skel}_1 P \subset M \subset \mathrm{skel}_2 P$.

Theorem 1 is proved by a local argument only considering the 3-ball

$$st(w,P) \cup st(x,P) \cup st(y,P) \cup st(z,P),$$

and part a) of it cannot be improved by a similar local argument as the proof of Theorem 2 shows.

PROBLEM. Is there a sewn 4-polytope on $n > 12$ vertices that admits a polyhedron M with $skel_1 P \subset M \subset skel_2 P$?

Although Theorem 1, b) strongly restricts the class of possible candidates for P it is not at all obvious that there is no such P in case of large n (see the proof of Theorem 2 below).

As will be shown in a forthcoming paper [12] Remark 3 of section 2 still holds in the more general case of neighborly triangulations of the 3-sphere, and thereby in Theorem 1 "neighborly 4-polytope" may be replaced by "neighborly 3-sphere". Theorem 2 however does not generalize to neighborly 3-spheres, since U. Brehm has found a neighborly 3-sphere S on 12 vertices containing a neighborly 2-manifold of genus 6 (private communication). S is not isomorphic to the boundary complex of a 4-polytope, and so does not yield a polyhedron with $g = 6$, $v = 12$ in E^3 by the projection argument mentioned in the introduction. Several attempts to realize S in E^4 in such a way that a modified projection argument may be applied have failed.

4. PROOFS

First we introduce some notations. As in the previous section P denotes a neighborly 4-polytope on $n > 7$ vertices. xy denotes a u.e. of P, and

$$D_{xy} = ast(y,link(x,P)),\ D_{yx} = ast(x,link(y,P)).$$

D_{xy} and D_{yx} are triangulated disks without interior vertices, and with common boundary link(xy,P). $D_{xy} \cup D_{yx}$ is the 2-sphere bounding the 3-ball $st(x,P) \cup st(y,P)$ (see section 2, Remark 3), and so, D_{xy} and D_{yx} do not have diagonals in common, $D_{xy} \cap D_{yx} = link(xy,P)$.

If M is a polyhedron with $skel_1 P \subset M \subset sekl_2 P$ then link(x,M), link(y,M) are Hamiltonian circuits on link(x,P), link(y,P), and

$$H_{xy} = D_{xy} \cap link(x,M),\ H_{yx} = D_{yx} \cap link(y,M)$$

are Hamiltonian paths on D_{xy}, D_{yx}.

LEMMA 1.

a) The endpoints of H_{xy} *and* H_{yx} *coincide.*

b) If a *is a common endpoint of* H_{xy} *and* H_{yx}, *and* ab *is one of the two edges of* link(xy,P) *containing* a *then* ab *may not belong to both* H_{xy} *and* H_{yx}.

d) There are no two consecutive edges ab, bc *in* $H_{xy} \cap H_{yx}$.

PROOF.

a) The endpoints a,b of H_{xy} correspond to the triangles axy, bxy of M, and so are also the endpoints of H_{yx}.

b) The assumption $ab \in H_{xy} \cap H_{yx}$ implies axy, abx, $aby \in M$, but this means that a is a 3-valent vertex of M contradicting $skel_1 P \subset M$.

c) ab, $bc \in H_{xy} \cap H_{yx}$ implies b to be a 4-valent vertex of M, again contradicting $skel_1 P \subset M$.

LEMMA 2. *If* link(x,P) *is cyclic then* H_{xy} *contains exactly one diagonal of* D_{xy}.

PROOF. Let z denote the universal vertex of link(x,P) different from y, and $x_1,\ldots,x_{n-3}$ the other vertices of link(xy,P) according to their order on link(xy,P). So, D_{xy} consists of the triangles zx_ix_{i+1}, $i = 1,\ldots,n-4$.

If H_{xy} does not contain any diagonal of D_{xy} it is just a Hamiltonian path on the circuit link(xy,P), and so its endpoints are neighbored in link(xy,P). By Lemma 1, a) H_{yx} has the same endpoints as H_{xy}. Thus H_{yx} may not contain any diagonal of D_{yx}, and we have $H_{xy} = H_{yx}$ contradicting Lemma 1, b) and c).

If H_{xy} contains two diagonals of D_{xy} its endpoints are the two vertices x_1 and x_{n-3} neighbored to z in link(xy,P). By Lemma 1, a) these are also the endpoints of H_{yx}. z is a universal vertex of D_{xy}, and so it is a 2-valent vertex of D_{yx} since D_{xy} and D_{yx} do not have diagonals in common. It follows that there is no Hamiltonian path H_{yx} from x_1 to x_{n-3} on D_{yx}.

PROOF OF THEOREM 1.

a) As mentioned in the introduction M is neighborly if and only if the quotient on the right hand side of equation (*) is integer. Together with $n > 7$ this yields $n \geq 12$.

Let us assume that P has three consecutive u.e.'s wx, xy, yz. By Remark 1 of section 2 link(x,P) and link(y,P) are cyclic. Now by Lemma 2 H_{xy} contains exactly one diagonal of D_{xy}, and H_{yx} contains exactly one diagonal of D_{yx}. So, $H_{xy} \cap$ link(xy,P) consists of two disjoint paths covering all the vertices of link(xy,P), and the same applies to $H_{yx} \cap$ link(xy,P). Since link(xy,P) is a circuit on $n-2 \geq 10$ vertices, this clearly contradicts Lemma 1, c).

b) link(x,P) is stacked, because P is a polytope [2] (or equivalently because it has the universal vertex y). So it has at least two 3-valent vertices. The 3-valent vertices of link(x,P) are exactly the 2-valent vertices of D_{xy}.

z is a universal vertex of D_{yx}, and since D_{xy} and D_{yx} do not have diagonals in common it is a 2-valent vertex of D_{xy}. Its neighbors a,b on link(xy,P) are joined by a diagonal of D_{xy}. By Lemma 2 H_{yx} contains exactly one diagonal zc of D_{yx}. So we may assume that one of the endpoints of H_{yx} is a while the other is d neighbored to c on link(xy,P).

Two 2-valent vertices of D_{xy} are not joined by an edge of D_{xy}. So if there would be two such vertices z_1, z_2 different from z they also would be different from a,b and at least one of them z_1 say would not be in {c,d}. Since H_{xy} passes through z_1 by two consecutive edges of link(xy,P) this contradicts Lemma 1, c).

In the proof of Theorem 2 we need another Lemma:

LEMMA 3. *If* xy *and* yz *are u.e.'s of* P *then* H_{yx} *is uniquely determined by* H_{yz} *(and vice versa), the relation of* H_{yx} *and* H_{yz} *being*

$$H_{yx} \cap \text{link}(xy,P) = H_{yz} \cap \text{link}(yz,P),$$

and a *being an endpoint of* H_{yx} *if and only if* ax *is an edge of* H_{yz}.

PROOF. link(y,P) is cyclic (Remark 1 of section 2) with universal vertices x,y, and $H_{yx} \cup H_{yz} = \text{link}(y,M)$ is a Hamiltonian circuit on link(y,P). From this the above assertions are following immediately.

PROOF OF THEOREM 2. The proof of Theorem 2 involves the consideration of a lot of subcases being not very difficult and rather tedious at the same time. We did part of the work by hand, and another part by computer. We shall only outline the proof leaving the veryifica-tion of some of the details to the reader.

Let xy and yz be u.e's of P (see Remark 2 of section 2), and let 1,...,9 be the other vertices of P according to their order on link(xy,P) (or equivalently on link(yz,P)). link(y,P) is cyclic with universal vertices x,z. So, z is a universal vertex of D_{yx}, and since D_{yx} and D_{xy} do not have diagonals in common z is a 2-valent vertex of D_{xy}.

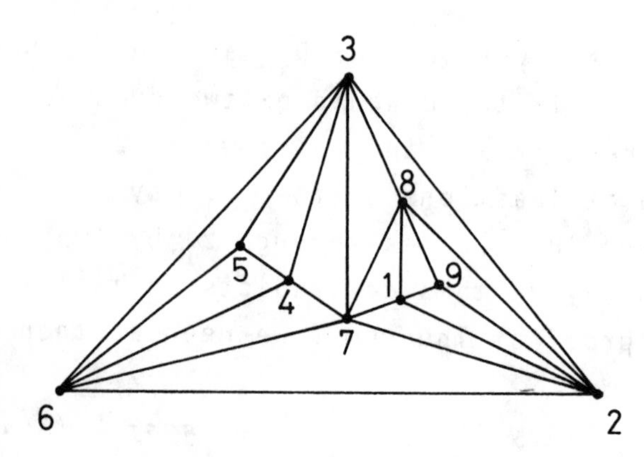

Figure 1

From this it follows that $19z \in D_{xy}$, and by the same argument one gets $19x \in D_{zy}$. Furthermore we observe that the 3-ball $st(x,P) \cup st(y,P) \cup st(z,P)$ (Remark 3 of section 2) is bounded by the 2-sphere $S = D_{xy} \cup D_{zy} - \{1x,9x,19x,1z,9z,19z\}$. Now using Lemmas 1 to 3 and Remark 3 of section 2 all possible quadruples $(H_{xy},H_{yx},H_{yz},H_{zy})$ may be listed. There exists exactly one such quadruple (up to obvious symmetries that just cause a change in notation), and moreover the disks D_{xy} and D_{zy} (and thereby the sphere S) are uniquely determined. Figure 1 shows S, the Hamiltonian circuit 1,...,9 separates the two disks $D_{xy} - \{1z, 9z, 19z\}$ and $D_{zy} - \{1x, 9x, 19x\}$.

The vertices of link(x,M), link(y,M), link(z,M) are (in cyclic order):

link(x,M): 1,y,5,4,6,7,3,8,2,9,z
link(y,M): 1,2,3,4,z,9,8,7,6,5,x
link(z,M): 1,x,9,y,4,5,3,6,2,7,8 .

This determines the subcomplex M' of M spanned by all triangles incident to x,y or z. M' has 27 triangles, and since (by Euler's Formula) M has 44 triangles there are 17 triangles left. By computer all complexes M_i spanned by 17 triangles with vertices from 1,...,9 such that $M = M' \cup M_i$ is a closed pseudomanifold were listed. There are 46 of them, and in exactly three cases none of the vertices of M is singular, i.e. M is a triangulated 2-manifold. The triangles of these M_i are:

M_1: 235,126,368,139,169,147,157,579,247,248,259,134, 379,158,469,489,568
M_2: 239,246,348,159,169,147,479,579,124,257,258,135, 137,369,168,489,568
M_3: 239,256,358,149,169,157,479,579,247,248,125,134, 137,369,168,468,589

At this point all possibilities for M and its position relative to the 2-sphere S of Figure 1 are determined. S has just one non-trivial automorphism induced by the permutation $p = (14)(23)(59)(68)$ of the vertices. So if there is a sewn 4-polytope P on 12 vertices admitting a polyhedron M with $\text{skel}_1 P \subset M \subset \text{skel}_2 P$ then $\text{skel}_2 P$

contains S, and one of the three complexes M_1, M_2, M_3 or an image of it under p. For the (neighborly) 4-polytope $Q = \mathrm{conv}(\mathrm{vert}\ P\setminus\{x,z\})$ we have $\mathrm{link}(y,Q) = S$, and $\mathrm{ast}(y,Q) = \{F \in \mathscr{B}(P) : F \cap \{x,y,z\} = \emptyset\}$. Thus ast(y,Q) still contains one of the M_i or an image of it under p.

According to Altshuler [2], Bokowski/Grams [3] and Bokowski/Sturmfels [4] there are exactly 334 neighborly 4-polytopes on 10 vertices. The 2-sphere S (it is the link of type 50 in Altshuler's notation) occurs exactly 35 times as link of a vertex x, and each time ast(x,Q) does not contain one of the complexes M_i, $p(M_i)$, $i = 1,2,3$. (In fact at least 5 triangles are always missing). This completes the proof of Theorem 2.

REFERENCES

[1] A. ALTSHULER, *Polyhedral Realizations in R^3 of Triangulations of the Torus and 2-Manifolds in Cyclic 4-Polytopes*, Discrete Math. 1(1971), 211-238.

[2] A. ALTSHULER, *Neighborly 4-Polytopes and Neighborly Combinatorial 3-Manifolds with 10 Vertices*, Can. J. Math. 29(1977), 400-420.

[3] J. BOKOWSKI and K. GARMS, *Altshuler's sphere M^{10}_{425} is not polytopal*, Preprint Darmstadt 1984.

[4] J. BOKOWSKI and B. STURMFELS, *Polytopal and non-polytopal spheres - an algorithmic approach*, Preprint Darmstadt 1985.

[5] U. BREHM, *Polyeder mit zehn Ecken vom Geschlecht drei*, Geometriae Dedicata 11(1981), 119-124.

[6] A. CSÁSZÁR, *A polyhedron without diagonals*, Acta Sci. Math. Szeged, 13(1949/50), 140.

[7] B. GRÜNBAUM, *Convex Polytopes*, Interscience Publishers, New York, 1967.

[8] P. McMULLEN, Ch. SCHULZ and J.M. WILLS, *Polyhedral 2-manifolds in E^3 with unusually large genus*, Israel J. Math. 46(1983), 12-144.

[9] A.F. MÖBIUS, Gesammelte Werke Bd. 2, 551-555.

[10] G. RINGEL, *Map Color Theorem*, Springer-Verlag, Berlin/Heidelberg/New York 1974.

[11] G. RINGEL and M. JUNGERMANN, *Minimale Dreieckszerlegungen geschlossener orientierbarer Flächen*, Tagungsbericht Oberwolfach 32/1977

[12] Ch. SCHULZ, *Construction of neighborly spheres* (in preparation)

[13] I. SHEMER, *Neighborly Polytopes*, Israel J. Math. 43(1982), 291-314.

Ch. SCHULZ
Fernuniversität (ZFE)
Postfach 940
D-5800 Hagen
Bundesrepublik Deutschland

COLLOQUIA MATHEMATICA SOCIETATIS JÁNOS BOLYAI
48. INTUITIVE GEOMETRY, SIÓFOK, 1985.

A STAR POLYHEDRON THAT TILES BUT NOT AS A FUNDAMENTAL DOMAIN

S. SZABÓ

To Prof. L. Fejes Tòth on the occasion of his 70th birthday

ABSTRACT. We shall construct a 3-dimensional centrally-symmetric star polyhedron whose translates tile 3-dimensional space but whose translates by a lattice do not tile it. In addition we will get a simple star polyhedron whose translates tile 3-dimensional space but whose congruent copies by a group of motions do not tile it. In particular, there is no lattice tiling by translates of this polyhedron.

1. THE BACKROUND. A system of congruent copies of an n-dimensional polyhedron, whose union is n-space and whose interiors are disjoint, is a *tiling*. D. Hilbert asked in the second part of his 18th problem: if congruent copies of a polyhedron P form a tiling whether has to be a group of motions whose fundamental domain is just P, or in other words whether then congruent copies of P by at least one group of motions can be a tiling.

In 1928 K. Reinhard constructed a 3-dimensional polyhedron which shows that the answer is "no". The answer remains "no" even if we restrict the polyhedron to be a star polyhedron, since J. Milnor gave a 2-dimensional star example.

To get sharper examples we may restrict the motions to the translations. In 1972 S.K. Stein constructed a 10-dimensional centrally-symmetric star example and he used only the translates of this polyhedron. In 1983 using his method [7] constructed a 5-dimensional centrally-symmetric star example. And now containing this investigation and using the same ideas we shall construct a 3-dimensional centrally-symmetric polyhedron whose translates tile space but translates by a lattice do not tile it. A slight modification of this method provides a 3-dimensional star polyhedron whose translates tile space but congruent copies by a group of motions do not tile it.

2. THE THREE DIMENSIONAL CASE

Since the 3-dimensional case seems to be the most interesting and since this is the intuitive base of the higher dimensional constructions we will start with it.

THEOREM 1. *There exists a centrally-symmetric star polyhedron whose translates tile space but translates by a lattice do not tile it.*

PROOF. To prove this statement we will divide space into centrally-symmetric star polyhedra which are translates of each other. If the translations do not form a group, or in other words a lattice, and we are able to force congruent copies of our polyhedron to tile space only in the previous way, then the proof will be complete.

1. Let e_1, e_2, e_3 be a fixed orthonormal basis in space. The union of translates of unit cube whose edges are parallel to e_1, e_2, e_3 respectively and whose centers are $-e_3$, $-e_2$, $-e_1$, 0, e_1, e_2, e_3 and 0, e_1, e_2, e_3 we call a *cross* and a *semicross* respectively (see Figure 1).

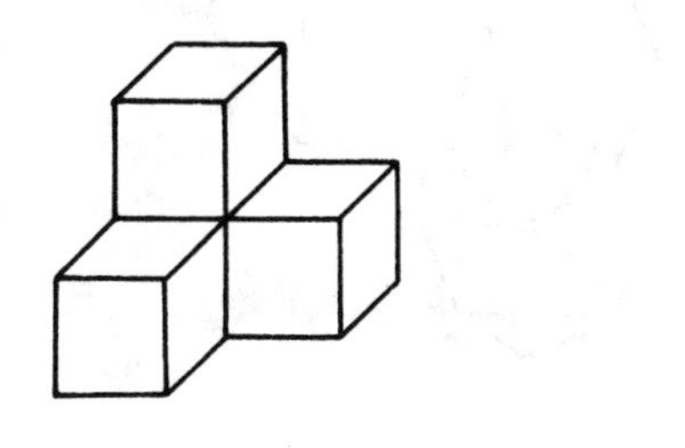
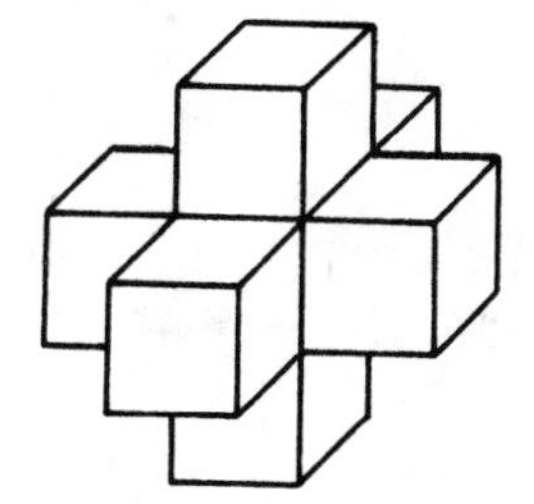

Figure 1

Consider the translates of a semicross by the vectors of the lattice L" spanned by $2e_1$, $2e_2$, $2e_3$. Since an end of an arm of a semicross always meets a face of a corner cube of another semicross, their union consists of infinitly many prisms shaped bars. For the sake of brevity call this structure a *semicross skeleton*. Translates of a semicross skeleton by the vectors 0 and $e_1 + e_2 + e_3$ tile space. Colour the corner cubes of the semicrosses of a semicross skeleton black and white like a chess board or more formally colour the corner cubes either black or white depending on whether the sum of its coordinates is a multiple of 4 or not. Place the central cube of a cross at every black corner cube. Note that the ends of the arms of the crosses always abut a white corner cube.

Divide every white cube into six pyramids whose common vertex is the centroid of the cube. We distribute these pyramids among the six crosses that abut the white cube, obtaining a centrally-symmetric star polyhedron whose translates tile space (see Figure 2, 3).

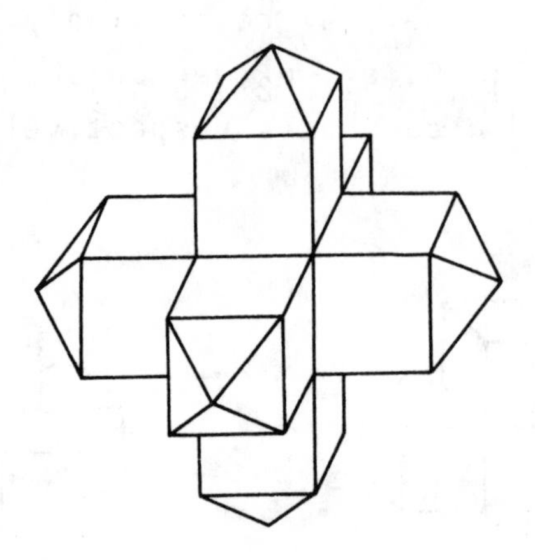

Figure 2

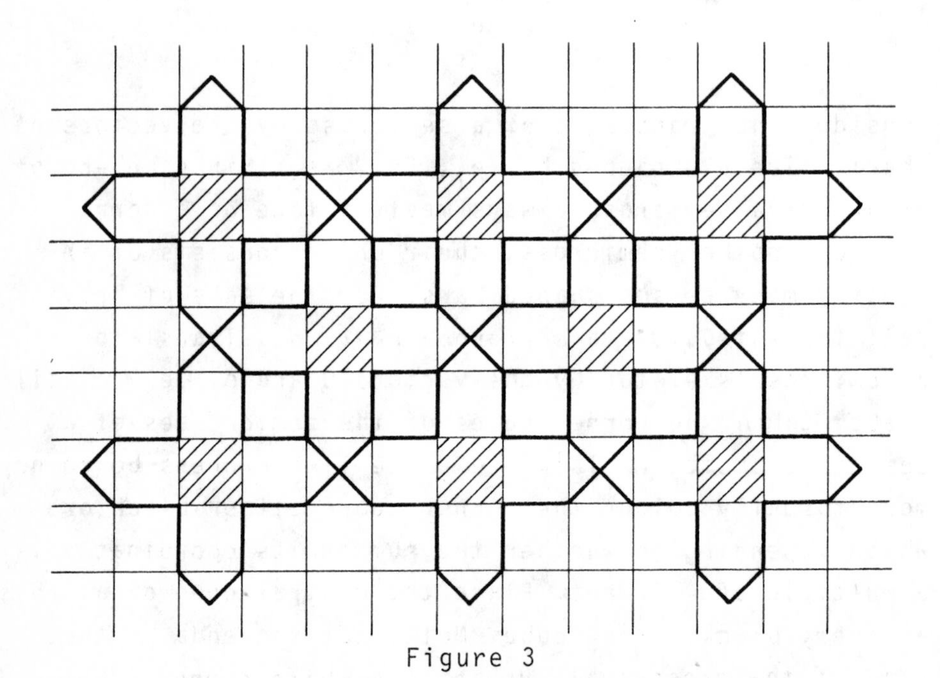

Figure 3

This tiling cannot be lattice like. Indeed, denote by L the translations of the crosses. If L were a lattice, then, since it contains $e_1 + e_2 + e_3$, it would contain $2(e_1 + e_2 + e_3)$. But $2(e_1 + e_2 + e_3)$ is not in L.

2. We would like to prove that if the congruent copies of our polyhedron tile space then they have to form two semicross skeletons. Since they are determined by each other this will complete the proof.

Divide the points of the surface of our polyhedron into 8 classes in the following way. The vertices of the central cube of the polyhedron belong to the 6th class; the internal points of its edges belong to the 8th class. The vertices at the ends of the arms belong to the 1st class; and the internal points of the edges meeting at these points belong to the 4th class. The vertices on the basis of the pyramids belong to the 2nd class and internal points of the edges of the bases belong to the 5th class. The internal points of the remaining edges belong to 3rd class and finally the internal points of the faces belong to the 7th class.

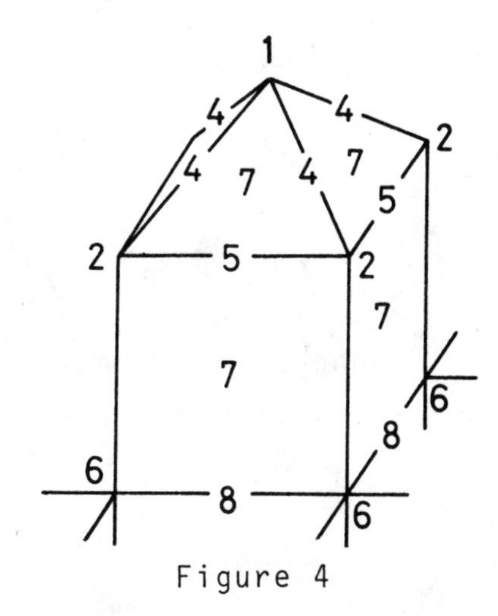

Figure 4

Take a point from a class and a small ball whose center is the point and see how large a part of the ball is covered by our polyhedron. The following table gives us the corresponding values and shows the type of that point: an edge point, a vertex, or a face point.

type	v	v	e	e	e	v	f	e
class	1	2	3	4	5	6	7	8
covered part	$\frac{1}{6}$	$\frac{1}{6}$	$\frac{1}{4}$	$\frac{1}{3}$	$\frac{3}{8}$	$\frac{1}{2}$	$\frac{1}{2}$	$\frac{3}{4}$

Consider a tiling by congruent copies of our polyhedron and take a point from the surface of one of the copies. Let $x_1,\ldots,x_8$ be the numbers of the different type of points which meet at this point. Since the small ball around point has to be covered,

$$x_1/6+x_2/6+x_3/4+x_4/3+3x_5/8+x_6/2+x_7/2+3x_8/4 = 1 ,$$

that is,

$$4x_1 + 4x_2 + 6x_3 + 8x_4 + 9x_5 + 12x_6 + 12x_7 + 18x_8 = 24 . \tag{1}$$

We are interested in nonnegative integer solutions of this equation. Since every term is even x_5 is even.

We shall obtain two consequences of (1).

(i) An 8th type edge has to meet with a 3rd type edge.

(ii) A 5th type edge has to meet with a 5th type and a 3rd type edge.

The proof is the following. If $x_8 \geq 1$, then since $x_8 > 1$ is impossible, $x_8 = 1$ and we have

$$4x_1 + 4x_2 + 6x_3 + 8x_4 + 9x_5 + 12x_6 + 12x_7 = 6 .$$

Now x_4, x_5, x_6, x_7 have to be zero and so $4x_1 + 4x_2 + 6x_3 = 6$. We can see that the only solution in nonnegative integers of this last equation is $x_1 = x_2 = 0$ and $x_3 = 1$.

If $x_5 \geq 1$, then since $x_5 \leq 2$ and is even $x_5 = 2$; hence (1) becomes

$$4x_1 + 4x_2 + 6x_3 + 8x_4 + 12x_6 + 12x_7 + 18x_8 = 6 .$$

Thus $x_4 = x_6 = x_7 = x_8 = 0$ and so $4x_1 + 4x_2 + 6x_3 = 6$ of which the only solution in nonnegative integers is $x_1 = x_2 = 0$ and $x_3 = 1$.

Try to construct a tiling by congruent copies of our polyhedron. Consider the first element of this tiling and the edges of its central cube (see Figure 5). According to (i) each free cube face of the arms has to meet an entire face of another cube. So locally the tiling appears in as figure 6. Consider the edges of the base the pyramid and apply (ii) , there will be a part of the tiling as figure 7. Let P be the common vertex of the five ends of arms. Since each member of the tiling has only finitely many neighbours and each of them has only

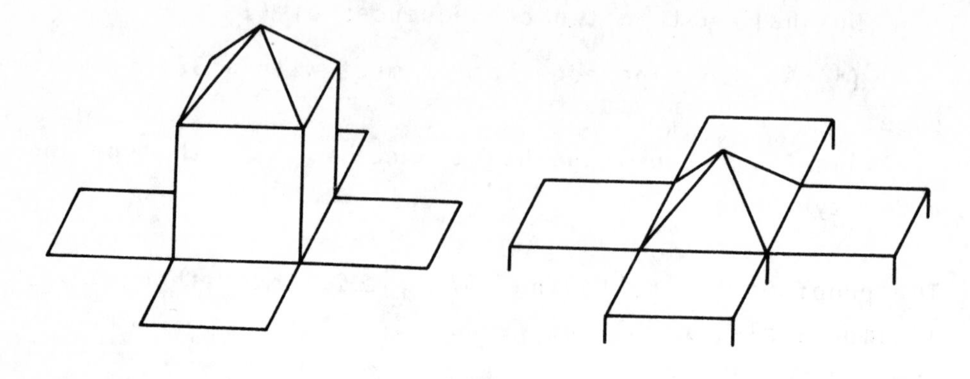

Figure 5

Figure 6

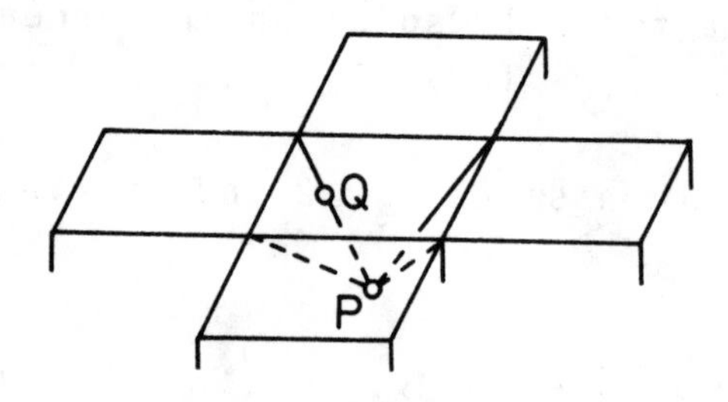

Figure 7

finitely many edges and vertices there is a point Q which is the closest to P in the edges joining to P and which is either a vertex of another polyhedron or a crossing point of the edges. Then consider an internal points of the section PQ and a small ball whose center is this point. Since our point is not a vertex of any polyhedron of the tiling x_1, x_2, x_6 have to be zero. It is clear that $x_4 \geq 2$. In this case equation (1) becomes

$$6x_3 + 8x_4 + 9x_5 + 12x_7 + 18x_8 = 24 \ .$$

We can see that x_4 is a multiple of 3 and $x_4 \leq 3$, that is, $x_4 = 3$ and the other x_i are zero.

Summing up our information on the tiling we get that the pyramids have to form cubes in every tiling and so the polyhedra have to form two semicross skeletons. This completes the proof.

REMARK

Translates of a modified cross tile 3-space but not by a group of translation. However, as S.K. Stein has pointed out, they do tile 3-space by a group of motions. Namely, let T_1, T_2, T_3, T_4 be translations given by $4e_1$, $2e_1 + 2e_2$, $2e_1 + 2e_3$, $e_1 + e_2 + e_3$ and let R be the reflection with respect to the origin. Then the modified cross is a fundamental domain of the group of motions generated by T_1, T_2, T_3, RT_4.

THEOREM 2. *There exists a star-polyhedron whose translates tile space but whose congruent copies by a group of motions do not tile it.*

PROOF. Consider again the semicross skeleton defined with the white cubes in the previous proof. Divide a white cube into six pyramids whose common vertex is *not* the centroid of the cube then translate this divided white cube to the other white cubes. Distribute these pyramids among the six crosses that abut the white cube, obtaining a star polyhedron whose translates tile space.

We can force that if the congruent copies of this polyhedron tile space then they must form two semicross skeletons. Note that the only symmetry of this polyhedron is the identity.

In this way conclude the translations of the polyhedra have to form a group which is impossible as we have already seen in the previous proof.

3. THE HIGHER DIMENSIONAL CASES

Using group theoretical considerations we will generalise the results of the previous section.

THEOREM 3. *If* $N = 2^t - 1$ *and* $t > 1$, *then there exists a centrally-symmetric star polyhedra whose translates tile* n-*space but whose translates by a lattice do not tile it.*

PROOF. Let $e_1, \ldots, e_n$ be a fixed orthonormal basis in R^n. The union of translates of an n-dimensional unit cube whose edges parallel to $e_1, \ldots, e_n$ respectively and whose centers are $-e_n, \ldots, -e_1, 0, e_1, \ldots, e_n$ and $0, e_1, \ldots, e_n$ we will call an n-dimensional *cross* and *semicross* respectively. A cross consists of a central cube and 2n arms; a semicross consists of a corner cube and n arms.

At first we construct a lattice semicross tiling. Let G be an abelian groups written additively of order $n + 1 = 2^t$ with generator elements $g_1, \ldots, g_t$ of order 2. Let L be the lattice spanned by the vector $e_1, \ldots, e_n$. Obviously, L is an abelian group under vector addition. Then consider the homomorphism f from L onto G defined by

$$f(x_1e_1 + \dots + x_ne_n) = x_1g_1 + \dots + x_ng_n ,$$

where $g_1,\dots,g_n$ are the nonzero elements of G and $x_1,\dots,x_n$ are integers. If L' is the kernel of f, then according to the homomorphism theorem every element of L uniquely expressible in the form $\ell = \ell' + x$, where $x \in \{0, e_1,\dots,e_n\}$. This means that the translates of a semicross by the elements of L' tile n-space. The vectors $2e_1,\dots,2e_n$ are elements of L' since $f(2e_i) = 2f(e_i) = 2g_i = 0$ because the orders of the nonzero elements of G are 2. Since the lattice L" spanned by the vectors $2e_1,\dots,2e_n$ is a sublattice of L' if follows that L' is a union of disjoint translates of L". The translates of a semicross by the elements of L" is a semicross skeleton. Colour its corner cubes black and white as in the preceding section. Then place the central cubes of the crosses at the black cubes. Between the ends of the arms of the crosses will be a white cube. We divide it into 2n pyramids whose common vertex is the centroid of the white cube. If we distribute the pyramids among the crosses we get a centrally-symmetric polyhedron whose translates tile n-space. For the sake of brevity we will modify this polyhedron in such a way that:

(i) The modified polyhedron remains centrally-symmetric starbody;

(ii) In every tiling by congruent copies of the modified polyhedron the polyhedra have to form semicross skeletons and these skeletons have to joint each other along entire (n-1)-dimensional faces of their cubes.

Using the benefits of the modification we prove that the congruent copies of a modified polyhedron by a group of motions cannot tile R^n. Assume the contrary: that it can and let L' be the set of the translations of the modified polyhedron. They are assumed to form a group or in other words a lattice. In virtue of (ii) each element $\ell \in L$ is uniquely expressible in the form $\ell = \ell' + x$, where $\ell' \in L'$ and $x \in \{-e_n, \ldots, -e_1, 0, e_1, 2e_1, e_2, \ldots, e_n\}$. Since L' is a subgroup of L we may consider the factor group $H = L/L'$. We have H is partitionable in the form

$$H = \{-h_n, \ldots, -h_1, 0, h_1, 2h_1, h_2, \ldots, h_n\},$$

where h_i is the coset containing the element e_i. According to (ii) $4e_1, \ldots, 4e_n \in L'$ so from it we have $4h_i = 0$, that is $|h_i| | 4$, and since from $h_i \neq -h_i$ we already know that $|h_i| > 2$, hence $|h_i| = 4$. Since $|H| = 2n+1+1 = 2^{t+1}$, H is a direct sum of cyclic groups of order of a power of 2. But H has at most one element of order two and so it has to be cyclic. Then the order of the generator element has to be four as well, so $|H| = 2^{t+1} = 4$, that is, $t = 1$ a contradiction.

The proof will be complete if we give the alteration. We divided the white corner cube into 2n n-dimensional pyramids. This figure is centrally-symmetric with respect to the centroid of the cube. Take an (n-1)-dimensional face which is a common border of two pyramids and place on it a small shallow n-dimensional pyramid. Throwing away the base of the small pyramid, we get a new modified face. Then reflect it with respect to the centre of the cube to get another new modified face. Repeat this process n times then translate this modified white cube to all of the others. (Figure 8 illustrated this in 2-dimension.)

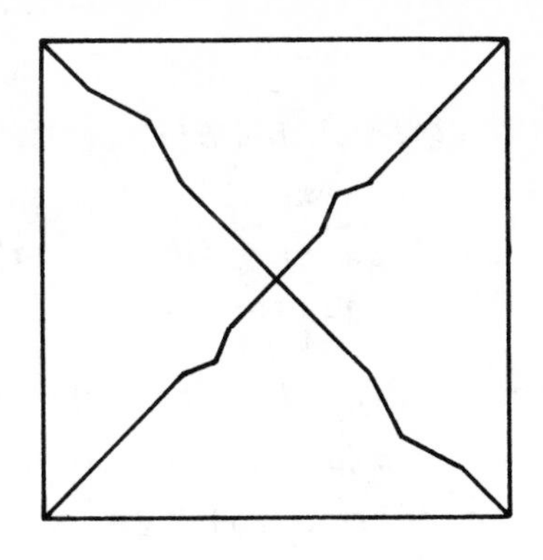

Figure 8

We can see that the modified polyhedra have to meet in a tiling in such a way that the modified n-dimensional pyramids form a cube. From this we see that the modified polyhedra have to form semicross skeletons. The skeletons are union of cubes but we do not know at this moment whether they have to join each other along entire (n-1)-dimensional cube faces, so we would like to force this by modifying the cubes of the skeletons. A face of a cube is an (n-1)-dimensional cube. Place a small shallow n-dimensional pyramids to a face and reflect it with respect to the center of the face. Throwing away the bases of the small pyramids we get a modified face. Reflect this modified face with respect to the centroid of the cube, we get a modified face pair of an original cube so we have to repeat the whole process n times. And if we have a modified cube then translate it to all of the other cubes. We can see that the modified skeletons have to join each other along entire (n-1)-dimensional cube faces.

ACKNOWLEDGEMENT. The author would like to thank the referee for his valuable suggestions.

REFERENCES

[1] D. HILBERT, *Mathematical problems, Lecture delivered before the International Congress of Mathematicians at Paris in 1980,* Bull. Amer. Math. Soc. 8 (1901-1902), 437-479.

[2] J. MILNOR, *Hilbert's problem 18: On crystalographic groups, fundamental domains, and sphere packing.* Proc. Sympos. Pure Math., Vol. 28. Amer. Math. Soc., Providence, R. I., p. 491-506.

[3] K. REINHARD, *Zur zerlegung der Euklidischen Raume in Kongruente Polytope,* Sitzberichte Preuss. Akad. Wiss. (1928), 150-155.

[4] S.K. STEIN, *A symmetric star body that tiles but not as a lattice,* Proc. Amer. Math. Soc. 36 (1972), 543-548.

[5] S.K. STEIN, *Tiling space by congruent polyhedra,* Bull. Amer. Math. Soc. 80 (1974), 819-820.

[6] S.K. STEIN, *Algebraic tiling,* Amer. Math. Monthly 81 (1974), 445-452.

[7] S. SZABÓ, *A symmetric star polyhedra that tiles but not as a fundamental domain,* Proc. Amer. Math. Soc. 89 (1983), 563-566.

S. SZABÓ
Dept. of Civil Engineering Math.
Techn. Univ. Budapest,
H-1111. Budapest, Stoczek u. 2.

COLLOQUIA MATHEMATICA SOCIETATIS JÁNOS BOLYAI
48. INTUITIVE GEOMETRY, SIÓFOK, 1985

COVERING THE SPHERE WITH EQUAL CIRCLES

T. TARNAI and Zs. GÁSPÁR

To Prof. L. Fejes Tóth on the occasion of his 70th birthday

How must the covering of a sphere be formed by n equal circles (spherical caps) so that the angular radius r_n of the circles will be as small as possible? The solution of this problem and also the extremal density of covering are known only for some values of n (n = 2,3,4,5,6,7,10,12,14). For some values of n there are only estimations of the extremal density (n=8,9,16,32) [3]. Lower bounds can be given, e.g., by Fejes Tóth's formula [3], but upper bounds can be most appropriately given by constructions. In the paper spherical circle-covering constructions are presented for n = 11 and 13.

The covering problem is discussed in such a way that to an arrangement of the circles a graph is associated. The graph contains two kinds of vertices. The vertices of the first kind are the centres of the spherical circles and the vertices of the second kind are the boundary points of the circles in which the spherical point is just covered (i.e. is not covered by the interiors of the circles). (In the figures, the vertices of the first kind will be marked by small circles but the vertices of the second kind will not have any special mark.) The edges of the graph are the shorter great circle arcs joining the centres and the boundary points of the circles just covered. Thus, all the edges of the graph are of equal length.

In our investigations the graph is modelled as a structure consisting of equal straight bars and frictionless pin joints lying on the surface of the sphere. In order to obtain a local minimum of the angular radius of the circles we decrease the temperature of the bars uniformly. Under the influence of a decrease in the temperature the bar-lengths can simultaneously decrease in the same proportion without inner forces. The graph moves until, at a certain temperature, the graph will tighten on the sphere. The graph becomes rigid but remains infinitesimally non-rigid. In this position the graph can be in a state of self-stress, but only tension may arise in the bars. Thus if there are bars in compression then those bars should be removed. For the modified structure we repeat the cooling process and removal of compressed bars until each bar will be in tension indicating that the length of the edges of the graph, that is, the radius of the circles has a local minimum. The details and the mechanical background of this method may be found in [4]. Now we apply the method for covering with 11 and 13 equal circles.

For $n = 12$ the solution is due to L. Fejes Tóth [2]. The centres of the circles form a regular icosahedron and the angular radius of the circles is $r_{12} = 37^{\circ}22'38.5''$. The graph of the arrangement of the circles can be seen in a simplified stereographic projection (Schlegel diagram) in Fig.1. The coverings for both 11 and 13 circles will be derived from the best covering with 12 circles.

$n = 11$. Consider the graph in Fig.1. Remove the second-kind vertices C and D together with the edges joining at these vertices, and shift the first-kind vertices A and B into coincidence under a proper increase in the edge-lengths, then apply the cooling process. So we obtain a covering with 11 circles, graph of which can be seen in Fig.2. However, for this arrangement $r_{11} > r_{10}$ where the angular radius of the circles of the best covering with 10 circles is $r_{10} = 42^{0}18'28.2''$ [1]. Fortunately, the graph contains some compressed bars. (The compressed bars are drawn by dashed lines in Fig.2 and also in all of the forthcoming pictures.) By removing the compressed bars and applying again the cooling process we can improve the arrangement. In this way we arrive at the configuration in Fig.3, where $r_{11} = 41^{0}30'47.0''$. Since in Fig. 3 $\alpha > 180^{0}$, this graph also contains compressed bars. By removing them and repeating the cooling process we obtain a locally extremal arrangement whose graph can be seen in Fig.4. The minimum of the angular radius of the circles was obtained at the angle $\beta \approx 153^{0}15'$ and its value is

$$r_{11} = 41^{0}29'28.0''.$$

The density of the covering (defined as the ratio of the total area of the surface of the spherical caps to the surface area of the sphere: $D_n = (1-\cos r_n)n/2$) is $D_{11} = 1.380$. We have also computed a lower bound for the extremal density by the formula of Fejes Tóth [3] and obtained: $D_{11}^{low} = 1.235$.

$n = 13$. Consider again the graph in Fig.1. Remove the edges CB and BD, and place an additional first-kind vertex between the vertices A and B. If we want to fix this new vertex by edges joining to the adjacent vertices, as seen in Fig. 5, then it needs an increase in the edge-lengths which results in $r_{13} > r_{12}$. Fortunately, by applying the cooling process we shall obtain some compressed bars in the graph of Fig.5. Thus by removing these bars and repeating the cooling process we can improve the arrangement so that we shall have $r_{13} < r_{12}$, but we omit the details (steps) of the improving procedure. After all we obtain a locally extremal covering for 13 equal circles whose graph can be seen in Fig.6. The minimum of the angular radius of the circles was obtained at the angle $\gamma = 120^{0}01'04.8''$ and its value is

$$r_{13} = 37^{0}04'06.7''.$$

The density of this covering is $D_{13} = 1.314$, and Fejes Tóth's lower bound for the extremal density is $D_{13}^{low} = 1.230$.

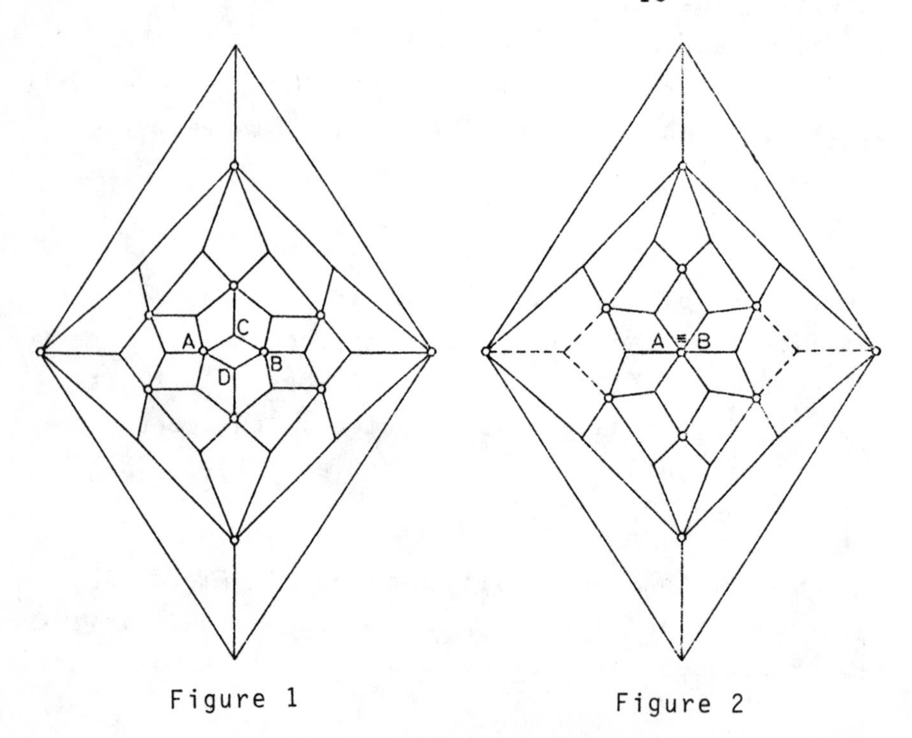

Figure 1 Figure 2

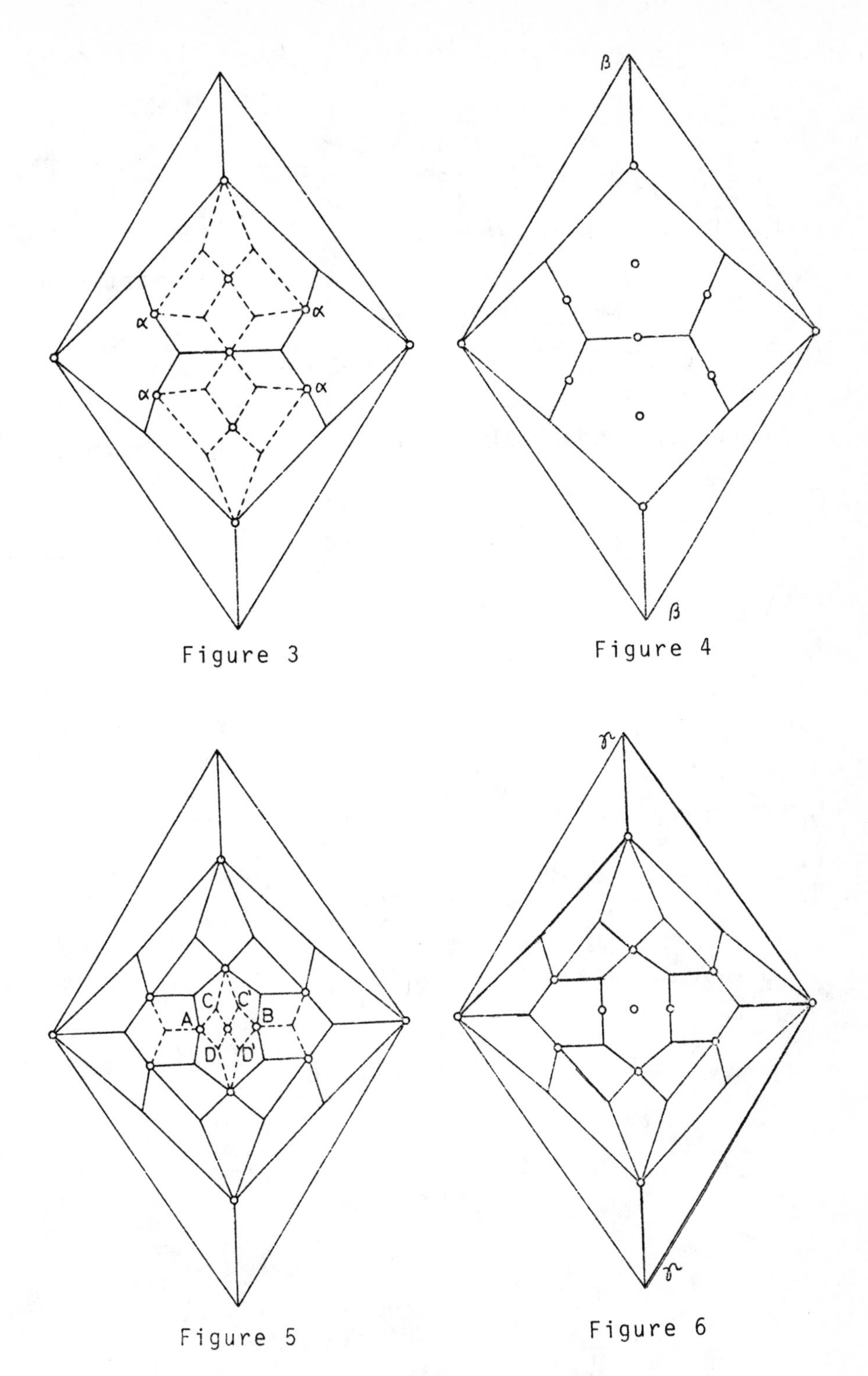

Figure 3

Figure 4

Figure 5

Figure 6

REFERENCES

[1] G. FEJES TÓTH, *Kreisüberdeckungen der Sphäre.* Studia Sci. Math. Hungar. 4(1969), 225-247.

[2] L. FEJES TÓTH, *Covering a spherical surface with equal spherical caps (in Hungarian).* Matematikai és Fizikai Lapok 50 (1943), 40-46.

[3] L. FEJES TÓTH, *Lagerungen in der Ebene auf der Kugel und im Raum.* Zweite Auflage. Springer-Verlag, Berlin, Heidelberg, New York 1972.

[4] T. TARNAI and Zs. GÁSPÁR, *Covering the Sphere with Equal Circles and Rigidity of its Graph.* (prepared for publication)

T. TARNAI
Hungarian Institute for Building Science,
Budapest, Dávid F.u.6., H-1113
Hungary

Zs. GÁSPÁR
Technical University of Budapest,
Department of Civil Engineering Mechanics,
Budapest, Müegyetem rkp.3.
H-1111
Hungary

COLLOQUIA MATHEMATICA SOCIETATIS JÁNOS BOLYAI
48. INTUITIVE GEOMETRY, SIÓFOK, 1985.

EXTENSIONS AND SHARPENINGS OF BRUNN-MINKOWSKI AND BONNESEN INEQUALITIES

B. UHRIN

0. INTRODUCTION

Denote by $A+B$ the algebraic sum of two sets $A,B\subset R^n$. The Brunn-Minkowski inequality asserts that

$$(0.1)\qquad \mu_{n*}(\lambda A + (1-\lambda)B) \geq (\lambda\mu_n(A)^{1/n}+(1-\lambda)\mu_n(B)^{1/n})^n,$$

where $0\leq\lambda\leq 1$, μ_n and μ_{n*} mean Lebesque (L)-measure and inner L-measure, respectively, and $A,B\subset R^n$ are L-measurable sets (see, e.g., [1]).

Bonnesen [1], [2] sharpened (0.1) and showed that

$$(0.2)\qquad \mu_{n*}(\lambda A+(1-\lambda)B) \geq (\lambda m(A)^{\frac{1}{n-1}} + (1-\lambda)\, m(B)^{\frac{1}{n-1}})^{n-1} \cdot \left(\lambda \frac{\mu_n(A)}{m(A)} + (1-\lambda)\frac{\mu_n(B)}{m(B)}\right)$$

where $m(A)$ and $m(B)$ are the maximal measures of sections of A and B by translates of R^{n-1} along the n-th coordinate.

Using the Hölder inequality we see that

$$(\lambda m(A)^{\frac{1}{n-1}} + (1-\lambda)\, m(B)^{\frac{1}{n-1}})^{n-1} \cdot \left(\lambda \frac{\mu_n(A)}{m(A)} + (1-\lambda) \frac{\mu_n(B)}{m(B)}\right) \geq$$

(0.3)

$$\geq (\lambda \mu_n(A)^{\frac{1}{n}} + (1-\lambda)\, \mu_n(B)^{\frac{1}{n}})^n .$$

The aim of this paper is to prove two type extensions of (0.1) and (0.2).

The first one sharpens and extends the inequality (0.2) from "both" sides in the following sense.

The sum $A + B$ is equal to $\{z \in R^n : A \cap (z - B) \neq \emptyset\}$ and define

$$(0.4) \qquad A \boxplus B := \{z \in R^n : \mu_n(A \cap (z - B)) > 0\} .$$

The definition is meaningful also if one of the sets is empty or has measure zero, in this case $A \boxplus B = \emptyset$ (the empty set). It is clear that $A + B = A \boxplus B$ for open sets A, B. The characteristic function of $A \boxplus B$ can be written as

$$(0.5) \qquad \chi_{A \boxplus B}(t) = \operatorname*{ess\text{-}sup}_{x \in R^n} \chi_A(x) \cdot \chi_B(t-x), \quad t \in R^n.$$

Taking in (0.5) "sup" instead of "ess-sup" we get χ_{A+B}. The idea to use "ess-sup" in functions of the above type is due to I. Dancs and had been first studied in [3]. Independently, Brascamp and Lieb [4] also proposed this idea and called $A \boxplus B$ "essential sum" of A and B. It is clear that $A \boxplus B$ remains the same if we change A or B by sets of measure zero. Further, $A \boxplus B$ is L-measurable (see [3], [4]), while $A \boxplus B$ is in general not (see, e.q., [5], [6]). It is also clear that

$$(0.6) \qquad \mu_{n*}(A+B) \geq \mu_n(A \boxplus B) .$$

The second extension of (0.1) gives a lower estimation for $\mu_{n*}(\lambda A \overset{\alpha}{+} (1-\lambda)B)$, where "$\overset{\alpha}{+}$" is a sort of curvilinear combination of two sets. The result gives (0.1) as a special case when the curves are straight lines.

1. A REDUCTION THEOREM

In what follows S is a k-dimensional linear subspace of R^n $(0 \leq k \leq n)$ and $T := S^{\perp}$ its orthogonal complement subspace. For $k = 0$, $S = \Theta$ (the zero vector) and $T = R^n$. By μ_k and μ_{n-k} we denote the L-measures in S and T, respectively. It is clear that $\mu_0(\Theta) = 1$, $\mu_0(\varphi) = 0$.

Given any L-measurable set $A \subset R^n$ the set $A \cap (S+x)$ is in general not L-measurable in S, but it is L-measurable for almost all $x \in T$. Using these sets we have

$$(1.1) \qquad \mu_n(A) = \int_T \mu_k(A \cap (S+x))dx .$$

In our investigations two more quantities related to the function $\mu_k(A \cap (S+x))$ will be useful.

The first one is

$$(1.2) \qquad m_S(A) := \operatorname*{ess\,sup}_{x \in T} \mu_k(A \cap (S+x)) .$$

Clearly $m_{\Theta}(A) = 1$.

The second one is the set of (normalized) upper level sets of the function, i.e.

$$(1.3) \qquad A_S(\xi) := \{x \in T : \mu_k(A \cap (S+x)) \geq \xi \cdot m_S(A)\},\ 0 \leq \xi \leq 1.$$

It is clear that $A_S(\xi) \subset A_S(\xi')$ if $\xi' < \xi$ and $A_\Theta(\xi) = A$, $A_{R^n}(\xi) = \{\Theta\}$ for all $0 \leq \xi \leq 1$.

One can easily see that

$$(1.4) \qquad \mu_n(A) = m_S(A) \cdot \int_0^1 \mu_{n-k}(A_S(\xi))d\xi \ .$$

Before the formulation of our results let us take some simple remarks.

Any L-measurable set can be approximated "from inside" by compact sets, hence when proving (0.1) one can prove it for compact sets. Both sides of (0.1) are invariant under the translations of the sets. For n=1 the proof of (0.1) is quite simple: we can assume that $A,B \subset R^1$ are compact sets such that $A \cap B = 0$ (the zero point) and A lies to the left of 0 and B to the right of 0. In this situation clearly $\lambda A + (1-\lambda)B \supset \lambda A \cup (1-\lambda)B$ and this proves

$$(1.5) \qquad \mu_{1*}(\lambda A + (1-\lambda)B) \geq \lambda\mu_1(A) + (1-\lambda)\ \mu_1(B)$$

for any L-measurable sets $A,B \subset R^1$.

In the proof of our results we shall need the set A^* of so called density points of $A \subset R^1$. A point $x \in R^1$ is a density point of A if

$$(1.6) \qquad \lim_{\delta \to 0+} \mu_1(A \cap [x-\delta,\ x+\delta])/\delta = 2\ .$$

It is clear that $A^* \subset A$ for compact sets A. It is well known that A^* is L-measurable and

$$(1.7) \qquad \mu_1(A^*) = \mu_1(A) .$$

Now, we shall prove the following reduction theorem.

THEOREM 1.1. *For any bounded* L-*measurable sets* $A,B \subset R^n$ *of positive* L-*measure and any* k-*dimensional linear subspace* $S \subset R^n$, $0 \le k \le n$, *we have*

$$(1.8) \qquad \begin{aligned} \mu_n(\lambda A \boxplus (1-\lambda)B) \ge \int_0^1 \mu_{n-k}(\lambda A_S(\xi) \boxplus (1-\lambda)B_S(\xi))d\xi \cdot \\ \cdot (\lambda m_S(A)^{\frac{1}{k}} + (1-\lambda)m_S(B)^{\frac{1}{k}})^k . \end{aligned}$$

It is clear that if, say, $\mu_n(A) = 0$, then (1.8) holds trivially as equality (both sides are zeros). If $k=0$ then (1.8) turns again to an equality.

If (1.8) is assumed to be true for $k=n=1$ then the case $k=n>1$ is a simple consequence of the case $k=n-1$, $n>1$: use first (1.8) ($k=n=1$) for the integrand, apply the identity (1.4) and after that the Hölder inequality (see (0.3)) gives (1.8) for $k=n$.

In the proof we shall need two more (almost trivial) inequalities of type (0.3). Namely, for $a,b \ge 0$, $-\infty < \alpha < +\infty$, $\alpha \ne 0$ denote

$$M(a,b) := \begin{cases} \max\{a,b\} & \text{if} \quad a \cdot b > 0 \\ 0 & \text{if} \quad a \cdot b = 0 , \end{cases}$$

$$M_\alpha^{(\lambda)}(a,b) := \begin{cases} (\lambda a^\alpha + (1-\lambda)b^\alpha)^{1/\alpha} & \text{if} \quad a \cdot b > 0 \\ 0 & \text{if} \quad a \cdot b = 0 . \end{cases}$$

Then for a, b, c, $d \geq 0$ we have

$$(1.9) \qquad M(a,b) \cdot M_\alpha^{(\lambda)}(c,d) \geq M_\alpha^{(\lambda)}(ac,bd)$$

$$(1.10) \qquad M_\alpha^{(\lambda)}(a,b) \cdot M_{-\alpha}^{(\lambda)}(c,d) \geq \min\{ac,bd\}$$

In fact, a more general inequality is true for the product of means (see, e.q., [7], [8]), but we shall need only these trivial cases.

PROOF OF THE THEOREM. First we prove the case k=n=1, i.e. the inequality

$$(1.11) \qquad \mu_1(A \boxplus B) \geq \mu_1(A) + \mu_1(B) .$$

As we have remarked above, we can assume without the loss of generality that $A,B \subset R^1$ are compact. The inequality (1.11) is a simple consequence of (1.5), (1.7) and the relation

$$(1.12) \qquad A \boxplus B \supset A^* + B^* .$$

Let $a \in A^*$, $b \in B^*$. For sufficiently small $\varepsilon > 0$ we have a $\delta > 0$ such that

$$\delta(2-\varepsilon) < \mu_1(A \cap [a-\delta,\ a+\delta]) \leq 2\delta$$

(1.13)

$$\delta(2-\varepsilon) < \mu_1(B \cap [b-\delta,\ b+\delta]) \leq 2\delta \quad .$$

Denote

$$A_1 = (A \cap [a-\delta, a)) - a, \quad A_2 = (A \cap (a, a+\delta]) - a$$

$$B_1 = b - (B \cap (b, b+\delta]), B_2 = b - (B \cap [b-\delta, b)) \quad .$$

Clearly

$$A_1 \cap B_2 = \emptyset, \quad A_2 \cap B_1 = \emptyset \quad .$$

The conditions (1.13) imply that

$$\mu_1((A_1 \cup A_2) \cap (B_1 \cup B_2)) \geq 2\delta(1-\varepsilon) \quad . \tag{1.14}$$

But the left hand side of (1.14) is equal to

$$\mu_1(A_1 \cap B_1) + \mu_1(A_2 \cap B_2) \quad , \tag{1.15}$$

consequently, one of the terms in (1.15) is positive. We can easily see that this implies

$$a + b \in A \boxplus B \quad . \tag{1.16}$$

By this (1.12), hence (1.11) is proved.

Now, we proceed with the proof of (1.8) by induction on the pairs (n,k), $0 \leq k \leq n$, $n = 1,2,\ldots$. Take these pairs into lexicographic order, i.e. $(n_1,k_1) < (n_2,k_2)$ if either $n_1 < n_2$ or $\{n_1 = n_2$ and $k_1 < k_2\}$. We get a sequence

$$(1.17) \qquad (1,0) < (1,1) < (2,0) < (2,1) < (2,2) < \dots$$

For k=n=1 the inequality (1.8) is true by (1.11). Assume that we have already proved (1.8) for all first N-1 members of the sequence (1.17) and let (n,k) be the N-th member. If k=0 we are ready. If n=k then we have already seen how to get the inequality (1.8), n=k, from the cases k=n=1 and n,k=n-1. So assume $1 \le k \le n-1$. Write the left hand side of (1.8) in the form of an integral

$$(1.18) \qquad \mu_n(\lambda A \boxplus (1-\lambda)B) = \int_{R^n} \operatorname*{ess\,sup}_{y \in R^n} M(\chi_A(\tfrac{y}{\lambda}),\ \chi_B(\tfrac{t-y}{1-\lambda}))dt\ .$$

Further, for $C, D \subset S$ and any positive numbers γ, δ we can write

$$(1.19) \qquad \begin{aligned} &\mu_k(\lambda C \boxplus (1-\lambda)D) \cdot (\lambda\gamma^{-1/k} + (1-\lambda)\delta^{-1/k})^{-k} = \\ &= \int_S \operatorname*{ess\,sup}_{u \in S} M^{(\lambda)}_{-1/k}(\gamma \cdot \chi_C(\tfrac{u}{\lambda}), \delta \cdot \chi_D(\tfrac{z-u}{1-\lambda}))dz\,. \end{aligned}$$

Denote $\gamma := m_S(A)^{-1}$, $\delta := m_S(B)^{-1}$. Using (1.18) and (1.9) we have

$$(1.20) \qquad \begin{aligned} &\mu_n(\lambda A \boxplus (1-\lambda)B)(\lambda\gamma^{-1/k} + (1-\lambda)\delta^{-1/k})^{-k} \ge \\ &\ge \int_{R^n} \operatorname*{ess\,sup}_{y \in R^n} M^{(\lambda)}_{-1/k}(\gamma \cdot \chi_A(\tfrac{y}{\lambda}),\ \delta \cdot \chi_B(\tfrac{t-y}{1-\lambda}))dt. \end{aligned}$$

The right hand side of (1.20) is not smaller than

$$(1.21)\qquad \int_T \operatorname{ess\,sup}_{x\in T}\Big(\int_S \operatorname{ess\,sup}_{u\in S} M_{-1/k}^{(\lambda)}\big(\gamma\chi_C(\tfrac{u}{\lambda}),\delta\chi_D(\tfrac{z-u}{1-\lambda})\big)dz\Big)d\tau,$$

where $C := (A - \frac{x}{\lambda}) \cap S$ and $D := (B - \frac{\tau - x}{1-\lambda}) \cap S$.

Using (1.19) and applying the induction hypothesis to $\mu_k(\lambda C \boxplus (1-\lambda)D)$ we can estimate the inner integral of (1.21) from below (in all points x,τ such that $\mu_k(C)$, $\mu_k(D)>0$) by the expression

$$(1.22)\qquad (\lambda\mu_k(C)^{1/k} + (1-\lambda)\mu_k(D)^{1/k})^k \cdot (\lambda\gamma^{-1/k} + (1-\lambda)\delta^{-1/k})^{-k}.$$

Using (1.10) we finaly get that the right hand side of (1.20) is not smaller than

$$(1.23)\qquad \int_T \operatorname{ess\,sup}_{x\in T} \min\left\{\frac{\mu_k((A-\frac{x}{\lambda})\cap S)}{m_S(A)}, \frac{\mu_k((B-\frac{\tau-x}{1-\lambda})\cap S)}{m_S(B)}\right\} d\tau$$

Denote by $h(\tau)$ the integrand in (1.23) and $C(\xi) := \{\tau\in T: h(\tau) \geq \xi\}$.

Then the integral (1.23) is equal to

$$(1.24)\qquad \int_0^1 \mu_{n-k}(C(\xi))d\xi .$$

Let $\tau \in T$ be such that

$$(1.25)\qquad \mu_{n-k}(\lambda A_S(\xi) \cap (\tau - (1-\lambda)B_S(\xi))) > 0 .$$

Then for any $x \in T$ belonging to the intersection in (1.25) we have

$$(1.26) \qquad \min\left\{\frac{\mu_k((A - \frac{x}{\lambda})\cap S)}{m_S(A)}, \frac{\mu_k((B - \frac{\tau-x}{1-\lambda})\cap S)}{m_S(B)}\right\} \geq \xi$$

that implies $h(\tau) \geq \xi$ i.e. $\tau \in C(\xi)$.

This proves that

$$(1.27) \qquad C(\xi) \supset \lambda A_S(\xi) \boxplus (1-\lambda)B_S(\xi)$$

and by this the theorem is proved.

By applying the appropriate case of (1.8) to $\mu_{n-k}(\ldots)$ in the integral of (1.8), we get

COROLLARY 1.1. *Under the assumptions of the Theorem 1.1 we have*

$$(1.28) \qquad \begin{aligned} \mu_n(\lambda A \boxplus (1-\lambda)B) &\geq (\lambda m_S(A)^{1/k} + (1-\lambda)m_S(B)^{1/k})^k \cdot \\ &\cdot (\lambda^{n-k}\frac{\mu_n(A)}{m_S(A)} + (1-\lambda)^{n-k}\frac{\mu_n(B)}{m_S(B)}) . \end{aligned}$$

For $k=n-1$ the inequality (1.28) is an extended form of (0.2).

Using (1.8) more cleverly, we get many other expressions estimating $\mu_n(\lambda A \boxplus (1-\lambda)B)$ from below. One of the hopeful possibilities is to use (1.8) successively for $S_1, S_2, \ldots, S_n$, mutually orthogonal 1-dimensional subspaces. We get a complicated expression on the right hand side of (1.8). The geometric contents of quantities in this expression are not quite clear yet. Let us demonstrate this approach for n=3 where S_1, S_2, S_3 are the usual coordinate axes.

Denote

$$m_1(A) := m_{S_1}(A), \; A_1(\xi) := A_{S_1}(\xi) \subset R^2, \; m_2(A_1(\xi)) := m_{S_2}(A_1(\xi))$$

and similarly for B. We get

$$\mu_3(\lambda A \boxplus (1-\lambda)B) \geq (\lambda m_1(A) + (1-\lambda)\, m_1(B)) \cdot$$

$$(1.29) \qquad \cdot (\lambda^2 \frac{\mu_3(A)}{m_1(A)} + (1-\lambda)^2 \frac{\mu_3(B)}{m_1(B)} + $$

$$+ \lambda(1-\lambda) \int_0^1 \Phi(\xi)\, d\xi) ,$$

where

$$\Phi(\xi) := \mu_2(A_1(\xi)) \frac{m_2(B_1(\xi))}{m_2(A_1(\xi))} + \mu_2(B_1(\xi)) \frac{m_2(A_1(\xi))}{m_2(B_1(\xi))} .$$

The righthand side of (1.29) can be compared with the "Bonnesen" right hand side of (1.28) i.e. when n=3, k=2 and S = {the coordinate plane of vectors (x_2, x_3)}.

Assume for simplicity that $m_S(A) = m_S(B)$ and $m_1(A) = m_1(B)$. Then we get on the right han sides of (1.28) and (1.29)

$$(1.30) \qquad \lambda\mu_3(A) + (1-\lambda)\mu_3(B) ,$$

$$(1.31) \qquad \lambda^2\mu_3(A) + (1-\lambda)^2\mu_3(B) + \lambda(1-\lambda)\, m_1(A) \int_0^1 \Phi(\xi) d\xi .$$

Dividing the both expressions by $\lambda(1-\lambda)\, m_1(A)$ and subtracting (1.30) from (1.31), the comparison boils down to the investigation of the sign of the following integral

$$\int_0^1 \Big(\mu_2(A_1(\xi))\Big(\frac{m_2(B_1(\xi))}{m_2(A_1(\xi))} - 1\Big) +$$

(1.32)

$$+ \mu_2(B_1(\xi))\Big(\frac{m_2(A_1(\xi))}{m_2(B_1(\xi))} - 1\Big)\Big)d\xi \quad .$$

If this integral is positive then (1.29) is "better" than (1.28) and vice versa. If the integral is zero then both right hand sides are equal.

It is clear that (1.29) is in all cases a better result than (1.28) applied for $n=3$, $k=1$, $S=S_1$.

2. A CURVILINEAR EXTENSION OF BRUNN-MINKOWSKI INEQUALITY

In this section all points and sets are assumed to be in R_+^n the non-negative orthant of R^n.

Let $-\infty \leq \alpha_i \leq +\infty$, $i=1,2,\ldots,n$, $x,y \in R_+^n$. We define

$$(2.1) \qquad (\lambda x \overset{\alpha}{+} (1-\lambda)y)_i := (\lambda x_i^{\alpha_i} + (1-\lambda)y_i^{\alpha_i})^{1/\alpha_i}, \qquad i=1,2,\ldots,n,$$

where for $\alpha_i = -\infty,\ 0,\ +\infty$ the expressions on the right hand sides are defined by limits as $\min\{x_i,y_i\}$, $x_i^{\lambda}y_i^{1-\lambda}$, $\max\{x_i,y_i\}$, respectively.

The combination (2.1) of two points is a sort of abstract convex combination that is sometimes studied in the literature. For $\alpha \neq 0, -\infty, +\infty$, the convexity of a set S with respect to (2.1) simply means that the set S^α is convex in the usual sense, i.e.

$$(2.2) \qquad \lambda S^\alpha + (1-\lambda)S^\alpha \subset S^\alpha .$$

Here by definition $S^\alpha := \{x^\alpha : x \in S\}$, where $x^\alpha := (x_1^{\alpha_1}, \ldots, \ldots, x_n^{\alpha_n})$. The relation (2.2) can be written in the form that is already meaningful also for $\alpha=0, -\infty, +\infty$. Namely, for any sets $A, B \subset R_+^n$ define

$$(2.3) \qquad \lambda A \overset{\alpha}{+} (1-\lambda)B := \{\lambda a \overset{\alpha}{+} (1-\lambda)b : a \in A,\ b \in B\}.$$

Then (2.2) means

$$(2.2') \qquad \lambda S \overset{\alpha}{+} (1-\lambda)S \subset S .$$

Using the function M(a,b) defined in the previous section we can write

$$(2.4) \qquad \chi_{\lambda A \overset{\alpha}{+} (1-\lambda)B}(t) = \sup_{\lambda x \overset{\alpha}{+} (1-\lambda)y = t} M(\chi_A(x), \chi_B(y)), \quad t \in R_+^n .$$

We remark that similarly to $\lambda A + (1-\lambda)B$, the set $\lambda A \overset{\alpha}{+} (1-\lambda)B$ is in general not L-measurable.
Now we have

THEOREM 2.1. *Let* $A, B \subset R_+^n$ *be bounded* L-*measurable sets of positive measure,* $0<\lambda<1$ *and* $0 \le \alpha_i \le 1$, $i=1,2,\ldots,n$.

Then

$$(2.5)\qquad \mu_{n*}(\lambda A \overset{\alpha}{+} (1-\lambda)B) \geq (\lambda\mu_n(A)^\gamma + (1-\lambda)\mu_n(B)^\gamma)^{1/\gamma}$$

where γ *is the harmonic mean of* α_i*-s, i.e.* $\gamma := (\sum_1^n \alpha_i^{-1})^{-1}$.

PROOF. The proof goes in two steps.
Assume first that (2.5) has been already proved for rectangular parallelotopes, i.e. the sets of the form

$$A := x + \{z : 0 \leq z_i \leq a_i,\ i=1,2,\dots,n\},\ B := y + \{z : 0 \leq z_i \leq b_i,\ i=1,2,\dots,n\},$$

where $x, y \in R_+^n$, $a_i, b_i > 0$, $i=1,2,\dots,n$.

Let

$$A := \bigcup_{i=1}^{p} A_i,\ B := \bigcup_{j=1}^{q} B_j\ ,$$

where A_i and B_j are parallelotopes of the above form such that A_k meats A_ℓ only on boundary (similarly for B_j). For such sets, (2.5) can be proved by induction on p+q using the usual trick known from a proof of Brunn-Minkowski inequality: we cut both A and B by translates of a hyperplane parallel to some R^{n-1}. Let $A^{(1)}$, $A^{(2)}$ be the parts of A being to the "left" and "right" of the hyperplane, respectively, and similarly $B^{(1)}$, $B^{(2)}$. If p_1 and p_2 (q_1 and q_2) are the numbers of parallelotopes unions of which are equal to and $A^{(1)}$ and $A^{(2)}$ ($B^{(1)}$ and $B^{(2)}$), then clearly

$$(2.6)\qquad p_1 + q_1 \leq p + q - 1,\ p_2 + q_2 \leq p + q - 1\ ,$$

and one can choose the hyperplanes such that

$$\mu_n(A^{(1)})/\mu_n(B^{(1)}) = \mu_n(A^{(2)})/\mu_n(B^{(2)}) = \omega > 0 . \tag{2.7}$$

One can also see easily that

$$\lambda A \overset{\alpha}{+} (1-\lambda)B \supset (\lambda A^{(1)} \overset{\alpha}{+} (1-\lambda)B^{(1)}) \cup (\lambda A^{(2)} \overset{\alpha}{+} (1-\lambda)B^{(2)}) \tag{2.8}$$

and that the intersection of the two sets on the right hand side of (2.8) is of measure zero.
Applying the induction hypothesis and using (2.7) we get (2.5) for A,B.
Now standard density considerations prove (2.5) for any bounded L-measurable sets A,B. The second step of the proof, i.e. that (2.5) is true for parallelotopes is detailed below.

For parallelotopes occuring above the left hand side of (2.5) is equal to

$$\prod_{i=1}^{n} \Big(((\lambda(a_i + x_i)^{\alpha_i} + (1-\lambda)(b_i+y_i)^{\alpha_i})^{1/\alpha_i} - (\lambda x_i^{\alpha_i} + (1-\lambda)y_i^{\alpha_i})^{1/\alpha_i} \Big) . \tag{2.9}$$

We have to prove that for $0 \le \alpha_i \le 1$, $i=1,2,\dots,n$, this expression is not smaller than

$$(\lambda(\prod_{i=1}^{n} a_i)^{\gamma} + (1-\lambda)(\prod_{i=1}^{n} b_i)^{\gamma})^{1/\gamma} , \tag{2.10}$$

where $\gamma = (\Sigma\alpha_i^{-1})^{-1}$.

Using the inverse Minkowski inequality if $0<\alpha_i\leq 1$ and the Hölder inequality if $\alpha_i = 0$, and taking the product of resulting inequalities (for $i=1,2,...,n$), we get that (2.9) is not smaller than

$$\prod_{i=1}^{n} (\lambda a_i^{\alpha_i} + (1-\lambda)b_i^{\alpha_i})^{1/\alpha_i} . \tag{2.11}$$

After that, the proof will be finished by the

LEMMA 2.1. *Let* $a_i,b_i>0$, $i=1,2,...,n$ *and let* α_i *be such numbers that*

$$\left(\sum_{i=1}^{j} \alpha_i^{-1}\right)^{-1} + \alpha_{j+1} > 0, \ j=1,2,...,n-1. \tag{2.12}$$

Then

$$\prod_{i=1}^{n} (\lambda a_i^{\alpha_i} + (1-\lambda)b_i^{\alpha_i})^{1/\alpha_i} \geq \left(\lambda\left(\prod_{i=1}^{n} a_i\right)^{\gamma} + (1-\lambda)\left(\prod_{i=1}^{n} b_i\right)^{\gamma}\right)^{1/\gamma}, \tag{2.13}$$

where

$$\gamma = (\Sigma\alpha_i^{-1})^{-1}.$$

PROOF. Using Hölder and inverse Hölder inequalities we have for $a,b,c,d>0$, $-\infty<\alpha,\beta<+\infty$ such that $\alpha + \beta > 0$

$$(2.14)\qquad \left(\lambda a^{\alpha} + (1-\lambda)b^{\alpha}\right)^{1/\alpha}\left(\lambda c^{\beta} + (1-\lambda)d^{\beta}\right)^{1/\beta} \geq$$

$$\geq \left(\lambda(ac)^{\frac{\alpha\beta}{\alpha+\beta}} + (1-\lambda)(bd)^{\frac{\alpha\beta}{\alpha+\beta}}\right)^{\frac{\alpha+\beta}{\alpha\beta}} .$$

Applying successively (2.14), we get (2.13).

The last lemma indicates that the fact (2.9) is greater than (2.10) might be true for a larger set of α_i than $0\leq\alpha_i\leq 1$. Indeed it can be shown that the relation (2.9) $\geq$ (2.10) is true also for α_i satisfying the conditions

$$(2.15)\qquad \left(\sum_{i=1}^{j} \alpha_i^{-1} - j\right)^{-1} + \frac{\alpha_{j+1}}{1-\alpha_{j+1}} > 0, \; j = 1,\dots n-1,$$

$$\left(\sum_{i=1}^{n} \alpha_i^{-1} - n\right)^{-1} \geq -\frac{1}{n}$$

The proof of this statement is more complicated, it uses an integral inequality extension of (1.28). In interesting problem would be to find some elementary proof analogously to the case $0\leq\alpha_i\leq 1$.

3. REMARKS

1. Given a k-dimensional affine space H (i.e. translate of a k-dimensional linear subspace), the quantity $\mu_k(A \cap H)$ is called in the literature "k-dimensional inner diagonal measure of A with respect to H" ("k-dimensional

inneren Quermasse" in German, see [1]). Integral geometry studies functions of these measures when H is varied (say, integral of $\mu_k(A \cap H)$ over some family of H-s). In our approach the family of H in question is $H = S + u$, $u \in S^{\perp}$ and the function is "ess-sup".
It is clear that Theorem 1.1 is true in the same form if instead of $T := S^{\perp}$ we take any subspace T such that $T \oplus S$ (the direct sum) is equal to R^n. The quantity $m_S(A)$ does not depend on T, but $A_S(\xi)$ depends on T, hence using general T, we get more "flexible" lower bounds on the right hand side of (1.8).

2. Both Brunn-Minkowski and Bonnesen inequalities have been intensively studied by many authors. One of the most important contributions is the paper of Henstock and Macbeath [9]. The connection of these inequalities with the general theory of "sum-sets" in locally compact groups has been first studied by Kneser [11]. The idea of using density points goes back also to Henstock and Macbeath [9], see also [4], [10].

3. The serious disadvantage of the curvilinear result (2.5) is that both sets should be in R^n_+. It is clear that we could choose any of 2^n orthants of R^n instead of R^n_+. In these cases use the following modification of (2.1):

$$(\lambda x \overset{\alpha}{+} (1-\lambda)y)_i := \begin{cases} (\lambda x_i^{\alpha_i} + (1-\lambda)y_i^{\alpha_i})^{1/\alpha_i} & \text{if } x_i, y_i \geq 0 \\ -(\lambda|x_i|^{\alpha_i} + (1-\lambda)|y_i|^{\alpha_i})^{1/\alpha_i} & \text{if } x_i, y_i \leq 0 . \end{cases}$$

The inequality (2.5) then still holds if A and B are in the same orthant.

The extension of (2.5) to the whole space depends on how to define $(\lambda x \overset{\alpha}{+} (1-\lambda)y)_i$ if say $x_i < 0$, $y_i > 0$. We have tried some plausible ideas (e.q. take one curve up to the boundary of the orthant and after that another curve), but serious technical difficulties arose.

4. In [8] we have proved an integral inequality for arbitrary L-measurable functions f, g of which (1.28) is a special case when $f := \chi_A$, $q := \chi_B$. An interesting problem would be to find an analogous extension of (1.8).

5. The identity (1.4) suggests that the ratio $\Phi_S(A) := \mu_n(A)/m_S(A)$ could behave like the (n-k)-dimensional L-measure. More precisely we can ask:

Under what assumptions is it true that

$$\Phi_S(\lambda A \boxplus (1-\lambda)B) \geq \tag{3.1}$$
$$\geq (\lambda \Phi_S(A)^{\frac{1}{n-k}} + (1-\lambda)\Phi_S(B)^{\frac{1}{n-k}})^{n-k} \quad ?$$

(3.1) for k=0 is the same as (1.8) for k=n, i.e. it is true. For k>0 clearly we have

$$\mu_n(\lambda A \boxplus (1-\lambda)B) \cdot (\lambda m_S(A)^{\frac{1}{k}} + (1-\lambda)m_S(B)^{\frac{1}{k}})^{-k} \geq \tag{3.2}$$
$$\geq \Phi_S(\lambda A \boxplus (1-\lambda)B).$$

If k=n-1 then obviously

$$\int_0^1 \mu_{n-k}(\lambda A_S(\xi) \boxplus (1-\lambda)B_S(\xi))d\xi \geq$$

(3.3)

$$\geq \lambda\Phi_S(A) + (1-\lambda)\Phi_S(B) \quad ,$$

hence in this case (3.1) might be also true. For $0<k\leq n-2$ the comparison of right hand sides of (3.1) and (1.8) is not so clear, we can state only that both expressions are greater than

$$\int_0^1 (\lambda\mu_{n-k}(A_S(\xi))^{\frac{1}{n-k}} + (1-\lambda)\mu_{n-k}(B_S(\xi))^{\frac{1}{n-k}})^{n-k} d\xi. \tag{3.4}$$

REFERENCES

[1] H. HADWIGER, "*Vorlesungen über Inhalt, Oberfläche und Isoperimetrie*", Springer, Berlin, 1957.

[2] T. BONNESEN, W. FENCHEL, "*Theorie der konvexer Körper*", Springer, Berlin, 1934.

[3] B. UHRIN, *On some inequalities of inverse Hölder type having some applications to stochastic programming problems*, Seminar Notes, Mathematics No. 2, Hungarian Committee for System Analysis, Budapest, 1975.

[4] H.J. BRASCAMP, E.H. LIEB, *On extensions of the Brunn-Minkowski and Prékopa-Leindler Theorems, including inequalities for logconcave functions, and with an application to the diffusion equation*, J. Funct. Anal., 22 (1976), 366-389.

[5] W. SIERPINSKI, *Sur la question de la measurabilité de la base de M. Hamel,* Fundamenta Math., 1 (1920), 105-111.

[6] P. ERDŐS, A.H. STONE, *On the sum of two Borel sets,* Proc. Am. Math. Soc., 25 (1970), 304-306.

[7] B. UHRIN, *Some remarks about the convolution of unimodal functions,* Annals of Prob., 12 (1984), 640-645.

[8] B. UHRIN, *Shapenings and extensions of Brunn-Minkowski-Lusternik inequality,* Stanford University, Dept. of Statistics, Techn. Report No. 203, Stanford, CA, November, 1984.

[9] R. HENSTOCK, A.M. MACBEATH, *On the measure of sum sets.* (I) *The theorems of Brunn, Minkowski and Lusternik,* Proc. London Math. Soc., Ser. III., 3 (1953), 182-194.

[10] H. FEDERER, *"Geometric measure theory",* Springer, Berlin, 1969.

[11] M. KNESER, *Summengen in lokalkompakten abelschen Gruppen,* Math. Zeitschr., 66 (1956), 88-110.

B. UHRIN
Computer and Automation Institute
Hungarian Academy of Sciences
1502 Budapest, P.f. 63, Hungary

COLLOQUIA MATHEMATICA SOCIETATIS JÁNOS BOLYAI
48. INTUITIVE GEOMETRY. SIÓFOK, 1985.

AN ISOPERIMETRIC PROBLEM FOR TILINGS

É. VÁSÁRHELYI

Dedicated to Prof. L. Fejes Tóth

Let us decompose a polygon of area A into N parts, say of perimeters $p_1,\dots,p_N$, and consider the supremum $S(q)$ of $A/\sum_{i=1}^{N} p_i^2$ extended over all decompositions of convex polygons with at most six sides, into parts of perimeters such that $\min p_i/\max p_i \geq q$. L. Fejes Tóth proved [1] that there is a constant $q_0 \approx 0{,}7713$ such that for $q \geq q_0$, we have $S(q) = \sqrt{3}/24$, and raised the question about the bounds of $S(q)$ when $q < q_0$. Presently we give an upper bound and a lower bound for $S(q)$ when $0 < q < q_0$.

Consider a finite set of polygons of areas $a_1,\dots,a_N$ and perimeters $p_1,\dots,p_N$. Let $\bar{S}(q)$ be the supremum of $\Sigma a_i/\Sigma p_i^2$ extended over all sets of polygons of average number of sides at most six such that $\min p_i/\max p_i \geq q$. The argument in [1] implies that $S(q) \leq \bar{S}(q)$.

The main result of this paper is the following explicit formula for $\bar{S}(q)$ for any $q \in (0,1]$.

THEOREM. *Let*

$$g_j = \frac{1}{4j}\cot\frac{\pi}{j}\ ,\quad S_k^\ell(q) = \frac{(\ell-6)g_k q^2 + (6-k)g_\ell}{(\ell-6)q^2 + (6-k)}$$

$$Q_m^\ell = \frac{(g_{\ell+1} - g_\ell)(6 - k)}{g_\ell - g_m - (\ell-6)(g_{\ell+1} - g_\ell)}$$

and

$$Q^\ell = \frac{g_\ell + g_4 - 2g_5}{(\ell-6)(g_5-g_4)}\ .$$

Then

$$\bar{S}(q) = \begin{cases} S_5^\ell(q) & \text{if}\quad q\in(\sqrt{Q_5^\ell}\ ;\ \sqrt{Q_5^{\ell-1}}]\ \ \ell=6,7,8,9,10;\ Q_5^5=1, \\ S_5^{11}(q) & \text{if}\quad q\in(\sqrt{Q^{11}};\ \sqrt{Q_5^{10}}\ ]\ , \\ S_4^{11}(q) & \text{if}\quad q\in(\sqrt{Q_4^{11}};\ \sqrt{Q^{11}}\ ]\ , \\ S_4^\ell(q) & \text{if}\quad q\in(\sqrt{Q_4^\ell}\ ;\ \sqrt{Q_4^{\ell-1}}]\ \ \ell=12,13,\dots\ . \end{cases}$$

To prove our theorem we need some lemmas.

Let $H = \{A_i\}_{i=1}^N$ be a finite set, where A_i is a polygon with perimeter p_i and with area a_i.

The quantity $\sum_{i=1}^N a_i / \sum_{i=1}^N p_i^2$ will be denoted by σ_H. We observe, that g_j is nothing else but an area-perimeter ratio a/p^2 of a regular j-gon. Thus $S_k^\ell(q)$ is the quotient σ_H, where H is a set of polygons consisting of $(\ell - 6)$ regular k-gons with perimeter q and $(6 - k)$ regular

ℓ-gons of unit perimeter. Firstly, we give some technical lemmas regarding the functions S_k^ℓ.

From the obvious formula

$$S_k^\ell(q) = g_k + (g_\ell - g_k)\frac{1}{\frac{\ell-6}{6-k}q^2+1} \quad (k=3,4,5;\ \ell>6),$$

we can see that $S_k^\ell(q)$ is a strictly monotone decreasing function of q, which is concave from below, and

$$\lim_{q\to 0} S_k^\ell = g_\ell .$$

(In case $\ell = 6$, $S_3^6(q) = S_4^6(q) = S_5^6(q) = g_6$.)

LEMMA 1. *We have*

$$S_3^\ell(q) \le S_4^\ell(q) \quad (\ell\ge 6;\ q\in(0,1]). \tag{1}$$

PROOF. In case $\ell = 6$ the statement is true. For $\ell > 6$, it is equivalent to the inequality

$$q^2(g_3 - g_4) \le 3g_4 - 2g_3 - g_\ell .$$

The left-hand side is negative and the right-hand side is positive because $g_\ell < \frac{1}{4\pi}$.

Thus (1) is proved.

LEMMA 2. *Let* $\ell \ge 6$, $q\in(0,1]$. *The inequalities*

$$S_4^\ell \lesseqgtr S_4^{\ell+1}(q) \tag{2}$$

hold if and only if

$$(2') \qquad q^2 \lesseqgtr Q_4^{\ell} = \frac{2}{\dfrac{g_\ell - g_4}{g_{\ell+1}-g_\ell} - (\ell-6)}.$$

PROOF. We observe that (2) is equivalent to

$$(2'') \qquad \frac{q^2}{2}\,[g_\ell - g_4 - (\ell-6)(g_{\ell+1} - g_\ell)] \lesseqgtr g_{\ell+1} - g_\ell .$$

In (2") the coefficient of q^2 is positive because g_j is a monotone increasing concave sequence. So (2") is equivalent to (2').

LEMMA 3. *The sequence* Q_4^ℓ, $\ell \geq 6$, *is strictly monotone decreasing, with limit* 0,

$$(3) \qquad Q_4^{\ell+1} < Q_4^{\ell}$$

and

$$(3') \qquad \lim_{\ell\to\infty} Q_4^{\ell} = 0 .$$

PROOF. Using the definition of Q_4^ℓ we have that

$$\frac{2}{Q_4^{\ell+1}} + (\ell-5) = \frac{g_{\ell+1} - g_4}{g_{\ell+2} - g_{\ell+1}}$$

and

$$\frac{2}{Q_4^{\ell}} + (\ell-5) = \frac{g_{\ell+1} - g_\ell + g_\ell - g_4}{g_{\ell+1} - g_\ell} .$$

Hence

$$\frac{1}{Q_4^{\ell+1}} > \frac{1}{Q_4^{\ell}}$$

follows and it is equivalent to (3).
(3') holds because of the properties of the sequence g_j.

LEMMA 4. *Let* $\ell \geq 6$, $q \in (0,1]$. *The inequalities*

$$(4) \qquad S_5^{\ell}(q) \lesseqqgtr S_5^{\ell+1}(q)$$

hold if and only if

$$(4') \qquad q^2 \lesseqqgtr Q_5^{\ell} = \frac{1}{\dfrac{g_\ell - g_5}{g_{\ell+1}-g_\ell} - (\ell-6)} \quad .$$

PROOF. The inequalities (4) are equivalent to

$$(4'') \qquad q^2 [g_\ell - g_5 - (\ell-6)(g_{\ell+1} - g_\ell)] \lesseqqgtr g_{\ell+1} - g_\ell \quad ,$$

where the coefficient of q^2 is positive. Consequently (4") is equivalent to (4').

LEMMA 5. *The sequence* Q_5^{ℓ}, $\ell \geq 6$, *is strictly monotone decreasing, with limit* 0; *that is*

$$(5) \qquad Q_5^{\ell+1} < Q_5^{\ell}$$

and

(5')
$$\lim_{\ell\to\infty} Q_5^{\ell} = 0 .$$

PROOF. Using the properties of the sequence g_j, one can easily see that the relations

$$\frac{g_{\ell+1} - g_5}{g_{\ell+2} - g_{\ell+1}} + 1 = \frac{g_{\ell+2} - g_5}{g_{\ell+2} - g_{\ell+1}} > \frac{g_{\ell+2} - g_5}{g_{\ell+1} - g_{\ell}} > \frac{g_{\ell} - g_5}{g_{\ell+1} - g_{\ell}}$$

hold. Adding $\ell - 6$ to the first and last terms and taking reciprocals, we obtain (5). (5') follows also from the properties of g_j.

LEMMA 6. *Let* $\ell > 6$ *be a fixed integer. The inequalities*

(6)
$$S_5^{\ell}(q) \lesseqqgtr S_4^{\ell}(q)$$

hold if and only if

(6')
$$q^2 \lesseqqgtr Q^{\ell} = \frac{g_{\ell} + g_4 - 2g_5}{(\ell-6)(g_5-g_4)} .$$

PROOF. The relations (6) are equivalent to

$$q^2(\ell-6)(g_5-g_4) \lesseqqgtr g_{\ell} + g_4 - 2g_5,$$

in which the coefficient of q^2 is positive. So it is equivalent to (6'). Hence (6) is equivalent to (6'). In case when $\ell \geq 8$, we have $Q^{\ell} \in (0;1]$. For $\ell = 7$ we have $Q^{\ell} < 0$. This means that $S_5^7 > S_4^7$.

LEMMA 7. *Let* $\ell \geq 12$ *be a fixed integer. Then we have*

$$Q^{\ell} > Q_5^{\ell-1}. \tag{7}$$

PROOF. The inequality (7) can be rewritten as

$$g_{\ell-1} + g_4 - 2g_5 > (\ell-7)(g_{\ell} - g_{\ell-1}) . \tag{8}$$

Adding $g_{\ell} - g_{\ell-1}$ to both sides and using the inequality

$$g_{\ell} - g_{\ell-1} > g_{\ell+1} - g_{\ell},$$

we obtain

$$g_{\ell} + g_4 - 2g_5 > (\ell-6)(g_{\ell} - g_{\ell-1}) > (\ell-6)(g_{\ell+1} - g_{\ell}).$$

Consequently if (8) holds for a certain ℓ, it holds for $\ell+1$ as well. In case $\ell = 12$, (8) is true. This can easily be verified by a calculation. So (8) holds for any $\ell \geq 12$.

The following lemma is fundamental.

LEMMA 8. *We have*

$$\bar{S}(q) = \sup_{k,\ell} S_k^{\ell}(q) \quad \textit{for} \quad q \in (0,1]\,, \ \ell = 6,7,\dots \quad \textit{and} \quad k = 3,4,5.$$

PROOF. Let $H^* = \left\{A_i^*\right\}_{i=1}^{N}$ be a set of polygons with average number of sides not exceeding 6, such that $\min p_i/\max p_i = q$. (Denote the area, perimeter and number of sides of A_i^* by a_i^*, p_i and s_i^*.) Next let H be a set of N regular polygons A_i with the perimeter p_i and the number $s_i \geq s_i^*$ of sides such that $\sum_{i=1}^{N} s_i = 6N$. For the area a_i of A_i and the area a_i^* of A_i^*, we have that

$$a_i = p_i^2 \, g_{si} \geq p_i^2 \, g_{s_i^*} \geq a_i^* .$$

From this we obtain the inequality

$$\sigma_H \geq \sigma_{H^*}.$$

Let us take the set H in a "sufficient" number of copies and consider their union $\mathcal{H}$.

Subdivide the union into subsets, each subset consisting of polygons of at most two kinds with average number of sides 6. (The condition $\sum_{i=1}^{N} s_i = 6N$ guarantees the existence of a "sufficient" number of copies.)

Select one of these subsets. Let us suppose for example that it contains u copies of A_i and v copies of A_j. Obviously we have

$$q' = \min\left\{\frac{p_i}{p_j}, \frac{p_j}{p_i}\right\} \geq q \quad \text{and}$$

(for this selected subset) $\sigma_{ij} = \dfrac{ua_i + va_j}{up_i^2 + vp_j^2}$.

If $s_i = s_j = 6$, then we obtain

$$\sigma_{ij} = S_k^6(q') = S_k^6(q) = g_6.$$

If the selected subset does not contain any hexagon, then we can choose the indices in such a way that $s_i < 6$ and $s_j > 6$ hold. In this case, the quantity σ_{ij} can be estimated by

$$\sigma_{ij} = \frac{(s_j-6)g_{s_i}p_i^2 + (6-s_i)g_{s_j}p_j^2}{(s_j-6)p_i^2 + (6-s_i)p_j^2} \leq S_{s_i}^{s_j}(q') \leq S_{s_i}^{s_j}(q) .$$

We note that $\sigma_H = \sigma_{\mathscr{H}}$ and that $\sigma_{\mathscr{H}}$ can be obtained as the weighted average of the values σ_{ij}. Thus

$$\sigma_H \leq \max \sigma_{ij}.$$

We have therefore for any set H^*, satisfying the condition described in the definition of S(q), the following chain of inequalities:

$$\sigma_{H^*} \leq \sigma_H \leq \max \sigma_{ij} \leq \sup_{k,\ell} S_k^{\ell}(q).$$

Hence $\bar{S}(q) \leq \sup_{k,\ell} S_k^{\ell}(q)$.

Since

$$\bar{S}(q) \geq \sup_{k,\ell} S_k^{\ell}(q)$$

by definition, we have that

$$\bar{S}(q) = \sup_{k,\ell} S_k^{\ell}(q).$$

PROOF OF THE THEOREM. Let $S_k(q) := \sup_{\ell} S_k^{\ell}(q)$ and call it the "k-gon bound". We show that the function $\overline{S}(q)$ coincides with the "square bound" $S_4(q)$ in case $q \leq q^*$, and with the "pentagon bound S_5 if $q > q^*$, that is

$$\text{(I)} \qquad \bar{S}(q) = \begin{cases} S_4(q), & q \in (0;q^*] \\ S_5(q), & q \in (q^*,1] \end{cases} \quad \text{where } q^* = \sqrt{Q^{11}} \approx 0,268 .$$

By Lemma 1, $S_3(q) \leq S_4(q)$ and consequently, the "triangle bound" $S_3(q)$ does not figure in $\bar{S}(q)$. Consider the intervals

$$I_\ell = (\sqrt{Q_5^\ell};\sqrt{Q_5^{\ell-1}}]$$

and

$$J_\ell = (\sqrt{Q_4^\ell},\sqrt{Q_4^{\ell-1}}] ,$$

where $Q_4^5 = 1$, $\ell \geq 6$. From Lemmas 3 and 5, we get

$$\bigcup_{\ell=6}^{\infty} I_\ell = \bigcup_{\ell=6}^{\infty} J_\ell = (0;1] .$$

By Lemmas 2 and 4, we obtain

$$\text{(II)} \qquad S_4(q) = S_4^\ell(q), \text{ if } q \in J_\ell,$$

and

$$\text{(III)} \qquad S_5(q) = S_5^\ell(q), \text{ if } q \in I_\ell .$$

From the numerical values,

$$q^* = \sqrt{Q^{11}} \in I_{11} \cap J_{11}$$

and this means that

$$S_4(q^*) = S_4^{11}(q^*) = S_5^{11}(q^*) = S_5(q^*) = \bar{S}(q^*).$$

Consider now the intervals

$$J_\ell \subset (0;\ q^*] .$$

By Lemma 3, we have the cases $\ell \geq 12$. From Lemma 7, we have that

$$I_\ell \subset (0; \sqrt{Q^\ell}] \text{ for } \ell \geq 12.$$

Applying Lemma 6, we obtain the relations

$$(9) \qquad S_5(q) = S_5^\ell(q) < S_4^\ell(q) \leq S_4(q) \quad (q \in I_\ell,\ \ell \geq 12),$$

$$(10) \qquad S_5(q) = S_5^{11}(q) \leq S_4^{11}(q) \leq S_4(q) \quad (q \in (\sqrt{Q_5^{11}}; \sqrt{Q^{11}})$$

and

$$(11) \qquad S_5(q) \geq S_5^{11}(q) > S_4^{11}(q) = S_4(q) \quad (q \in (\sqrt{Q^{11}}; \sqrt{Q_4^{10}}),$$

respectively.

Let us consider now the intervals

$$J_\ell \subset (q^*;\ 1] .$$

This means that $6 \leq \ell \leq 10$. The relations

$$(12) \qquad S_4(q) = S_4^6(q) = g_6 \leq S_5(q) \quad (q \in J_6)$$

and

(13) $$S_4(q) = S_4^7(q) < S_5^7(q) \leq S_5(q) \quad (q \in J_7)$$

are obvious. We can easily verify the inequalities

$$Q^\ell < Q_4^\ell \qquad (\ell = 8,9,10)$$

by calculation. Using this and Lemma 6, we get

(14) $$S_4(q) = S_4^\ell(q) < S_5^\ell(q) \leq S_5(q) \quad (q \in J_\ell)$$

From (9) - (14), we have (I).

Comparing (I) with the relations (II) and (III) we obtain the Theorem.

We remark that

$$\lim_{\ell\to\infty} S_4(q) = \lim_{\ell\to\infty} S_5(q) = \lim_{q\to 0} \bar{S}(q) = \frac{1}{4\pi},$$

and that this value was obtained as a trivial bound by using the isoperimetric inequality.

GENERALIZATION FOR NORMAL CONVEX TILINGS

A tiling is said to be *normal* if there are two positive numbers ρ_1 and ρ_2 such that each face of the tiling contains a circle of radius ρ_1 and is contained in a circle of radius ρ_2.

The set of all normal tilings with convex faces is denoted by $\mathcal{T}$, the set of convex domains by $\mathcal{D}$ and the set of convex polygons by $\mathcal{P}$.

We consider a tiling $T \in \mathcal{T}$ and a domain $D \in \mathcal{D}$, and let O be a fixed inner point of D.

Let λD be the domain obtained from D by a homothety with quotient λ and centre O.

We consider those faces of T which are contained in λD and let $\bar{s}_{\lambda D}$ [$\sigma_{\lambda D}$] be the average number of sides [an area-perimeter ratio $\Sigma a_i / \Sigma p_i^2$] of the set of these faces, respectively.

Then $\bar{s}_T$ and σ_T of a tiling $T \in \mathcal{T}$ are defined by

$$\bar{s}_T = \sup_{D \in \mathcal{D}} \quad \sup_{O \in \operatorname{int} D} \limsup_{\lambda \to \infty} \bar{s}_{\lambda D}$$

and

$$\sigma_T = \sup_{D \in \mathcal{D}} \quad \sup_{O \in \operatorname{int} D} \limsup_{\lambda \to \infty} \sigma_{\lambda D} .$$

The homogeneity of a set $\left\{A_i\right\}_{i \in I}$ *of polygons* (I is any set of indices) is defined by

$$q = \frac{\inf p_i}{\sup p_i}$$

(p_i is the perimeter of A_i).

We observe the following three facts for a tiling $T \in \mathcal{T}$.

1.) $\bar{s}_T \leq 6$ by Euler's formula for polyhedra.

2.) A convex domain may be approximated by polygons. Thus to determine σ_T, it is sufficient to consider $P \in \mathscr{P}$; that is

$$\sigma_T = \sup_{P \in \mathscr{P}} \sup_{0 \in \text{int } P} \limsup_{\lambda \to \infty} \sigma_{\lambda P}.$$

3.) It is obvious that for the homogeneity q of a tiling $T \in \mathscr{T}$, we have

$$q \in (0;1] .$$

Finally, normality readily yields, that σ_T of a tiling from $\mathscr{T}$, with prescribed homogeneity q, can be estimated from above by the area-perimeter ratio $\Sigma a_i / \Sigma p_i^2$ of a set of polygons with average number of sides not exceeding 6.

The argument of $\bar{S}(q)$ implies that

$$\sigma_T \leq \bar{S}(q) .$$

A LOWER BOUND AND CONSTRUCTIONS

The previously given bound $\bar{S}$ for σ_T is obviously not sharp. For example, $\bar{S}(q) = S_5^7(q)$ on an interval but there does not exist a tiling in the Euclidean plane consisting of regular pentagons and heptagons.

In order to demonstrate the deviation of our upper bound $\bar{S}$ from the least upper bound, we shall find a lower bound for S(q).

Such a bound $s(q)$ can be given by finding (for each $q_0 \in (0;1]$) a tiling $T \in \mathcal{T}$ of homogeneity q_0 such that $s(q_0) = \sigma_T$.

We shall deal with nine kinds of initial tilings. These will be modified to families of tilings. The area-perimeter ratio of the i-th family ($i = 0,1,2,...,8$) will be described by a function $\sigma_i(q)$ on a homogeneity interval $[a_i,b_i] \subset (0;1]$.

By an appropriate modification of a single face and by that of its neighbours in a tiling $T \in \mathcal{T}$ of homogeneity q_0, we can construct a tiling of arbitrary homogeneity $q < q_0$ without changing the area-perimeter ratio. It follows that the function σ_i can be extended to the interval $(0;1]$ by setting

$$s_i(q) = \begin{cases} \sigma_i(a_i), & q \in (0;a_i) \\ \sigma_i(q), & q \in [a_i,b_i] \\ 0, & q \in (b_i,1] \end{cases}.$$

The lower bound $s(q)$ can be given in the form

$$s(q) = \max_i s_i(q), \quad (i = 0,...8).$$

We shall describe now the details of the construction.

C0. Let T_0 denote the regular tiling of symbol $\{6;3\}$. Then $\sigma_0 = s_0(q) = 1/8\sqrt{3}$ $(q \in (0;1])$.

C1. In order to define the first family of tilings consider the following system of circles (Figure 1).

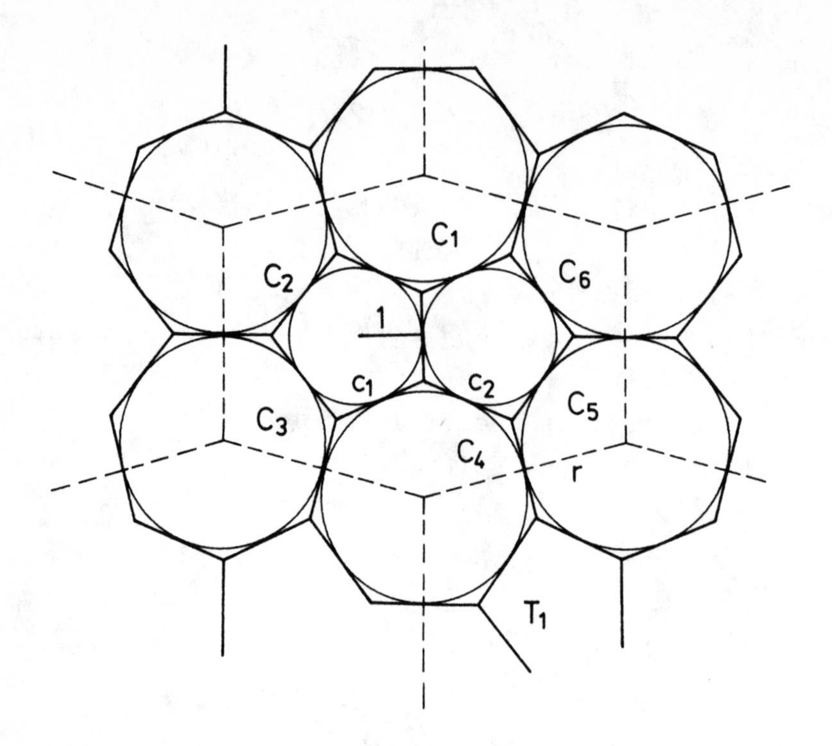

Figure 1

Let c_1 and c_2 be unit circles touching one another. Let C_1 and C_4 be circles of radius r touching both c_1 and c_2. Let C_2, C_3, C_5 and C_6 be further circles of radius r touching both C_1 and c_1, c_1 and C_4, C_4 and c_2, c_2 and C_1, respectively. We choose r so that C_2 and C_3 also touch one another. In this case r is a root, of value greater than 1, of the equation

$$9x^4 - 8x^3 - 10x^2 + 1 = 0.$$

(This value is about 1,568.)

By translating the hexagon of side 2r, spanned by the centres of the big circles, we obtain a tiling T'. The translated circles will produce a packing of circles.

The system of cells of this packing* will be chosen as the initial tiling T_1.

The tiling T_1 is now modified as follows. The centres of the circles are held fixed. Circles of radius μ replace the unit circles and circles of radius $1+r-\mu$ replace those of radius r. The modified tiling $T_1(\mu)$ is the system of cells of the modified system of circles. If

$$\mu \in \left(\frac{1}{1+r} ; \frac{1+2r}{1+r}\right)$$

then the faces of $T_1(\mu)$ are pentagons and heptagons. If

$$\mu \leq \frac{1}{1+r}\left(1+\frac{6r\sqrt{r^2+2r}}{\sqrt{1+2r}+3\sqrt{r^2+2r}}\right)$$

then the perimeter of the pentagons is not greater than that of the heptagons; consequently, the family $T_1(\mu)$ will be discussed for values

$$\mu \in \left(\frac{1}{1+r} ; \frac{1}{1+r}\left(1+\frac{6r\sqrt{r^2+2r}}{\sqrt{1+2r}+3\sqrt{r^2+2r}}\right)\right).$$

In this case, the homogeneity of $T_1(\mu)$ is a continuous, strictly monotone increasing function of μ. With the inverse of this function, the area-perimeter ratio of the family $T_1(\mu)$ can be expressed by

$$\sigma_1(q) = \frac{(r+q)^2}{1+q^2} \cdot \frac{1}{12\sqrt{r^2+2r^3}+4\sqrt{r^2+2r}}.$$

* Using the radical axes of the circles we get the system of generalized Dirichlet cells of a system of circles, which we call in this paper system of cells.

The function $\sigma_1(q)$ takes its maximum at $q = \frac{1}{r}$ and for values $q > \frac{1}{r}$, it is strictly monotone decreasing. The μ belonging to $q = \frac{1}{r}$ lies in the admissible interval and for any $q \in \left[\frac{1}{r}; 1\right]$, there is a tiling of area-perimeter ratio $\sigma_1(q)$ in the family $T_1(\mu)$.

Summing up, the function $s_1(q)$ is of the form

$$s_1(q) = \begin{cases} \sigma_1(\frac{1}{r}), & q < \frac{1}{r} \\ \sigma_1(q), & q \in \left[\frac{1}{r}; 1\right] \end{cases} .$$

C2. For constructing the second family we consider the following system of circles (Figure 2). C_1 and C_2 are unit circles touching each other, and a circle c_1 of radius r will touch both. The circle c_2 is of radius r, it touches c_1 and C_1. The value of r will now be chosen by requiring that the segment connecting the centres of C_1 and c_2 will be perpendicular to the segment connecting the centres of C_1 and C_2. In this case r is a root, smaller than 1, of equation

$$8x^3 + 3x^2 - 2x - 1 = 0, \quad r \approx 0{,}533.$$

By reflecting the circle c_1 about the line through the centres of the circles C_1 and c_2, we obtain the circle c_3. Reflecting c_3, c_2 and c_1 about the line through the centres of C_1 and C_2, we get the circles c_4, c_5 and c_6.

Now we translate the intersection of these circles with the hexagon spanned by the centres of the small circles so that the translates of the hexagon tile the plane, and thus obtain a packing of circles.

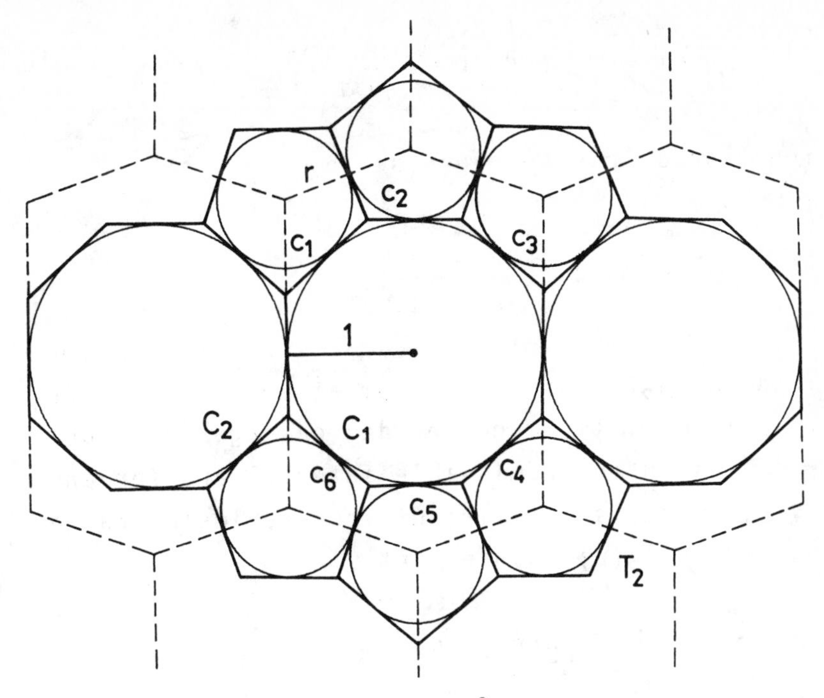

Figure 2

Let T_2 be the system of cells of this packing. The family $T_2(\mu)$ is defined in a similar way: we draw circles of radius μ around the centres of the circles of radius r, and circles of radius $1+r-\mu$ will be drawn around the centres of the unit circles;

$$\mu \in \left[\frac{1}{1+r}, \frac{2r+1}{r+1}\right] .$$

The family $T_2(\mu)$ is the system of cells of this set of circles.

By introducing the notion

$$c_2 = \frac{1}{8[r(a+4b)+2(a+b)]} \approx 0,04637$$

where

$$a = \sqrt{\frac{r}{2+r}} \quad \text{and} \quad b = \frac{r}{\sqrt{2r+1}} ,$$

the function σ_2 can be expressed by

$$\sigma_2(q) = c_2 \frac{(2rq + 1)^2}{2q^2 + 1} .$$

The function $\sigma_2(q)$ has a maximum at $q = r$. For values $q > r$, it is strictly monotone decreasing. The μ belonging to $q = r$ lies in the admissible interval and for any $q \in [r,1]$, there is a tiling of area-perimeter ratio $\sigma_2(q)$ in the family $T_2(\mu)$. Consequently,

$$s_2(q) = \begin{cases} \sigma_2(r), & q < r \\ \sigma_2(q), & q \in [r;1] \end{cases} .$$

C3. Consider the regular tiling with symbol {4;4} of side length $1 + \sqrt{2}$. Cut off isosceles rectangular triangles with hypothenuse $\mu \in (0;\ 1 + \frac{\sqrt{2}}{2})$ from each corner of the squares. The four small triangles, which are cut off around a vertex, constitute a square of side μ. $T_3(\mu)$ is defined as those squares and the resulting octagons (Fig. 3).

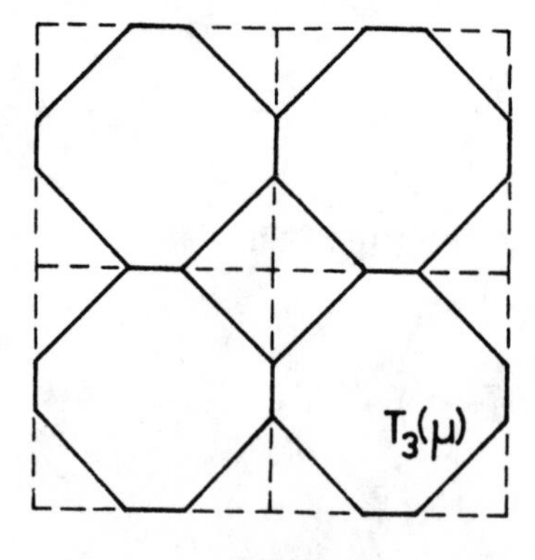

Figure 3

The area-perimeter ratio of the family $T_3(\mu)$ can be expressed by

$$\sigma_3(q) = \frac{(1 + q(\sqrt{2} - 1))^2}{16(1 + q^2)} .$$

The function $\sigma_3(q)$ has a maximum at $q = \sqrt{2} - 1$. For any $q \in [\sqrt{2} - 1, 1]$, it is a strictly monotone decreasing function.
Hence

$$s_3(q) = \begin{cases} \sigma_3(\sqrt{2} - 1), & q < \sqrt{2} - 1 \\ \sigma_3(q), & q \in [\sqrt{2} - 1, 1] \end{cases}$$

C4. Consider the system of circles consisting of unit circles and of circles of radius

$$r = \frac{\sin 15^0}{1 - \sin 15^0} .$$

In this system any unit circle touches 12 small circles, and the neighbouring small circles also touch one another. We note that any small circle has 3 small and 2 big neighbours (Fig. 4). The initial tiling T_4 is the system of cells of this packing of circles.*

The tiling T_4 is now modified as follows: we draw circles of radius μ $(\mu < \frac{2r + 1}{r+1} \cdot \frac{1}{\sin 15^0})$ around the centres of the unit circles, and circles of radius

---

* The packings of circles in C1, C2 and C4 were constructed by J. Molnár [3] (see also [2], p. 185).

$1 + r - \mu$ around the centres of the circles of radius r.

The modified tiling $T_4(\mu)$ in the system of cells of the modified set of circles. The area-perimeter ratio of $T_4(\mu)$ can be expressed by

$$\sigma_4(q) = \frac{\sqrt{3}}{96r^2} \cdot \frac{(r+\sqrt{3}\sqrt{1+2r})^2}{(r+2\sqrt{3}\sqrt{1+2r})^2} \cdot \frac{(6rq+1)^2}{6q^2+1} .$$

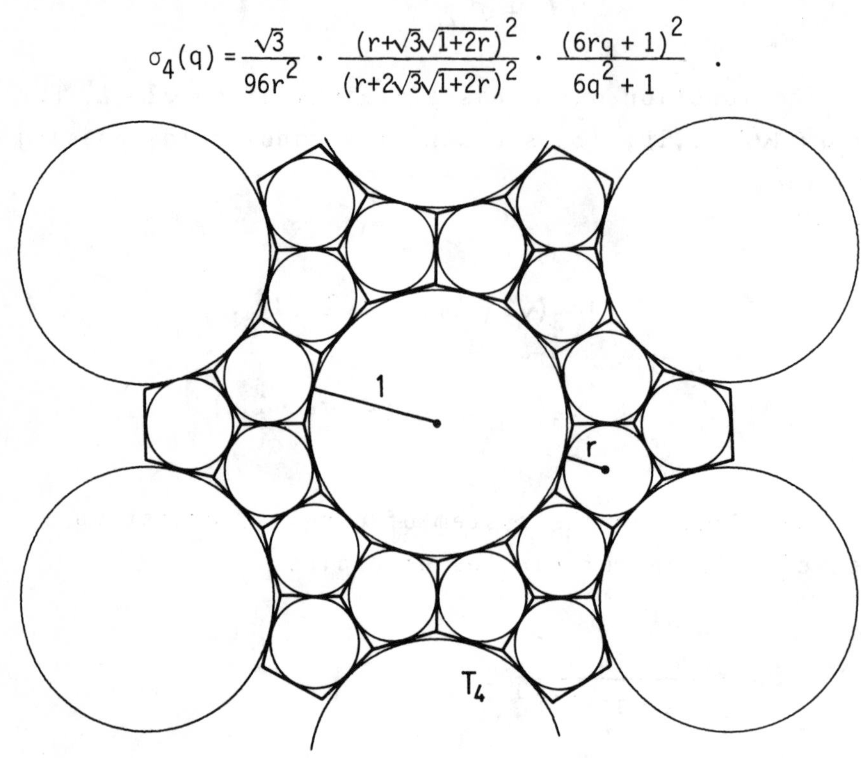

Figure 4

$\sigma_4(q)$ has a maximum at $q = r$. For $q \in [r,1]$, it is strictly monotone decreasing function. Finally,

$$s_4(q) = \begin{cases} \sigma_4(r), & q < r \\ \sigma_4(q), & q \in [r,1] \end{cases} .$$

C5. The following initial tiling T_5 is an Archimedean tiling with symbol (4,6,12). Consider the system of incircles of its faces and let their radii be 1, $\sqrt{3}$, $2+\sqrt{3}$, respectively. The centres of incircles are held fixed and draw circles of radius μ ($\mu \leq 3\sqrt{3} - 4{,}5$) around the centres of the squares, circles of radius $1+\sqrt{3}-\mu$ around the centres of the hexagons, and circles of radius $3+\sqrt{3}-\mu$ around the centres of 12-gons. Let $T_5(\mu)$ be the system of cells of this set of circles (Fig. 5 for $\mu > 1$).

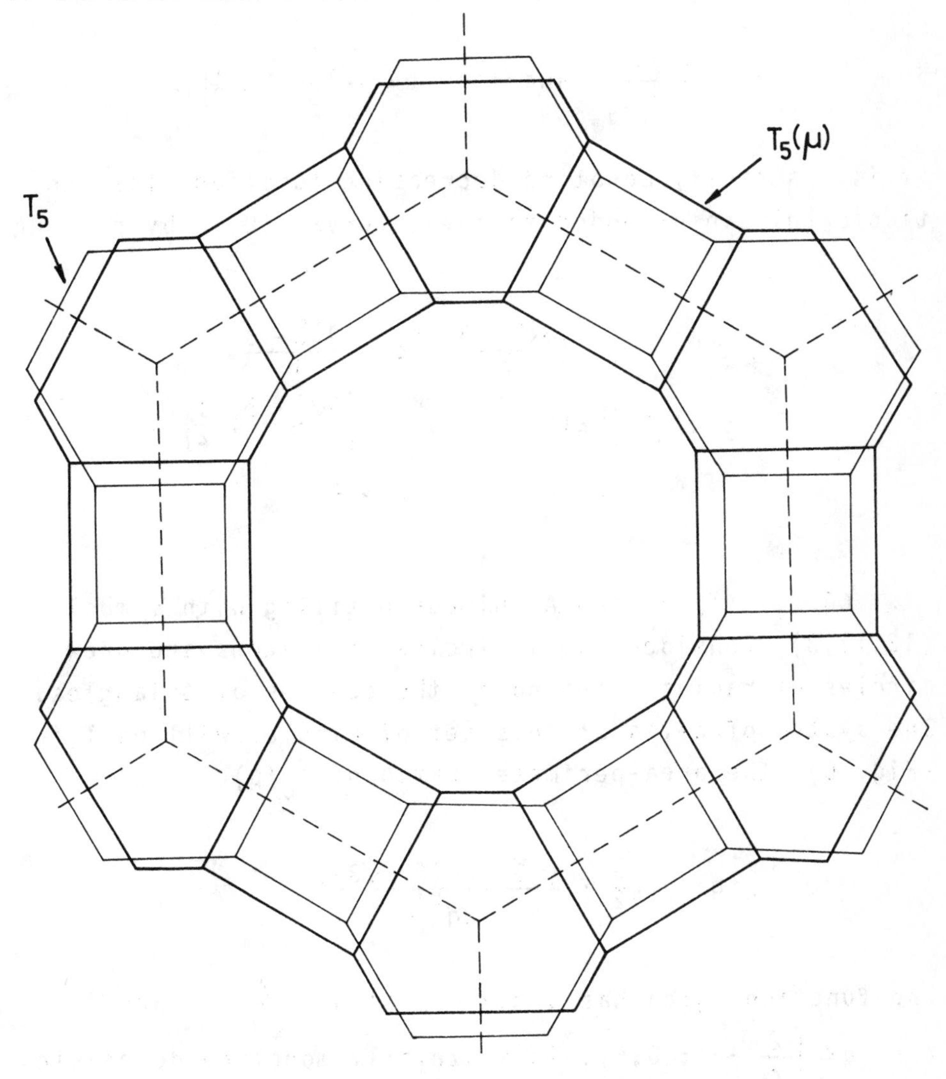

Figure 5

The area-perimeter ratio of $T_5(\mu)$ can be expressed by

$$\sigma_5(q) = \frac{\sqrt{3}}{432} \cdot \frac{(4 + 3q(\sqrt{3} - 1))^2}{2q^2 + 1} .$$

It has a maximum at

$$q = \frac{3(\sqrt{3} - 1)}{8} \text{ and for } q \in \left[\frac{3(\sqrt{3} - 1)}{8}; \frac{1}{2}\right] ,$$

it is a strictly monotone decreasing function. The function $\sigma_5(q)$ can extended to the interval (0;1] by putting

$$s_5(q) = \begin{cases} \sigma_5(\frac{3\sqrt{3} - 3}{8}), & q < \frac{3\sqrt{3} - 3}{8} \\ \sigma_5(q) & , q \in \left[\frac{3\sqrt{3} - 3}{8}; \frac{1}{2}\right] \\ 0 & , q > \frac{1}{2} \end{cases} .$$

C6. Let T_6 be the Archimedean tiling with symbol (12,12,3). Consider the incircles of 12-gons and draw circles of radius μ around of the centres of triangles. The system of cells of this set of circles will be $T_6(\mu)$ (Fig. 6). The area-perimeter ratio of $T_6(\mu)$

$$\sigma_6(q) = \frac{\sqrt{3}}{72} \cdot \frac{(\sqrt{3} + 2q(2 - \sqrt{3}))^2}{2q^2 + 1}$$

The function $\sigma_6(q)$ has a maximum at $q = \frac{2}{\sqrt{3}} - 1$ and for any $q \in \left[\frac{2}{\sqrt{3}} - 1; 0,5\right]$, it a strictly monotone decreasing function. The function $s_6(q)$ is given by

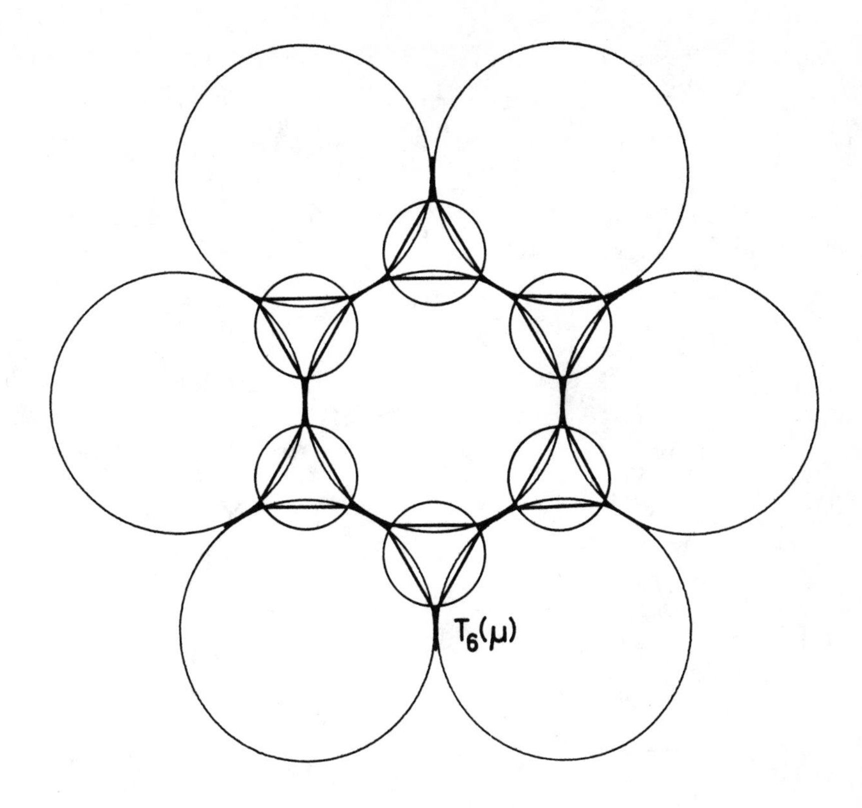

Figure 6

$$s_6(q) = \begin{cases} \sigma_6(\frac{2}{\sqrt{3}} - 1), & q < \frac{2}{\sqrt{3}} - 1 \\ \sigma_6(q) & , q \in \left[\frac{2}{\sqrt{3}} - 1;\ 0,5\right] \\ 0 & , q > 0,5 \end{cases} .$$

C7. Consider the incircles of the faces of T_6. Then we obtain a circle packing. To this packing we add the incircles of the holes. These incircles have radius r.

We now define T_7 as the system of cells associated with this new set of circles.

If circles of radii μ, around the centres of circles of radii r, replace the latter circles then we get a modified set of circles. Let $T_7(\mu)$ be the system of cells of this modified set of circles (Fig. 7).

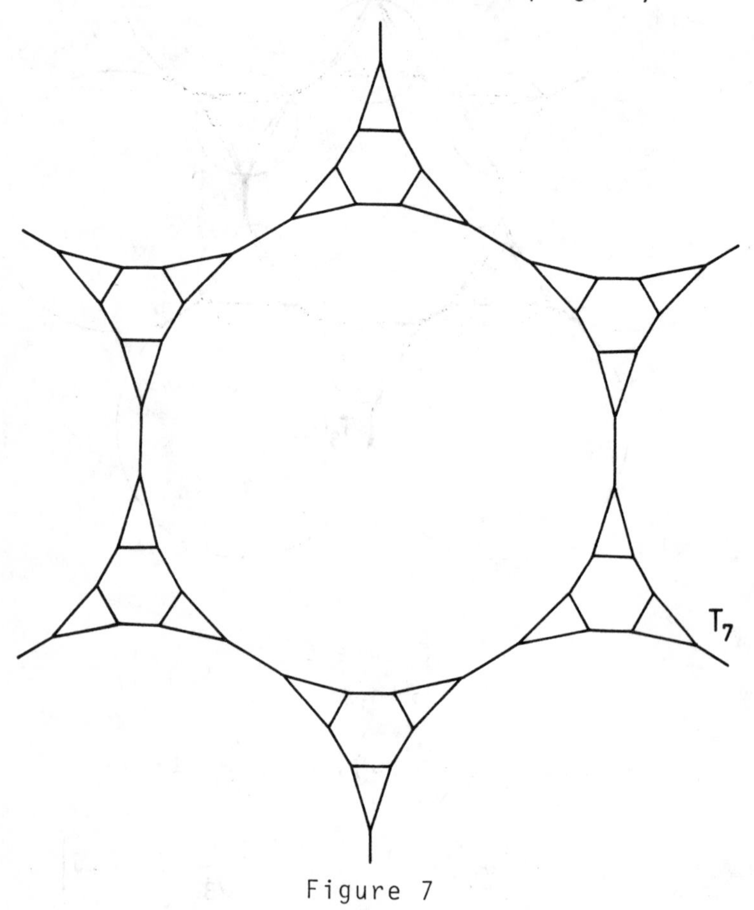

Figure 7

The area-perimeter ratio of $T_7(\mu)$ is

$$\sigma_7(q) = \frac{7\sqrt{3} + 12}{36} \cdot \frac{(c_1 + 12\, q\, c_2)^2}{8\, c_1^2(6q^2 + 1) + (c_1 + 12q\, c_3)^2}$$

where

$$c_1 = 64 - 29\sqrt{3}$$

$$c_2 = 40 - 23\sqrt{3}$$

$$c_3 = 14 - 10\sqrt{3}.$$

For the value $0{,}0734 = q_0 \leq q \leq q_1 = \dfrac{64 - 29\sqrt{3}}{4(22+\sqrt{3})}$, $\sigma_7(q)$ is a strictly monotone decreasing function.

The function $s_7(q)$, we obtain by:

$$s_7(q) = \begin{cases} \sigma_7(q_0), & q < q_0 \\ \sigma_7(q), & q \in [q_0, q_1] \\ 0, & q > q_1 \end{cases} .$$

C8. Now starting with T_7, we adjoin the incircles of the large gaps to obtain T_8.

We draw circles of radii μ around the centres of smallest circles replace the latter circles, and let $T_8(\mu)$ be the system of cells of this modified set of circles (Fig. 8).
We obtain the area-perimeter ratio of $T_8(\mu)$ by

$$\sigma_8(q) = \frac{2\sqrt{3}(c_1 + 12q\, c_2)^2}{(6q^2+1)24^2[c_1(2-\sqrt{3})+36q\, c_3]^2 + 8(2-\sqrt{3})^2}$$

where

$$c_1 = \frac{28 - 10\sqrt{3}}{121} + \frac{6 - \sqrt{3}}{33};$$

$$c_2 = \frac{4 - 2\sqrt{3}}{3} + \frac{4 - \sqrt{3}}{13};$$

$$c_3 = \frac{28 - 10\sqrt{3}}{121} - \frac{4 - \sqrt{3}}{13}.$$

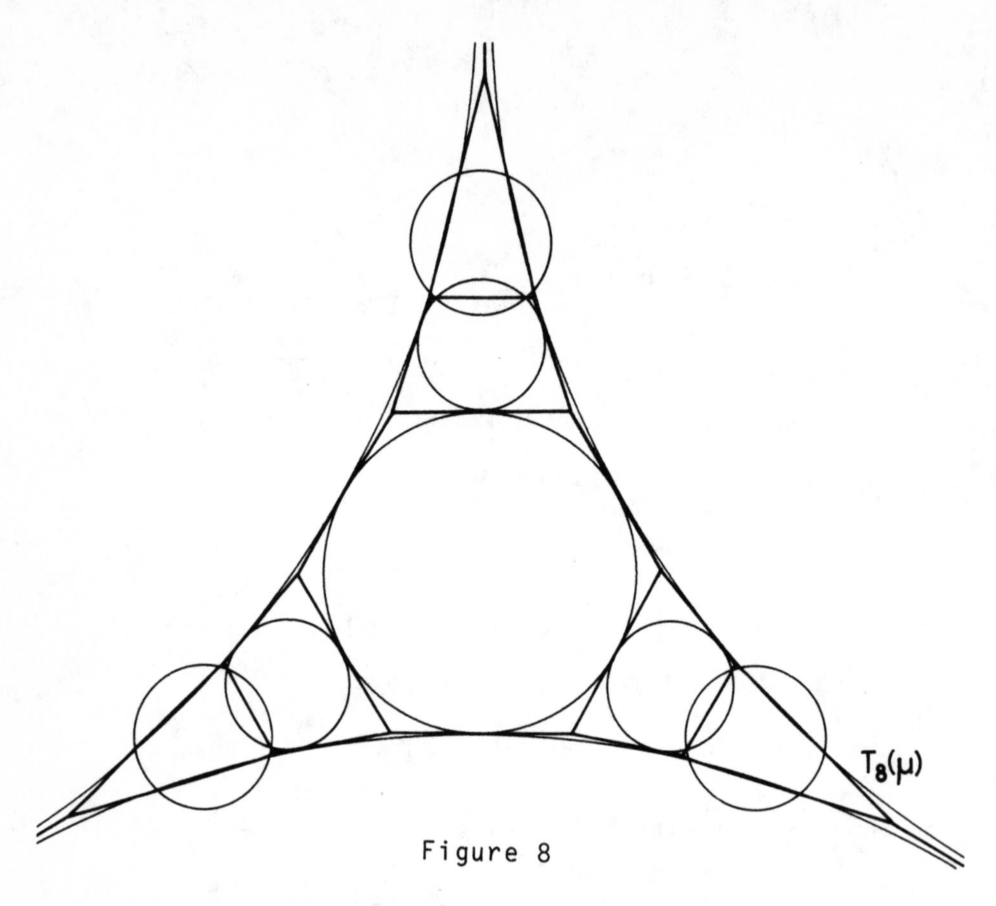

Figure 8

The function $\sigma_8(q)$, for $q \in [0,046;\ 0,0734]$ is strictly monotone decreasing.

Hence we have

$$s_8(q) = \begin{cases} \sigma_8(0,046), & q < 0,046 \\ \sigma_8(q) & , q \in [0,046;\ 0,0734] \\ 0 & , q > 0,0734 \end{cases}.$$

We summarize our results in the following:

$$s(q) \le S(q) \le \bar{S}(q)$$

where

$$s(q) = \max_i \ s_i(q)$$

and $s_i(q)$, $0 \leq i \leq 8$, are defined in the i-th construction. Our bounds $\bar{S}(q)$ and $s(q)$ are represented in the following diagram.

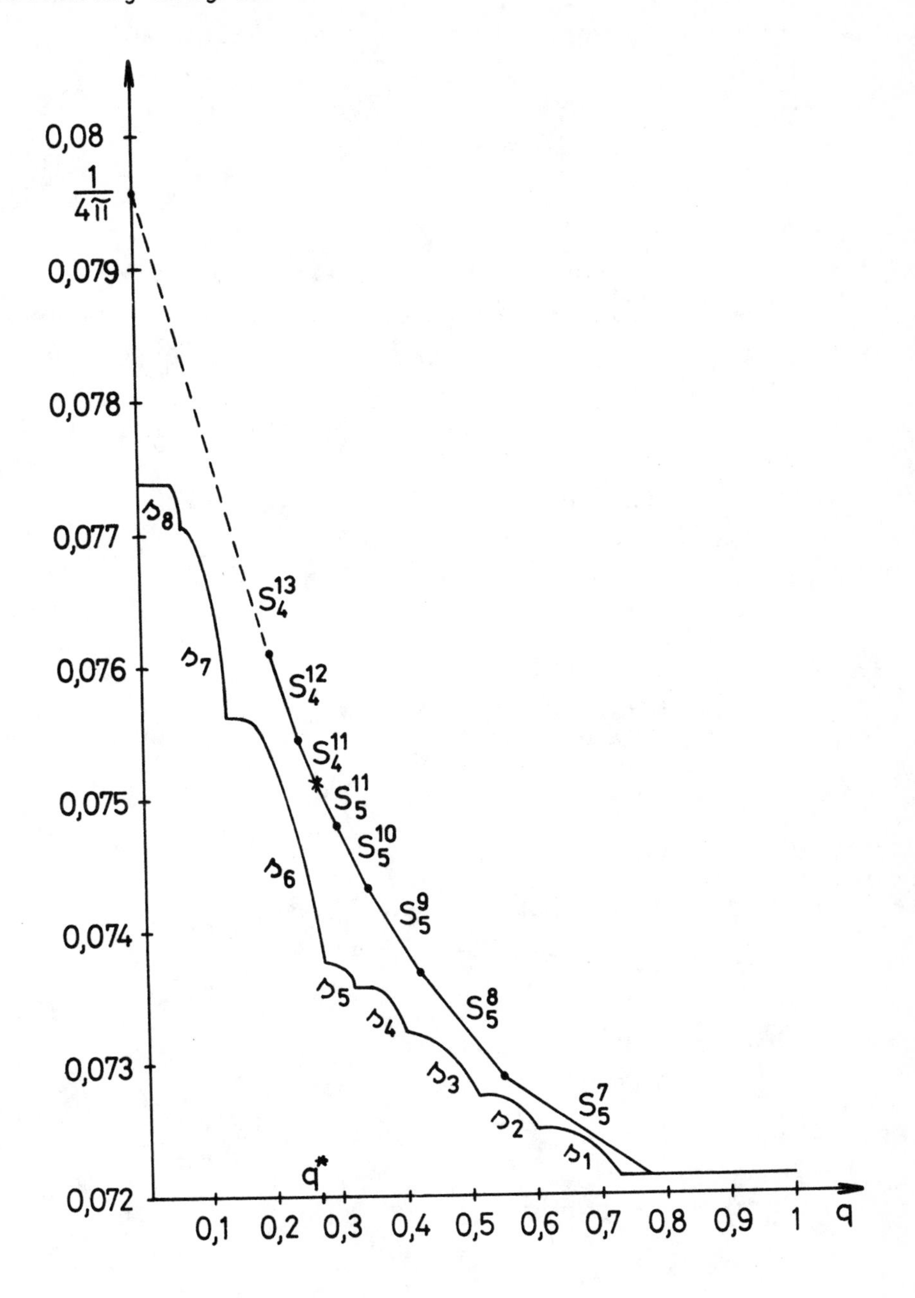

ACKNOWLEDGEMENT. The author would like to thank the referee for his valuable suggestions. She also thanks the Hungarian National Foundation for Scientific Research for supporting this work under grant no. 1238.

REFERENCES

[1] L. FEJES TÓTH, *Isoperimetric problem for tilings*, *3. Kolloquium über Diskrete Geometrie*, Institut für Mathematik der Univ. Salzburg, Salzburg, 1985.

[2] L. FEJES TÓTH, *Regular figures*, Pergamon Press, 1964.

[3] J. MOLNÁR, *Unterdeckung und Überdeckung der Ebene durch Kreise*, Ann. Univ. Sci. Budapestiensis, 2, 33-40.

É. VÁSÁRHELYI
Eötvös Loránd University,
Institute of Mathematics,
Department of Geometry,
Budapest, VIII. Rákóczi út 5.
H-1088.

COLLOQUIA MATHEMATICA SOCIETATIS JÁNOS BOLYAI
48. INTUITIVE GEOMETRY, SIÓFOK, 1985.

ON SYSTEMS OF CIRCLES SURROUNDING A CIRCLE

I. VINCZE

1. INTRODUCTION

We shall consider N (> 3) circles in the plane denoting the i-th of them by $\mathbf{C}_i$, its center by $\mathbf{O}_i$ and its radius by r_i. In our main statements we shall assume that the circles $\mathbf{C}_i$ are not congruent and the set of r_i-s is fixed. (Equality of the radii will occur in some examples.) The ordered set of circles

$$(1.1) \qquad \underline{\mathbf{C}} = \{\mathbf{C}_1, \mathbf{C}_2, \ldots, \mathbf{C}_N; \mathbf{C}\}$$

will be called a surrounding system of a circle **C** (**SSC**), if no two of them have inner points in common, further $\mathbf{C}_i$ touches outside exactly $\mathbf{C}_{i-1}$, $\mathbf{C}_{i+1}$ and a circle **C** with center **O** and with radius **R**; $i=1,2,\ldots,N$, $\mathbf{C}_0 = \mathbf{C}_N$, $\mathbf{C}_{N+1} = \mathbf{C}_1$. - The starting point of our investigations is the observation, that when $\underline{\mathbf{C}}$ form an **SSC**, then the rearranged circles $(\mathbf{C}_{j_1}, \mathbf{C}_{j_2}, \ldots, \mathbf{C}_{j_N})$, where $(j_1, j_2, \ldots, j_N) = \underline{\mathbf{P}}$ is a permutation of $(1,2,\ldots,N)$, either does not yield an **SSC**, or when it does, then with a circle $\mathbf{C}(\mathbf{P})$ in the middle, having a radius $\mathbf{R}(\mathbf{P})$ different from the initial $\mathbf{R}$. We shall denote the new system by $\underline{\mathbf{C}}(\mathbf{P})$.

In §.2. conditions will be considered under which for all permutations $\underline{\mathbf{P}}$ the system

$$(1.2) \qquad \underline{\mathbf{C}}(\mathbf{P}) = \{\mathbf{C}_1(\mathbf{P}), \mathbf{C}_2(\mathbf{P}), \dots, \mathbf{C}_N(\mathbf{P}); \mathbf{C}(\mathbf{P})\}$$

built also an **SSC**; here the notation $\mathbf{C}_i(\mathbf{P})$ means $\mathbf{C}_{j_i}$ when $\mathbf{P} = (j_1, j_2, \dots, j_i, \dots, j_N)$.

In §.3. the corresponding system of angles $(\varphi_1, \varphi_2, \dots, \varphi_N)$, where $\varphi_i = \sphericalangle \mathbf{O}_i \mathbf{O} \mathbf{O}_{i+1}$, will be treated considering the map $(r_1, r_2, \dots, r_N) \to (\varphi_1, \varphi_2, \dots, \varphi_N)$ and its inverse. We shall turn to the main result of our considerations in §.4. determining the arrangement $\mathbf{P}_{max}$ and $\mathbf{P}_{min}$ for which $\mathbf{R}(\mathbf{P})$ attains its maximum and minimum, respectively. To obtain this result we shall apply the following theorem proved in [1]:

THEOREM 1.1. *Let* $f(x,y)$ *be a real valued, symmetric function of two variables defined on the* (x,y)*-plane monotonously increasing in its variables and satisfying also the relation*

$$(1.3) \qquad \frac{\partial^2 f}{\partial x \partial y} \geq 0 \ .$$

Having a sequence $a_1 \geq a_2 \geq \dots \geq a_n$ *of real numbers and denoting by* $\mathbf{P}$ *a permutation* $(j_1 = 1, j_2, j_3, \dots, j_n)$ *of the first* n *positive integers, the function*

$$\mathbf{A}_n(\mathbf{P}) = \sum_{i=1}^{n} f(a_{j_i}, a_{j_{i+1}}), \ a_{j_{n+1}} = a_1$$

takes on its extreme values $\min_{\mathbf{P}} \mathbf{A}_n(\mathbf{P})$ *and* $\max_{\mathbf{P}} \mathbf{A}_n(\mathbf{P})$ *among all permutations for*

$$(1.4) \qquad \mathbf{P}_{min} = (1,n,2,n-2,n-4,\ldots,n-5,5,n-3,3,n-1)$$

and

$$(1.5) \qquad \mathbf{P}_{max} = (1,2,3,4,\ldots,2i,2(i+1),\ldots,n,\ldots,\ldots,(2i+1),(2i-1),\ldots,7,5,3),$$

respectively.

2. ON SURROUNDING SYSTEMS OF A CIRCLE

Let $\underline{\mathbf{C}} = \{\mathbf{C}_1,\mathbf{C}_2,\ldots,\mathbf{C}_N;\ \mathbf{C}\}$ be an **SSC** with the corresponding system of radii $\underline{r} = (r_1,r_2,\ldots,r_N)$. From now on $\underline{\mathbf{C}}$ will denote that system in which the radii are in descending order:

$$(2.1) \qquad \infty > r_1 > r_2 > \ldots > r_N > 0 .$$

The value $r_1 = \infty$ will also occur in an example. We shall denote by P the set of permutations of the first N natural numbers having first term 1 and leading to essentially different configurations; this means that $(1,j_2,j_3,\ldots,j_N) = (1,j_N,j_{N-1},\ldots,j_2)$, the corresponding configurations being congruent. Hence the number of different elements of P is $N_1 = (N-1)!/2$. Having a permutation $\mathbf{P} = (1,j_2,\ldots,j_N)$ we come to the system given in (1,2) while (1.1) belongs to the identity permutation $\mathbf{I} = (1,2,\ldots,N)$. Our first remark concerning **SSC**-s is contained in the following

PROPOSITION 2.1. *When* $\underline{\mathbf{C}}(\mathbf{P})$ *is an* **SSC** *and* $\mathbf{P}' \in P$ *is different from* **P**, *then* $\underline{\mathbf{C}}(\mathbf{P}')$ *is in general not an* **SSC**, *or at least with* $\mathbf{R}(\mathbf{P}') \neq \mathbf{R}(\mathbf{P})$.

PROOF. Let N=4 and $r_1 = r_2 = \infty$, $r_3 = r_4 = r > 0$. Then to $\mathbf{P} = (r_1, r_3, r_2, r_4)$ belongs an **SSC** with $\mathbf{R}(\mathbf{P}) = r$, but we cannot change the order of the circles at all. - If we diminish r_1 and r_2 having a large common value $\bar{r}$ and for r_3 and r_4 we keep the value r, then in the new situation $\mathbf{R}(\mathbf{P})$ will be smaller than r, at the same time to the order $(r_1, r_2, r_3, r_4) = (\bar{r}, \bar{r}, r, r)$ belongs an **SSC** but with **R** larger than r.

Another kind of the violation of the **SSC** property, when taking a permutation of the circles, is the following. Let $\underline{\mathbf{C}} = \{\mathbf{C}_1, \mathbf{C}_2, \ldots, \mathbf{C}_N; \mathbf{C}\}$ be an **SSC** with r_N small r_1 and r_2 large in such a way that in $\underline{\mathbf{C}}(\mathbf{P}) = \{\mathbf{C}_1, \mathbf{C}_N, \mathbf{C}_2, \mathbf{C}_3, \ldots, \mathbf{C}_{N-1}; \mathbf{C}(\mathbf{P})\}$ $\mathbf{C}_1$ and $\mathbf{C}_2$ will touch each other but $\mathbf{C}_N$ may move between them. This also contradicts to the **SSC** property: $\underline{\mathbf{C}}$ is an **SSC** but $\underline{\mathbf{C}}(\mathbf{P})$ not.

In the sequel we shall consider mainly such kind of systems $\underline{\mathbf{C}}$ only which are along with all $\underline{\mathbf{C}}(\mathbf{P})$, $\mathbf{P} \in P$ surrounding systems of corresponding circles $\mathbf{C}(\mathbf{P})$, $\mathbf{P} \in P$.

Given a set of system $\underline{\mathbf{C}}(\mathbf{P})$, $\mathbf{P} \in P$ the corresponding $\mathbf{R}(\mathbf{P})$-s may have N_1 different values. The permutations to which the smallest and the largest radius belong will be denoted by $\mathbf{P}_{min}$ and $\mathbf{P}_{max}$ respectively, the corresponding radii will be denoted by $R_{min} = \mathbf{R}(\mathbf{P}_{min})$, $R_{max} = \mathbf{R}(\mathbf{P}_{max})$. To guarantee that $\underline{\mathbf{C}}(\mathbf{P})$ form an **SSC** for all permutations $\mathbf{P} \in P$ we have to assume that the smallest radius r_N is larger than the radius of the circle having room inside the three circles $\mathbf{C}_1, \mathbf{C}_2$ and $\mathbf{C}(\mathbf{P}_{max})$ when they touch each other. *This condition is sufficient*. Replacing $\mathbf{C}(\mathbf{P}_{max})$

determine the system $\underline{t} = (t_1, t_2, \ldots, t_N)$ uniquely or not.

Let first $N = 2n+1$, $n > 1$ and let us introduce the following notations

$$A_0^{(N)} = 1,\quad A_i^{(N)} = \prod_{j=0}^{i-1} \sin \frac{\varphi_{2j+1}}{2},\quad i=1,2,\ldots,n$$

$$\bar{A}_i^{(N)} = \prod_{j=i}^{n} \sin \frac{\varphi_{2j+1}}{2},\quad i=0,1,2,\ldots,n$$

(3.4)

$$B_0^{(N)} = 1,\quad B_i^{(N)} = \prod_{j=1}^{i} \sin \frac{\varphi_{2j}}{2},\quad i=1,2,\ldots,n$$

$$\bar{B}_i^{(N)} = \prod_{j=i+1}^{n} \sin \frac{\varphi_{2j}}{2},\quad i=0,1,\ldots,n-1,\quad \bar{B}_n^{(N)} = 1\ .$$

Then the following relations hold:

$$t_{2i} = \frac{A_i^{(N)}\ \bar{B}_{i-1}^{(N)}}{\bar{A}_i^{(N)}\ B_{i-1}^{(N)}},\quad i=1,2,\ldots,n$$

(3.5)

$$t_{2i+1} = \frac{\bar{A}_i^{(N)}\ B_i^{(N)}}{A_i^{(N)}\ \bar{B}_i^{(N)}},\quad i=0,1,2,\ldots,n$$

To justify the solution (3.5) of the system (3.3) we express t_2 from the first equation ($i=1$) by means of t_1, then t_3, $t_4, \ldots$ from the consecutive equations; we obtain

$$t_{2i} = \frac{\sin^2 \frac{\varphi_{2i-1}}{2}}{t_{2i-1}} = \frac{\prod_{j=0}^{i-1} \sin \frac{\varphi_{2j+1}}{2}}{\prod_{j=1}^{i-1} \sin^2 \frac{\varphi_{2j}}{2}} \cdot \frac{1}{t_1}$$

(3.6)

$$t_{2i+1} = \frac{\sin^2 \frac{\varphi_{2i}}{2}}{t_{2i}} = \frac{\prod_{j=1}^{i} \sin^2 \frac{\varphi_{2j}}{2}}{\prod_{j=0}^{i-1} \sin^2 \frac{\varphi_{2j+1}}{2}} \cdot t_1$$

$i=1,2,\ldots,n$. For $i=n$, t_{2n+1} can be derived from the $N-1 = 2n$-th equation; considering the last equation in (3.3) we can determine t_1:

$$(3.7) \qquad t_1 = \frac{\prod_{j=0}^{n} \sin \frac{\varphi_{2j+1}}{2}}{\prod_{j=1}^{n} \sin \frac{\varphi_{2j}}{2}} = \frac{\bar{\mathbf{A}}_0^{(N)}}{\bar{\mathbf{B}}_0^{(N)}}$$

where the notations of (3.4) are applied. (3.7) along with (3.6) yields the solution given in (3.5). - This way we uniquely determined the system $\underline{t}$ with the aid of the system $\underline{\varphi}$. The uniqueness follows doubtless from Remark 2.2 below.

In the case $N = 2n$ let $(t_1, t_2, \ldots, t_N)$ be a solution of the system of equations (2.3); thus corresponding to the initial system $\underline{\mathbf{C}}$ we have: $1 > t_1 > t_2 > \ldots > t_{2n} > 0$. Let ε be a real value for which $1 < \varepsilon < \varepsilon_1$, where

$$\varepsilon_1 = \sup\{\varepsilon:\ 1 > \varepsilon t_{2i-1} > \frac{1}{\varepsilon}\, t_{2i} > \varepsilon t_{2i+1},\quad i=1,2,\ldots,n-1\}.$$

Then for any ε $(1 < \varepsilon < \varepsilon_1)$ the set

$$\underline{t}_\varepsilon = \{\varepsilon t_1,\ \frac{1}{\varepsilon}\, t_2,\ \varepsilon t_3, \ldots, \varepsilon t_{2n-1},\ \varepsilon t_{2n}\}$$

satisfies (3.3) as well; consequently with an arbitrary chosen $\mathbf{R}$ the corresponding system

$$r_{i\varepsilon} = \mathbf{R}\,\frac{t_{i\varepsilon}}{1-t_{i\varepsilon}}\ ,\quad i=1,2,\ldots,N$$

determines an **SSC** which can be denoted by $\underline{\mathbf{C}}_\varepsilon$.

REMARK 3.2. Turning to the logarithms in the system of equations (3.3) a system of linear equations will be obtained with the $N \times N$ determinant

$$\begin{vmatrix} 1 & 1 & 0 & 0 & \ldots & 0 \\ 0 & 1 & 1 & 0 & \ldots & 0 \\ 0 & 0 & 1 & 1 & \ldots & 0 \\ & & & & & \\ 1 & 0 & 0 & 0 & \ldots & 1 \end{vmatrix} = 1 - (-1)^N = \begin{cases} 0 \text{ if } N = 2n\ , \\ 2 \text{ if } N = 2n+1\ ; \end{cases}$$

this justifies the unicity of the solution in the case of an odd N, while the infinity of solution-systems in the case if N is even. -

We are now in the position to answer the question whether a system of angles $\underline{\varphi} = (\varphi_1, \varphi_2, \ldots, \varphi_N)$, satisfying the conditions (3.3) is compatible with an **SSC** $\underline{\mathbf{C}}$, the next theorem settles the case of an odd N only.

THEOREM 3.2. *Let* N *be an odd integer': 2n+1. A system of angles* $\varphi = (\varphi_1, \varphi_2, \ldots, \varphi_N)$ *satisfying the relation (3.2) belongs to an* **SSC** $\underline{\mathbf{C}}$ *if and only if the following inequalities hold:*

$$(3.8) \qquad \frac{\mathbf{A}_i^{(N)}}{\bar{\mathbf{A}}_i^{(N)}} > \frac{\mathbf{B}_{i-1}^{(N)}\ \mathbf{B}_i^{(N)}}{\bar{\mathbf{B}}_{i-1}^{(N)}\ \bar{\mathbf{B}}_i^{(N)}} \quad , \ i=1,2,\ldots,n$$

$$(3.9) \qquad \frac{\mathbf{B}_i^{(N)}}{\bar{\mathbf{B}}_i^{(N)}} > \frac{\mathbf{A}_i^{(N)}\ \mathbf{A}_{i+1}^{(N)}}{\bar{\mathbf{A}}_i^{(N)}\ \bar{\mathbf{A}}_{i+1}^{(N)}} \quad , \ i=1,2,\ldots,n-1$$

$$(3.10) \qquad \frac{\bar{\mathbf{A}}_0^{(N)}}{\bar{\mathbf{B}}_0^{(N)}} \cdot \frac{\bar{\mathbf{A}}_{n-s}^{(N)}\ \mathbf{B}_{n-s}^{(N)}}{\mathbf{A}_{n-s}^{(N)}\ \bar{\mathbf{B}}_{n-s}^{(N)}} > \; > \sin^2\left(\frac{\varphi_{N-2s}}{2} + \varphi_{r-2s+1} + \ldots + \varphi_{N-1} + \frac{\varphi_r}{2}\right)$$

$$\frac{\bar{\mathbf{A}}_0^{(N)}}{\bar{\mathbf{B}}_0^{(N)}} \cdot \frac{\mathbf{A}_{n-s+1}^{(N)} \; \bar{\mathbf{B}}_{n-s+1}^{(N)}}{\bar{\mathbf{A}}_{n-s+1}^{(N)} \; \mathbf{B}_{n-s+1}^{(N)}} >$$

(3.11)

$$> \sin^2\left(\frac{\varphi_{N-2s+1}}{2} + \varphi_{N-2s+1} + \ldots + \varphi_{N-1} + \frac{\varphi_N}{2}\right)$$

$s=1,2,\ldots$.

PROOF. Having an **SSC** $\underline{\mathbf{C}}$, the system of angles can be calculated by means of $\mathbf{R}$ and of $r_1 > r_2 > \ldots > r_N$. Then the conditions $t_{2i} > t_{2i+1}$ and $t_{2i+1} > t_{2i+2}$ have to be satisfied, which is expressed in (3.8) and in (3.9), respectively. (3.10) and (3.11) must also be valid, which means that the circles close to $\mathbf{C}_1$ do not touch it, except $\mathbf{C}_N$. - On the other hand: having the system $\underline{\varphi}$, (3.5) yields the t_i values and due to (3.8) and (3.9) these form a monotone sequence: consequently the circles ith radii $r_i = \frac{t_i}{1-t_i}\mathbf{R}$, $i=1,2,\ldots,N$, with an arbitrarily chosen $\mathbf{R}$, define an **SSC**. The latter property is guaranteed by the validity of (3.10) and (3.11).

4. ON THE LARGEST AND SMALLEST CIRCLES WHICH CAN BE SURROUNDED BY A GIVEN SYSTEM OF CIRCLES

We are looking now for those permutations $\mathbf{P}_{min}$ and $\mathbf{P}_{max}$ for which $\mathbf{R}(\mathbf{P})$ takes on their extreme values.

THEOREM 4.1. *Let* $\underline{\mathbf{C}}(\mathbf{P})$ *be* **SSC** *for all* $\mathbf{P} \in P$. *Then the minimum and maximum values of the radii* $\mathbf{R}(\mathbf{P})$, $\mathbf{P} \in P$ *will be taken on for the permutations given in* (1.4) *and* (1.5) *respectively.*

PROOF. Let us rewrite system (3.3) in the following form:

$$\sin^2 \frac{\varphi_i(\mathbf{P})}{2} = \frac{r_i(\mathbf{P})}{r_i(\mathbf{P})+\mathbf{R}(\mathbf{P})} \frac{r_{i+1}(\mathbf{P})}{r_{i+1}(\mathbf{P})+\mathbf{R}(\mathbf{P})}, \quad i=1,2,\ldots,N$$

where the notation $\varphi_i(\mathbf{P})$, $t_i(\mathbf{P})$, $r_i(\mathbf{P})$ means φ_{j_i}, t_{j_i}, r_{j_i} respectively when $\mathbf{P} = \{j_1=1, j_2, \ldots, j_i, \ldots, j_N\}$. From this we have

$$\varphi_i(\mathbf{P}) = 2 \arcsin \sqrt{\frac{r_i(\mathbf{P})}{r_i(\mathbf{P})+\mathbf{R}(\mathbf{P})} \frac{r_{i+1}(\mathbf{P})}{r_{i+1}(\mathbf{P})+\mathbf{R}(\mathbf{P})}}.$$

Introducing the notation

$$A_N(z,\mathbf{P}) = \sum_{i=1}^{N} \arcsin \sqrt{\frac{r_i(\mathbf{P})}{r_i(\mathbf{P})+z} \frac{r_{i+1}(\mathbf{P})}{r_{i+1}(\mathbf{P})+z}},$$

obviously

$$A_N(\mathbf{R}(\mathbf{P}),\mathbf{P}) = \pi, \text{ for all } \mathbf{P} \in \mathcal{P}. \tag{4.1}$$

We shall apply now Theorem 1.1 with

$$f(x,y) = \arcsin \sqrt{\frac{x}{x+c} \frac{y}{y+c}}, \quad c > 0,\ x > 0,\ y > 0,$$

for which

$$\frac{\partial f}{\partial x} = \frac{c}{2} \sqrt{\frac{y}{x}} \frac{1}{(c+x)\sqrt{(x+y)c+c^2}} > 0,$$

$$\frac{\partial f}{\partial y} = \frac{c}{2}\sqrt{\frac{x}{y}}\,\frac{1}{(c+y)\sqrt{(x+y)c+c^2}} > 0 ,$$

$$\frac{\partial^2 f}{\partial x \partial y} = \frac{c^2}{4\quad xy[(x+y)c+c^2]^{3/2}} > 0$$

hold.

This means that for any value of $z > 0$

$$\mathbf{A}_N(z,\mathbf{P}_{min}) \le \mathbf{A}_N(z,\mathbf{P}) \le \mathbf{A}_N(z,\mathbf{P}_{max}), \quad \mathbf{P} \in \mathcal{P} .$$

Since the function $\mathbf{A}_N(z,\mathbf{P})$ is monotonically decreasing when z increases, the value π can be taken on by it when the corresponding $\mathbf{R}(\mathbf{P})$ values have the same increasing property as the $\mathbf{A}_N(z,\mathbf{P})$-s, as function of $\mathbf{P}$. Hence the relation

$$\mathbf{R}(\mathbf{P}_{min}) \le \mathbf{R}(\mathbf{P}) \le \mathbf{R}(\mathbf{P}_{max})$$

has to be valid.

ACKNOWLEDGEMENT

The author is indebted to Professor L. Fejes Tóth, F. Kárteszi and P.P. Pálfy and last but not least to the referee for their valuable remarks.

REFERENCE

[1] VINCZE, I., *Combinatorial inequalities*. To be published in the Proceedings of the Haaz Memorial Conference (Series of Colloquia Matematica Societatis János Bolyai).

I. VINCZE
Mathematical Institute of the
Hungarian Academy of Sciences
Budapest, Reàltanoda u. 13-15.
H-1053.
HUNGARY

COLLOQUIA MATHEMATICA SOCIETATIS JÁNOS BOLYAI
48. INTUITIVE GEOMETRY, SIÓFOK, 1985.

ON REFLECTIONS IN MÖBIUS PLANES

B. WERNICKE

Let $\mathfrak{P} = \{P, Q, \dots\}$ be a set of points and $\mathfrak{K} = \{k, l, \dots\}$ a subset of the power set of $\mathfrak{P}$. The elements of $\mathfrak{K}$ are called circles. An incidence structure $(\mathfrak{P}, \mathfrak{K}, \in)$ is called a Möbius plane if the following axioms are satisfied [2].

M1. Any three distinct points lie on exactly one circle.

M2. If a point P is on a circle k and Q is not on k, there is exactly one circle through Q whose only common point with k is P.

M3. Every circle contains at least one point; there are four points which do not lie on a circle.

With reference to the ordinary Euclidean plane $\mathfrak{E}$ with one point at infinity, to the complete Gauß plane (geometry of complex numbers) and the Riemannian sphere of numbers we only remark, that investigations of Möbius planes can be made in very different manners: In $\mathfrak{E}$ for all $P \in \mathfrak{P}$ is

$$\mathfrak{E}_P := (\mathfrak{P} \setminus \{P\}, \{k \setminus \{P\} : P \in k \in \mathfrak{K}\}, \in)$$

a Euclidean plane (in general an affine plane). In the Gauß plane $\mathbb{C} \cup \{\infty\}$ is the set of points and the set of circles is the set of images of the projective line $\mathbb{R} \cup \{\infty\}$ under the projective general linear group $PGL_2(\mathbb{C})$. In this sense, the geometry over a pair of fields (K,L) in the place of $(\mathbb{R}, \mathbb{C})$ can be studied. (K,L)-planes were investigated in [14]. They are Möbius planes with the theorem of Miquel (M4) and the theorem of tangent pencil (M5):

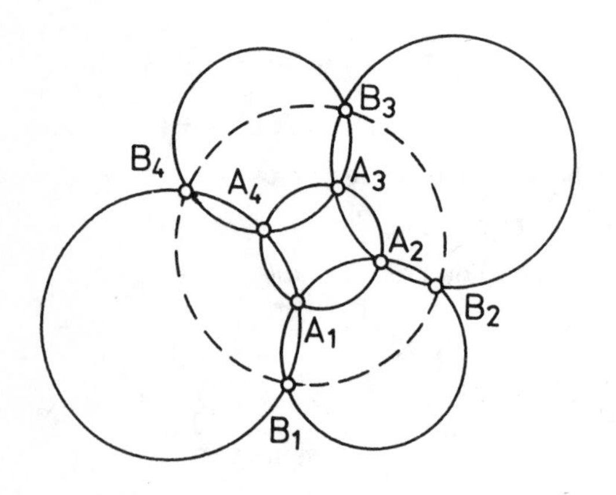

Figure 1

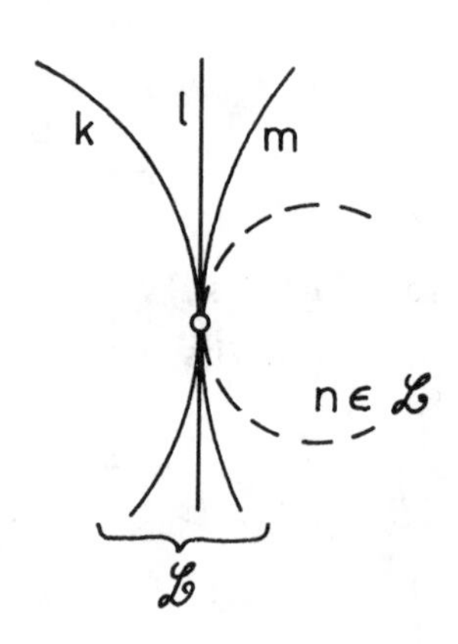

Figure 2

M4. Let A_1, A_2, A_3, A_4, B_1, B_2, B_3, B_4 be eight distinct points. If the quadrupels (A_1,A_2,B_1,B_2), (A_2,A_3,B_2,B_3), (A_3,A_4,B_3,B_4), (A_4,A_1,B_4,B_1), (A_1,A_2,A_3,A_4) are, in each case, elements of a circle, then the points B_1,B_2,B_3,B_4 also lie on a circle.

M5. If k, l, m are three distinct circles of a tangent pencil and circle n touches each of them, then n is a circle of this tangent pencil.

In the present paper a group theoretical axiom system is given for miquelian Möbius planes with the theorem of tangent pencil: It is possible to characterize these Möbius planes in a theory of reflections like "Spiegelungsgeometrie" by Bachmann [1].

At the beginning we refer to results of Dembowski, Karzel and Mäurer [4,8,10]. Reflection geometry in Möbius planes was also investigated by Molnár [11], Lang [9] and in work [13].

In this note we weaken our "Grundannahme" [13], see also [11], and we also give a group theoretical axiom system for miquelian Möbius planes whose characteristics are $\neq 2$. One can say that in this approach the concept "reflection" is a basic concept. In the group theoretical language our axiom system consist of three simple incidence axioms and one orthogonality axiom. With a modification of the "orthogonality" axiom the condition of characteristic $\neq 2$ can be neglected, see [13].

1. REFLECTION, ORTHOGONALITY AND TOUCH

Each transformation, which maps points one to one onto points and circles one to one onto circles and preserves incidence, is called a circle-preserving transformation, in short a circle-transformation. The results of Dembowski [4] taken in consideration a reflection in a circle k can be defined as a non- identical circle-transformation which leaves invariant each point of the circle k.

(1) LEMMA (Dembowski [4]). *If* σ_k *is a reflection in* k, *then* σ_k *leaves fixed all the points on* k *and no other points. For each point* $P \in k$ *exists exactly one tangent pencil* $\mathfrak{B}'$ *with* P *as its point of contact, so that* $\mathfrak{B}' \cup \{k\}$ *is the set of invariant circles under* σ_k *through*

P. *Every reflection is an involution, and in a circle there exists one reflection at most.*

(2) THEOREM (Karzel and Mäurer [10,8]). *A Möbius plane with the properties*

(i) *For any circle* k *there exists a reflection in* k.

(ii) *If* σ *is a reflection and* P, Q *are two points, then* P, Q, $\sigma(P)$ *and* $\sigma(Q)$ *lie on just one circle.*

is isomorphic to a miquelian Möbius plane satisfying the theorem of tangent pencil.

We consider the case that for each point $P \in k$ the tangent pencil $\mathfrak{B}'$ in (1) does not contain the circle k. Möbius planes with (i), (ii) and the conditions $k \notin \mathfrak{B}'$ in the sense of (1) are miquelian Möbius planes whose characteristics $\neq 2$ [10]. In these Möbius planes one can define an orthogonality in the set of pairs of intersecting circles with the relation of touch and vice versa. First we define more generally

$$k \perp l \; :\Longleftrightarrow \sigma_k(l) = l \wedge k \neq l.$$

For all $k, l \in \mathfrak{K}$ we have

$$\sigma_k \sigma_l \sigma_k = \sigma_{\sigma_k(l)}$$

and

$$\sigma_k(l) = l \Longleftrightarrow \sigma_k \sigma_l = \sigma_l \sigma_k.$$

It is easy to prove that in Möbius planes with (i) the condition (ii) in Theorem (2) is equivalent to one of the following properties [10,13].

(iii) If a point P is not on a circle k, then for all circles l with P, $\sigma_k(P) \in l$ the orthogonality $l \perp k$ holds.

(iv) For any circle k and for any points P,Q there is a circle l with P, $Q \in l$ and $\sigma_k(l) = l$.

In miquelian Möbius planes whose characteristics are $\neq 2$ we have now a *criterion for orthogonality* in the set of pairs of intersecting circles.

The intersecting circles k, l *are orthogonal if and only if* k *belongs to a triad of mutually tangent circles, touching one another at three distinct points, while* l *is the connecting circle of these three points.*

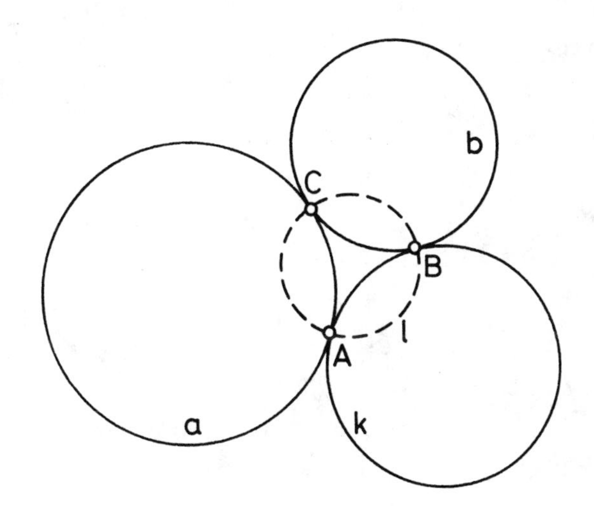

Figure 3

It is easy to see that the condition for the orthogonality $k \perp l$ is necessary: If the two points A,B lie on the circles k,l, than we regard the two circles through a third point C on l, touching k in A and B, respectively. With the properties of reflections in lemma (1) of Dembowski it follows that these two circles are tangent, with C as their point of contact.

Figure 4

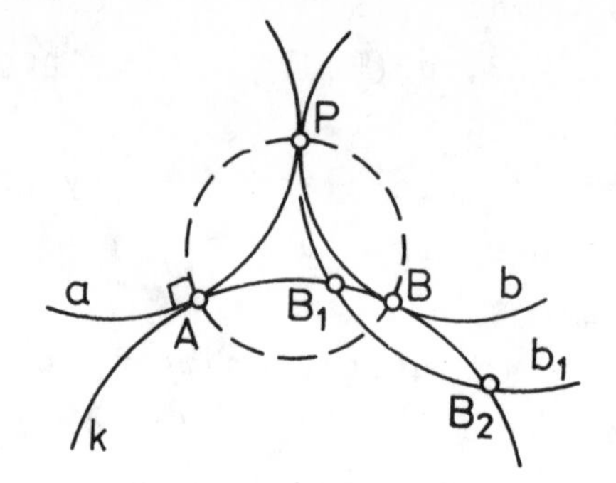

The condition is also sufficient: Let l be the connecting circle of the three distinct points A,B,C and k be a circle of a triad (k,a,b) of mutually tangent circles with A,B,C as points of contact and $A,B \in k$. Consider the connecting circle l' of C,A and $\sigma_k(C)$. From property (iii) it follows that $\sigma_k(l') = l'$ i.e. $k \perp l'$. Therefore $\sigma_{l'}(k) = k$, and we have $\sigma_{l'}(a) = a$ and $\sigma_{l'}(b) = b$ by lemma (1) of Dembowski. Now, since B is the only point of k and b, $\sigma_{l'}(B) = B$, and it follows $B \in l'$ by lemma (1). It is $l' = l$ and $k \perp l$.

The criterion for orthogonality is proved.

If point P lies not on the circle k, we can now construct the point $\sigma_k(P)$.

We regard a point A on k and the circle a through P touching k in A (M2). Let B_1 be a point on k with $B_1 \neq A$. If the circle b_1 through B_1 touching a in P has a second point B_2 in common with k, then we consider the mid-point B of B_1, B_2 in the affine plane with respect to $A = \infty$. (This mid-point B can be constructed as intersection of two diagonals in a parallelogram.) The circle b through B touching a in P has only the point B in common with k. The connecting circle of A,B,P is orthogonal to k by the criterion. Two such orthogonal circles through P to k describe the point $\sigma_k(P)$.

Figure 5

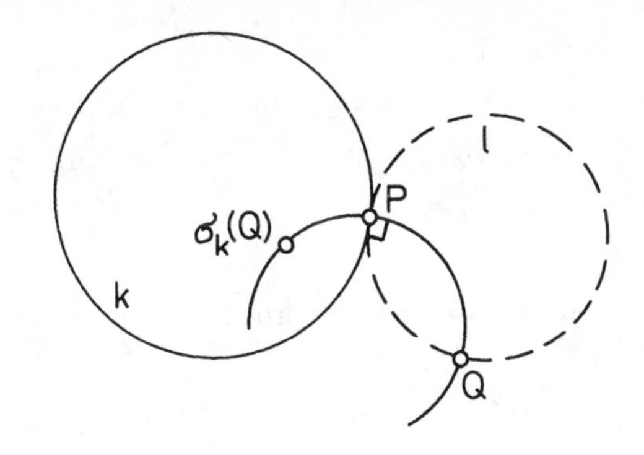

Figure 5 shows, how the relation of touch can be defined by the orthogonality. The point P lies on circle k and Q is not on k. Consider the circle l through P, Q which is orthogonal to the connecting circle of P, Q, $\sigma_k(Q)$. Clearly, the circle l has only the point P in common with k.

2. A GROUP THEORETICAL AXIOM SYSTEM

Consider a group G with a generating set $\mathcal{S} = \{\alpha, \beta, \ldots\}$ and a subset $\mathcal{P} = \{A, B, \ldots\}$ of the power set of $\mathcal{S}$ satisfying the following conditions.

(1) For any $A \in \mathcal{P}$ and any $\alpha \in \mathcal{S}$ it holds

$$\alpha \in A \iff A^{\alpha} := \alpha^{-1} A \alpha = A .$$

(2) For any $\alpha \in \mathcal{S}$ and any $A \in \mathcal{P}$ it is

$$A^{\alpha} \in \mathcal{P} .$$

(3) For any $\alpha \in \mathcal{S}$ there exists a $A \in \mathcal{P}$ with

$$\alpha \in A.$$

The conditions (1), (2), (3) say that $\mathscr{P}$, consisting of self-invariant sets, is invariant under inner automorphisms and is a covering of $\mathscr{S}$. In the group G we define the $|$-relation:

$$a|b \ :\Longleftrightarrow a,\ b,\ ab \text{ are involutions.}$$

The elements of $\mathscr{P}$ are called points, those of $\mathscr{S}$ circles, and with the relation I in $\mathscr{P} \times \mathscr{S}$:

$$A I \alpha \ :\Longleftrightarrow \alpha \in A$$

we have a group plane ($\mathscr{P}$, $\mathscr{S}$, I) of G.

The group G with the set $\mathscr{S}$ of generators and the set $\mathscr{P}$ of points shall satisfy the following axioms.

A1. There are four points not on a generator.
A2. Any three distinct points lie on exactly one generator.
A3. If α, β, γ are any distinct generators and A, B are any distinct points with A, B I α, β and A Iγ, then there exists a third point C on γ lying on α or β.
A4. For any generator α and any points A, B there exists a generator β with A, B I β and $\beta|\alpha$.

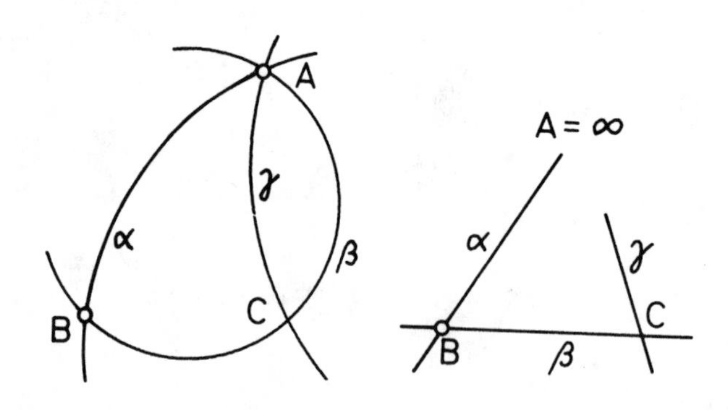

Figure 6

Figure 7

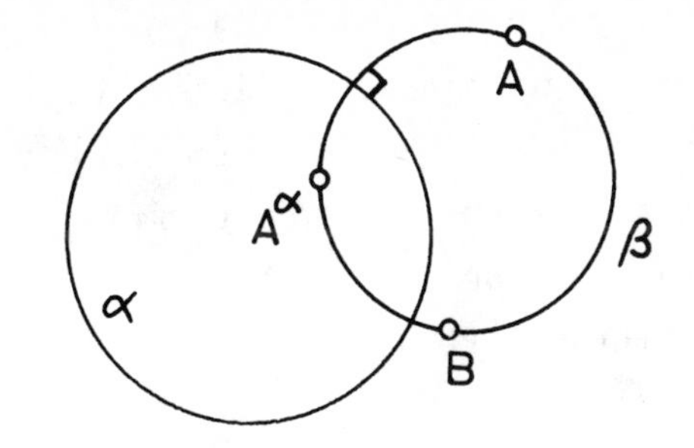

REMARK 1. Since $\mathcal{P}$ is a covering of $\mathcal{S}$ (3) and $\mathcal{P}$ is invariant under inner automorphisms (2) it follows that $\mathcal{S}$ is invariant under inner automorphisms.

Axiom A4 shows that the set $\mathcal{S}$ of generators consists of involutions.

Therefore the I-relation is also invariant under inner automorphisms.

Now we can prove that the group plans ($\mathcal{P}$, $\mathcal{S}$, I) is a miquelian Möbius plane the characteristic of which is $\neq 2$.

PROPOSITION 1. If a point A is on a circle α and B is not on α, there is exactly one circle β through B whose only common point with α is A (touch axiom M2).

PROOF. EXISTENCE. Let A be a point on a circle α and B a point not on α. From conditions (2) and (1) follows that B^{α} is a point different from B and A. Since A2 and A4 hold there exist first a circle γ through A, B, B^{α} with $\gamma|\alpha$ and second a circle β through A, B with $\beta|\gamma$. The circle β has exactly the point A in common with α: If there is a point $C \neq A$ with C I α,β, then C does not lie on γ and $C^{\gamma} \neq C$, A. Since $\gamma|\alpha,\beta$ therefore C^{γ} Iα,β holds and for the circle δ through C, A, C^{γ} (A2) we have $\alpha = \delta = \beta$.

UNIQUENESS. Suppose that β is a circle through B whose only common point with α is A. We regard a circle $\beta' \neq \beta$ through A and B. We have now A, B I β, β'; A Iα and α, β have exactly point A in common. From axiom A3 it follows that β' has the point A and a second point in common with α. Therefore β is the only circle through B touching α in A.

Since axiom A2 (M1), proposition 1 (M2), condition (3) and axiom A1 (M3) hold, the group plane $(\mathcal{P}, \mathcal{S}, I)$ is an Möbius plane. It is easy to see that in a Möbius plane a circle has at least three points.

From lemma (1) of Dembowski in 1. follows

PROPOSITION 2. The mapping

$$\bar{\alpha} : \mathcal{P} \to \mathcal{P} \quad \text{with} \quad \bar{\alpha}(A) = A^{\alpha}$$

determined by a generator α of the group G is a reflection in the circle α of the group plane. Hence, for any circle there exists a reflection.

PROPOSITION 3. If a point A is not on a circle α, then for all circles β with A, A^{α} I β is $\beta|\alpha$.

PROOF. Let α be a circle and A a point not on α. Then $A^{\alpha} \neq A$. Consider a circle β with A, A^{α} I β and a point $B \neq A$, A^{α} with B I β. Since axiom A4 hold there exists a circle β' through A, B with $\beta'|\alpha$. We have now $(A^{\alpha})^{\beta'} = A^{\beta'\alpha} = A^{\alpha}$ and also A^{α} I β'. The axiom A2 says that $\beta' = \beta$.

From theorem of Karzel and Mäurer with respect to the equivalence of (ii) and (iii) in 1. it follows that the group plane $(\mathcal{P}, \mathcal{S}, I)$ is a miquelian Möbius plane whose characteristic is $\neq 2$ (A4 implies that char. $\neq 2$, see lemma of Dembowski and remarks in 1.).

On the other hand, the supposition about the group G and the axioms A1 to A4 are true for the group generated by all reflections in the circles of a miquelian Möbius plane (char. $\neq 2$).

Since in a miquelian Möbius plane every circle-transformation which leaves each point of a circle invariant is exactly determined, we can say that for a group G with the conditions (1) to (3) and axioms A1 to A4 about $\mathcal{S}$ and $\mathcal{P}$ there exists exactly one (up to an isomorphism) miquelian Möbius plane M, so that G is isomorphic to the group of circle-transformations of M.

We have now the main result of this paper.

THEOREM. *A group* G *is the group generated by circle-reflections of a miquelian Möbius plane* (char. $\neq 2$) *if and only if* G *satisfies the conditions* (1) *to* (3) *and axioms* A1 *to* A4 *about* $\mathcal{S}$ *and* $\mathcal{P}$.

3. APPENDIX

Instead of axiom A4 in the group theoretical axiom system in 2. we suppose the following axioms:

A4'. For any generator α and any points A, B there exists a generator β with A, B I β and $\alpha\beta = \beta\alpha$.

A5 . If A, B are two distinct points on generators α, β, γ with $\alpha | \beta, \gamma$, then $\beta = \gamma$.

The condition $\alpha\beta = \beta\alpha$ in A4' is more general than $\alpha | \beta$ in A4. In A4' here $\alpha = \beta$ also allowed, which holds, if A, B I α and $A \neq B$.

In [13] we proved that this more general axiom system characterizes the group of circle-transformations of a miquelian Möbius plane satisfying the theorem of tangent pencil.

REFERENCES

[1] BACHMANN, F., *Aufbau der Geometrie aus dem Spiegelungsbegriff*. 2. Aufl., Springer-Verlag, Berlin-Heidelberg-New York 1973.

[2] BENZ, W., *Vorlesungen über Geometrie der Algebren*. Springer-Verlag, Berlin-Heidelberg-New York 1973.

[3] COXETER, H.S.M., *Inverse geometry*. Educational Studies in Mathematics 3 (1971), 310-321.

[4] DEMBOWSKI, P., *Möbiusebenen gerader Ordnung*. Math. Ann. 157 (1964), 175-205.

[5] EWALD, G., *Axiomatischer Aufbau der Kreisgeometrie*. Math. Ann. 131 (1956), 354-371.

[6] EWALD, G., *Über den Begriff der Orthogonalität in der Kreisgeometrie*. Math. Ann. 131 (1956), 463-469.

[7] FEJES TÓTH, L., *Regular figures*. Pergamon Press, Oxford 1964.

[8] KARZEL, H., and H. Mäurer, *Eine Kennzeichnung miquelscher Möbiusebenen durch eine Eigenschaft der Kreisspiegelungen*. Resultate d. Math. 5 (1982), 52-56.

[9] LANG, K., *Kennzeichnung von Möbiusebenen durch Spiegelungsrelationen*. J. of Geometry 16 (1981), 5-18.

[10] MÄURER, H., *Kreisspiegelungen in Möbiusebenen.* Geom. Ded. 2 (1973), 261-268.

[11] MOLNÁR, E., *Begründung der Möbiusschen Kreisebene aus dem Spiegelungsbegriff.* Ann. Univ. Sci. Budapest, Sect. Math. 18 (1975), 143-170.

[12] QUAISSER, E., *Zum Aufbau affiner Ebenen aus dem Spiegelungsbegriff.* Z. Math. Logik u. Grundlagen der Math. 27 (1981), 131-140.

[13] WERNICKE, B., *Möbiusebenen in spiegelungsgeometrischer Darstellung.* Studia Sci. Math. Hungarica (to appear).

[14] ZEITLER, H., *Über (K,L)-Ebenen.* Dissertation, Kassel 1977.

B. WERNICKE
Pädagogische Hochschule
Sektion Mathematik/Physik
DDR-5064 Erfurt
Nordhäuser Str. 63.

COLLOQUIA MATHEMATICA SOCIETATIS JÁNOS BOLYAI
48. INTUITIVE GEOMETRY, SIÓFOK, 1985.

ON ISOPERIMETRIC TETRAHEDRA

G. WEIß

Dedicated to Prof. Dr. F. Hohenberg on the occasion of his 80th birthday

L. FEJES TÓTH suggested in [3, p. 284] to destinate the convex polyhedron P of given character and volume V minimizing the sum of the lenghts of its edges. (Furtheron this sum shortly will be named the "perimeter" U of P). Whilst this general problem still is unsolved, several spezial cases and modifications of that problem have been attacked successfully, (c.f. [8], [10], [11], [12], nad [1], [2], [5], [6], [7], [9, p. 567]). Already in 1928 G. SANSONE [11] proved

THEOREM 1. *Among all tetrahedra* T *of equal volume* V *the regular one* T_0 *has minimal perimeter.*

Meanwhile theorem 1 has been proved independently and in different ways by Z.A. MELZAK 8 , G. RIEGER*, and the author**.

---

*Lecture given at the DMV-Tagung 1984, Kaiserlautern (BRD). RIEGER discusses a formula of HERON-type [4, p. 1056] to describe the volume V of T as a function of U and the lenghts of the tetrahedron's edges.

**Lecture given at the Conference on Intuitive Geometry 1985, Siofok (Hungary). The authors's elementary proof is based on STEINER-symmetrization of T, so that T gets two symmetry-planes, and discusses the "isoperimetric quotient" U^3/V of such symmetrizised tetrahedra.

In the following we give generalizations of Theorem 1 with regard to the dimension of T.

Let T $\mathbb{R}^n$ be an n-dimensional simplex with vertices $A_0,\dots,A_n$. Any k+1 vertices of T, $(1 \le k < n)$, form a k-dimensional face $T_{(k);j_k}$, $(j_k \in \{1,\dots,\binom{n+1}{k+1}\})$, the k-measure of which we abbreviate by $V_{(k);j_k}$. The sum of the k-measures of all the $\binom{n+1}{k+1}$ k-faces $T_{(k);j_k}$ is usually named "k-measure" of T. We state

THEOREM 2. *Among all* n*-simplices* T *of equal volume* $V_{(n)}$ *the regular one has minimal* k*-measure.*

PROOF. Following an idea of KORCHMÁROS G. [13] we first look at the (n-1)-faces $T_{(n-1);i}$ of T, $(i=0,\dots,n)$. As T is convex, it is possible to orientate the unit-vectors e_i, normal to the face-hyperplanes of T, "outwards" with respect to T. Let be E_i the convex hull of $\{0;\, e_0,\dots,e_{i-1},\, e_{i+1},\dots,e_n\}$ and let be $\mathrm{int}E_i$ the set of inner points of E_i, then convexity of T causes $\mathrm{int}E_i \cap \mathrm{int}E_j = \emptyset \quad \forall i \ne j \in \{0,\dots,n\}$, and the convex hull E of $\{0,\, e_0,\dots,\, e_n\}$ is the union of all simplices E_i. For such a simplex E, inscribed in the unit-sphere of $\mathbb{R}^n$, the volume W is limited according to [4, p. 191] by

$$(1) \qquad W \le \frac{1}{n!}\,(n+1)^{\frac{n+1}{2}} \cdot n^{-\frac{n}{2}} \quad .$$

In (1) equality holds exactly for regular E. (It is immediately obvious that E is regular iff T is regular.) On the other hand, using the n-volumes W_i of E_i and the (n-1)-measures $V_{(n-1);i} =: V_i$ of $T_{(n-1);i} \subset T$, BARTOS P. (c.f. [3]) gave the following formulae for the volume $V_{(n)}$ of an n-simplex:

$$V_{(n)} = \frac{1}{n} \left(n!(n-1)!\; V_0 \cdot V_1 \cdots V_{i-1} \cdot W_i \cdot V_{i+1} \cdots V_n\right)^{\frac{1}{n-1}}, \tag{2}$$

$$(i=0,\ldots,n).$$

By multiplying these n+1 equations, and because of

$$(W_0 \cdots W_n)^{\frac{1}{n+1}} \le \frac{1}{n+1} (W_0 + \ldots + W_n) =: \frac{1}{n+1} \cdot W, \tag{3}$$

we get, by replacing W according to (1),

$$\begin{aligned} V_{(n)} \le n!^{\frac{1}{n+1}} \cdot \; n+1 \cdot n^{-\frac{3n}{2(n-1)}} \, . \\ \cdot \left(V_{(n-1);0}, \ldots, V_{(n-1);n}\right)^{\frac{n}{n^2-1}} . \end{aligned} \tag{4}$$

Again equality holds if and only if $W_i = W_j$, that is for a regular n-simplex T. Now (4) can be taken as the initial term of a recursive process, the q-th step of which we assume to deliver

$$\begin{aligned} V_{(n)} \le S_{(n)} \cdot S_{(n-1)}^{\frac{n}{n-1}}, \ldots, S_{(n-q+1)}^{\frac{n}{n-q+1}} \, . \\ \cdot \left(\prod_{1}^{\binom{n+1}{q}} V_{(n-q);j_q} \right)^{\frac{n}{(n-q)\binom{n+1}{q}}} . \end{aligned} \tag{5}$$

Within (5) we made the abbreviation

$$(6) \qquad S_{(\alpha)} := \sqrt{\alpha+1} \cdot (\alpha! \, \alpha^{-\frac{3\alpha}{2}})^{\frac{1}{\alpha-1}}, \quad \alpha \in \mathbb{N}$$

We point out that equality holds iff T is regular. After replacing n in (4) by n-q and combining (4) with (5) we get as the (q+1)-th step

$$(7) \qquad V_{(n)} \leq S_{(n)}^{\frac{n}{n}} \cdots S_{(n-q)}^{\frac{n}{n-q}} \cdot \left(\prod_{1}^{\binom{n+1}{q}} \prod_{1}^{(n-q-1)} V_{(n-q-1);j_{q+1}} \right)^{\frac{n}{(n-q-1)(q+1)\binom{n+1}{q+1}}} .$$

Each (n-q-1)-face $T_{(n-q-1);j_{q+1}}$ belongs to exactly q+1 faces $T_{(n-q);j_q}$ and therefore in (7) the twofold product within the brackets can be written as

$$(8) \qquad \prod_{1}^{\binom{n+1}{q}} \prod_{1}^{(n-q-1)} V_{(n-q-1);j_{q+1}} = \prod_{1}^{\binom{n+1}{q+1}} V_{(n-q-1);j_{q+1}}^{q+1} .$$

Thus follows

$$(9) \qquad V_{(n)} \leq S_{(n)}^{\frac{n}{n}} \cdots S_{(n-q)}^{\frac{n}{n-q}} \left(\prod_{1}^{\binom{n+1}{q+1}} V_{n-q-1);j_{q+1}} \right)^{\frac{n}{(n-q-1)\binom{n+1}{q+1}}}$$

as we had to show. For the product $\gamma_{(n;q)}$ of the terms $S_{(\alpha)}^{n/\alpha}$ in (9) a little reckoning delivers

$$(10)\qquad \gamma_{(n;q)} = \frac{1}{n!}\sqrt{n+1}\cdot (n-q)!^{\frac{n}{n-q-1}} \cdot (n-q)^{\frac{-3n}{2(n-q-1)}} .$$

Together with the mean value inequality

$$(11)\qquad \prod_{1}^{\binom{n+1}{q+1}} V_{(n-q-1);j_{q+1}} \le \left(\frac{1}{\binom{n+1}{q+1}} \sum_{1}^{\binom{n+1}{q+1}} V_{(n-q-1);j_{q+1}} \right)^{\frac{1}{\binom{q+1}{n+1}}} ,$$

and with k instead of n-q-1 follows from (9)

$$(12)\qquad V_{(n)} \le \frac{\sqrt{n+1}\cdot (k+1)!^{\frac{n}{k}}}{n!(k+1)^{\frac{3n}{2k}} \cdot \binom{n+1}{k+1}^{\frac{n}{k}}} \left(\sum_{1}^{\binom{n+1}{k+1}} V_{(k);j_k} \right)^{\frac{n}{k}} , \quad (1 \le k \le n-1) .$$

In the first instance equality in (12) holds exactly if all k-faces $T_{(k);j}$ of T have equal k-volume. But this implies regularity of the n-simplex T, as it obviously follows from the inequality (9), which can be received from (4) by recursion.

REMARK. The case k=1 has been treated by J.N. LILLINGTON [7, p. 522]; his proof is based on relations between the volume, the radii of circumsphere and insphere, and the perimeter of T.

Let us now replace n,k in (12) by k,ℓ ($\ell \le k-1$), such that for any $V_{(k);j_k}$ in (12) holds an analogues inequality to (12), limiting the k-volume of $V_{(k);j_k}$ by its ℓ-measure.

Thus follows an inequality between the k-measure of $V_{(n)}$ and a function of its ℓ-measure, whereby equality occurs only if $V_{(n)}$ is regular. By reversing the arguments we get the same result also for $\ell > k$. In extension of theorem 2 we now have shown

THEOREM 3. *Among all* n-*simplices* T *of equal* k-*measure the regular one has extremal* ℓ-*measure.*

REMARK. For $n=3$, $k=2$, $\ell=1$ the result above is a consequence of a theorem due to M. KÖMHOFF [5]. The author is indepted to her for numerous references.

REFERENCES

[1] ABERTH, O., *An isoperimetric inequality for polyhedra and its application to an extremal problem.* Proc. London Math. Soc. (3), 13 (1963), p. 322-336.

[2] ABERTH, O., *An inequality for convex polyhedra.* J. London Math. Soc. (2), 6 (1973), p. 382-384.

[3] BARTOS, P., *Sinusova veta o simplexoch v* E_n. Casopis Pest Math. 93 (1968), p. 273-277.

[4] FEJES TÓTH, L., *Reguläre Figuren.* Akademiai Kiadó, Budapest, 1965.

[5] KÖMHOFF, M., *An isoperimetric inequality for convex polyhedra with triangular faces.* Canad. Math. Bull. 11 (1968), p. 723-727.

[6] KÖMHOFF, M., *On a 3-dimensional isoperimetric problem.* Canad. Math. Bull. 13 (4), (1970), p. 447-449.

[7] LILLINGTON, J.N., *Some extremal properties of convex sets*. Math. Proc. Camb. Phil. Soc. 77 (1975), p. 515-524.

[8] MELZAK, Z.A., *An isoperimetric inequality for tetrahedra*. Canad. Math. Bull. 9 (1966), p. 667-669.

[9] MELZAK, Z.A., *Problems connected with convexity*. Canad. Math. Bull. 8 (1965), p. 565-573.

[10] PROCISSI, A., *Sul massimo volume di alcune specie di poliedri dei quali e data la somma degli spigoli*. Period. Matem. IV, 12 (1932), p. 159-172.

[11] SANSONE, G., *Sulle espressioni del volume del tetraedro e su qualche problema di massimi*. Period. Matem. IV, 3 (1928), p. 20-50.

[12] SANSONE, G., *Su una proprieta di massimo dell' ottaedro regolare e del cubo*. Boll. Unione Mat. Ital. II, 3 (1940), p. 140-146.

[13] VELJAN, D. - KORCHMÁROS, G., *Problem 629A*, Elemente d. Math. 27 (1972).

[14] ZACHARIAS, M., *Elementargeometrie*. Enzykl. d. math. Wiss. III A B 9; Teubner, Leipzig, 1914-1931.

G. WEIß
Institut für Geometrie
Technische Universität Wien
A-1040. WIEN
Wiedner Hauptstr. 8-10.

COLLOQUIA MATHEMATICA SOCIETATIS JÁNOS BOLYAI
48. INTUITIVE GEOMETRY, SIÓFOK, 1985.

POLYHEDRAL COVERS

B. WEIßBACH

1. Let $\mathfrak{M} = \{M_i : i \in I\}$ be a family of sets in the d-dimensional euclidean space E^d, and let $\mathfrak{G}$ be a group of isometric maps in E^d. A futher set $D \subset E^d$ is then called a cover to $\mathfrak{M}$ relative to $\mathfrak{G}$ if and only if for every $M_i \in \mathfrak{M}$, a $\varphi_i \in \mathfrak{G}$ exists, such that $\varphi_i(D) \supset M_i$ holds. If $\mathfrak{G}$ is the group of all motions in E^d and $\mathfrak{M}$ is the family of all bounded sets with diameter not greater then a given number $\delta > 0$ (the value of δ is not essential) then D is mostly called simply a cover or a universal cover.
This definition is unsatisfying in a certain way. Therefore, a sharper definition is to be used here. A set $D \subset E^d$ is called a "strong" cover if and only if D contains a congruent copy of every set with diameter not greater than 2, and there is only a ball with radius 1 contained in D.
Obviously a parallelotope circumscribed to the unit ball is a strong cover.
We ask the question if further polytopes are such covers. In the following we summarize some results known at this time.

2. We restrict our attention to polytopes with midpoint. Especially let M(m) be the set of all convex polytopes circumscribed to the unit ball $B^d := \{x : \|x\| = 1\}$, which are symmetric to the origin and have 2m facets. If the polytope D is a cover in E^d than it is clear that the $(d+1)$-dimensional cylinder of height 2 with D as base is a cover in E^{d+1}. We are talking only about such covers - called essential - which cannot be constructed in this way.

In 1920 I. Pål [9] showed, that for $d = 2$ a (undoubtably strong) cover exists in M(3), - for instance a regular hexagon is a such cover. D. Gale [4] in 1953 stated, that for $d = 3$ a regular octahedron belonging to M(4) is a cover.

Indeed Gale has not proved this statement. The first proof of Gales conjecture was given presumably by H. Meschkowski in 1960 ([8]). Another proof was given by M. G. Boltjanski and J. Z. Gochberg in 1965 ([2]).

The theorems of Pål and Gale are special cases of the following: If T is a d-dimensional simplex circumscribed to B^d and there exists a orientation - preserving motion φ in E^d with $\varphi(T) = -T$, then $T \cap -T$ (belonging to M(d+1)) is a cover.

This statement is to be found without proof in the known report "Helly's theorem and it's relatives" by Danzer, Grünbaum and Klee ([3]).

Proofs are given by the author in 1977 ([10]) and by V. Makejev in 1979 ([7]). Lately V. Makejev stated remarkable results on covers. He showed that for $d = 3$ covers exist in M(5), and that for $d = 4$ even all polytopes in M(7) and for $d = 8$ all polytopes in M(15) are (strong) covers.

In the following first of all we will give sufficient conditions for a polytope $D \in M(m)$ which ensure D is a cover. The way to such conditons seems to be simpler than the approach of Makejev. According to that we want to show that all the polytopes in E^d, $d \geq 3$, belonging to $M(d+2)$, which satisfy a special condition are covers. If d is even this condition is valid for every polytope in $M(d+2)$, i. e. all these polytopes are covers in E^d.

3. To prove that a polytope D circumscribed to the unit ball is a strong cover, it is sufficient to show, that D contains a congruent copy of every body of constant width 2. This will hold if for such a body K there exists a point $x \in E^d$ and a motion φ with the origin as fixed point such that

$$(1) \qquad K + x \subset \varphi(D) \quad .$$

If K is a convex body with the support function $h(v)$, then the support function $h_x(v)$ of a translate $K + x$ of K satisfies the equation

$$(2) \qquad h_x(v) = h(v) + \langle v, x \rangle \quad .$$

If K is of constant width 2, then

$$(3) \qquad h(v) + h(-v) = 2$$

for every point v on the unit sphere $S^{d-1} := \{v : \|v\| = 1\}$. Suppose the polytope D touches the unit ball B^d in the points $\pm v_i$, $i=1,\ldots,m$. Then it follows from (2) and (3) that $K + x$ lies in $\varphi(D)$ if and only if the m equations

$$(4) \qquad h_x(\varphi(v_i)) = h(\varphi(v_i)) + \langle \varphi(v_i), x \rangle = 1$$

are satisfied. Instead of (4) it is appropriate to write

$$(5) \qquad \langle\varphi(v_i),\ x\rangle = k(\varphi(v_i));\ i=1,\dots,m\ .$$

Here $k(v) := 1 - h(v)$ is a continuous odd function on S^{d-1}. Therefore one obtains the sufficient condition: A polytope with midpoint, touching the unit ball in the points $\pm v_i$, is a strong cover, if to every odd continuous function $k(v)$ on S^{d-1} a motion φ in the rotation group $SO(d)$ exists such that the system $\langle\varphi(v_i),x\rangle = k(\varphi(v_i))$ has a solution $x \in E^d$.

4. If $k(v)$ is given and φ is known, than (5) is a system of linear equations for the coordinates of x. One may suppose that $\{v_i,\ i=1,\dots,d\}$ is a base in E^d. Then

$$(6) \qquad v_{d+r} - \sum_{i=1}^{d} \lambda_{ir} v_i = 0;\ r=1,\dots,n;\ n+d=m$$

holds with uniquely determined numbers λ_{ir}, and (5) has a (unique) solution x, if and only if

$$(7) \qquad F_r(\varphi) := k(\varphi(v_{d+r})) - \sum_{i=1}^{d} \lambda_{ir}\, k(\varphi(v_i)) = 0;$$

$$r=1,\dots,n$$

holds too. To prove that a polytope $D \in M(d+n)$ is a cover in E^d, it is therefore enough to show that (independently of the choice of the function $k(v)$) by the continuous map

$$(8) \qquad f : SO(d) \to R^n,\quad \varphi \to (F_1,\dots,F_n)$$

the origin in R^n is the image of a motion $\varphi \in SO(d)$. This succeeds surely by additional suppositions about the structure of the polytope D in the cases $n = 1$ and $n = 2$. We shall suppose that there is a involution $\varphi_1 \in SO(d)$, (i. e. $\varphi_1 \circ \varphi_1 = \varphi_o$, where in the sequel φ_o stands for the identity in E^d) such that the conditions

$$(9) \qquad \begin{aligned} &\varphi_1(v_{d+r}) = -v_{d+r}, \ r=1,...,n \\ &\varphi_1(\{v_1,...,v_d\}) = \{-v_1,..., -v_d\} \end{aligned}$$

are valid.

This demand is superfluous if d is even, because then the map $\varphi_1 : E^d \to E^d$, $x \to -x$ lies in SO(d). If d is odd also a great number of essential polytopes exist to which one can find such a map - in the case $n = 1$ for instance if the convex hull of the points v_i is a regular simplex. From the conditions (9) by means of the uniqueness of the coordinates λ_{ir} it can be seen easily that the relations

$$(10) \qquad F_r(\varphi \circ \varphi_1) = -F_r(\varphi \circ \varphi_o); \ r=1,...,n$$

are valid for every $\varphi \in SO(d)$.

(From $0 = v_{d+r} - \sum_{i=1}^{d} \lambda_{ir} v_i = -v_{d+r} + \sum_{i=1}^{d} \lambda_{j_i r} v_{j_i}$

and $0 = \varphi_1(0) = -v_{d+r} + \sum_{i=1}^{d} \lambda_{ir} v_{j_i}$ follows $\lambda_{ir} = \lambda_{j_i r}$

and consequently

$$F_r(\varphi \circ \varphi_1) = k(\varphi(\varphi_1(v_{d+r}))) - \sum_{i=1}^{d} \lambda_{ir} k(\varphi(\varphi_1(v_i)))$$

$$= -k(\varphi(v_{d+r})) + \sum_{i=1}^{d} \lambda_{j_i r}\, k(\varphi(v_{j_i})) =$$

$$= -F_r(\varphi \circ \varphi_0).)$$

In the case $m = d+1$ only the condition $F_1(\varphi) = 0$ has to be satisfied. Because the rotation group $SO(d)$ is connedted one finds a motion φ_t with $F_1(\varphi_t) = 0$ on a path $w := \{\varphi_t : 0 \leq t \leq 1\}$ which connects the identity with the given map φ_1, since $F(\varphi_t)$ is a continuous function of t and $F_1(\varphi_1) = -F_1(\varphi_o)$ according to (10). Because the Betti number B_o of a polyhedron states the number of its components one can say that on account of $B_o(SO(d)) = 1$ in E^d there are essential covers in $M(d+1)$.
In the case $m = d+2$ the Betti number B_1 of $SO(d)$ has analogous importance. Suppose that there is no motion $\varphi \in SO(d)$ with $F_1(\varphi) = 0$ and $F_2(\varphi) = 0$.
Then with $\varphi \to (\xi_1, \xi_2)$

$$(11) \qquad \xi_i := F_i(F_1{}^2 + F_2{}^2)^{-\frac{1}{2}}, \quad i=1,2$$

a continuous map f^* from $SO(d)$ in the circle S^1 is given. It follows from (10) that $f^*(\varphi \circ \varphi_1)$ and $f^*(\varphi \circ \varphi_o)$ are antipodal points of S^1 for every $\varphi \in SO(d)$.
Hence the image of the path $w = \{\varphi_t : 0 \leq t \leq 1\}$ by f^* covers at least a closed half circle.
Further one now sets $\bar{w} := \{\varphi_t \circ \varphi_1 : \varphi_t \in w\}$, then the image of the closed path $w \cup \bar{w}$ by f^* covers S^1, and obviously it is not continuously contractible on S^1 in a single point, i. e. the map f^* is not homotopic to zero. Now it is known (see Alexandroff-Hopf [1] p. 517) that there is

only one class of continuous maps from a polyhedron in the circle, if the Betti number B_1 of this polyhedron is zero. L. Pontrjagin and other authors showed in the thirties, that $B_1(SO(d)) = 0$ for every $d \geq 3$ (see [5] p. 238). Therefore any continuous map from SO(d) into S^1 is then homotopic to zero, and the assumption turns out as false. Consequently in E^d, $d \geq 3$, there exists essential covers in M(d+2).

It is to be conjectured that also for $d + 3 \leq m \leq \frac{1}{2} d(d+1)$ there are convers in M(m). In order to prove it, one presumably has to apply besides the Betti numbers more topological invariants of SO(d) e. g. the degree of acyclicity in the sense of Jaworowski [6].

REFERENCES

[1] ALEXANDROFF, P., HOPF, H., *Topologie*. Bd 1 Berlin, 1935.

[2] BOLTJANSKI, W. G., GOCHBERG, I. Z., *Sätze und Probleme der kombinatorischen Geometrie*. Berlin, 1972.

[3] DANZER, L., GRÜNBAUM, B., and KLEE, H., *Helly's theorem and its relatives*. Proc. Symp. Pure Math. Vol. VII. (Amer. Math. Soc.) 1963; 101-180.

[4] GALE, D., *On inscribing n-dimensional sets in a regular n-simplex*. Proc. Amer. Math. Soc. 4 (1953); 222-225.

[5] HODGE, W. V. D., *Theorie und Anwendungen harmonischer Integrale*. Leipzig, 1958.

[6] JAWOROWSKI, J. W., *On antipodal sets on the sphere and on continuous involutions*. Fundamenta Math. XLIII. (1956); 241-254.

[7] MAKAJEV, V., *Universal coverings I and II.* (in Russian). Ukrainsk. Geometr. Sbornik 24 (1981); 70-73 and 25 (1982); 82-86.

[8] MESCHKOWSKI, H., *Ungelöste und unlösbare Probleme der Geometrie.* Braunschweig, 1960.

[9] PÁL, J., *Über ein elementares Variationsproblem.* Det. Kgl. Danske Vidensk. Selskab. Math. - Fys. Medd. III/2 (1920).

[10] WEIßBACH, B., *Pálsche Deckel im* E^n, Materialien 2. Tagung Geometrie und Anwendungen Math. Gesell. DDR, 1977.

B. Weißbach
Sektion Math.-Physik der Techn. Hochschule
Magdeburg, PSF 124.
DDR-3010.

COLLOQUIA MATHEMATICA SOCIETATIS JÁNOS BOLYAI
48. INTUITIVE GEOMETRY, SIÓFOK, 1985.

PERIODIC ISOCLINAL SEQUENCES

J.B. WILKER

1. INTRODUCTION

An isoclinal sequence of n-balls in inversive n-space $\mathbb{R}^n \cup \{\infty\}$ is a doubly infinite sequence B_k, $k = 0, \pm 1, \pm 2, \ldots$ in which each pair of n-balls B_i, B_j with indices i,j satisfying $|i-j| \leq n+1$ has the same conformal invariant $\gamma = B_i * B_j$. The geometric meaning of γ is a follows. When $|\gamma| \leq 1$, the bounding (n-1)-spheres ∂B_i, ∂B_j intersect and $\gamma = \cos\varphi$ where φ is the dihedral angle measured between their inward pointing normals. When $|\gamma| > 1$, $\partial B_i \cap \partial B_j \neq \emptyset$ and $\gamma = \pm\cosh\delta$ where δ is the inversive distance between ∂B_i and ∂B_j and the sign is + or - as the n-balls are nested or not.

Isoclinal sequences exist if and only if $\gamma < -\frac{1}{n+1}$, $\gamma \neq -\frac{1}{n}$, and each instance determines a conformal transformation that advances the sequence of bounding (n-1)-spheres by one notch $\partial B_k \to \partial B_{k+1}$. The conjugacy class of these conformal transformations is determined by n and γ and usually it is of loxodromic type. There are exceptional cases however and these occur in odd dimensions. When n is odd and $\gamma = -\frac{n+2}{n^2+3n+1}$, the transformation is an $(\frac{n-1}{2}, 0)$-parabolic; when n is odd and

$$-\frac{n+2}{n^2+3n+1} < \gamma < -\frac{1}{n+1},$$

the transformation is an $(\frac{n+1}{2}, 0)$-elliptic.

The occurence of elliptic transformations raises the possibility of periodic isoclinal sequences. These will occur whenever there are $\frac{n+1}{2}$ distinct rational multiples of π in $(0, \frac{\pi}{2})$ that share the same value in $(0, \frac{1}{n+2})$ of the function $\frac{\tan(n+2)\theta}{\tan\theta}$. When $n = 1$ there is only one angle to consider and hence there are, in a trivial way, infinitely many periodic sequences; when $n = 3$ it turns out that there is a single instance of a periodic sequence and it occurs at $\gamma = \frac{1-2\sqrt{2}}{7}$; when $n > 3$ there are no known examples of periodic sequences but the question of existence is still open. We shall develop these results and relate the periodic sequence in dimension $n = 3$ to the Petrie polygons of two dual {3,4,3}'s.

For previous work on isoclinal sequences see Coxeter [3], Weiss [6], [7], Wilker [10] and the references they cite, especially [8] and [9]. In the next section we shall proceed by recalling futher details of this previous work to set the present material in context. I would like to thank A. Baker, H.S.M. Coxeter and P.J. Leah for very helpful conversations in connection with periodic isoclinal sequences.

2. ISOCLINAL SEQUENCES OF LOXODROMIC TYPE

Consider dimension $n=2$ and let B_0, B_1, B_2, B_3 be four mutually tangent disks. There is a unique disk $B_4 \neq B_0$ tangent to B_1, B_2, B_3 and if we drop B_0 and add B_4 we have taken the first step forward in defining a doubly infinite sequence of disks $\ldots, B_{-2}, B_{-1}, B_0, B_1, B_2, \ldots$ with the property that any four successive disks are mutually tangent. This is the case $n = 2$, $\gamma = -1$ of an isoclinal sequence and, apart from the possibility of applying a conformal transformation or of simultaneously replacing each disk by its complement, the sequence is uniquely determined by its parameters. The points of contact of successive disks in this sequence lie on a loxodrome and if we invert one end of this loxodrome to ∞ and translate the other end to 0 we see that for suitable δ and θ the (1,0)-loxodromic transformation or spiral similarity $z \to e^{2(\delta+i\theta)}z$ advances our sequence by one notch.

For other values of $\gamma < -\frac{1}{2}$ the story is qualitatively the same. The γ-clinal sequence is essentially unique; there is a conformal transformation advancing it by one notch; this conformal transformation is a (1,0)-loxodromic and all that really changes with γ is the conjugacy class of the driving transformation determined by $\delta(\gamma)$ and $\theta(\gamma)$. For $-\frac{1}{2} < \gamma < -\frac{1}{3}$ the story is slightly different. The γ-clinal sequence is unique up to conformal transformations alone and there is no conformal transformation mapping $B_k \to B_{k+1}$. Instead there is a conformal transformation that maps $\partial B_k \to \partial B_{k+1}$ and $B_k \to -B_{k+1}$ where $-B_{k+1}$ is the complement of B_{k+1}. This driving transformation is again a (1,0)-loxodromic whose conjugacy class depends on γ.

The two-dimensional situation gives a very good indication of what happens in all the even dimensions. If $n = 2m$ and $\gamma < \frac{-1}{n+1}$, $\gamma \neq -\frac{1}{n}$, there is a γ-clinal sequence B_k unique up to conformal transformations and simultaneous complementation if $\gamma < -\frac{1}{n}$ and unique up to conformal transformations alone if $-\frac{1}{n} < \gamma < -\frac{1}{n+1}$. This sequence determines a conformal transformation that maps $\partial B_k \to \partial B_{k+1}$ and sends $B_k \to B_{k+1}$ or $B_k \to -B_{k+1}$ as $\gamma < -\frac{1}{n}$ or $-\frac{1}{n} < \gamma < -\frac{1}{n+1}$. The conjugacy class of this conformal transformation depends on γ but it is always an $(m,0)$-loxodromic and can be conjugated to

$$(z_1, z_2, \ldots, z_m) \to e^{2\delta}(e^{2i\theta_1} z_1, e^{2i\theta_2} z_2, \ldots,$$

$$\ldots, e^{2i\theta_m} z_m).$$

In odd dimensions the γ-clinal sequence and driving transformation are determined exactly as they are in even dimensions. The thing that is quite different in the odd dimensions is the dependence of the conjugacy class of the driving transformation on γ. If $n = 2m+1$ and $\gamma < -\frac{1}{n}$ it is an $(m,1)$-loxodromic:

$$(z_1, z_2, \ldots, z_m, x_{2m+1}) \to$$

$$\to e^{2\delta}(e^{2i\theta_1} z_1, e^{2i\theta_2} z_2, \ldots, e^{2i\theta_m} z_m, -x_{2m+1}).$$

This is a sense-reversing transformation and it advances $B_k \to B_{k+1}$. All the other transformations to be mentioned here are sense-preserving and advance $B_k \to -B_{k+1}$. When

$$-\frac{1}{n} < \gamma < -\frac{n+2}{n^2+3n+1}$$

the driving transformation is an (m,0)-loxodromic:

$$(z_1, z_2, \ldots, z_m, x_{2m+1}) \to$$

$$\to e^{2\delta}(e^{2i\theta_1} z_1, e^{2i\theta_2} z_2, \ldots, e^{2i\theta_m} z_m, x_{2m+1}).$$

For the single value $\gamma = -\frac{n+2}{n^2+3n+1}$ it is an (m,0)-parabolic:

$$(z_1, z_2, \ldots, z_m, x_{2m+1}) \to$$

$$\to (e^{2i\theta_1} z_1, e^{2i\theta_2} z_2, \ldots, e^{2i\theta_m} z_m, x_{2m+1}+d).$$

Then for

$$-\frac{n+2}{n^2+3n+1} < \gamma < -\frac{1}{n+1}$$

it is an (m+1,0)-elliptic most conveniently written as a transformation of the n-sphere in $\mathbb{R}^{n+1}$:

$$(z_1, z_2, \ldots, z_{m+1}) \to$$

$$\to (e^{2i\theta_1} z_1, e^{2i\theta_2} z_2, \ldots, e^{2i\theta_{m+1}} z_{m+1}) .$$

The one-dimensional situation described in [10] gives some feeling for the odd-dimensional cases. Our main purpose here is to further clarify the situation by fleshing out the details of the parabolic and elliptic cases in dimension $n = 3$. These enhance our intuition for parabolic and elliptic isoclinal sequences and provide one example (perhaps the only nontrivial example) of a periodic isoclinal sequence.

3. THE PARABOLIC CASE $n = 3$, $\gamma = -\frac{n+2}{n^2+3n+1} = -\frac{5}{19}$

The general results quoted above are derived with the aid of inversive coordinates. In terms of these coordinates a γ-clinal sequence B_k is given by a linear transformation $B_k = B_0 M^k$, $k = 0, \pm 1, \pm 2, \ldots$, where M has the characteristic equation

$$\lambda^{n+2} - \frac{2}{n+\gamma^{-1}} (\lambda^{n+1} + \ldots + \lambda) + 1 = 0 .$$

The Möbius transformation that drives the sequence is given by M if $\gamma < -\frac{1}{n}$ and by -M if $-\frac{1}{n} < \gamma < -\frac{1}{n+1}$. With this observation we can determine the conjugacy class of the driving transformation from a careful analysis of the roots of the characteristic equation. However these considerations do not yield immediately certain other details about the relation between the initial n-ball B_0 and the driving transformation. Once the general picture has been established it may be better to compute these details by making a fresh start along more elementary lines.

For the case in hand, general considerations guarantee that repeated applications of a (1,0)-parabolic or twist of the form $(z, x_3)^T = (e^{2i\theta} z, x_3 + d)$ will map a ball B_0 with centre $(a,0)$ and radius r into the proper balls or complements of the improper balls in an isoclinal sequence

$B_k = (-1)^k B_0 T^k$ with inclination $\gamma = -\frac{5}{19}$. General considerations do not determine the relations among d, r and a and we now proceed to do this. Since we can arrange to have a real and in fact $a = 1$ by conjugating with an appropriate similarity, we assume henceforth that this has been done.

Since B_0 is inclined at γ to B_0T^2 and B_0T^4, the distance from the centre of B_0 to the centres of these balls must be equal:

$$d_1^2 = 4\sin^2 2\theta + 4d^2 = 4\sin^2 4\theta + 16d^2 \ .$$

Since B_0 is inclined at $-\gamma$ to B_0T and B_0T^3, the distance from the centre of B_0 to the centres of these balls must also be equal:

$$d_2^2 = 4\sin^2\theta + d^2 = 4\sin^2 3\theta + 9d^2 \ .$$

These equations yield

$$2(\sin^2 2\theta - \sin^2 4\theta) = 6d^2 = 3(\sin^2\theta - \sin^2 3\theta)$$

hence

$$\begin{aligned} 2\sin^2 2\theta(1-4\cos^2 2\theta) &= 3(\sin^2\theta - \sin^2\theta\,\cos^2 2\theta - 2\sin\theta\cos\theta\sin 2\theta\cos 2\theta - \\ &\quad - \sin^2 2\theta\,\cos^2\theta) = \\ &= 3\sin^2 2\theta(\sin^2\theta - \cos 2\theta - \cos^2\theta) = -6\sin^2 2\theta\,\cos 2\theta \end{aligned}$$

hence

$$4\cos^2 2\theta - 3\cos 2\theta - 1 = 0 \ .$$

It follows that

$$\cos 2\theta = -\frac{1}{4} \qquad \sin 2\theta = \frac{\sqrt{15}}{4}$$

and also that

$$d = \frac{\sqrt{15}}{8} \quad .$$

The fact that B_0 and B_0T^2 are inclined at γ while B_0 and B_0T are inclined at $-\gamma$ yields

$$\frac{d_1^2}{2r^2} = 1 - \gamma \qquad \text{and} \qquad \frac{d_2^2}{2r^2} = 1 + \gamma$$

hence

$$\gamma = -\frac{5}{19} \qquad \text{and} \qquad r = \frac{5\sqrt{19}}{16} \quad .$$

We note that B_0 meets the axis of the twist and so the balls B_0T^k look like tomatoes on a kebab skewer which have been pierced somewhat off centre and have somehow kept their shape even though they have been squeezed into overlapping.

We also note that when $n = 3$ and $\gamma = -\frac{5}{19}$ the characteristic equation of M reduces to

$$(\lambda + 1)^3(\lambda^2 - \frac{1}{2}\lambda + 1) = 0$$

with a pair of complex roots $\lambda = e^{\pm 2i\bar{\theta}}$, $\bar{\theta} = \frac{\pi}{2} - \theta$. This agrees with the fact that -M represents a twist with rotation angle 2θ.

4. THE ELLIPTIC CASE $n = 3$, $-\frac{5}{19} = -\frac{n+2}{n^2+3n+1} < \gamma < -\frac{1}{n+1} = -\frac{5}{20}$

In odd dimensions $n = 2m+1$ the characteristic equation of M always has one root equal to -1. The remaining n+1 roots come in m+1 reciprocal pairs and when

$$-\frac{n+2}{n^2+2n+1} < \gamma < -\frac{1}{n+1}$$

they are all complex numbers of modulus 1 given by $\lambda = e^{\pm 2i\bar{\theta}_j}$ where the m+1 angles $\bar{\theta}_j$ are the roots of

$$\tan(n+2)\bar{\theta} = (n+1+\gamma^{-1})\tan\bar{\theta}, \quad 0 < \bar{\theta} < \frac{\pi}{2} .$$

Indeed the condition on γ can be understood as the condition for this trigonometric form of the characteristic equation to have a full complement of m+1 roots; for smaller values of γ it has only m roots. By inspecting the graphs of $\tan(n+1)\bar{\theta}$ and $\tan\bar{\theta}$ we see that these m+1 roots satisfy

$$\frac{2j\pi}{2(n+2)} < \bar{\theta}_j < \frac{(2j+1)\pi}{2(n+2)} , \qquad j = 1,2,\dots,m+1 .$$

Consideration of the eigenvector corresponding to -1 indicates [10] that the driving transformation of these sequences is given by -M. It is an $(\frac{n+1}{2}, 0)$-elliptic which can be conjugated to an $\frac{n+1}{2}$-fold rotation of the n-sphere in $\mathbb{R}^{n+1}$ with rotation angles $2\theta_j = \pi - 2\bar{\theta}_j$. This means that the driving transformation is periodic if and only if the m+1 angles $\theta_j(\gamma)$ are all rational multiples of π. Considerations involving inversive coordinates also indicate [10] that when the driving transformation is expressed as an $\frac{n+1}{2}$-fold rotation (periodic or not) the n-balls of the corresponding isoclinal sequence have angular radius ψ or $\pi-\psi$ where

$$\cot \psi = \sqrt{\frac{(n+1)\gamma^2 - n\gamma - 1}{(n^2+3n+1)\gamma + (n+2)}} \quad .$$

When $n = 3$ the characteristic equation for M can be written

$$(\lambda+1)(\lambda^4 - \frac{5\gamma+1}{3\gamma+1}\lambda^3 + \lambda^2 - \frac{5\gamma+1}{3\gamma+1}\lambda + 1) = 0$$

with roots

$$\lambda = -1$$

$$\lambda = e^{\pm 2i\bar{\theta}_1} , \qquad \frac{\pi}{5} < \bar{\theta}_1 < \frac{3\pi}{10}$$

$$\lambda = e^{\pm 2i\bar{\theta}_2} , \qquad \frac{2\pi}{5} < \bar{\theta}_2 < \frac{\pi}{2} \quad .$$

The coefficient of λ (or λ^3) in the second factor of the characteristic equation leads to the equation

$$\cos 2\bar{\theta}_1 + \cos 2\bar{\theta}_2 = 2\,\frac{5\gamma+1}{3\gamma+1}$$

and the coefficient of λ^2, to the equation

$$\cos 2\bar{\theta}_1 \cos 2\bar{\theta}_2 = -\frac{1}{4} \quad .$$

To pursue the second of these equations we write

$$\phi_1 = 2\bar{\theta}_1 \qquad \text{and} \qquad \phi_2 = \pi - 2\bar{\theta}_2$$

so that

$$\cos \phi_1 \cos \phi_2 = \frac{1}{4}$$

and ϕ_2 decreases from $\frac{\pi}{5}$ to 0 as ϕ_1 increases from $\frac{2\pi}{5}$ to $\cos^{-1} \frac{1}{4}$. By differentiating this equation with respect to ϕ_1 we see that

$\sigma(\phi_1) = \phi_1 + \phi_2$ decreases steadily from $\frac{3\pi}{5}$ to $\cos^{-1} \frac{1}{4}$ and

$\delta(\phi_1) = \phi_1 - \phi_2$ increases steadily from $\frac{\pi}{5}$ to $\cos^{-1} \frac{1}{4}$

so the equation in ϕ_1 and ϕ_2 is equivalent to

$$\cos \sigma + \cos \delta = \frac{1}{2} \quad .$$

By making use of P. Gordan's solution to a similar trigonometric Diophantine equation ([5] and [2] p. 109; see also Crosby [4] and Conway and Jones [1]) we see that the obvious solution

$$\sigma = \frac{\pi}{2} \, , \; \delta = \frac{\pi}{3}$$

is the only one relevant to our question. This shows that there is a unique periodic isoclinal sequence in dimension $n = 3$ and it has parameters

$$\bar{\theta}_1 = \frac{5\pi}{24}\ , \ \bar{\theta}_2 = \frac{11\pi}{24}\ , \ \gamma = \frac{1-2\sqrt{2}}{7}\ .$$

The driving transformation of this sequence is a double rotation of period 24 with rotation angles

$$2\theta_1 = \frac{7\pi}{12} \quad \text{and} \quad 2\theta_2 = \frac{\pi}{12}\ .$$

We note that with $\gamma = \frac{1-2\sqrt{2}}{7}$ the characteristic polynomial for M reduces to

$$(\lambda + 1)(\lambda^4 + \sqrt{2}\lambda^3 + \lambda^2 + \sqrt{2}\lambda + 1)$$

and this is a factor of $\lambda^{24}-1$. At an early stage in our investigation Prof. A. Baker spotted this directly and thereby assured us of the existence of at least one periodic isoclinal sequence in dimension $n = 3$.

We also note that the general formula for the size of the n-balls involved in a spherical isoclinal sequence shows us that when $n = 3$ and $\gamma = \frac{1-2\sqrt{2}}{7}$ the angular radius ψ satisfies $\sin\psi = \sqrt{\frac{3-\sqrt{2}}{4}}$.

5. THE PERIODIC ISOCLINAL SEQUENCE IN DIMENSION n=3

Here we seek to relate the periodic isoclinal sequence in dimension n=3 to a more familiar geometric object. We coordinatize the 3-sphere in $\mathbb{R}^4$ with pairs of complex numbers and accept the fact that a (2,0)-elliptic or double rotation of the form

$$(z_1,z_2)^R = (e^{2i\theta_1}z_1,\ e^{2i\theta_2}z_2)$$

will map a ball B_0 with centre (a,b) and angular radius ψ into the balls or complements of the balls in an isoclinal sequence $B_k = (-1)^k B_0 R^k$ with inclination $\gamma = \frac{1-2\sqrt{2}}{7}$. General considerations do not determine a and b but we can arrange to have them both real by conjugating with an appropriate isometry that commutes with R and we assume henceforth that this has been done.

It is convenient to write $a^2 = A$ and $b^2 = B$. Then the fact that (a,b) lies on the unit sphere gives

$$A + B = 1 \ ;$$

the fact that B_0 is inclined at γ to B_0R^2 and B_0R^4 gives

$$A\sin^2 2\theta_1 + B\sin^2 2\theta_2 = A\sin^2 4\theta_1 + B\sin^2 4\theta_2 = \\ = \frac{1}{2}\sin^2\psi(1-\gamma) \ ;$$

and the fact that B_0 is inclined at $-\gamma$ to B_0R and B_0R^3 gives

$$A\sin^2\theta_1 + B\sin^2\theta_2 = A\sin^2 3\theta_1 + B\sin^2 3\theta_2 = \\ = \frac{1}{2}\sin^2\psi(1+\gamma) \ .$$

It should be possible to use these five equations to express A, B, θ_1, θ_2 and ψ as functions of γ for $-\frac{5}{19} < \gamma < -\frac{5}{20}$ and thereby make this the starting point of an elementary treatment similar to the one we have given for the parabolic case, albeit with the addition of some trigonometric Diophantine analysis. But this does not seem to be altogether straightforward and so we shall use all the results of the previous section and simply verify that when

$$\gamma = \frac{1-2\sqrt{2}}{7}, \; \sin^2\psi = \frac{3-\sqrt{2}}{4}, \; \theta_1 = \frac{7\pi}{24} \text{ and } \theta_2 = \frac{\pi}{24}$$

the overdetermined linear system in A and B is consistent and admits the unique solution

$$A = \frac{3-\sqrt{3}}{6} \quad \text{and} \quad B = \frac{3+\sqrt{3}}{6}.$$

For vertification it sufficies to mention that

$$\sin^2\psi(1-\gamma) = \frac{1}{2}, \; \sin^2\psi(1+\gamma) = \frac{2-\sqrt{2}}{2}$$

and the values of $(\sin^2 k\theta_1, \sin^2 k\theta_2)$ for k=1,2,3,4 are respectively

$$\left(\frac{2+\sqrt{2-\sqrt{3}}}{4}, \frac{2-\sqrt{2+\sqrt{3}}}{4}\right), \left(\frac{2+\sqrt{3}}{4}, \frac{2-\sqrt{3}}{4}\right),$$

$$\left(\frac{2-\sqrt{2}}{4}, \frac{2-\sqrt{2}}{4}\right) \text{ and } \left(\frac{1}{4}, \frac{1}{4}\right).$$

From this it follows that the centres of the 24 balls $B_k = B_0R^k$ are the points

$$(z_1,z_2) = \left(\sqrt{\frac{3-\sqrt{3}}{6}}\left(\frac{-\sqrt{2-\sqrt{3}}+i\sqrt{2+\sqrt{3}}}{2}\right)^k, \sqrt{\frac{3+\sqrt{3}}{6}}\left(\frac{\sqrt{2+\sqrt{3}}+i\sqrt{2-\sqrt{3}}}{2}\right)^k\right)$$

The square of our double rotation is conjugate to the square of the one given by Coxeter [2] p. 245 and this shows that the even points are the vertices of a Petrie polygon of one {3,4,3} while the odd points are the vertices of a Petrie polygon of the {3,4,3} dual to the first.

With this insight it may be more appealing to simplity the coordinates of the centres rather than the formula for the driving transformation. In terms of real coordinates we can take the centres to be the 24 points

$$(1,0,0,0)\begin{pmatrix} \frac{1}{\sqrt{2}} & 0 & \frac{1}{\sqrt{2}} & 0 \\ 0 & \frac{1}{\sqrt{2}} & 0 & \frac{-1}{\sqrt{2}} \\ 0 & \frac{1}{\sqrt{2}} & 0 & \frac{1}{\sqrt{2}} \\ \frac{1}{\sqrt{2}} & 0 & \frac{-1}{\sqrt{2}} & 0 \end{pmatrix}^k$$

$k=1,2,\ldots,24$. By way of vertification we note that the orthogonal transformation has characteristic equation

$$\lambda^4 - \sqrt{2}\lambda^3 + \lambda^2 - \sqrt{2}\lambda + 1 = 0$$

so it does represent a double rotation with angles $2\theta_1$ and $2\theta_2$. Moreover the even points are of the form $(\pm 1,0,0,0)$ or $(\pm \frac{1}{2}, \pm \frac{1}{2}, \pm \frac{1}{2}, \pm \frac{1}{2})$ while the odd points the of form $(\pm \frac{1}{\sqrt{2}}, \pm \frac{1}{\sqrt{2}}, 0, 0)$ and these are the most familiar coordinates for the vertices of a {3,4,3} and its dual (cf. Coxeter [2] p. 156).

REFERENCES

[1] J.H. CONWAY - A.J. JONES, *Trigonometric diophantine equations (On vanishing sums of roots of unity)*, Acta Arith. 30 (1976) 229-240.

[2] H.S.M. COXETER, *Regular Polytopes*, 2nd ed., Macmillan, New York (1963).

[3] H.S.M. COXETER, *Loxodromic sequences of tangent spheres*, Aequationes Math. 1 (1968) 104-121.

[4] W.J.R. CROSBY, *Solution to problem 4136*, Amer. Math. Monthly 53 (1946) 103-107.

[5] P. GORDAN, *Über endliche Gruppen linearer Transformationen einer Veränderlichen*, Math. Ann. 12 (1877) 23-46.

[6] A. WEISS, *On Coxeter's loxodromic sequences of tangent spheres*, The Geometric Vein, Springer-Verlag, New York (1981) 243-250.

[7] A. WEISS, *On isoclinal sequences of spheres*, Proc. Amer. Math. Soc. 88 (1983) 665-671.

[8] J.B. EILKER, *Inversive geometry*, The Geometric Vein, Springer-Verlag, New York (1981) 379-442.

[9] J.B. WILKER, *Möbius transformations in dimension n*, Period. Math. Hungar. 14 (1983) 93-99.

[10] J.B. WILKER, *Möbius transformations and isoclinal sequences of spheres*, Aequationes Math., 30 (1986) 161-179.

J.B. WILKER
Scarborough College
University of Toronto
West Hill, Ontario M1C 1A4
Canada

COLLOQUIA MATHEMATICA SOCIETATIS JÁNOS BOLYAI
48. INTUITIVE GEOMETRY, SIÓFOK, 1985.

A NECESSARY CONDITION FOR BEST HAUSDORFF APPROXIMATION OF PLANE CONVEX COMPACTA BY POLYGONS

N.V. ZHIVKOV

1. INTRODUCTION

Let $\mathbb{P}$ be a two-dimensional Minkowskian plane, i.e. the unit ball U of $\mathbb{P}$'s metric d is a planer convex and symmetric body. Hence d is induced by a norm $\|.\|$. In the particular case of U being a circle, $\mathbb{P}$ coincides with the usual Euclidean plane.

An n-gon is a convex polygon with at most n vertices. A polygon Δ is a non-degenerate n-gon provided that the number of its vertices is equal to n. Denote the class of n-gons by $\mathscr{P}_n$ and the class of planar convex compacta by $\mathscr{C}$.

Let $<.,.>$ be the inner product on $\mathbb{P}$. Given a convex compact C define its support function h(C.,) as follows

$$h(C,x) = \max\{<x,z>: z \in C\}.$$

The set $U^* = \{x^* \in \mathbb{P} : h<U,x^*> \leq 1\}$ is closed, convex, bounded and symmetric. Hence U^* generates a new norm $\|.\|_*$

on $\mathbb{P}$. Denote by $\mathbb{P}^*$ the Minkowskian plane equipped with this norm. $\mathbb{P}^*$ is usually referred to as the dual (conjugate) plane of $\mathbb{P}$.

The Hausdorff distance $D(.,.)$ on $\mathbb{P}$ is defined for any two members C_1 and C_2 of by

$$D(C_1,C_2) = \inf\{\varepsilon > 0 : C_1 + \varepsilon U \supset C_2 \text{ and } C_2 + \varepsilon U \supset C_1\},$$

Consider the following approximation problem:

Given a convex compact C. Find for $n > 2$ an n-gon

(*) Δ_0 on minimal Hausdorff distance from C, i.e.

$$D(C, \Delta_0) = \min\{D(C, \Delta) : \Delta \in \mathscr{P}_n\}$$

The Blaschke selection theorem [1] implies existence of a solution to the problem (*) but the solution may not always be unique.

The problems of Hausdorff approximations of planar convex compacta by polygons have been treated by many authors, for instance [1-8], [10-11], [14]. The reader may refer to the book [14] of Bl. Sendov for a detailed discussion on numerous approximation problems in Hausdorff metrics generated by non-Euclidean metrics, and to the survey paper [6] of Gruber where various types of approximation with convex bodies are considered.

In [8] Kenderov reveals the connection between best approximating n-gons and the alternating polygons in the case of Hausdorff metric generated by the Euclidean norm. Here, a polygon Δ is alternating for a given convex compact C if the support functions $h(C,.)$ and $h(\Delta,.)$, viewed as functions on the unit sphere S^* of $\mathbb{P}^*$, have Chebyshev alternance at 2n different points. Later in [18] this connection has been traced for arbitrary norm generated Hausdorff metric in the plane. The analogy with best Chebyshev approximations of continuous functions by polynomials of a given degree n is apparent, especialy if is regarded as a subset of $C(S^*)$, but in the present situation the alternance condition is only necessary ([8] theorem 3.1, [18] theorem 3.2), and not sufficient. In order to ensure both necessity and sufficiency, as was shown by Kenderov, we have to approximate by polygons which have a common outward normal. More precisely:

Associate with any non-degenerate n-gon Δ n unit vectors called normals or side-directions which are perpendicular to the sides of Δ and directed "outward". For any $e^* \in S^*$ and $n > 1$ define a class of polygons $\mathcal{P}_n(e^*)$ as follows

$$\mathcal{P}_n(e^*) = \mathcal{P}_{n-1} \cup \{\Delta \in \mathcal{P}_n \setminus \mathcal{P}_{n-1} : e^* \text{ is a normal vector of } \Delta\},$$

and consider the approximation problem:

Given $C \in \mathcal{C}$, $n > 2$ and $e^* \in S^*$. Find a polygon Δ_0 in (**) $\mathcal{P}_n(e^*)$ on minimal Hausdorff distance from C, i.e.

$$D(C,\Delta) = \min\{D(C,\Delta) : \Delta \in \mathcal{P}_n(e^*)\}$$

According to Kenderov's result ([8] theorem 4.12) the uniqueness of the solution to the problem (**) is completely determined by the alternance condition. Extentions of this result for Minkowskian planes are given in [18].

In the present paper a stronger necessary condition for best approximation of a plane convex figure by n-gons is obtained. To this end we consider the class of approximation problems in $\mathcal{P}_n(e^*)$ when e^* runs over the unit sphere S^*. To each $e^* \in S^*$ the value of the distance from C to

$\mathscr{P}_n(e^*)$ is assigned, and the function, thus defined, is subjected to optimization. Recall from [18] its definition.

Given $C \in \mathscr{C}$ and $n > 2$. The *normal projection function* is defined for every $e^* \in S^*$ by

$$\rho_n(C,e^*) = \min\{D(C,\Delta) : \Delta \in \mathscr{P}_n(e^*)\}.$$

Under an appropriate transformation the function $\rho_n(C,.)$ might be viewed as a function on the interval $[0, 2\pi]$, and in such a form we are interested in its differentiability. For instance, if D is generated by the Euclidean norm $|.|$ and C is strictly convex then $\rho_n(C.,)$ is continuously differentiable on $[0,2\pi]$.

Thus, the necessary condition for best approximation, which we claim, expresses the fact that ρ_n's derivative is zero at a point $\omega \in [0,2\pi]$ corresponding to some $e^* \in S^*$, and then the best approximating polygon in $\mathscr{P}_n(e^*)$ is a candidate for a best approximation in $\mathscr{P}_n$. The same result in Minkowskian planes is true if both U and C are strictly convex and at least one of them is smooth. Under less restrictive assumptions on C and U the function ρ_n has left-hand and right-hand derivatives.

Some of the results in the paper were announced in [17].

2. BASIC GEOMETRICAL CONSTRUCTION

Let $\Delta = [V_1,\ldots,V_n] \in \mathscr{P}_n \setminus \mathscr{P}_{n-1}$ be a polygon and $n > 2$. For the sake of convenience we assume that Δ's vertices follow the counter-clockwise (c.c.w.) circle ordering and $V_i = V_j$ if and only if $i \equiv j \pmod n$. Suppose that a sequence of points $(Q_i)_{i=1}^n$ such that for any i $Q_i \in (V_i, V_{i+1})$ and a sequence of axes $(l_i)_{i=1}^n$ with $l_i \cap \Delta = \{V_i\}$ are given. The l_i-directions e_i, $|e_i| = 1$, are defined so that when "running" in positive direction on l_i the polygon Δ remains on the "left side". Here $|.|$ designates the Euclidean norm.

Let $a_i = V_i - Q_{i-1}$, $b_i = Q_i - V_i$ and

$$\alpha_i = \text{mes}\sphericalangle(a_i, e_i), \quad \beta_i = \text{mes}\sphericalangle(e_i, b_i), \quad i = 1, \ldots, n,$$

where $\sphericalangle(x,y)$ is the elementary angle between two vectors x and y and mes stands for the radian angle measure.

Let also $(c_i)_{i=1}^n$ be a sequence of vectors so that $|c_i| = 1$, $\langle e_i, c_i \rangle = 0$ and for every i the triad $\{V_i; e_i, c_i\}$ forms a left coordinate system.

A curve r is defined in a neighbourhood of V_1 as follows (see Figure 1):

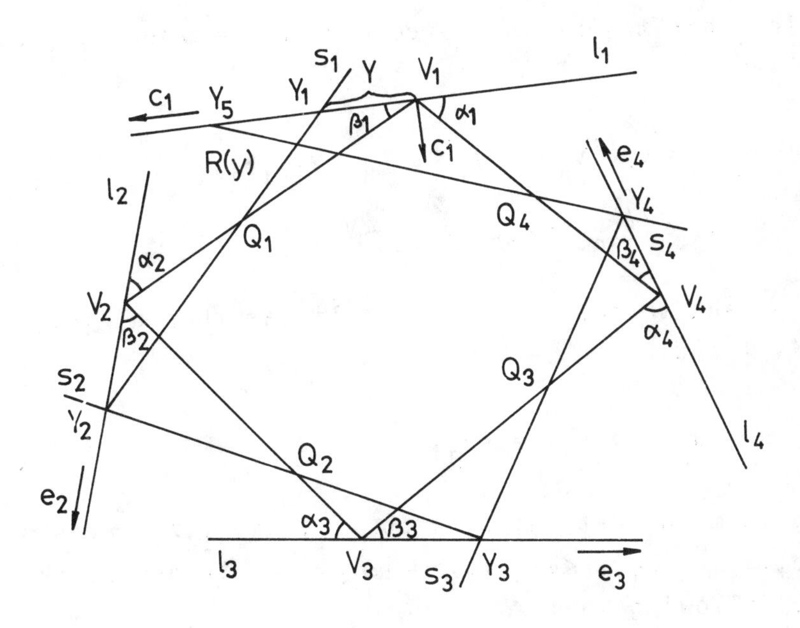

Figure 1

Let Y_1 belong to l_1 and Y_2 be the intersection point of the straight line s_1 (passing through Y_1 and Q_1) with the axis l_2. Analogously, by induction, define s_2 (through Y_2 and Q_2), Y_3 and so on ... s_{n-1}, Y_n and s_n (through Y_n and Q_n). The line s_n meets l_1 at Y_{n+1} and s_1 at a point R. The set of all R thus defined is denoted by r. Let $y_i = \langle Y_i - V_i, e_i \rangle$ for every i, and let also $y = y_1$. The point R depends on Y_1 and hence r admits a parametric representation of the form $r = \{R(y) \in \mathbb{P} : y \in (-\delta, \delta)\}$ for some $\delta > 0$. It is to be shown that r is actually a curve which belongs to the class C^∞.

Continuity arguments yield that for every $i \in \{1,\ldots,n\}$ there are relative neighbourhoods $\mathcal{O}_i$ of V_i and $\mathcal{O}_{i+1}$ of V_{i+1} such that the mapping $Y_i \to Y_{i+1}$ is well defined and projects $\mathcal{O}_i$ onto $\mathcal{O}_{i+1}$.

In order to obtain the connection between y_{i+1} and y_i we let

$$u_i = b_i - y_i e_i, \quad v_i = a_{i+1} + y_{i+1} e_{i+1},$$

$$z_i = (V_{i+1} - V_i)/|V_{i+1} - V_i|.$$

Since u_i and v_i are positively colinear then $u_i/|u_i| = v_i/|v_i|$ and

$$\langle u_i, z_i \rangle |v_i| = \langle v_i, z_i \rangle |u_i| \tag{2.1}$$

Having in mind that $\langle e_i, z_i \rangle = \cos \beta_i$, $\langle e_{i+1}, z_i \rangle = \cos \alpha_{i+1}$ and $\langle b_i, z_i \rangle = |b_i|$, $\langle a_{i+1}, z_i \rangle = |a_{i+1}|$ we transform (2.1) in the following way:

$$(\langle u_i, z_i \rangle)^2 (y_{i+1}^2 \sin^2 \alpha_{i+1} + (\langle v_i, z_i \rangle)^2) =$$

$$= (\langle u_i, z_i \rangle)^2 (|a_{i+1}|^2 + 2|a_{i+1}| y_{i+1} \cos\alpha_{i+1} +$$

$$+ y_{i+1}^2(\sin^2\alpha_{i+1} + \cos^2\beta_{i+1})) = (\langle u_i,z_i\rangle)^2\langle v_i,v_i\rangle =$$

$$= (\langle v_i,z_i\rangle)^2\langle u_i,u_i\rangle = (\langle v_i,z_i\rangle)^2(|b_i|^2-2|b_i|y_i\cos\beta_i +$$

$$+ y_i^2(\sin^2\beta_i + \cos^2\beta_i)) = (\langle v_i,z_i\rangle)^2(y_i^2\sin^2\beta_i +$$

$$+ (\langle u_i,z_i\rangle)^2),$$

whence

$$(|b_i| - y_i\cos\beta_i)^2 y_{i+1}^2\sin^2\alpha_{i+1} = \tag{2.2}$$

$$= (|a_{i+1}| + y_{i+1}\cos\alpha_{i+1})^2 y_i^2\sin^2\beta_i$$

The expressions $|a_{i+1}| + y_{i+1}\cos\alpha_{i+1}$ and $|b_i| - y_i\cos\beta_i$ are positive provided that $\mathcal{C}_i$ and $\mathcal{C}_{i+1}$ have been taken sufficiently small. Therefore (2.2) implies

$$y_{i+1} = \frac{|a_{i+1}|\sin\beta_i\cdot y_i}{-\sin(\alpha_{i+1} + \beta_i)y_i + |b_i|\sin\alpha_{i+1}} \tag{2.3}$$

The last formula indicates that the connection between y_{n+1} and y_i is rational, i.e.

$$y_{n+1} = \kappa y/(-\lambda y + \mu) =: \Psi(y), \tag{2.4}$$

where $\kappa = \prod_{i=1}^{n} |a_i|.\sin\beta_i$, λ is a constant, $\mu = \prod_{i=1}^{n} |b_i|.\sin\alpha_i$ and $y = y_1$.

Let r be represented parametrically (with parameter y) with respect to the local coordinate system $\{V_1;e_1,c_1\}$. If Q_1 and Q_n have local coordinates (p_1,q_1) and (p_n,q_n) respectively, then it is easily checked for any point (ξ,η) on r that

$$r := \begin{cases} \xi = \dfrac{(q_1-q_n)y\Psi(y) + p_1q_n\Psi(y) - p_nq_1y}{q_1\Psi(y) - q_ny + p_1q_n - p_nq_1} = F(y) \\ \\ \eta = \dfrac{q_1q_n(\Psi(y) - y)}{q_1\Psi(y) - q_ny + p_1q_n - p_nq_1} = G(y)(\Psi(y)-y) \end{cases}$$

The functions $F(y)$, $G(y)$ (and $\Psi(y)$) are rational and finite at $y = 0$, since $p_1q_n - p_nq_1 = 0$ would imply that Q_1, V_1 and Q_n belong to a straight line.

Hence r is a curve from the class C^∞ in a neighbourhood of V_1. Now, the line l_1 is a tangent to r at V_1 if and only if $d\eta(0)/dy = 0$, and since $q_1q_n > 0$ this implies $d\Psi(0)/dy = 1$. Thus we have

PROPOSITION 1. *The line* l_1 *is a tangent to the curve* r *at* V_1 *if and only if the condition* $\prod_{i=1}^{n} \dfrac{|a_i|.\sin\beta_i}{|b_i|.\sin\alpha_i} = 1$ *is fulfilled.*

It might be shown that in such a case r is a graph of a strictly convex function with respect to $\{V_1;e_1,c_1\}$, but we will make no use of that result.

The n-gons $\Delta(y) = [R(y),Y_2(y),\ldots,Y_n(y)]$ defined in a small Hausdorff neighbourhood of Δ are called *variational with respect to* Δ, $(Q_i)_{i=1}^n$, $(l_i)_{i=1}^n$ provided that l_1 is a tangent to r. Further we shall see that the variational n-gons provide smooth variations of the alternating ones.

3. A NECESSARY CONDITION FOR EXTREMUM.

Let $C \in \mathscr{C}$ and $n > 2$. For $e^* \in S^*$ the support function $e^* \to h(C,e^*)$ and the normal projection function $e^* \to \rho(C,e^*) = D(C, \mathscr{P}_n(e^*))$ might be viewed as defined on the segment

$[0,2\pi]$ in which case, for ω corresponding to e^*, we simply write $h(C,\omega)$ and $\rho_n(C,\omega)$ instead of $h(C,e^*)$ and $\rho_n(C,e^*)$ respectively.

Recall some definitions from [18]. Let C be a convex compact with non-empty interior. A polygon Δ is said to be *oscillating* for C provided that Δ's vertices do not belong to C and each side of Δ meets the interior of C.

For any convex figure $C \in \mathscr{C}$ and polygon $\Delta \in \mathscr{P}_n \setminus \mathscr{P}_{n-1}$ with side-directions $e_1^*, e_2^*, \ldots, e_n^*$ the numbers

$$\lambda_i = h(C,e_i^*) - h(\Delta,e_i^*), \quad \nu_i = \min(\|V_i - Z\| : Z \in C\}$$

are called *oscillation numbers* of Δ with respect to C.

A polygon Δ is *alternating* for C if all the 2n oscillation numbers of Δ to C are equal.

It was established in [18] (theorem 5.3) that for any non-degenerate n-gon Δ which is oscillating for $C \in \mathscr{C}$ the following estimations hold:

$$(3.1) \qquad \min_{1 \le j \le n} \{\lambda_j, \nu_j\} \le \rho_n(C,e_i^*) \le \max_{1 \le j \le n} \{\lambda_j, \nu_j\}$$

where λ_j, ν_j are the oscillation numbers and e_i^* are the normals of Δ, $i = 1,2,\ldots,n$.

A convex figure C is *strictly convex* provided that its boundary does not contain any non-degenerate line segments.

A set $C \in \mathscr{C}$ with int $C \neq$ is said to be *smooth* if for any $x \in \text{bd } C$ there is only one line touching C at x.

Let now C and U be strictly convex, C or U be smooth and $n > 2$ be fixed. According to more general results from [18] (theorems 3.3 and 4.2), for every $\omega \in [0,2\pi)$ there exists a unique non-degenerate n-gon $\Delta(\omega)$ which is alternating for C and $\rho_n(C,\omega) = D(C,\Delta(\omega))$. Let the vertices

$(V_i(\omega))_{i=1}^n$ and the normals $(e_i^*(\omega))_{i=1}^n$ of $\Delta(\omega)$ be indexed in such a way that ω corresponds to e_i^*. Then the mappings $\omega \to V_i(\omega)$ and $\omega \to e_i^*(\omega)$ are well defined on $[0,2\pi)$.

Consider the following construction for arbitrary ω (for the sake of brevity we drop the argument ω whenever it is understood):

For every i the maximal value of $\langle e_i^*,.\rangle$ over C is attained at a single point Z_i on C's boundary. There is a unique point $Q_i \in [V_i,V_{i+1}]$ such that $\|Z_i - Q_i\| = D(C,\Delta)$. It follows from the smoothness condition on C or U that Q_i actually belongs to (V_i,V_{i+1}) (because if for some i Q_i is a vertice of Δ, i.e. $Q_i = V_i$, then $Z_i \in V_i + D(C,\Delta)U$ and as in the proof of lemma 2.4 from [18] neither C nor U would be smooth).

Let l_i be the axes supporting the set $C+D(C,\Delta)U$ at V_i, $i = 1,\dots,n$. Further, the l_i-directions e_i, the vectors a_i, b_i and the angle-measures α_i and β_i are determined by l_i, V_i and Q_i as in section 2. (see Figure 2 in the case of a triangle):

Define $\Phi(C,\Delta(\omega)) = \prod_{i=1}^{n} \frac{\|a_i\|.\sin\beta_i}{\|b_i\|.\sin\alpha_i}$. It is easily seen, from $\|a_i\|/\|b_{i-1}\| = |a_i|/|b_{i-1}|$ for each i, that $\Phi(C,\Delta(\omega))$ is also equal to $\prod_{i=1}^{n} \frac{|a_i|.\sin\beta_i}{|b_i|.\sin\alpha_i}$.

THEOREM 2. *Let* C *and* U *be strictly convex and* C *or* U *be smooth. Then for every* $n > 2$ *the normal projection function* $\rho_n(C,.)$ *is continuously differentiable on* $[0,2\pi]$. *If* $\Delta_0(\omega_0)$ *is a non-degenerate* n*-gon which is alternating for* C *then the sign of* $\rho_n'(\omega_0)$ *is the same as the sign of* $1-\Phi(C,\Delta_0(\omega_0))$. *In particular* $\rho_n'(\omega) = 0$ *if and only if*

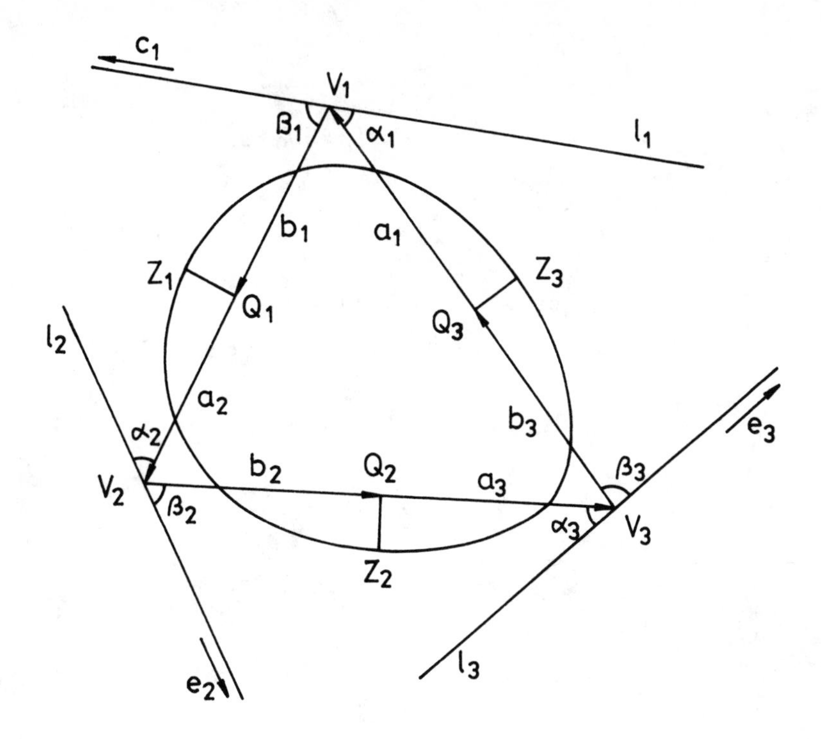

Figure 2

$$\prod_{i=1}^{n} \frac{\|a_i\| . \sin \beta_i}{\|b_i\| . \sin \alpha_i} = 1.$$

PROOF. Consider several separate cases:

(I) Differentiability of $\rho_n(C,.)$ at ω_0 in the case $\Phi(C,\Delta_0) > 1$.

Define a function φ as follows

$$\varphi(t_1,\Theta_1,\ldots,t_n,\Theta_n) = \prod_{i=1}^{n} \frac{|a_i|-t_i}{|b_i|+t_i} \cdot \frac{\sin(\beta_i-\Theta_i)}{\sin(\alpha_i+\Theta_i)} \tag{3.2}$$

where $t_i \in (-|b_i|,|a_{i+1}|)$, $\Theta_i \in (-\alpha_i,\beta_i)$ for each i. Obviously $\varphi(0,0,\ldots,0) = \Phi(C,\Delta_0)$. It is not difficult to verify that φ is strictly decreasing from $+\infty$ to 0 with respect to any t_i as well as to any Θ_i. Let $\bar{t}_i$ and $\bar{\Theta}_i$ be such that

$$\varphi(0,\dots,\bar{t}_i,0,\dots,0) = 1 = \varphi(0,\dots,0,\bar{\Theta}_i,\dots,0).$$

For every i such $\bar{t}_i \in (0,|a_{i+1}|)$ and $\bar{\Theta}_i \in (0,\beta_i)$ are uniquely determined.

Let $H^{2n-1} \subset \mathbb{R}^{2n}$ be a set defined by

$$H^{2n-1} = \left(\prod_{i=1}^{n} [0,\bar{t}_i]\times[0,\bar{\Theta}_i]\right) \cap \{(t_1,\Theta_1,\dots,t_n,\Theta_n) \in \mathbb{R}^{2n} : \tag{3.3}$$
$$\varphi(t_1,\Theta_1,\dots,t_n,\Theta_n) = 1\}.$$

For any fixed $\chi = (t_1,\Theta_1,\dots,t_n,\Theta_n) \in H^{2n-1}$ let $Q_{i\chi} := Q_i + (t_i/|b_i|).b_i$ and $l_{i\chi}$ be the axes carried by $e_{i\chi}$, $|e_{i\chi}| = 1$ with $\text{mes}\angle(a_i,e_{i\chi}) = \alpha_i + \Theta_i$ for $i = 1,\dots,n$. Denote by r_χ the curve defined by Δ_0, $(Q_{i\chi})_{i=1}^{n}$ and $(l_{i\chi})_{i=1}^{n}$ as in section 2 and consider the variational n-gons $\Delta_\chi(\omega) \in \mathscr{P}_n$ with vertices $V_{i\chi}(\omega)$ and side-directions $e^*_{i\chi}(\omega)$ such that $\omega = \omega(e^*_{1\chi}(\omega))$, $V_{1\chi}(\omega) \in r_\chi$, $V_{i\chi}(\omega) \in l_{i\chi}$ for $i > 1$ and $Q_{i\chi} \in (V_{i\chi}(\omega),V_{i+1,\chi}(\omega))$ for $i = 1,2,\dots,n$, where $\omega \in (\omega_0-\delta,\omega_0+\delta)$ for some $\delta > 0$.

The metric projection function $d(X,C) = \min\{\|X-Z\|:Z\in C\}$ is convex and differentiable at every point $X \notin C$, its subdifferential at X being singleton [13]. For each i denote the subdifferential $\partial d(V_i,C)$ by $\{c^*_i\}$. It is well known (for instance [12]) that c^*_i attains its maximum on $C + \rho_n(C,\Delta_0)U$ at V_i and $\|c^*_i\|_* = 1$. The points $V_{i\chi}(\omega)$ have local coordinates $(y_i(\omega),0)$ with respect to $\{V_i;e_{i\chi},c_{i\chi}\}$, $i \neq 1$, and $V_{1\chi}(\omega) = (y_1(\omega),z(\omega))$ with respect to $\{V_i;e_{1\chi},c_{1\chi}\}$. The functions $y_i(\omega)$ are continuous and satisfy $y_i(\omega_0) = 0$. Then according to Proposition 1 $dz(\omega_0)/d\omega = 0$.

Let $\lambda_{i\chi}(\omega)$ and $\nu_{i\chi}(\omega)$ be the oscillation numbers of $\Delta_\chi(\omega)$ with respect to C. Define for each i

$$p_i(\chi) = \lim_{\omega\to\omega_0} (\omega-\omega_0)^{-1}(\lambda_{i\chi}(\omega)-\rho_n(C,\Delta_0))$$

$$q_i(\chi) = \lim_{\omega\to\omega_0} (\omega-\omega_0)^{-1}(\nu_{i\chi}(\omega)-\rho_n(C,\Delta_0)).$$

In order to calculate $p_i(\chi)$ and $p_i(\chi)$ and $\Theta_i(\chi)$ denote at first $\mathrm{mes}\sphericalangle(e_i^*,e_{i\chi}^*(\omega))$ by $|\sigma_i(\omega)|$, where $\mathrm{sign}\ \sigma_i(\omega) = \mathrm{sign}\ y_i(\omega)$. In particular $\sigma_1(\omega) = \omega-\omega_0$ (see Figure 3):

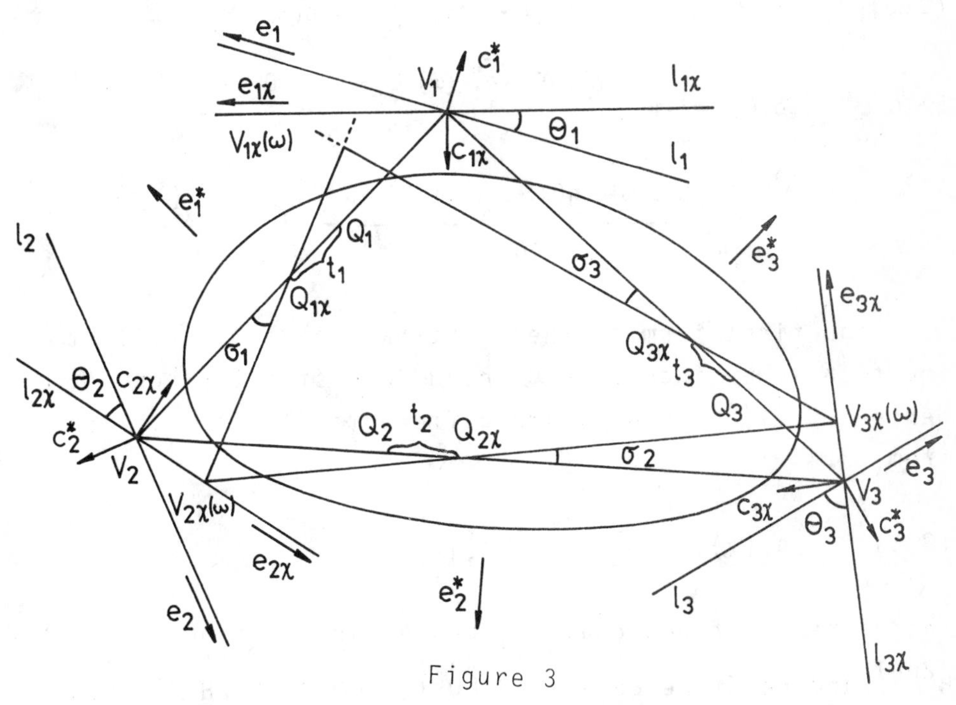

Figure 3

It follows from the elementary sine theorem applied to the triangles $[Q_{i-1,\chi},V_i,V_{i\chi}(\omega)]$ and $[Q_{i\chi},V_i,V_{i\chi}(\omega)]$ that

$$\frac{\sin\sigma_i(\omega)}{\sin\sigma_{i-1}(\omega)} = \frac{(|a_i| - t_{i-1}).\sin(\beta_i-\Theta_i+\sigma_i(\omega))}{(|b_i| + t_i).\sin(\alpha_i+\Theta_i-\sigma_{i-1}(\omega))}$$

whence for $i \neq 1$

$$(3.4) \qquad \lim_{\omega \to \omega_0} \frac{\sigma_i(\omega)}{\omega - \omega_0} = \prod_{j=2}^{i} \frac{(|a_j| - t_{j-1}).\sin(\beta_j - \Theta_j)}{(|b_j| + t_j).\sin(\alpha_j + \Theta_j)}$$

Denote the last expression by $A_i(\chi)$ for all $i \neq j$ and put $A_1(\chi) = 1$. Thus for $i = 1,2,\ldots,n$

$$p_i(\chi) = \lim_{\omega \to \omega_0} \frac{d(V_{i\chi}(\omega),C) - \rho_n(C,\Delta_0)}{y_i(\omega)} \cdot \frac{y_i(\omega)}{\sin \sigma_i(\omega)} \cdot \frac{\sigma_i(\omega)}{\omega - \omega_0} =$$

$$= \langle c_i^*, e_{i\chi}^* \rangle \cdot A_i(\chi) \cdot (|b_i| + t_i)/\sin(\beta_i - \Theta_i) =$$

$$(3.5) \qquad = -A_i(\chi) \cdot |c_i^*| \cdot \sin \Theta_i \cdot (|b_i| + t_i)/\sin(\beta_i - \Theta_i).$$

$$q_i(\chi) = \lim_{\omega \to \omega_0} \frac{h(C,\omega) - \langle e_{i\chi}^*(\omega), Q_i + (Q_{i\chi} - Q_i) \rangle - \rho_n(C,\Delta_0)}{\omega - \omega_0} =$$

$$= \lim_{\omega \to \omega_0} \frac{h(C,\omega) - h(Q_i + \rho_n(C,\Delta_0)U,\omega)}{\omega - \omega_0} - \lim_{\omega \to \omega_0} \frac{\langle e_{i\chi}^*(\omega), Q_{i\chi} - Q_i \rangle}{\omega - \omega_0}.$$

The first term on the right-hand side of the last equation is 0 since both C and U are strictly convex, hence $h(C,.)$ and $h(U,.)$ are differentiable at ω_0. Then from (3.4)

$$(3.6) \qquad q_i(\chi) = -A_i(\chi).t_i |e_i^*|$$

Therefore the functionals p_i and q_i are continuous on H^{2n-1} and negative on $\mathrm{int}(\prod_{j=1}^{n} [0,\bar{t}_j] \times [0,\bar{\Theta}_j]) \Pi H^{2n-1}$, for $i = 1,2,\ldots,n$.

Our next goal is to show the existence of a point $\tilde{\chi}$ in the relative interior of H^{2n-1} such that

$$(3.7) \qquad p_1(\tilde{\chi}) = \ldots = p_n(\tilde{\chi}) = q_i(\tilde{\chi}) = \ldots = q_n(\tilde{\chi}) < 0$$

In proving (3.7) we need the following variant of the well known Sperner lemma [15], due to Knaster-Kuratowski-Mazurkiewicz [9]:

LEMMA. *Let* $S^m = [V_0, V_1, \ldots, V_m]$ *be an* m*-dimensional simplex covered by the closed sets* $F_0, F_1, \ldots, F_m$ *such that* $F_j \cap [V_0, \ldots, V_{j-1}, V_{j+1}, \ldots, V_m] = \emptyset$ *for each* $j = 0,1,\ldots,m$. *Then* $(\operatorname{int} S^m) \cap (\bigcap_{j=0}^{m} F_j) \neq \emptyset$.

In order to apply the Sperner lemma in the present context observe at first that H^{2n-1} is homeomorhic to the (2n-1)-dimensional simplex

$$S^{2n-1} = \left\{(x_1, x_2, \ldots, x_{2n}) \in \mathbb{R}_+^{2n} : \sum_{i=1}^{2n} x_i = 1\right\}.$$

Indeed, let

$$f_i(t_i) = \frac{|b_i| + t_i}{|a_i| - t_i} \cdot \frac{|a_i|}{|b_i|}, \text{ where } t_i \in [0, \bar{t}_i], \quad i=1,2,\ldots,n$$

$$g_i(\Theta_i) = \frac{\sin(\alpha_i + \Theta_i)}{\sin(\beta_i - \Theta_i)} \cdot \frac{\sin \beta_i}{\sin \alpha_i}, \text{ where } \Theta_i \in [0, \bar{\Theta}_i], \ i=1,\ldots,n$$

These functions are strictly increasing and for every i $f_i(0) = 1 = g_i(0)$.

Consider the mapping $\mathscr{L} : H^{2n-1} \to S^{2n-1}$ defined by

$$\mathscr{L}(t_1, \Theta_1, \ldots, t_n, \Theta_n) = (\ln\Phi(C, \Delta_0))^{-1} . (\ln f_1(t_i),$$

$$\ln g_1(\Theta_1), \ldots, \ln f_n(t_n), \ln g_n(\Theta_n))$$

It is a routine matter to verify that $\mathscr{L}$ is one-to-one, onto and both $\mathscr{L}$ and $\mathscr{L}^{-1}$ are continuous. Moreover, the sets $H^{2n-1}\cap\{t_i=0\}$ and $H^{2n-1}\cap\{\Theta_i=0\}$ are mapped onto the facets $S^{2n-1}\cap\{x_{2n-1}=0\}$ and $S^{2n-1}\cap\{x_{2i}=0\}$ respectively.

Now, the closet sets

$$F_{2i} = \{X\in H^{2n-1} : p_i(X)\le p_j(X) \text{ and } p_i(X)\le q_j(X),\ \forall j\},$$

$$F_{2i-1} = \{X\in H^{2n-1} : q_i(X)\le p_j(X) \text{ and } q_i(X)\le q_j(X),\ \forall j\}$$

constitute a covering of H^{2n-1} which satisfies all the assumptions of Sperner's lemma. Then there exists a point $\tilde{X}$ satisfying (3.7).

Finally, consider the variational n-gons $\Delta_{\tilde{X}}(\omega)$ generated by Δ_0 and $\tilde{X}$. Since Δ_0 is alternating, there is some $\delta>0$ such that $\Delta_{\tilde{X}}(\omega)$ is oscillating for C whenever $\omega\in(\omega_0-\delta,\omega_0+\delta)$. Then (3.1) is fulfilled. Let $m_{\tilde{X}}(\omega)$ and $M_{\tilde{X}}(\omega)$ be the minimal and the maximal value of all the oscillation numbers $\lambda_{i\tilde{X}}(\omega)$ and $\nu_{i\tilde{X}}(\omega)$ respectively. It follows from (3.1) and (3.7) that

$$q_1(\tilde{X}) = \lim_{\omega\to\omega_0}\frac{m_{\tilde{X}}(\omega)-\rho_n(C,\Delta_0)}{\omega-\omega_0}\le\lim_{\omega\to\omega_0}\frac{\rho_n(C,\omega)-\rho_n(C,\omega_0)}{\omega-\omega_0}\le$$

$$\le\lim_{\omega\to\omega_0}\frac{M_{\tilde{X}}(\omega)-\rho_n(C,\omega_0)}{\omega-\omega_0}<0,$$

which completes the proof of part (I).

(II) Differentiability of $\rho_n(C,.)$ at ω_0 in the case $\Phi(C,\Delta_0)<1$.

It is reduced to the former one. We have to consider a symmetric image of C with respect to a certain line in the plane. In this case $\rho_n'(C,\omega_0)>0$.

(III) Differentiability of $\rho_n(C.,)$ at ω_0 in the case $\Phi(C,\Delta_0) = 1$.

This is the most interesting case since then it follows that ρ_n's derivative is equal to 0. In that situation we need not consider any set H^{2n-1}. Instead, apply (3.1) directly to the variational n-gons generated by Δ_0, $(Q_i)_{i=1}^n$ and $(l_i)_{i=1}^n$.

Thus $\rho_n(c,.)$ is a differentiable function on $[0,2\pi]$ and sign $d\rho_n(C,\omega_0)/d\omega = \text{sign}(1 - \Psi(C,\Delta(\omega_0)))$.

(IV) Continuity of $\rho_n'(C,.)$ at ω_0 in the case $\Phi(C,\Delta_0) > 1$.

It was pointed above that the elements associated with the alternating n-gons $\Delta(\omega)$, $\omega \in [0,2\pi)$ such as vertices and sidedirections might be considered separately as mappings defined on $[0,2\pi)$. The mappings $\omega \to V_i(\omega)$, $\omega \to e_i^*(\omega)$ are well defined. So are $\omega \to Q_i(\omega)$, $\omega \to a_i(\omega)$, $\omega \to e_i(\omega)$ etc. It is a consequence of the Blaschke selection theorem that $\lim_{\omega\to\omega_0} D(\Delta(\omega),\Delta(\omega_0)) = 0$, hence $\omega \to V_i(\omega)$ are continuous. The smoothness of $C + \rho_n(C,\Delta_0).U$ and the continuity of $\rho_n(C,\Delta(.))$ imply continuity of $\omega \to e_i(\omega)$, whence for $i = 1,2,\ldots,n$ the mappings $\omega \to \alpha_i(\omega)$ and $\omega \to b_i(\omega)$ are continuous too. In order to demonstrate the continuity of $\omega \to Q_i(\omega)$ let $(Q_i(\omega_k))_{k=1}^\infty$ be a sequence such that $\lim_{\omega\to\omega_0} (Q_i(\omega_k)) = Q$ and $\lim_{\omega\to\omega_0} \omega_k = \omega_0$. By Blaschke's theorem again $Q \in [V_i(\omega_0),V_{i+1}(\omega_0)]$. A sequence of points $(Z_i(\omega_k))_{k=1}^\infty$ corresponds to $(Q_i(\omega_k))_{k=1}^\infty$ such that $e_i^*(\omega_k)$ attains its maximum on C at $Z_i(\omega_k)$ and $\|Z_i(\omega_k)-Q_i(\omega_k)\| = \rho_n(C,\Delta(\omega_k))$. Let Z be a cluster point of the latter sequence. It follows from the continuity of the mappings involved that e_i^* attains maximal value on C at Z and $\|Z-Q\| = D(C,\Delta_0)$. Since $\omega \to Q_i(\omega)$ is single-valued then $Z = Z_i(\omega_0)$ and

$Q = Q_i(\omega_0)$. Therefore, for every i the mappings $\omega \to \alpha_i(\omega)$, $\omega \to b_i(\omega)$ and the function $\Phi(C,\Delta(.))$ are continuous.

Let $\delta > 0$ be sufficiently small such that $\Phi(C,\Delta(\omega)) > 1$ whenever $\omega \in [\omega_0-\delta,\omega_0+\delta]$. Define $\varphi(X,\omega)$ as in (3.2), where a_i, b_i, α_i, β_i depend on ω. The functions $\bar{t}_i(\omega)$, $\bar{\Theta}_i(\omega)$ are easily seen to be continuous on $[\omega_0-\delta,\omega_0+\delta]$. Define now by (3.3), for $\omega \in [\omega_0-\delta,\omega_0+\delta]$, the family of sets $H^{2n-1}(\omega)$. Suppose (ω_k) is an arbitrary sequence which converges to ω_0. As in the proof of (I), for each k there exists $X_k \in H^{2n-1}(\omega)$ such that

$$(3.8) \qquad p_i(\omega,X_k) = \rho_n(C,\omega_k) = q_j(\omega,X_k); \; i,j \in \{1,2,...,n\}$$

where p_i and q_j are defined by (3.5-3.6) and depend continuously on both ω and X.

The sequence (X_k) is bounded. Hence it has a convergent subsequence. Without any change of the indexation, let $X_0 = \lim_{k-\infty} X_k$. Then $X_0 \in H^{2n-1}(\omega_0)$ and by (3.8)

$$(3.9) \qquad p_i(\omega_0,X_0) = \lim_{k\to\infty} \rho_n(C,\omega_k) = q_j(\omega_0,X_0); \; i,j \in \{1,...,n\}$$

It was proved in (I) that (3.9) implies $\rho_n'(C,\omega_0) = p_1(\omega_0,X_0)$ and consequently the continuity of $\rho_n'(C,\omega_0)$ at ω_0.

(V) Continuity of $\rho_n'(C,.)$ at ω_0 in the case $\Phi(C,\Delta_0) < 1$.

The same argument as for (II) is applied in this case too.

(VI) Continuity of $\rho'(C,.)$ at ω_0 in the case $\Phi(C,\Delta_0) = 1$.

Let $\lim_{k\to\infty} \omega_k = \omega_0$ for a sequence (ω_k). With no loss of generality suppose $\Phi(C,\Delta(\omega_k)) > 1$ for each k. According to

(3.6) and (3.8) it is sufficient to prove $\lim_{k\to\infty} t_1(\omega_k) = 0$, which follows immediately from the continuity of $\Phi(C,\Delta(.))$. The theorem is proved.

COROLLARY. *For a Hausdorff metric generated by the Euclidean norm in* $\mathbb{R}^2$, *the above result holds under the strict convexity assumption on* C *only.*

Combining results from [8], [18] and theorem 2 we obtain the following condition for a best Hausdorff approximation:

Let both C and U be strictly convex and at least one of them is smooth. If Δ_0 is a best Hausdorff approximation for C in $\mathscr{P}_n$ then

(a) Δ_0 is a non-degenerate n-gon,

(b) Δ_0 is alternating w.r.t C,

(c) $\prod_{i=1}^{n} (\|a_i\|.\sin \beta_i)/(\|b_i\|.\sin \alpha_i) = 1.$

4. EQUIVALENT FORMULATIONS

In this section, two equivalent formulations of the condition (c) are presented.

Let U,C and Δ satisfy the assumptions of theorem 2 and the points $(Q_i)_{i=1}^n$, the vectors $(a_i)_{i=1}^n$, $(b_i)_{i=1}^n$ and the angle-measures $(\alpha_i)_{i=1}^n$, $(\beta_i)_{i=1}^n$ are defined as in section 3. Let also $(e_i^*)_{i=1}^n$ be the normals of Δ and $(c_i^*)_{i=1}^n$ satisfy $\{c_i^*\} = \partial d(V_i C)$, $i = 1,2,\dots,n$, i.e. e_i^* and c_i^* are the alternance vectors. Consider Figure 4:

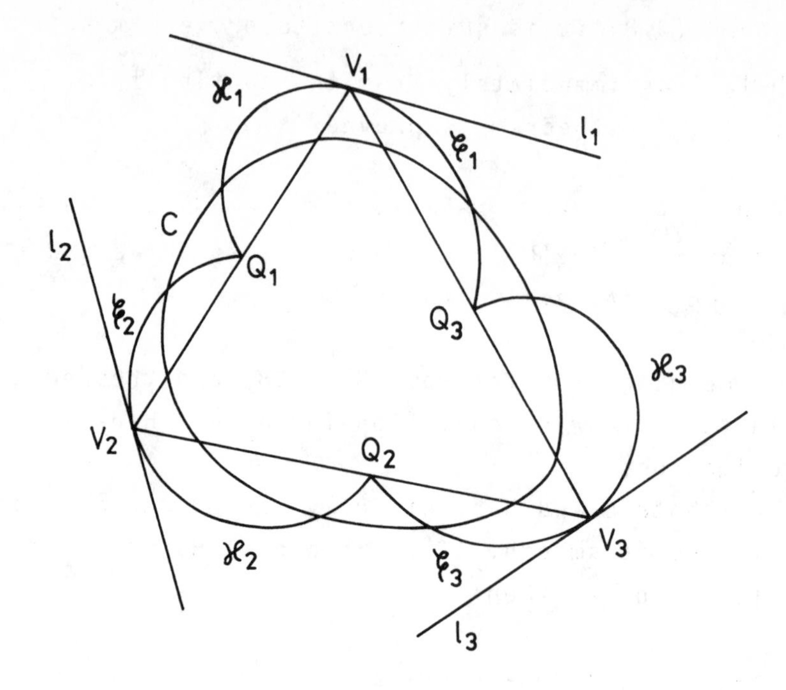

Figure 4

The circumferences κ_i pass through V_i and Q_i and have common tangents l_i at V_i with the circumferences ζ_i through V_i and Q_{i-1}, $i = 1,2,\ldots,n$. For the radii r_i of κ_i and s_i of ζ_i, we calculate:

$$r_i = |b_i|/2\sin\beta_i, \quad s_i = |a_i|/2\sin\alpha_i.$$

Since for every i $\|a_i\|/\|b_{i-1}\| = |a_i|/|b_{i-1}|$ then (c) is equivalent to

$$\prod_{i=1}^{n} r_i = \prod_{i=1}^{n} s_i \tag{4.1}$$

The next form expresses (c) in terms of derivatives of the support functions h_C and h_Δ.

Let e_i^* corresponds to $\omega_i \in [0,2\pi)$ and c_i^* corresponds to $\Theta_i \in [0,2\pi)$, $i = 1,2,\dots,n$. For $\omega \in (\omega_i, \omega_{i+1})$, $\omega = \omega(e^*)$ and $\rho_0 = \rho_n(C,\Delta)$ we have

$$<e^*, V_{i+1} - Q_i> = |e^*|.|a_{i+1}|.\cos(\pi/2 - (\omega - \omega_i)),$$

whence $|e^*|.|a_{i+1}|.\sin(\omega-\omega_i) = h(\Delta,e^*) - <e^*,Q_i>$ and

$$\frac{|e^*|.|a_{i+1}|.\sin(\omega-\omega_i)}{\omega - \omega_i} = \frac{h(\Delta,\omega) - h(\Delta,\omega_i)}{\omega - \omega_i} -$$

$$- \frac{h(Q_i+\rho_0.U,\omega) - h(Q_i+\rho_0.U,\omega_i)}{\omega - \omega_i}$$

Therefore

$$(4.2) \qquad |e_i^*|.|a_{i+1}| = h'_+(\Delta,\omega_i) - h'(C,\omega_i),$$

since $h(C,\omega_i) = h(Q_i+\rho_0 U,\omega_i)$.
Analogously, for $\omega \in (\omega_{i-1},\omega_i)$ we have

$$<e^*, V_i - Q_i> = |e^*|.|b_i|.\sin(\omega_i - \omega) \quad \text{and}$$

$$(4.3) \qquad |e^*|.|b_i| = h'(C,\omega_i) - h'_-(\Delta,\omega_i)$$

On the other hand for every i

$$(4.4) \qquad \alpha_{i+1} = \Theta_{i+1} - \omega_i \quad \text{and} \quad \beta_i = \omega_i - \Theta_i$$

Now (4.2-4.4) entail the next equivalent form of (c):

$$\prod_{i=1}^{n} \frac{h'_+(\Delta,\omega_i) - h'(C,\omega_i)}{h'(C,\omega_i) - h'_-(\Delta,\omega_i)} \cdot \frac{\sin(\omega_i - \Theta_i)}{\sin(\Theta_{i+1} - \omega_i)} = 1,$$

where ω_i, $\Theta_i \in [0,2\pi)$ are the alternance points of the support functions h_C and h_Δ, and ω_i correspond to the normals of Δ.

5. MORE GENERAL CASE

Throughout the sequel a relaxation of the strict convexity assumption on C and U in theorem 2 is discussed. It is supposed, everywhere in this section, that C is not an n-gon and at least one of the convex compacta C and U is smooth.

Let $\Delta(\omega) \in \mathscr{P}_n$ be alternating for $C, \omega \in [0,2\pi]$. Define $Z_i = \{z \in C : h(C,e_i^*) = \langle e_i^*, z\rangle\}$, $i = 1,2,\ldots,n$. Obviously, Z_i are compact intervals (possibly degenerate) which are parallel to the sides of $\Delta(=\Delta(\omega))$. Let then $Z_i = [z_i^l, z_i^r]$ where $z_i^r - z_i^l$ and $V_{i+1} - V_i$ are positively colinear. For $\rho = \rho_n(C,\Delta)$, denote by I_i^l and I_i^r the intervals $(z_i^l + \rho.U) \cap [V_i, V_{i+1}]$ and $(z_i^r + \rho.U) \cap [V_i, V_{i+1}]$ respectively. Let also Q_i^l and Q_i^r be such that $|Q_i^l - V_i| = \max\{|Y - V_i| : Y \in I_i^l\}$ and $|Q_i^r - V_i| = \min\{|Y - V_i| : Y \in I_i^r\}$ where $i = 1,2,\ldots,n$. It follows from the smoothness hypothesis that Q_i^l and Q_i^r belong to (V_i, V_{i+1}), and also the axes l_i supporting $C + \rho U$ at V_i are uniquely determined. Define then with respect to $(l_i)_{i=1}^n$, $(Q_i^l)_{i=1}^n$ and $(Q_i^r)_{i=1}^n$ the functions

$$\Phi_-(C,\Delta) = \prod_{i=1}^{n} \frac{\|a_i^l\| \sin \beta_i}{\|b_i^l\| \sin \alpha_i} \quad \text{and} \quad \Phi_+(C,\Delta) = \prod_{i=1}^{n} \frac{\|a_i^r\| \sin \beta_i}{\|b_i^r\| \sin \alpha_i}$$

where $a_i^l = V_i - Q_{i-1}^l$, $b_i^l = V_i - Q_i^l$ and

$a_i^r = V_i - Q_{i-1}^r$, $b_i^r = V_i - Q_i^r$.

THEOREM 3. *Under the above assumptions on* C *and* U *the function* $\rho_n(C,.)$ *has left-hand derivative and right-hand derivative at every point* $\omega \in [0,2\pi]$. *Moreover*

$$\operatorname{sign} \rho'_{n-}(\omega) = \operatorname{sign}(1 - \Phi_-(C,\Delta(\omega))),$$

$$\operatorname{sign} \rho'_{n+}(\omega) = \operatorname{sign}(1 - \Phi_+(C,\Delta(\omega))).$$

The proof is quite analogous with the proof of theorem 2. The only difference is that for the existence of ρ'_{n-} and ρ'_{n+} we have to consider left-hand and right-hand neighbourhoods of ω respectively.

REMARKS

Obviously $\rho_n(C,.)$ is differentiable at ω if for every i $Q_i^l = Q_i^r$.

Under the smoothness assumption on C or U and $\Delta = \Delta(\omega)$ an alternating n-gon for C, consider two separate cases:

(I) U is strictly convex,

(II) C is strictly convex.

In the case (I) for every i I_i^l and I_i^r are singletons and $\langle Q_i^r - Q_i^l, V_{i+1} - V_i \rangle \geq 0$. Then it is easily seen that

$$\Phi_+(C,\Delta) \leq \Phi_-(C,\Delta) \quad \text{and} \quad \rho'_{n+}(C,\omega) \geq \rho'_{n-}(C,\omega).$$

In the case (II) for every i Z_i are degenerate intervals and $\langle Q_i^r - Q_i^l, V_{i+1} - V_i \rangle \leq 0$ which implies $\Phi_+(C,\Delta) \geq \Phi_-(C,\Delta)$ and $\rho'_{n-}(C,\omega) \geq \rho'_{n+}(C,\omega)$.

The next result follows immediately from Theorem 3 and the above remarks.

PROPOSITION 4. *Suppose* $\Delta = \Delta(\omega)$ *is alternating for* C *and at least one of the intervals* $[Q_i^l, Q_i^r]$ *is non-degenerate. Then the following condition:*

there exists a sequence $(Q_i)_{i=1}^n$, $Q_i \in [Q_i^l, Q_i^r]$ *such that*

$$\prod_{i=1}^{n} \frac{\|a_i\| \cdot \sin \beta_i}{\|b_i\| \cdot \sin \alpha_i} = 1 \quad \textit{for} \quad a_i = V_i - Q_{i-1}, \; b_i = V_i - Q_i$$

is necessary and sufficient for $\rho_n(C,.)$ *to have:*

a (strict) local minimum at ω *in the case* (I),
a (strict) local maximum at ω *in the case* (II),

provided that for some j $Q_j \in (Q_j^l, Q_j^r)$.

REFERENCES

[1] W. BLASCHKE, *Kreis und Kugel*. Berlin: de Gruyter 1956.

[2] D.E. McCLURE and R.A. VITALE, *Polygonal Approximation of Plane Convex Bodies*. J. Math. Anal. Appl. 51, 1975, 326-358.

[3] P.J. DAVIS, R.A. VITALE, and E. BEN-SABAR, *On the Deterministic and Stochastic Approximation of Regions*. J. of Approx. Theory 21, 1977, 60-88.

[4] P.M. GRUBER, *Approximation of Convex Bodies*. Compt. rend. Acad. bulg. Sci. 34, 1981, 621-622.

[5] P.M. GRUBER and P.S. KENDEROV, *Approximation of Convex Bodies by Polytopes*. Rendiconti del Circolo Math. di Palermo 31, 1982, 195-225.

[6] P.M. GRUBER, *Approximation of Convex Bodies. Convexity and Its Applications* (ed. by P.M. Gruber, J.M. Wills), 131-162, Birkhauser, Basel 1983.

[7] P.S. KENDEROV, *Approximation of Plane Convex Compacta by Polygons*. Compt. rend. Acad. bulg. Sci. 33, 1980, 889-891.

[8] P.S. KENDEROV, *Polygonal Approximation of Plane Convex Compacta*. J. Approx. Theory 38, 1983, 221-239.

[9] B. KNASTER, K. KURATOWSKI, and S. MAZURKIEWICZ, *Ein Beweis des Fixpunktsatzes fur n-dimensionale Simplexe*. Fund. Math. 14, 1929, 132-137.

[10] V.A. POPOV, *Approximation of Convex Figures*. Compt. rend. Acad. bulg. Sci. 21, 1968, 993-995 (in russian).

[11] V.A. POPOV, *Approximation of Convex Sets*. Bulgar. Akad. Nauk Otd. Mat. Fiz. Nauk Izv. Mat. Inst. 11, 1970, 67-80 (in bulgarian).

[12] B.N. PSHENICHNII, *Convex Analysis and Extremal Problems*. Moscow: "Nauka", 1980 (in russian).

[13] R.T. ROCAFELLAR, *Convex Analysis*. Princeton Univ. Press, 1970.

[14] Bl. SENDOV, *Hausdorff Approximations*. Sofia: Bulg. Akad. Nauk, 1979 (in russian).

[15] E. SPERNER, *Neuer Beweis fur die Invarianz der Dimensionszahl und des Gebietes*. Abh. Math. Sem. Univ. Hamburg 6, 1928, 265-272.

[16] F. TOTH, *Approximation by Polygons and Polyhedra*. Bull. Amer. Math. Soc. 54, 1948, 431-438.

[17] N.V. ZHIVKOV, *Plane Polygonal Approximations of Bounded Convex Sets*. Compt. rend. l'Acad. bulg. Sci. 35, 1982, 1631-1634.

I. BÁRÁNY, K. BÖRÖCZKY, E. MAKAI Jr. and J. PACH

Let P be a non-degenerate self-intersecting oriented polytope in R^n. Let P^* denote the unique convex polytope which has the same set of outward normals of facets and the same (sum of) areas of facets with any given outward normal, as P. (The existence and uniqueness of P^* follows from a theorem of Minkowski, cf. [2].) Is it true that the total volume V of the regions enclosed by the facets of P is at most Vol P^*? For $n = 2$ this follows from a result of Pach [3]. For $n \geq 3$ we know only $V/\mathrm{Vol}\ P^* < 2^{n/(n-1)}$ (see [1]).

REFERENCES

[1] I. BÁRÁNY, K. BÖRÖCZKY, E. MAKAI Jr. and J. PACH, *Maximal volume enclosed by plates and proof of the chessboard conjecture*, Dicrete Math. 60 (1986), 101-120.

[2] T. BONNESEN and W. FENCHEL, *Theorie der konvexen Körper*, Springer, Berlin 1934.

[3] J. PACH, *On an isoperimetric problem*, Stud. Sci. Math. Hungar. 13 (1978), 43-45.

M. BLANK

Let K be a finite subset of the boundary of a centrally symmetric convex polytope P in E^d, $d \geq 1$. We say that a facet F of P is represented by K if $K \cap F \neq \emptyset$. Let K represent all the facets and be minimal with respect to inclusion. The question is what is the smallest number of points which must be taken for all the facets to be represented. More precisely, we aim at finding the numbers

$$L_d = \sup_{P \subset E^d} \min_{K \subset \partial P} |K|/|P|, \quad d \geq 1, \text{ where}$$

$|K|$ denotes the number of points of K, and $|P|$ denotes the number of facets of P.

M. BLEICHER

Given N points in a bounded region we consider the following operation:

1) Division: The region is divided into cells, each given point being in the cell of all those points closer to it than to any other of the given points (the Dirichlet or Voronoi cell).

2) Centralizing: Each given point is moved to the centroid of the corresponding cell. (The inceter, ex-center, Steiner point replacing the centroid are variants of the problem.)

The above two steps are iterated. Is there a limiting position or is cyclic behavior possible? What do the limiting positions (if they exist) look like for special cases.

The one dimensional case is solved for segments, rays, lines and Jordan arcs. The problem can be generalized to other compact surfaces, e.g. sphere, torus, to higher dimensions, or other metrics or planes.

R. CONNELLY

If $\mathscr{P}$ is a packing of congruent circles in Euclidean plane, we say $\mathscr{P}$ is n-*stable* if every subset of n circles from $\mathscr{P}$ is held fixed by the rest. If $\mathscr{P}$ is n-stable for every $n = 1,2,\ldots,$ is there a positive lower bound for the density of $\mathscr{P}$?

In [1] there are examples of circle packings that are 1-stable and have density 0. However all these packings are not 9-stable. The notion of 1-stable here is what L. Fejes Tóth has previously defined as stable [2].

REFERENCES

[1] K. BÖRÖCZKY, *Über stabile Kreis- und Kugelsysteme*. Annal. Univ. Sci Budapest. Sect. Math. 7 (1964), 79-82.

[2] L. FEJES TÓTH, *On the stability of a circle packing*, Annal. Univ. Sci. Budapest. Sect. Math. 3-4 (1960-1961), 63-66.

L. DANZER

Let J be a Jordan-domain, which tiles R^2 by translation. Does this imply that there exists even a lattice that tiles R^2 with translates of J?

P. ENGEL

Consider a lattice $L := \{\mathbf{t} = m_1\mathbf{a}_1 + \ldots + m_n\mathbf{a}_n, m_i \in \mathbb{Z}\}$ in E^n, and let $|\mathbf{a}_1| = \ldots = |\mathbf{a}_n| = 1$. At every lattice point $\mathbf{t}$ take a ball $B(\mathbf{t})$ of unit radius. Under which conditions do we have

$$\bigcup_{\mathbf{t} \in L} B(\mathbf{t}) = E^n?$$

I. FÁRY and E. MAKAI Jr.

Let C_1 and C_2 be centrally symmetric convex bodies in R^n. Suppose that any affine image of C_1 has a surface area not greater than the corresponding affine image of C_2. Does this imply

$$\text{Vol } C_1 \leq \text{Vol } C_2?$$

For $n = 2$ this holds (see [1]). That some caution has to be taken, is shown by some examples by Larman and Rogers [2] and Schneider [3].

REFERENCES

[1] I. FÁRY and E. MAKAI Jr., *Problem 31*, Period. Math. Hungar. 14 (1983), 111-114.

[2] D.G. LARMAN and C.A. ROGERS, *The existence of a centrally symmetric convex body with central sections that are unexpectedly small*, Mathematika 22 (1975), 164-175.

[3] R. SCHNEIDER, *Zu einem Problem von Shephard über die Projectionen konvexer Körper*, Math. Z. 101 (1967), 71-82.

G. FEJES TÓTH and L. FEJES TÓTH

A packing of circles is called *totally separable* if to any pair of circles from the packing there exists a straight line which separates these circles and does not meet any other member of the packing [2]. It is conjectured that the (upper) density of a totally separable packing of circles is at most $11\pi/24\sqrt{3} = 0.831\ldots$. The incircles of the faces of the Archimedean tiling (3,6,3,6) constitute a totally separable packing attaining this bound. In [1] it is shown that the density of a totally separable packing of circles is at most 0.98. Give improvements of this bound.

REFERENCES

[1] G. FEJES TÓTH, *Totally separable packing and covering with circles*, Studia Sci. Math. Hungar. 21 (1986) (to appear).

[2] G. FEJES TÓTH and L. FEJES TÓTH, *On totally separable domains*, Acta Math. Acad. Sci. Hungar. 24 (1973), 229-232.

R.K. GUY

Is there a point at a rational distance from each of the four corners of a unit square?

The point (x,y) is at a rational distance from $(0,1)$, $(1,0)$ and $(1,1)$ just if each of

$$X = \frac{x}{1-y}, \quad Y = \frac{y}{1-x} \quad \text{and} \quad M = \frac{1-y}{1-x}$$

is a Pythagorean number, i.e. the ratio of the rational sides of a rectangle whose diagonal is also rational. Infinitely many such points are known for each Pythagorean number

$$M = \pm 3/4, \ \pm 5/12, \ \pm 8/15, \ldots .$$

The distance from (x,y) to $(0,0)$ will also be rational if MX/Y is Pythagorean. No such number is known, nor has its impossibility been established.

REFERENCES

[1] R.B. EGGLESTON, A.S. FRAENKEL, R.K. GUY and J.L. SELFRIDGE, *Tiling the square with rational triangles* (to appear).

[2] R.K. GUY, *Problem D 19, Unsolved Problems in Number Theory,* Springer, Berlin-Heidelberg-New York 1981.

W. JANK

Find all types of tiles in the Euclidean plane, which admit a regular tiling (with a plane ornament group transitive on the tiles) and have at least a second tiling.

W. KUPERBERG

I. A packing $\{K_i\}$ with congruent copies of a convex body K is said to be *uniform* if there exists a tiling $\{T_i\}$ with all tiles T_i congruent to the model-tile T, such that $K_i \subset T_i$ for all i. Is it true that every convex body admits its highest density of packing through a uniform packing?

II. A p-*hexagon* is a hexagon with a pair of parallel opposite sides of equal length.

a) Find the smallest number H with the property that every convex plane body of area 1 is contained in a p-hexagon of area H.

b) Find the greatest number h with the property that every convex plane body of area 1 contains a p-hexagon of area h.

Since every p-hexagon tiles the plane, the answers to these questions will produce results on packing and covering densities, respectively.

III. Given a plane convex body K, let pd(K) denote the greatest density with which K can pack the plane, and let cd(K) denote the smallest density with which K can cover the plane. It is known [1] that $pd(K)/cd(K) \geq 3/4$

for every K. It is also known that if K is an ellipse, then $pd(K)/cd(K) = 3/4$. Find all convex bodies K for which the equality occurs.

REFERENCE

[1] W. KUPERBERG, *An inequality linking packing and covering densities of plane convex bodies*, Geom. Dedicata (to appear).

M. LASSAK

I. Can every set of diameter 1 in the Minkowski plane be covered by 4 balls of diameter $\sqrt{2}/2$?

II. No plane convex body can be covered by 4 homothetic copies with a positive ratio smaller than 1/2. Is this true for 5 (for 6) copies?

C. LINDERHOLM

Let K be a convex body in R^2 of area 1. Can K be enclosed in some right triangle of area 2?

J.B. WILKER

I. J. Schaer (private communication) and A. Szulkin [4] have constructed tilings of R^3 by proper circles. Both of these tilings involve arbitrarily large circles and both involve pairs of linked circles. (Two circles are linked if they have a unique common chord which they meet alternately.) If circles are replaced by arbitrary simple closed curves Bankston and McGovern [2] have shown that tiles of diameter at most 1 suffice. Note however that links occur in the tiling they exhibit. On the other hand Bankston and Fox [1] have described a tiling constructed by S. Kakutani that avoids links but includes simple closed curves of arbitrary large diameter.

We ask whether there is a tiling of R^3 by simple closed curves that (1) uses tiles of diameter at most 1 and also (2) avoids links. In addition we ask for explicit constructions of tilings by proper circles with either or both of these properties. This last question is made more tantalizing because Conway and Croft [3] have observed that the axiom of choice implies R^3 can be tiled with congruent circles.

REFERENCES

[1] P. BABKSTON and R. FOX, *Topological partitions of Euclidean space by spheres*, Amer. Math. Monthly 92(1985), 423-424.

[2] P. BANKSTON and R.J. McGOVERN, *Topological partitions*, General Topology and Appl. 10 (1979), 215-229.

[3] J.H. CONWAY and H.T. CROFT, *Covering a sphere with congruent great-circle arcs*, Proc. Camb. Phil. Soc. 60 (1964), 787-800.

[4] A. SZULKIN, *R^3 is the union of disjoint circles*, Amer. Math. Monthly 90 (1983), 640-641.

II. What is the densest packing (thinnest cocering) of R^3 by congruent circles? Vary the problem by packing or covering with other natural lower dimensional objects.

A. Blokhuis (private communication) has pointed out that a packing (covering) of R^2 by ellipses with major axis equal to 1 yields cylinders in R^3 that can be tiled with circular discs of diameter 1. Thus one approach to our problem raises an intriguing question of more traditional character: can the efficiency of packing (covering) the plane by ellipses of diameter 1 enhanced by allowing variable eccentricity?

If in the original problem we allow disks of arbitrarily small radius instead of congruent ones then it is possible to pack the plane and hence space so as to leave uncovered a set of points of Lebesgue measure 0. The Hausdorff dimension of this residual set is then an appropriate measure of efficiency. Hausdorff dimension may be useful also in assessing the thinness of coverings of R^3 by congruent disks. If it should turn out that points covered to various multiplicities constitute interesting fractal sets the appropriate measure of efficiency of a covering might be a function from cardinal numbers (the multiplicities) to real numbers (the Hausdorff dimension of the point set covered to that multiplicity) and the appropriate definition of "thinnest" open to interesting debate.

LIST OF PARTICIPANTS

J. BARACS, Faculté de l'Aménagement Université de Montreal, Montreal, Que. C.P. 6128 Svcc.A Canada H3C 3JC

A. BARAGAR, 6608-84 street, Edmonton, Alberta T6E 2W9 Canada

I. BÁRÁNY, MTA MKI, Budapest, Reáltanoda u. 13-15. H-1053

G. BARON, Institut für Algebra und Diskrete Mathematik Abteilung für Diskrete Mathematik (118.4) Technische Universität Wien, Wiedner Hauptstr. 8-10, A-1040 Wien

Sz. BÉRCZI, ELTE TTK Ált. Technika Tanszék, Budapest, Rákóczi út 5. H-1088

K. BEZDEK, ELTE TTK Geometria Tanszék, Budapest, Rákóczi út 5. H-1088

S. BILINSKY, Svibovac 10, 41000 Zagreb, Yugoslavia

T. BISZTRICZKY, Dept. of Mathematics University of Calgary, Calgary, Alberta T2N 1N4, Canada

M. BLANK, Institut Matematyki i Fizyki ATR, ul.Kaliskiego 7, 85-791, Bydgoszcz, Poland

M.N. BLEICHER, University of Wisconsin, Math. Dept., Madison, Wisc. 53706, USA

A. BLOKHUIS, Dept. of Math. Techn. University Eindhoven, Den Dolech 1, Eindhoven, Netherlands

K. BOGNÁR MÁTHÉ, Ybl Miklós Építőipari Műszaki Főiskola, Budapest, Thököly ut 74. H-1146

J. BOKOWSKI, Technische Hochschule, Darmstadt, Fachbereich Mathematik, Schloßgartenstr. 7. D-6100, Darmstadt

U. BOLLE, Dangenstorf 71, D-3131 Lübbow 1

J. BÖHM, Friedrich-Schiller-Universität, Jena, Sektion Mathematik Universitätschochaus, 17 OG. DDR - 6900 Jena

K. BÖRÖCZKY, ELTE TTK Geometria Tanszék, Budapest, Rákóczi út 5. H-1088

U. BREHM, Technische Universität Berlin, FB Mathematik, Straße des 17. Juni 135, D-1000 Berlin 12

C. BUCHTA, Mathematisches Institut der Universität, Albertstraße 23b, D-7800 Freiburg im Breisgau

G. CETKOVIC, Institute for Mathematics Physics and Social Sciences University of Beograd, Faculty of Civil Engineering. Bulevar revolucije 73, 11000 Belgrade, Yugoslavia

R. CONNELLY, Dept. of Mathematics Cornell University, Ithaca, NY 14853, USA

J. VAN DE CRAATS, Marinus de Jongstraat 12, 4904 PL Oosterhout -NB, The Netherlands

G. CSÓKA, ELTE TTK Geometria Tanszék, Budapest, Rákóczi ut 5. H-1088

L.W. DANZER, Math. Inst. der Univ. Dortmund, Postfach 50 0500, D-46 Dortmund 50

A. DRESS, Universität Bielefeld Fakultät für Mathematik, Postfach 8640, D-4800 Bielefeld

Z. DZIECHCINSKA-HALAMODA, Institut Matematyki Wyzsza Szkola Pedagogiczna, 45-056 Opole, Poland

D-G. EMMERICH, Unité Pedagogique d'Architecture N°6 144, Rue de Flandre, 75019 Paris, France

P. ENGEL, University of Bern Lab. for Crystallography, Freiestr. 3, CH-3012 Bern

P. ERDŐS, MTA MKI, Budapest, Reáltanoda u. 13-15. H-1053

I. FÁBIÁN, BHG Hiradástechnikai Vállalat, Budapest, Fehérvári ut 70. H-1119

G. FEJES TÓTH, MTA MKI, Budapest, Reáltanoda u. 13-15. H-1053

L. FEJES TÓTH, MTA MKI, Budapest, Reáltanoda u. 13-15 H-1053

J. FLACHSMEYER, Sektion Mathematik Universität Greifswald Ludwig-Jahn, Str, 15a, DDR-22 Greifswald

A. FLORIAN, Mathem. Institut Univ. Salzburg, Hellbrunner-str. 34, A-5020 Salzburg

S. FUDALI, Wyzsza Szkola Pedagogiczna Zaklad Matematyki, ul. Wielkopolska 15, 70-451 Szczecin, Poland

Z. FÜREDI, MTA MKI, Budapest, Reáltanoda u. 13-15. H-1053

Zs. GÁSPÁR, BME Épitömérnöki Kar Mech. Tsz., Budapest, Müegyetem rkp. 3. H-1111

J.E. GOODMAN, City College, C.U.N.Y. New York, NY 10023 USA

P. GOOSSENS, Institut de Mathematique AV. des Tilleuls, 15, B-4000 Liege

P.M! GRUBER, Techn. Univ. Vienna, Inst. für Analysis Techn.Math.u.Veis.-Math. Wiedner Hauptstraße 8-10, A-1040 Wien

R.K. GUY, Department of Mathematics & Statistics, The University of Calgary, Calgary, Alberta T2N 1N4, Canada

H. HARBORTH, Bienroder Weg 47, D-3300 Braunschweig

I. HERBURT, Institute of Math. Warsaw Technical University, pl. Jednosci Rob.1, 00-661 Warszawa, Poland

M. HOLLAI, ELTE TTK Matematikai Szakmódszertani Csoport, Budapest, Rákóczi út 5. H-1088

I. HORTOBÁGYI, ELTE TTK Matematikai Szakmódszertani Csoport, Budapest, Rákóczi út 5. H-1088

M. HUSTY, Montanuniversität Leoben, Franz Josefstraße 18, A-8700 Leoben

V. IGNATENKO, 333044 Simferopol, October 60th Str. 10
USSR

H-C. IM HOF, Mathematisches Institut der Universität, Rheinsprung 21, CH-4051 Basel

A. IVIC WEISS, Department of Mathematics York University, 4700 Keele st. Downsview, Ont. M35 IP3, Canada

W. JANK, Insitut für Geometrie, TU Wien, Wiedner Hauptstraße 8-10., A-1040 Wien

E. JUCOVIČ, P.J. Šafárik University, Dept. of Geometry and Algebra, Jesenná ul. 5, 04154 Košice, CSSR

I. JUHÁSZ, Országos Vezetőképző Központ, Budapest, Könyves Kálmán krt. 48-52. H-1087

H. KAISER, Friedrich-Schiller-Universität, Sekt. Mathematik, Universitäts Hochhaus, DDR-6900 Jena

A. KEMNITZ, Wümmeweg 10, D-3300 Braunschweig

J. KINCSES, JATE Bolyai Intézet, Szeged, Aradi vértanuk tere 1. H-6720

B. KLOTZEK, Pädagogische Hochschule "Karl Liebknecht", DDR-1500, Potsdam, Am Neuen Palais

A. KUBA, JATE Kibernetikai Lab., Szeged, Árpád tér 2. H-6720

W. KUPERBERG, Auburn University, Auburn, Alabama 36849 USA

D. LARMAN, Mathematics Dept. University College London, Gower Street, London WC1E 6BT, England

M. LASSAK, Institut Matematyki i Fizyki ATR, ul. Kaliskiego 7 85-790 Bydgoszcz, Poland

C. LEE, Institut für Mathematik Ruhr Universität-Bochum, D-4630 Bochum

Á. SZILÁGYI LENGYEL, Budapesti Tanítóképző Főiskola, Budapest, Kiss J. altb. u. 40., H-1126

C.E. LINDERHOLM, Auburn University, Auburn, Alabama 36849, USA

J. LINHART, Mathematisches Institut der Universität, Hellbrunnerstr. 34., A-5020 Salzburg

D. LJUBIC, Institute of Mathematics University of Belgrade, Studentski TRG 16 P.B. 550, 11000 Beograd, Yugoslavia

Cs. LOZANOV, Bul. Hriszto Kobakcsiev 63, 1111 Sofia, Bulgaria

Z. LUČIC, Department of Mathematics Faculty of Sciences, Studentski TRG 16 pp 550, 11000 Beograd, Yugoslavia

A.L. MACKAY, Dept. of Crystallography Birkbeck College University of London, Malet Street, London WC1 E6HX, England

Z. MAJOR, ELTE TTK Matematikai Szakmódszertani Csoport, Budapest, Rákóczi út 5. H-1088

E. MAKAI, MTA MKI, Budapest, Reáltanoda u. 13-15. H-1053

P. MANI, University of Bern, Sidlerstrasse 5, CH-3012, Bern

H. MARTINI, Pädagogische Hochschule Dresden Sektion Mathematik, Wigardstraße 17, DDR-8060 Dresden

P. McMULLEN, Department of Mathematics University College, Gower Street, London WC1E 6BT, England

F. MÉSZÁROS, Berzsenyi D. Tanárképző Főiskola, Szombathely, Szabadság tér 4. H-9701

E. MOLNÁR, ELTE TTK Geometria Tanszék, Budapest, Rákóczi út 5. H-1088

B.R. MONSON, University of New Brunswick Dept. of Mathematics & Statistics, Post Office 4400, Frederiction, N.B. E3B 5A3, Canada

W. MÖGLING, Pädagogosche Hochschule Erfurt/Mühlhausen, Nordhäuser Str. 63, DDR-5064 Erfurt

J. MÜLLER, Technische Universität Wien Abteilung für Analysis, Wiedner Hauptstr. 8-10, A-1040 Wien

D. NAGY, ELTE TTK Ált. Technika Tanszék, Budapest, Rákóczi út 5. H-1088

P. NAGY, JATE Bolyai Intézet, Szeged, Aradi vértanuk tere 1. H-6720

W. NOWACKI, Tulpenweg 6, CH-3004 Bern

J. PACH, MTA MKI, Budapest, Reáltanoda u. 13-15. H-1053

I. PALÁSTI, MTA MKI, Budapest, Reáltanoda u. 13-15. H-1053

R. POLLACK, Courant Institute, NYU 251 Mercer St., New York, NY 10012, USA

G.J. RIEGER, Universität Hannover Institut für Mathematik, Welfengarten 1, D-3000 Hannover

H. SACHS, Montan Universität, Franz-Josefstr. 18, A-8700 Leoben

J. SCHAER, Deopt. of Mathematics Univ. of Calgary, Calgary, Alberta T2N 1N4, Canada

R. SCHNEIDER, Math. Institut Universität Freiburg, Albertstr. 23b, D-7800 Freiburg

E. SCHULTE, Universität Dortmund Math. Insitut, D-46 Dortmund

C. SCHULZ, Fernuniversität (ZFE), Postfach 940, D-5800 Hagen

J. SEBESTYÉN, Berzsenyi D. Tanárképző Főiskola, Szombathely, Szabadság tér 4. H-9701

J.J. SEIDEL, Vesaliuslaan 26, 5644 HK Eindhowen, The Netherlands

V.Sh. SHEKHTMAN, Insitute of Solid State Physics Academy of Sciences of the USSR, Moscow district 142432, Chernogolovka, USSR

I.M. SHMIT'KO, Institute of Solid State Physics, Academy of Sciences Moscow district 142432, Chernogolovka, USSR

J. SIMONIS, Delft University of Technology, P.O.Box 356, 2600 AJ Delft, The Netherlands

H. SORGER, Institut für Analysis Technische Universität Wien, Wiedner Hauptstraße 8-10, A-1040 Wien

H. STACHEL, Technische Universität Wien, Institut für Geometrie, Wiedner Hauptstraße 8-10, A-1040 Wien

L. STAMMLER, Martin-Luther-Universität, Universitätsplatz 6, DDR-4010 Halle

Gy. STROMMER, Budapest, Munkácy Mihály u. 23. II.4. H-1063

S. SZABÓ, BME Építőmérnöki Kar Mat. Tsz., Budapest, Stoczek u. 2. H-1111

J. SZÉKELY, ELTE TKFK Mat. Tsz., Budapest, Kazinczy u. 17. H-1075

J. SZENTHE, BME Geomezría Tsz., Budap-st, Stoczek u.2. H-1111

I. SZEPESVÁRI, MTA SZTAKI, Budapest, Kende u. 13-17. H-1111

L. TAMÁSSY, KLTE Geometria Tsz, Debrecen 10, Pf. 12. H-4010

T. TARNAI, Építéstudományi Intézet, Budapest, Dávid Ferenc u. 6. H-1113

Y.G. TETERIN, Leningrad Dept. of Math. Institut Acad. Sci. (LOMI) Fontanka 27, Leningrad 191011, USSR

A.C. THOMPSON, Dept. of Mathematics, Dalhousie University, Halifax, N.S. B3H 3T5, Canada

B. UHRIN, MTA SZTAKI, Budapest Victor Hugo u. 18-22. H-1132

E. VÁRADI, ELTE TTK Matematikai Szakmódszertani Csoport Budapest, Rákóczi út 5. H-1088

É. VÁSÁRHELYI, ELTE TTK Geometria Tsz., Budapest, Rákóczi út 5. H-1088

I. VERMES, BME Geometria Tsz., Budapest, Stoczek u. 2. H-1111

I. VINCZE, MTA MKI, Budapest, Reáltanoda u. 13-15. H-1053

A. VOLČIČ, Instituto di Matematica Applicata, Piazzale Europa 1, I-34100, Trieste